The Future of Post-Human Accounting

The Future of Post-Human Accounting

Towards a New Theory of Addition and Subtraction in Information Management

By
Peter Baofu

IAP
INFORMATION AGE PUBLISHING, INC.
Charlotte, NC • www.infoagepub.com

Library of Congress Cataloging-in-Publication Data

A CIP record for this book is available from the Library of Congress
http://www.loc.gov

ISBN: 978-1-62396-682-9 (Paperback)
978-1-62396-683-6 (Hardcover)
978-1-62396-684-3 (ebook)

Printed and bound by CPI Group (UK) Ltd, Croydon, CR0 4YY

To Those in the Future World beyond Addition and Subtraction

Books also by Peter Baofu

Category I: Social Sciences and Related Fields

- 65. *The Future of Post-Human Waste* (2012)
- 64. *The Future of Post-Human Migration* (2012)
- 63. *The Future of Post-Human Criminality* (2012)
- 62. *The Future of Post-Human Geography* (2011)
- 61. *The Future of Post-Human Gambling* (2011)
- 60. *The Future of Post-Human Education* (2011)
- 59. *The Future of Post-Human War and Peace* (2010)
- 58. *The Future of Post-Human Law* (2010)
- 57. *The Future of Post-Human Organization* (2009)
- 56. *The Future of Post-Human Mass Media* (2009)
- 55. *The Rise of Authoritarian Liberal Democracy* (2007)
- 54. *Beyond the World of Titans, and the Remaking of World Order* (2007)
- 53. *Beyond Capitalism to Post-Capitalism* (2005)
- 52. Volume 1: *Beyond Democracy to Post-Democracy* (2004)
- 51. Volume 2: *Beyond Democracy to Post-Democracy* (2004)
- 50. *The Future of Capitalism and Democracy* (2002)

Category II: Natural Sciences and Related Fields

- 49. Volume 1: *The Future of Post-Human Aerology* (2014)
- 48. Volume 2: *The Future of Post-Human Aerology* (2014)
- 47. *The Future of Post-Human Health Care* (2013)
- 46. *Beyond Natural Resources to Post-Human Resources* (2013)
- 45. *The Future of Post-Human Transportation* (2012)
- 44. *The Future of Post-Human Architecture* (2012)
- 43. *The Future of Post-Human Chemistry* (2011)
- 42. *The Future of Post-Human Acoustics* (2011)
- 41. *The Future of Post-Human Geology* (2010)
- 40. *Beyond Cosmology to Post-Cosmology* (2010)
- 39. *The Future of Post-Human Personality* (2010)
- 38. *The Future of Post-Human Sexuality* (2010)
- 37. *The Future of Post-Human Urban Planning* (2009)
- 36. *The Future of Post-Human Engineering* (2009)
- 35. *The Future of Post-Human Unconsciousness* (2008)
- 34. *The Future of Complexity* (2007)
- 33. *Beyond Nature and Nurture* (2006)
- 32. *The Future of Post-Human Space-Time* (2006)
- 31. *The Future of Post-Human Consciousness* (2004)

Category III: The Humanities and Related Fields

- 30. Volume 2: *The Future of Post-Human Etiology* (2014)
- 29. Volume 1: *The Future of Post-Human Etiology* (2014)
- 28. *The Future of Post-Human Thanatology* (2013)
- 27. *The Future of Post-Human Sports* (2013)
- 26. *The Future of Post-Human Culinary Art* (2012)
- 25. *The Future of Post-Human History* (2012)
- 24. *The Future of Post-Human Performing Arts* (2012)
- 23. *The Future of Post-Human Literature* (2011)
- 22. *The Future of Post-Human Humor* (2011)
- 21. *Beyond Ethics to Post-Ethics* (2011)
- 20. *The Future of Post-Human Religion* (2010)
- 19. *The Future of Post-Human Chess* (2010)
- 18. *The Future of Post-Human Martial Arts* (2009)
- 17. *The Future of Post-Human Creative Thinking* (2009)
- 16. *The Future of Post-Human Knowledge* (2008)
- 15. *The Future of Aesthetic Experience* (2007)
- 14. *Beyond Civilization to Post-Civilization* (2006)
- 13. Volume 1: *The Future of Human Civilization* (2000)
- 12. Volume 2: *The Future of Human Civilization* (2000)

Category IV: Formal Sciences and Related Fields

- 11. *The Future of Post-Human Phonology* (2014)
- 10. *The Future of Post-Human Probability* (2013)
- 09. *The Future of Post-Human Morphology* (2013)
- 08. *The Future of Post-Human Semantics* (2012)
- 07. *The Future of Post-Human Data Analysis* (2011)
- 06. *The Future of Post-Human Computing* (2011)
- 05. *The Future of Post-Human Formal Science* (2010)
- 04. *The Future of Post-Human Language* (2009)
- 03. *The Future of Post-Human Geometry* (2009)
- 02. *The Future of Post-Human Mathematical Logic* (2008)
- 01. *The Future of Information Architecture* (2008)

CONTENTS

Part One: Introduction

Part Two: Addition

Part Three: Subtraction

Part Four: Conclusion

TABLES

Category III: Visions on the Mind

Category IV: Visions on History

Category VII. Visions on Society (Socio-Political)

Category VIII. Visions on Society (Socio-Economic)

Category IX: Visions on Culture

ACKNOWLEDGMENTS

Like all other previous books of mine, this one here is no exception and is written to challenge conventional wisdom or, more precisely, in the current context, the one on accounting—and to replace it with an original way of thinking about its future fate (especially, though not exclusively, in relation to addition and subtraction).

Consequently, this book receives no external funding nor help from any formal organization or institution, because of its political incorrectness—as this is something that I often stressed (and repeated) in all my previous books.

My only reward is the joy of coming up with something original in the intellectual history of accounting.

In any event, I bear the sole responsibility for the ideas presented in this book.

ABBREVIATIONS

ALD = Peter Baofu. 2007. *The Rise of Authoritarian Liberal Democracy: A Preface to a New Theory of Comparative Political Systems*. Cambridge, England: Cambridge Scholars Publishing, Ltd.

BCIV = Peter Baofu. 2006. *Beyond Civilization to Post-Civilization: Conceiving a Better Model of Life Settlement to Supersede Civilization*. NY: Peter Lang Publishing, Inc.

BCPC = Peter Baofu. 2005. *Beyond Capitalism to Post-Capitalism: Conceiving a Better Model of Wealth Acquisition to Supersede Capitalism*. NY: The Edwin Mellen Press.

BCOS = Peter Baofu. 2010. *Beyond Cosmology to Post-Cosmology: A Preface to a New Theory of Different Worlds*. Cambridge, England: Cambridge International Science Publishing, Ltd.

BDPD1 = Peter Baofu. 2004. Volume 1. *Beyond Democracy to Post-Democracy: Conceiving a Better Model of Governance to Supersede Democracy*. NY: The Edwin Mellen Press.

BDPD2 = Peter Baofu. 2004. Volume 2. *Beyond Democracy to Post-Democracy: Conceiving a Better Model of Governance to Supersede Democracy*. NY: The Edwin Mellen Press.

BEPE = Peter Baofu. 2011. *Beyond Ethics to Post-Ethics: A Preface to a New Theory of Morality and Immorality*. Charlotte, NC: Infomration Age Publishing.

BNN = Peter Baofu. 2006. *Beyond Nature and Nurture: Conceivng a Better Way to Understand Genes and Memes*. Cambridge, England: Cambridge Scholars Publishing, Ltd.

BNR = Peter Baofu. 2013. *Beyond Natural Resources to Post-Human Resources: Towards a New Theory of Diversity and Discontinuity*. Cambridge, England: Cambridge Scholars Publishing, Ltd.

BWT = Peter Baofu. 2007. *Beyond the World of Titans, and the Renaking of World Order: A Preface to a New Logic of Empire-Building*. Cambridge, England: Cambridge Scholars Publishing, Ltd.

FAE = Peter Baofu. 2007. *The Future of Aesthetic Experience: Conceiving a Better Way to Understand Beauty, Ugliness and the Rest.* Cambridge, England: Cambridge Scholars Publishing, Ltd.

FC = Peter Baofu. 2007. *The Future of Complexity: Conceiving a Better Way to Understand Order and Chaos.* London, United Kingdom: World Scientific Publishing Co.

FCD = Peter Baofu. 2002. *The Future of Capitalism and Democracy.* MD: The University Press of America.

FHC1 = Peter Baofu. 2000. Volume 1. *The Future of Human Civilization.* NY: The Edwin Mellen Press.

FHC2 = Peter Baofu. 2000. Volume 2. *The Future of Human Civilization.* NY: The Edwin Mellen Press.

FIA = Peter Baofu. 2008. *The Future of Information Architecture: Conceiving a Better Way to Understand Taxonomy, Network, and Intelligence.* Oxford, England: Chandos Publishing (Oxford) Limited.

FPHA = Peter Baofu. 2011. *The Future of Post-Human Acoustics: A Preface to a New Theory of Sound and Silence.* Cambridge, England: Cambridge International Science Publishing, Ltd.

FPHACCO = Peter Baofu. 2014. *The Future of Post-Human Accounting: Towards a New Theory of Addition and Subtraction in Information Management.* Charlotte, NC: Information Age Publishing.

FPHAERO1 = Peter Baofu. 2014. Volume 1. *The Future of Post-Human Aerology: A Preface to a New Theory of Predictability and Nonpredictability.* Hauppauge, NY: Nova Science Publishers, Inc.

FPHAERO2 = Peter Baofu. 2014. Volume 2. *The Future of Post-Human Aerology: A Preface to a New Theory of Predictability and Nonpredictability.* Hauppauge, NY: Nova Science Publishers, Inc.

FPHARCH = Peter Baofu. 2012. *The Future of Post-Human Architecture: A Preface to a New Theory of Form and Function.* Cambridge, England: Cambridge International Science Publishing, Ltd.

FPHC = Peter Baofu. 2004. *The Future of Post-Human Consciousness.* NY: The Edwin Mellen Press.

FPHCA = Peter Baofu. 2012. *The Future of Post-Human Culinary Art: Towards a New Theory of Ingredients and Techniques.* England: Cambridge Scholars Publishing, Ltd.

FPHCHEM = Peter Baofu. 2011. *The Future of Post-Human Chemistry: A Preface to a New Theory of Substances and their Changes.* England: Cambridge Scholars Publishing, Ltd.

FPHCHESS = Peter Baofu. 2010. *The Future of Post-Human Chess: A Preface to a New Theory of Tactics and Strategy.* Cambridge, England: Cambridge International Science Publishing, Ltd.

FPHCOM = Peter Baofu. 2011. *The Future of Post-Human Computing: A Preface to a New Theory of Hardware and Software.* Cambridge, England: Cambridge International Science Publishing, Ltd.

FPHCRIM = Peter Baofu. 2012. *The Future of Post-Human Criminality: A Preface to a New Theory of Heroes and Villains.* Cambridge, England: Cambridge International Science Publishing, Ltd.

FPHCT = Peter Baofu. 2009. *The Future of Post-Human Creative Thinking: A Preface to a New Theory of Invention and Innovation.* Cambridge, England: Cambridge Scholars Publishing, Ltd.

FPHDA = Peter Baofu. 2011. *The Future of Post-Human Data Analysis: A Preface to a New Theory of Quantitative and Qualitative Research.* Cambridge, England: Cambridge International Science Publishing, Ltd.

FPHE = Peter Baofu. 2009. *The Future of Post-Human Engineering: A Preface to a New Theory of Technology.* Cambridge, England: Cambridge Scholars Publishing, Ltd.

FPHEDU = Peter Baofu. 2011. *The Future of Post-Human Education: A Preface to a New Theory of Teaching and Learning.* Cambridge, England: Cambridge International Science Publishing, Ltd.

FPHETIO1 = Peter Baofu. 2014. Volume 1. *The Future of Post-Human Etiology: Towards a New Theory of Cause and Effect.* Hauppauge, NY: Nova Science Publishers, Inc.

FPHETIO2 = Peter Baofu. 2014. Volume 2. *The Future of Post-Human Etiology: Towards a New Theory of Cause and Effect.* Hauppauge, NY: Nova Science Publishers, Inc.

FPHFS = Peter Baofu. 2010. *The Future of Post-Human Formal Science: A Preface to a New Theory of Abstraction and Application.* Cambridge, England: Cambridge Scholars Publishing, Ltd.

FPHG = Peter Baofu. 2009. *The Future of Post-Human Geometry: A Preface to a New Theory of Infinity, Symmetry, and Dimensionality.* Cambridge, England: Cambridge Scholars Publishing, Ltd.

FPHGAM = Peter Baofu. 2011. *The Future of Post-Human Gambling: A Preface to a New Theory of Risk and Caution.* Cambridge, England: Cambridge International Science Publishing, Ltd.

FPHGEOG = Peter Baofu. 2011. *The Future of Post-Human Geography: A Preface to a New Theory of Environments and their Interactions.* Cambridge, England: Cambridge International Science Publishing, Ltd.

FPHGEOL = Peter Baofu. 2010. *The Future of Post-Human Geology: A Preface to a New Theory of Statics and Dynamics.* Cambridge, England: Cambridge International Science Publishing, Ltd.

FPHH = Peter Baofu. 2011. *The Future of Post-Human Humor: A Preface to a New Theory of Joking and Laughing.* Cambridge, England: Cambridge International Science Publishing, Ltd.

FPHHC = Peter Baofu. 2013. *The Future of Post-Human Health Care: Towards a New Theory of Mind and Body.* Hauppauge, NY: Nova Science Publishers, Inc.

FPHHIST = Peter Baofu. 2012. *The Future of Post-Human History: A Preface to a New Theory of Universality and Relativity.* Cambridge, England: Cambridge Scholars Publishing, Ltd.

FPHK = Peter Baofu. 2008. *The Future of Post-Human Knowledge: A Preface to a New Theory of Methodology and Ontology.* Oxford, England: Chandos Publishing (Oxford) Limited.

FPHL = Peter Baofu. 2009. *The Future of Post-Human Language: A Preface to a New Theory of Structure, Context, and Learning.* Cambridge, England: Cambridge Scholars Publishing, Ltd.

FPHLAW = Peter Baofu. 2010. *The Future of Post-Human Law: A Preface to a New Theory of Necessity, Contingency, and Justice.* Cambridge, England: Cambridge Scholars Publishing, Ltd.

FPHLIT = Peter Baofu. 2011. *The Future of Post-Human Literature: A Preface to a New Theory of Fiction and Non-Fiction.* Cambridge, England: Cambridge International Science Publishing, Ltd.

FPHMA = Peter Baofu. 2009. *The Future of Post-Human Martial Arts: A Preface to a New Theory of the Body and Spirit of Warriors.* Cambridge, England: Cambridge Scholars Publishing, Ltd.

FPHMIG = Peter Baofu. 2012. *The Future of Post-Human Migration: A Preface to a New Theory of Sameness, Otherness, and Identity.* Cambridge, England: Cambridge Scholars Publishing, Ltd.

FPHML = Peter Baofu. 2008. *The Future of Post-Human Mathematical Logic: A Preface to a New Theory of Rationality.* Cambridge, England: Cambridge Scholars Publishing, Ltd.

FPHMM = Peter Baofu. 2009. *The Future of Post-Human Mass Media: A Preface to a New Theory of Technology.* Cambridge, England: Cambridge Scholars Publishing, Ltd.

FPHMORP = Peter Baofu. 2013. *The Future of Post-Human Morphology: Towards a New Theory of Typologies and Rules.* Hauppauge, NY: Nova Science Publishers, Inc.

FPHO = Peter Baofu. 2009. *The Future of Post-Human Organization: A Preface to a New Theory of Communication, Decision-Making, and Leadership.*.Cambridge, England: Cambridge Scholars Publishing, Ltd.

FPHP = Peter Baofu. 2010. *The Future of Post-Human Personality: A Preface to a New Theory of Normality and Abnormality.* Cambridge, England: Cambridge International Science Publishing, Ltd.

FPHPHON = Peter Baofu. 2014. *The Future of Post-Human Phonology: Towards a New Theory of Generality and Specifcity.* Hauppauge, NY: Nova Science Publishers, Inc.

FPHPA = Peter Baofu. 2012. *The Future of Post-Human Performing Arts: A Preface to a New Theory of the Body and its Presence.* Cambridge, England: Cambridge Scholars Publishing, Ltd.

FPHPROB = Peter Baofu. 2013. *The Future of Post-Human Probability: Towards a New Theory of Objectivity and Subjectivity.* Hauppauge, NY: Nova Science Publishers, Inc.

FPHR = Peter Baofu. 2010. *The Future of Post-Human Religion: A Preface to a New Theory of Spirituality.* Cambridge, England: Cambridge International Science Publishing, Ltd.

FPHS = Peter Baofu. 2010. *The Future of Post-Human Sexuality: A Preface to a New Theory of the Body and Spirit of Love-Makers.* Cambridge, England: Cambridge Scholars Publishing, Ltd.

FPHSEM = Peter Baofu. 2012. *The Future of Post-Human Semantics: A Preface to a New Theory of Internality and Externality.* Cambridge, England: Cambridge Scholars Publishing, Ltd.

FPHSPORT = Peter Baofu. 2013. *The Future of Post-Human Sports: Towards a New Theory of Training and Winning.* Cambridge, England: Cambridge Scholars Publishing, Ltd.

FPHST = Peter Baofu. 2006. *The Future of Post-Human Space-Time: Conceivng a Better Way to Understand Space and Time.* New York: Peter Lang Publishing, Inc.

FPHTHAN = Peter Baofu. 2013. *The Future of Post-Human Thanatology: Towards a New Theory of Death and After-Death.* Berlin, Germany: Logos Verlag Berlin.

FPHTRAN = Peter Baofu. 2012. *The Future of Post-Human Transportation: A Preface to a New Theory of Networks and Operations.* Cambridge, England: Cambridge Scholars Publishing, Ltd.

FPHU = Peter Baofu. 2008. *The Future of Post-Human Unconsciousness: A Preface to a New Theory of Anomalous Experience.* Cambridge, England: Cambridge Scholars Publishing, Ltd.

FPHUP = Peter Baofu. 2009. *The Future of Post-Human Urban Planning: A Preface to a New Theory of Density, Void, and Sustainability.* Cambridge, England: Cambridge Scholars Publishing, Ltd.

FPHW = Peter Baofu. 2012. *The Future of Post-Human Waste: Towards a New Theory of Uselessness and Usefulness.* Cambridge, England: Cambridge Scholars Publishing, Ltd.

FPHWP = Peter Baofu. 2010. *The Future of Post-Human War and Peace: A Preface to a New Theory of Aggression and Pacificity.* Cambridge, England: Cambridge Scholars Publishing, Ltd.

• Part One •

Introduction

CHAPTER 1
INTRODUCTION—THE INVENTION OF ACCOUNTING

> You have to know accounting. It's the language of practical business life. It was a very useful thing to deliver to civilization. I've heard it came to civilization through Venice which of course was once the great commercial power in the Mediterranean.
> —Charlie Munger (WOO 2013)

A Quarrel about an Invention

Is the invention of accounting so useful that, as Charlie Munger once said, "you have to know accounting. It's the language of practical business life. It was a very useful thing to deliver to civilization. I've heard it came to civilization through Venice which of course was once the great commercial power in the Mediterranean"? (WOO 2013)

This positive view on accounting can be contrasted with an opposing view by Paul Browne that "the recent [accounting] scandals have brought a new level of attention to the accounting profession as gatekeepers and custodians of social interest." (DUM 2013)

Contrary to these opposing views (and other ones as will be discussed in the book), accounting (in relation to addition and subtraction) are neither possible (or impossible) nor desirable (or undesirable) to the extent that the respective ideologues (on different sides) would like us to believe.

Of course, this reexamination of different opposing views on accounting does not mean that the study of addition and subtraction is useless, or that those fields (related to accounting)—like bookkeeping, auditing, forensics, info management, finance, philosophy of accounting, accounting ethics, lean accounting, mental accounting, environmental audit, creative accounting, carbon accounting, social accounting, and so on—are unimportant. (WK 2013) In fact, neither of these extreme views is plausible.

Rather, this book offers an alternative (better) way to understand the future of accounting in regard to the dialectic relationship between addition and subtraction—while learning from different approaches in the literature but without favoring any one of them (nor integrating them, since they are not necessarily compatible with each other).

More specifically, this book offers a new theory (that is, *the double-sided theory of accounting*) to go beyond the existing approaches in a novel way.

This seminal project will fundamentally change the way that we think about accounting in relation to addition and subtraction from the combined perspectives of the mind, nature, society, and culture, with enormous implications for the human future and what I originally called its "post-human" fate.

Definition and History of Accounting

A good point of departure for the present inquiry is to provide a working definition of the term "accounting" and a brief history of the field.

Definition

The term "accounting" is "from the French word *Compter*, which took its origin from the Latin word *Computare*. The word was formerly written in English as 'Accomptant,' but in process of time the word, which was always pronounced by dropping the 'p,' became gradually changed both in pronunciation and in orthography to its present form." (WK 2013; F. Pixley 1990)

History

Historically, accounting can be traced back to antiquity.

For instance, "early accounting records were found amongst the ruins of ancient Babylon, Assyria and Sumeria, which date back more than 7,000 years. The people of that time relied on primitive accounting methods to record the growth of crops and herds. Because there is a natural season to farming and herding, it is easy to count and determine if a surplus had been gained after the crops had been harvested or the young animals weaned." (WK 2013; T. Friedlob 1996)

Later, "between the 4th millennium BC and the 3rd millennium BC, in ancient Iran, new socioeconomic situations resulted in unequal distribution of wealth and in such conditions leaders and priests started to rule. They had people to look after the financial matters. In Godin Tepe (buildings were discovered ,)تپه یحیی(and also Tepe Yahya)گدین تپه with large rooms for storage of crops. Cylindrical tokens were found in these buildings, which were used for bookkeeping on clay scripts. In Godin Tepe's findings, the scripts only contained tables with figures. In Tepe Yahya's findings, the scripts also contained graphical representations." (WK 2013)

Then, in the Roman Empire, "the *Res Gestae Divi Augusti* (Latin: 'The Deeds of the Divine Augustus') is a remarkable account to the Roman people of the Emperor Augustus' stewardship. It listed and quantified his public expenditure, which encompassed distributions to the people, grants of land or money to army veterans, subsidies to the *aerarium* (treasury), building of temples, religious offerings, and expenditures on theatrical shows and gladiatorial games. It was not an account of state revenue and expenditure, but was designed to demonstrate Augustus' munificence. The significance of the *Res Gestae Divi Augusti* from an accounting perspective lies in the fact that it illustrates that the executive authority had access to detailed financial information, covering a period of some forty years, which was still retrievable after the event. The scope of the accounting information at the emperor's disposal suggests that its purpose encompassed planning and decision-making." (WK 2013; D. Oldroyd 1995)

By the 13th century, "one important breakthrough took place, and that was "the introduction of double-entry bookkeeping, which is defined as any bookkeeping system in which there was a debit and credit entry for each transaction, or for which the majority of transactions were intended to be of this form. The historical origin of the use of the words 'debit' and 'credit' in accounting goes back to the days of single-entry bookkeeping in which the chief objective was to keep track of amounts owed by customers (debtors) and amounts owed to creditors....The earliest extant evidence of full double-entry bookkeeping is the Farolfi ledger of 1299-1300....The oldest discovered record of a complete double-entry system is the *Messari* (Italian: Treasurer's) accounts of the city of Genoa in 1340." (WK 2013; A. Heeffer 2009; G. Mills 1993)

But "Luca Pacioli's *'Summa de Arithmetica, Geometria, Proportioni et Proportionalità*' (early Italian: 'Review of Arithmetic, Geometry, Ratio and Proportion') was first printed and published in Venice in 1494. It included a 27-page treatise on bookkeeping, *'Particularis de Computis et Scripturis'* (Latin: 'Details of Calculation and Recording')....It represents the first known printed treatise on bookkeeping; and it is widely believed to be the forerunner of modern bookkeeping practice. In *Summa Arithmetica*, Pacioli introduced symbols for plus and minus for the first time in a printed book, symbols that became standard notation in Italian Renaissance mathematics....Although Luca Pacioli did not invent double-entry bookkeeping, his 27-page treatise on bookkeeping contained the first known published work on that topic, and is said to have laid the foundation for double-entry bookkeeping as it is practiced today." (WK 2013; B. Carruthers 1991; A. Sangster 2007)

Later, in the 19th century, "the modern profession of the chartered accountant originated in Scotland..., as well as the first public accountant societies. Early accountants tended to regard themselves as solicitors and would subsume their accounting activities within the rubric of their legal profession, often calling themselves 'Writers' as well as 'Accountants'. Several members of the Society of Writers to the Signet, the leading Solicitor Society in Scotland, practiced as accountants....To improve their status and combat criticism of low standards, local professional bodies in England amalgamated to form the Institute of Chartered Accountants in England and Wales, established by royal charter in 1880," and "it soon drew up standards of conduct and examinations for admission and members were authorised to use the professional designations 'FCA' (Fellow Chartered Accountant), for a firm partner and 'ACA' (Associate Chartered Accountant) for a qualified member of an accountant's staff. In the United States the American Institute of Certified Public Accountants was established in 1887." (WK 2013; R. Perks 1993)

In our time, "national accountancy and trade administrative vocabulary definitions are at present date influenced by international treaties and especially within Europe the EU VAT directive defining a legal terminology set of expressions. Most countries have had a very unregulated business administrative vocabulary and there is in practice for instance different English Business administrative vocabulary in different areas of the world. The VAT directive is published in every official language within the EU, and make an impact on each business administrative language. The chart of accounts are to a large extent dependent on VAT and tax law and an important issue of understanding the accounting and analyse reports. This has importance in the development of accounting." (WK 2013)

And more recently, there is "electronic accounting," such that "we see a world where banks starts to refuse cash, and internet banking and mobile phone banking apps are making huge change in the common life in accounting....The projects of electronic trade documents like UN/CEFACT and previously UBL will not only change the daily world of accounting, the work tools but also to a large extent accounting logics itself. Accounting definitions will be tighter and the national accounting culture differences will decrease in a process being able to use modern tools and information sources, in order to save work, costs and increase quality of information." (WK 2013)

Addition and Subtraction in Accounting

With this definition and brief history in mind—accounting can be re-interpreted in this book in terms of two major processes concerning *addition* (like debit in asset, credit in liability, quality audit to increase effectiveness, etc.) and subtraction (like credit in expense, debit in income, lean accounting to decrease waste, etc.).

Of course, there can be other different re-interpretations (like debit vs. credit, expense vs. income, etc.), so the one in this book is illustrative only, for *the double-sided theory of accounting*.

In other words, the purpose of this book is to re-interpret accounting on the basis of the dialectic relationship between "addition" and "subtraction," as explained below (and summarized in *Table 1.1*), with the ultimate goal to transcend both addition and subtraction, for a different approach to the study of accounting altogether.

Addition

The word "addition," in accordance to the Merriam-Webster dictionary, refers to "the act or process of joining something to something else," such that "something or someone more...is included." (MWD 2013)

In the context of accounting, "addition" can be illustrated with the following examples, namely, (a) debit (in asset), (b) credit (in liability), and (c) quality audit (to increase efficiency).

Debit (in Asset)

The first example of addition in accounting to be illustrated here has to do with "debit" (in "asset").

In "financial accounting," the word "debit" refers to "an entry…made on the left side" of the account's ledger. (WK 2013f) And an "asset," by definition, refers to "an economic resource. Anything tangible or intangible that is capable of being owned or controlled to produce value and that is held to have positive economic value is considered an asset. Simply stated, assets represent value of ownership that can be converted into cash (although cash itself is also considered an asset)." (WK 2013a & 2013b) Good examples of "tangible" assets include "buildings and equipment"—and those of "intangible" assets include "copyrights, trademarks, patents and computer programs," as well as "bonds and stocks." (WK 2013b)

The "general accounting equation" to understand the arrangement of "assets" is shown below: (WK 2013a)

Assets = Liabilities + Owners' Equity (or Net Assets)

This then means that "assets" normally have "debit balances," so that "debiting these types of accounts increases them," whereas crediting them decreases them. (WK 2013a)

Credit (in Liability)

The second example of addition in accounting to be illustrated here has to do with "credit" (in "liability").

In "financial accounting," the word "credit" refers to "an entry…made on the right side" of the account's ledger. (WK 2013f) And a "liability" refers to "a current obligation of an entity arising from past transactions or events," and good examples are "mortgages and loans that are payable" over a period of time. (WK 2013c & 2013d)

The "general accounting equation" to understand the arrangement of "liability" is shown below: (WK 2013a)

Assets = Liabilities + Owners' Equity (or Net Assets)

This then means that "liabilities" normally have "credit balances," so that "crediting these increases them," whereas debiting them decreases them. (WK 2013a)

In other words, this difference between "credit balances" (in the case of "liabilities") and "debit balances" (in the case of "assets") is due to the general accounting equation which places assets and liabilities on the opposite sides.

Quality Audit (to Increase Effectiveness)

And the third example of addition in accounting to be illustrated here has to do with "quality audit" (to increase efficiency).

The term "quality audit" refers to "the process of systematic examination of a quality system carried out by an internal or external quality auditor or an audit team. It is an important part of organization's quality management system and is a key element in the ISO quality system standard, ISO 9001." (WK 2013e)

In so doing, "quality audits are typically performed at predefined time intervals and ensure that the institution has clearly defined internal system monitoring procedures linked to effective action"; for instance, "audits are an essential management tool to be used for verifying objective evidence of processes, to assess how successfully processes have been implemented, for judging the effectiveness of achieving any defined target levels," and so on. (WK 2013e)

Subtraction

By contrast—"subtraction," in accordance to the Merriam-Webster dictionary, refers to an act or operation "to take (a number or amount) from another number or amount." (MWD 2013a)

In the context of accounting, "subtraction" can be illustrated with the following examples, namely, (a) credit (in expense), (b) debit (in income), and (c) lean accounting (to decrease waste).

Credit (in Expense)

The first example of subtraction in accounting to be illustrated here has to do with "credit" (in "expense").

In "financial accounting," the term "expense" has "a very specific meaning. It is an outflow of cash or other valuable assets from a person or company to another person or company," and good examples include "rent," "tuition," "dining," "refreshments," and so on. (WK 2013g)

The "extended accounting equation" to understand the arrangement of "expense" is shown below: (WK 2013f)

Assets + Expenses = Liabilities + Equity + Income

This then means that "expenses" normally have "debit balances," so that crediting these decreases (or subtracts from) them (whereas debiting them increases them). (WK 2013f)

Debit (in Income)

The second example of subtraction in accounting to be illustrated here has to do with "debit" (in "income").

In economics, the term "income" refers to, as is the case for households and individuals, "the sum of all the wages, salaries, profits, interests payments, rents and other forms of earnings received... in a given period of time." (WK 2013h; K. Case 2007)

The "extended accounting equation" to understand the arrangement of "income" is shown below: (WK 2013f)

Assets + Expenses = Liabilities + Equity + Income

This then means that "income" normally has "credit balances," so that debiting these decreases (or subtracts from) them (whereas crediting them increases them). (WK 2013f)

Again, like the difference in the context of "liabilities" and "assets"—the difference between "credit balances" (in the case of "expenses") and "debit balances" (in the case of "income") is due to the extended accounting equation which places expenses and income on the opposite sides.

Lean Accounting (to Decrease Waste)

And the third example of subtraction in accounting to be illustrated here has to do with "lean accounting" (to decrease waste).

The term "lean accounting" refers to "lean manufacturing...developed by Toyota and other Japanese companies....As pioneer American and European companies embraced lean manufacturing methods in the late 1980s, they discovered that lean thinking must be applied to every aspect of the company including the financial and management accounting processes." (WK 2013i; B. Emiliani 2007)

One important objective of "lean accounting" is "to eliminate waste, free up capacity, speed up the process, eliminate errors and defects, and make the process clear and understandable." (WK 2013i)

Diverse Aspects of Accounting

This introduction to "addition" and "subtraction" (in the context of accounting) thus shows its more complicated nature.

To illustrate this more complicated nature further, consider some diverse aspects of accounting below (as summarized in *Table 1.1*).

Debit vs. Credit

The first aspect of accounting to be shown here in regard to its more complicated nature has to do with the distinction between "debit" and "credit."

On the one hand, as pointed out earlier, in "financial accounting," the word "debit" refers to "an entry…made on the left side" of the account's ledger. (WK 2013f)

On the other hand, the word "credit" refers to "an entry…made on the right side" of the account's ledger. (WK 2013f)

However, "debiting" can mean increasing (adding) or decreasing (subtracting), just as "crediting" can imply increasing (adding) or decreasing (subtracting), as shown in the following "summary table of standard increasing and decreasing attributes for the five accounting elements": (WK 2013f)

	Debit	*Credit*
Asset	+	−
Expense	+	−
Liability	−	+
Income	−	+
Equity	−	+

This variation is due to the "extended accounting equation" which places assets and expenses on the left side and liabilities, income, and equity on the right side, as shown earlier.

A Single Point in Time vs. a Period over Time

The second aspect of accounting to be shown here in regard to its more complicated nature has to do with the distinction between "a single point in time" and "a period over time."

On the one hand, in financial accounting, "a single point in time" is static, because it refers to "a balance sheet or statement of financial position" which is "a summary of the financial balances of a sole proprietorship, a business partnership, a corporation or other business organization, such as an LLC or an LLP. Assets, liabilities and ownership equity are listed as of a specific date, such as the end of its financial year. A balance sheet is often described as a 'snapshot of a company's financial condition.' Of the four basic financial statements, the balance sheet is the only statement which applies to a single point in time of a business' calendar year." (WK 2013j)

On the other hand, in financial accounting, "a peirod over time" is dynamic, because it refers to, say, a "statement of comprehensive income" or an "income statement," which "reports on a company's income, expenses, and profits over a period of time. A profit and loss statement provides information on the operation of the enterprise. These include sales and the various expenses incurred during the processing state." (WK 2013k)

Bookkeeping vs. Accounting

The third aspect of accountng to be shown here in regard to its more complicated nature has to do with the distinction between "bookkeeping" and "accounting."

On the one hand, "bookkeeping" refers to "the recording of financial transactions. Transactions include purchases, sales, receipts and payments by an individual or organization." (WK 2013l)

On the other hand, "accounting" is more than just "bookkeeping," even though it makes good use of bookkeeping, and the reason is that "the accounting process includes the bookkeeping function, but is just one part of the accounting process. The accountant creates reports from the recorded financial transactions recorded by the bookkeeper and files forms with government agencies." (WK 2013l)

Purchasing Power Accounting vs. Historical Cost Accounting

And the fourth aspect of accounting to be shown here in regard to its more complicated nature has to do with the distinction between "purchasing power accounting" and "historical cost accounting."

On the one hand, the term "purchasing power accounting" refers to an "accounting model originally authorized in IFRS in 1989," such that "financial capital maintenance is always and everywhere measured in units of constant purchasing power in terms of a Daily CPI (consumer price index) or monetized daily indexed unit of account (e.g. the Unidad de Fomento in Chile) during low inflation and deflation, and in terms of a daily relatively stable foreign currency parallel rate or daily index during high inflation and hyperinflation." (WK 2013m)

On the other hand, the term "historical cost accounting" refers to an accounting model based on "the original monetary value of an economic item. Historical cost is based on the stable measuring unit assumption. In some circumstances, assets and liabilities may be shown at their historical cost, as if there had been no change in value since the date of acquisition. The balance sheet value of the item may therefore differ from the 'true' value." (WK 2013n)

The Theoretical Debate

These diverse aspects of accounting, together with the introduction to "addition" and "subtraction," then lead us to a more important question concerning the theoretical debate on the very nature of accounting. Indeed, this theoretical debate can also say much about the nature of accounting as a discipline over time.

Although there can be many different theoretical approaches in the debate—there are five main ones to be summarized hereafter, and they can be called, in the absence of better words, namely, (a) the *mental* argument, (b) the *natural* argument, (c) the *societal* argument, (d) the *cultural* argument, and (e) the *double-sided* argument (which is my original contribution to the debate), to be elabroated hereafter, respectively.

Lest misunderstanding occurs, four clarifications should be made here, as this is something that I regularly emphasize in my previous books, whenever a theoretical debate is summarized for a topic, *almost verbatim*.

Firstly, there can be other arguments besides the first four here, that is, the "mental," the "natural," the "societal," and the "cultural" arguments (as indicated above). The advantage to select these four here has to do with their illustrative representation of the diverse theories in the literature (which are deemed sufficient for the purpose at hand in this book).

Secondly, the examples (theories) given to illustrate each argument are also supposed to be illustrative, not exhaustive. After all, the arguments will be addressed again in the rest of the book, with other examples (theories) to be introduced.

Thirdly, the four arguments are not necessarily mutually exclusive. For instance, those who advocate the "mental" argument can also consider the "cultural" argument, although they do not focus on the latter. The reverse is also true, in that those who make the "cultural" argument can also consider the "mental" argument, although they do not focus on the latter. In other words, their disagreement is often one in degree, not in kind.

And fourthly, any given case study for one of the two dialectical opposites (in one chapter) is not completely unrelated to the other opposite (in another chapter)—depending on the particular meaning and context in which the case study is used; therefore, the same case study can be used to analyze the other opposite, though in a different context altogether.

The Mental Argument

With these clarifications in mind—the first major theoretical approach is called, in the absence of better words, the *mental* argument, which focuses, relatively speaking of course, more on the physiological, biological and psychological aspects of accounting.

Thus, the word "mental" in the title of the argument is suggestive, because it refers to the focus on the physiological, biological and psychological aspects of accounting so understood.

But the mental argument has different issues of concern, and different defenders of the argument offer their own distinctive stands on the issues, depending on their own interests.

For illustration, consider two major versions of the mental argument (with the first on addition and the second on subtraction), namely, (a) addition, deception, and creative accounting, and (b) subtraction, cognitive biases, and mental accounting—to be addressed below, in that order.

Addition, Deception, and Creative Accounting

The first version of the mental argument has to do with addition, deception, and creative accounting (in the context of addition) in accounting.

An argument in the literature is that people use "creative accounting" to complicate (or add) things (not necessary) so as to deceive others for private gains.

In other words, "creative accounting and earnings management are euphemisms referring to accounting practices that may follow the letter of the rules of standard accounting practices, but certainly deviate from the spirit of those rules. They are characterized by excessive complication and the use of novel ways of characterizing income, assets, or liabilities and the intent to influence readers towards the interpretations desired by the authors." (WK 2013q)

However, to the critics, this argument on addition, deception, and creative accounting has some major problems. Consider, for illustration, a few examples below.

Firstly, one main criticism of of the argument on addition, deception, and creative accounting is that the problem with accounting scandals is not wholly the fault of creative accountants.

For instance, to the critics, although "earnings management is seen as a pressing issue in current accounting practice," one needs to understand that "part of the difficulty lies in the accepted recognition that there is no such thing as a single 'right' earnings figure," so this tempts "creative accountants" to offer alternative interpretations of earnings figures. (WK 2013q)

And secondly, another main criticism of the argument on addition, deception, and creative accounting is that the nature of business cycles in the free market makes it easier for creative accountants to cheat.

For instance, as the critics are right to point out, "conformance to acceptable accounting practices is a matter of personal integrity. Aggressive earnings management becomes more probable when a company is affected by a downturn in business." (WK 2013q)

Subtraction, Cognitive Biases, and Mental Accounting

And the second version of the mental argument has to do with subtraction, cognitive biases, and mental accounting (in the context of subtraction) in accounting.

An argument in the literature is that people use "mental accounting" to "code, categorize and evaluate economic outcomes" so as to simplify (or subtract things from) their decision-making process.

Richard Thaler (1980) first proposed this "mental accounting" theory, and "one detailed application of mental accounting, the behavioral life cycle hypothesis (Shefrin & Thaler, 1988), posits that people mentally frame assets as belonging to either current income, current wealth or future income and this has implications for their behavior as the accounts are largely non-fungible and marginal propensity to consume out of each account is different." (WK 2013o)

However, to the critics, the argument on subtraction, cognitive biases, and mental accounting has some major problems. Consider, for illustration, a few examples below.

Firstly, one main criticism of the argument on subtraction, cognitive biases, and mental accounting is that the issue of "rationality" in the debate on mental accounting can be relative.

For instance, in the fourth example about "bonuses," spending them (on "boats, new cars or lavish vacations" instead of something else more productive if the money comes from a different source) may appear questionable from the "rational" perspective of the principle of "fungibility," since "different mental accounts violates one of the basic rules of economics—that money is fungible, or interchangeable" (and should be treated the same in different mental accounts). (S. Vedantam

2007) But why should the economic principle of "fungibility" be so god-like that no violation is allowed or that it must be so desirable? As the poet Walt Whitman well put it, "Do I contradict myself? Very well then I contradict myself, (I am large, I contain multitudes.)" (QUO 2013)

And secondly, another main criticism of the argument on subtraction, cognitive biases, and mental accounting is that cognitive biases in mental accounting, if they exist in some cases, can be more lasting than previously thought.

For instance, to the critics, even when those with cognitive biases in "mental accounting" receive treatment from a CBM [Cognitive Bias Modification] program, "one concern is whether CBM modification procedures will achieve lasting benefits. This is not yet clear from research. Another concern is that participants with cognitive vulnerability are not able to maintain their new learning even within a CBM course, with their negative cognitive bias beginning to re-emerge over time." (WK 2013p)

The Natural Argument

The second theoretical approach is called, in the absence of better words, the *natural* argument, which focuses, relatively speaking of course, more on the natural aspects of accounting, especially in relation to the state of nature.

Therefore, the word "natural" in the title of the argument is suggestive, because it refers to the focus on the natural aspects of accounting so understood.

Like the mental argument—the natural argument has different issues of concern, and different defenders of the argument offer their own distinctive stands on the issues, depending on their own interests.

For illustration, consider two major versions of the natural argument (with the first on addition and the second on subtraction), namely, (a) addition, eco-management, and audit scheme, and (b) subtraction, carbon accounting, and environmental sustainability—to be summarized below, in that order.

Addition, Eco-Management, and Audit scheme

The first version of the natural argument has to do with addition, eco-management, and audit scheme (in the context of addition) in accounting.

An argument in the literature is that a voluntary environmental management instrument like an "audit scheme" for "eco-management" can be used to increase "environmental performance." (WK 2013)

A good example is "the Eco-Management and Audit Scheme (EMAS)...developed in 1993 by the European Commission. It enables organizations to assess, manage and continuously improve their environmental performance. The scheme is globally applicable and open to all types of private and public organizations. In order to register with EMAS, organisations must meet the requirements of the EU EMAS-Regulation. Currently, more than 4,600 organisations and more than 7,900 sites are EMAS registered." (WK 2013r)

However, to the critics, this argument on addition, eco-management, and audit scheme has some major problems. Consider, for illustration, a few examples below.

Firstly, one main criticism of the argument on addition, eco-management, and audit scheme is that the required measurement to implement EMAS is not practical.

For instance, to the critics, "given the heterogeneity of registered organisations" (with "more than 4,600 organisations and more than 7,900 sites"), "calculating average benefits and costs of EMAS is nearly unfeasible." (WK 2013r)

And secondly, another main criticism of the argument on addition, eco-management, and audit scheme is that the benefits of EMAs vary from country to country and from organization to organization.

For instance, to the critics, "the financial benefits of an EMAS registered organisation can differ per country. For example, some countries offer substantial reductions of waste fees, lower permitting costs, faster licensing procedures, etc."—but others do not. (WK 2013r)

Subtraction, Carbon Accounting, and Environmental Sustainability

And the second version of the natural argument has to do with subtraction, carbon accounting, and environmental sustainability (in the context of subtraction) in accounting.

An argument in the literature is that "carbon accounting" can be used to "'measure' amounts of carbon dioxide equivalents emitted by an entity," so as to contribute to the reduction of carbon emissions.

For example, "carbon accounting...is used *inter alia* by nation states, corporations and individuals. Correspondingly, examples for products based upon forms of carbon accounting can be found in national inventories, corporate environmental reports or carbon footprint calculators. Likening sustainability measurement, as an instance of ecological modernisation discourses and policy, carbon accounting is hoped to provide a factual ground for carbon-related decision-making," so as to further reduce carbon emissions in practice. (WK 2013s)

However, to the critics, this argument on subtraction, carbon accounting, and environmental sustainability has some major problems. Consider, for illustration, a few examples below.

Firstly, one main criticism of the argument on subtraction, carbon accounting, and environmental sustainability is that the practice of carbon accounting is not objective at all.

For instance, to the critics, "social scientific studies of accounting challenge this hope, pointing to the socially constructed character of carbon conversion factors or of the accountants' work practice which cannot implement abstract accounting schemes into reality." (WK 2013s; L. Lohmann 2009; I. Lippert 2012; D. MacKenzie 2009)

And secondly, another main criticism of the argument on subtraction, carbon accounting, and environmental sustainability is that different carbon accountants can produce different results (in addition to the problem of "abstraction" in accounting schemes as mentioned above).

For instance, to the critics, "while natural sciences claim to know and measure carbon, for organisations it is usually easier to employ forms of carbon accounting to represent carbon. The trustworthiness of accounts of carbon emissions can easily be contested....Science and Technology Studies scholar Donna Haraway's pluralised concept of knowledge, i.e. knowledges, can well be used to understand better the status of knowledge produced by carbon accounting: carbon accounting produced a version of understanding of carbon emissions. Other carbon accountants would produce other results." (WK 2013s)

The Societal Argument

The third theoretical approach is called, in the absence of better words, the *societal* argument, which focuses, relatively speaking of course, more on the societal aspects of accounting (e.g., organizational, institutional, structural, and technological issues).

In this sense, the word "societal" in the title of the argument is suggestive, because it refers to the societal aspects of accounting so understood.

Like the mental and natural arguments—the societal argument has different issues of concern, and different defenders of the argument offer their own distinctive stands on the issues, depending on their own interests.

For illustration, consider two major versions of the societal argument (with the first on addition and the second on subtraction), namely, (a) addition, financial performance, and auditing, and (b) subtraction, computer programming, and bookkeeping—to be summarized in what follows, in that order.

Addition, Financial Performance, and Auditing

The first version of the societal argument has to do with addition, financial performance, and auditing (in the context of addition) in the study of accounting.

An argument in the literature is that an "audit" is important for a business to increase financial performance.

In other words, "auditing is a vital part of accounting. Traditionally, audits were mainly associated with gaining information about financial systems and the financial records of a company or a business. Financial audits are performed to ascertain the validity and reliability of information, as well as to provide an assessment of a system's internal control." (WK 2013t)

However, to the critics, this argument on addition, financial performance, and auditing has some major problems. Consider, for illustration, a few examples below.

Firstly, one main criticism of the argument on addition, financial performance, and auditing is that an audit cannot guarantee a financial statement to be "free from material error."

For instance, as the critics are quick to point out, "due to constraints, an audit seeks to provide only reasonable assurance that the statements are free from material error. Hence, statistical sampling is often adopted in audits." (WK 2013t)

And secondly, another main criticism of the argument on addition, financial performance, and auditing is that even the conventional standard of "reasonable assurance" is no longer sufficient for some people. .

For instance, to the critics, "recently, the argument that auditing should go beyond just true and fair is gaining momentum. And the US Public Company Accounting Oversight Board has come out with a concept release on the same." (WK 2013t; F. McKenna 2011)

Subtraction, Computer Programming, and Bookkeeping

And the second version of the societal argument has to do with subtraction, computer programming, and bookkeeping (in the context of subtraction) in accounting.

An argument in the literature is that computer programming can be used to make bookkeeping easier, especially with the reduction of the need for "boilerplate." (WK 2013u)

In other words, "in computer programming, boilerplate code or boilerplate is the sections of code that have to be included in many places with little or no alteration. It is more often used when referring to languages that are considered verbose, i.e. the programmer must write a lot of code to do minimal jobs. The need for boilerplate can be reduced through high-level mechanisms such as metaprogramming (which has the computer automatically write the needed boilerplate text), convention over configuration (which provides good default values, reducing the need to specify program details in every project) and model-driven engineering (which uses models and model-to-code generators, eliminating the need for boilerplate manual code)." (WK 2013u)

And a good example is the use of "object-oriented programming" (or "OOP" in short), and this reduction of the need for boilerplate is important for bookkeeping…." (WK 2013u)

However, to the critics, this argument on subtraction, computer programming, and bookkeeping has some major problems. Consider, for illustration, a few examples below.

Firstly, one main criticism of this argument on subtraction, computer programming, and bookkeeping is that computer programming often does not model the "real world."

For instance, to the critics, "not everyone agrees that OOP facilitates direct real-world mapping…; Bertrand Meyer argues in *Object-Oriented Software Construction* that a program is not a model of the world but a model of some part of the world; 'Reality is a cousin twice removed.' At the same time, some principal limitations of OOP had been noted. For example, the Circle-ellipse problem is difficult to handle using OOP's concept of inheritance." (WK 2013v; M. Trofimov 1993; B. Meyer 1997)

Secondly, another main criticism of this argument on subtraction, computer programming, and bookkeeping is that OOP is too obsessed with "objects" (or "nouns") but not other things (like "verbs" or "actions," etc.).

For instance, the critic Alexander Stepanov (2010) "compares object orientation unfavourably to multimethods," because "OOP…attempts to decompose the world in terms of interfaces that vary on a single type [for objects]…. It claims that everything is an object. Even if it is true it is not very interesting—saying that everything is an object is saying nothing at all…." (WK 2013v)

The Cultural Argument

And the fourth theoretical approach is called, in the absence of better words, the *cultural* argument, which focuses, relatively speaking of course, more on the cultural aspects of accounting (e.g., issues in relation to values and beliefs).

Therefore, the word "cultural" in the title of the argument is suggestive, because it refers to the focus on the cultural aspects of accounting so understood.

Like the mental, natural, and societal arguments—the cultural argument has different issues of concern, and different defenders of the argument offer their own distinctive stands on the issues, depending on their own interests.

For illustration, consider two major versions of the cultural argument (with the first on addition and the second on subtraction), namely, (a) addition, accounting, and principles, and (b) subtraction, accounting scandals, and ethics—to be summarized in what follows, in that order.

Addition, Accounting, and Principles

The first version of the cultural argument has to do with addition, accounting, and principles (in the context of addition) in accounting.

An argument in the literature is that certain accounting principles which have developed over the last few hundreds of years govern the convention of doing accounting.

For instance, "accounts are prepared by reference to accounting principles. Of these, three key ones are: (1) the accounting entity, (2) the unit of measure and (3) the historic cost convention." (ACCO 2013) The first one, "the principle of the accounting entity," means that "it is separate and distinct from its owners and other firms"; the second one, "the principle of the unit of measure," means that "transactions should be included using a standard unit of measure and that the monetary unit is used"; and the third one, "the historical costs convention," means that "costs incurred in the past…is the basis for how items are valued." (ACCO 2013)

However, to the critics, the argument on addition, accounting, and principles has some major problems. Consider, for illustration, a few examples below.

Firstly, one main criticism of the argument on addition, accounting, and principles is that the princples are too past-oriented.

For instance, to the critics, "the historical costs convention…has always been difficult since significant areas of financial accounting are not based on historical costs. Many require an assessment of cost (or value) based on a forecast of the future where costs have not yet been incurred. Think of the provision for bad debts, fair value and depreciation." (ACCO 2013)

And secondly, another main criticism of the argument on addition, accounting, and principles is that the princples can be non-objectively implemented.

For instance, "the principle of the unit of measure is that transactions should be included using a standard unit of measure and that the monetary unit is used, which means that accounts will be restricted to those transactions that can be expressed using monetary units." (ACCO 2013) But there are other transactions which cannot be so easily expressed using monetary units, and good examples include "carbon trading," because of "the socially constructed character of carbon conversion factors or of the accountants' work practice which cannot implement abstract accounting schemes into reality." (ACCO 2013; WK 2013s; L. Lohmann 2009; I. Lippert 2012; D. MacKenzie 2009)

Subtraction, Accounting Scandals, and Ethics

And the second version of the cultural argument has to do with subtraction, accounting scandals, and ethics (in the context of subtraction) in accounting.

An argument in the literature is that there is accounting ethics which sets up standards for accountants to follow, so as to minimize or reduce accounting scandals.

For instance, "accounting ethics is primarily a field of applied ethics, the study of moral values and judgments as they apply to accountancy. It is an example of professional ethics. Accounting ethics were first introduced by Luca Pacioli, and later expanded by government groups, professional organizations, and independent companies. To combat...and prevent fraudulent accounting, various accounting organizations and governments have developed regulations and remedies for improved ethics among the accounting profession." (WK 2013w)

However, to the critics, this argument on subtraction, accounting scandals, and ethics has some major problems. Consider, for illustration, a few examples below.

Firstly, one main criticism of this argument on subtraction, accounting scandals, and ethics is that there are business scandals in spite of the existence of accounting ethics.

For instance, to the critics, "due to range of accounting services and recent corporate collapses, attention has been drawn to ethical standards accepted within the accounting profession. These collapses have resulted in a widespread disregard for the reputation of the accounting profession." (WK 2013w; B. Jackling 2007; S. Dellaportas 2006)

And secondly, another main criticism of this argument on subtraction, accounting scandals, and ethics is that different countries have different standards and regulations in regard to the enforcement of accounting practices.

For instance, "most countries have differing focuses on enforcing accounting laws. In Germany, accounting legislation is governed by 'tax law'; in Sweden, by 'accounting law'; and in the United Kingdom, by the 'company law.' In addition, countries have their own organizations which regulate accounting. For example, Sweden has the Bokföringsnämden (BFN - Accounting Standards Board), Spain the Instituto de Comtabilidad y Auditoria de Cuentas (ICAC), and the United States the Financial Accounting Standards Board (FASB)." (WK 2013w; C. Gowthorpe 1998)

The Double-Sided Argument

And fifthly, unlike the previous four arguments (viz., the mental, the natural, the societal, and the cultural)—the fifth one to be introduced here can be called, in the absence of better words again, the *double-sided* argument, which learns from all other theoretical approaches (which include the previous four arguments and also other examples as will be introduced in the rest of this book), without, however, favoring any of them, nor trying to integrate them (as they are not necessarily compatible with each other, as the introduction of the theoretical debate in the previous sub-sections has shown).

The Double-Sided Theory of Accounting

My *double-sided* argument can be more precisely called *the double-sided theory of accounting*, since it is to propose a new theory of addition and subtraction (especially, though not solely, in the context of accounting).

My *double-sided theory of accounting* has four major distinct features.

Firstly, my theory makes good use of all theoretical approaches to accounting in relation to addition and subtraction, be they about the mental argument, the natural argument, the societal argument, or the cultural argument—especially from the most comprehensive combined perspectives of the mind, nature, society, and culture (as will be dealt with in Chapter Two and Chapter Three).

Secondly, just like many other theories of mine in my previous books, my theory here does not heavily favor any specific theory over others in the literature, nor trying to integrate them (as they are not necessarily compatible with each other).

Thirdly, my theory re-interprets accounting in the distinctive dialectic context of no analysis of addition without that of subtraction, and vice versa—to be eventually transcended altogether later.

And fourthly, it contains thirty-six major theses, namely, (a) the first thesis: the formalness-informalness principle, (b) the second thesis: the absoluteness-relativeness principle, (c) the third thesis: the partiality-totality principle, (d) the fourth thesis: the predictability-unpredictability principle, (e) the fifth thesis: the explicability-inexplicability principle, (f) the sixth thesis: the fiction-reality principle, (g) the seventh thesis: the cognitiveness-noncognitiveness principle, (h) the eighth thesis: the finiteness-transfiniteness principle, (i) the ninth thesis: the preciseness-vagueness principle, (j) the tenth thesis: the simpleness-complicatedness principle, (k) the eleventh thesis: the openness-hiddenness principle, (l) the twelfth thesis: the denseness-emptiness principle, (m) the thirteenth thesis: the rule-exception principle, (n) the fourteenth thesis: the prototypicality-variation principle, (o) the fifteenth thesis: the change-constancy principle, (p) the sixteenth thesis: the order-chaos principle, (q) the seventeenth thesis: the slowness-quickness principle, (r) the eighteenth thesis: the expansion-contraction principle, (s) the nineteenth thesis: the optimality-nonoptimality principle, (t) the twentieth thesis: the simultaneity-nonsimultaneity principle, (u) the twenty-first thesis: the isolation-interaction principle, (v) the twenty-second thesis: the theory-praxis principle, (w) the twenty-third thesis: the convention-novelty principle, (x) the twenty-fourth thesis: the evolution-transformation principle, (y) the twenty-fifth thesis: the symmetry-asymmetry principle, (z) the twenty-sixth thesis: the softness-hardness principle, (aa) the twenty-seventh thesis: the seriousness-playfulness principle, (bb) the twenty-eighth thesis: the activeness-inactiveness principle, (cc) the twenty-ninth thesis: the selfness-otherness principle, (dd) the thirtieth thesis: the regression-progression principle, (ee) the thirty-first thesis: the sameness-difference principle, (ff) the thirty-second thesis: the stability-reaction principle, (gg) the thirty-third thesis: the functionality-nonfunctionality principle, (hh) the thirty-fourth thesis: the intentionality-nonintentionality principle, (ii) the thirty-fifth thesis: the survivability-non-survivability principle, and (jj) the thirty-sixth thesis: the post-human double-sidedness—to be elaborated in the rest of the book.

Because of the last thesis on "the post-human double-sidedness," the word "double-sidedness" in the title of my theory is suggestive, because it refers to the dual analyses of both addition and subtraction in the history of accounting —which then require a multi-faceted inquiry of the dialectic relationship between addition and subtraction, as well as an understanding of the different views regarding the relationships in accounting (together with other debates as will be discussed in the rest of the book)—and in the end it calls for a need to transcend both addition and subtraction by future advanced humans and later post-humans in the context of the great transformations of accounting in the future worlds (as will be discussed in the concluding chapter, Chapter Four).

With this clarification in mind—the seminal project here will fundamentally change the way that we think about addition and subtraction (especially, though not solely, in the context of accounting), together with other debates as will be discussed in the rest of the book, from the combined perspectives of the mind, nature, society, and culture, with enormous implications for the human future and what I originally called its "post-human" fate.

Theory and Meta-Theory

The *double-sided* argument, or *the double-sided theory of accounting*, presupposes some kind of meta-theory, in special relation to methodology and ontology—just as all theories, either explicitly or implicitly, require their own versions of meta-theory.

Meta-Theory

As so often clarified in my previous books—theory and meta-theory, broadly speaking, can be further classified into different categories, as shown below:

- On Theory
  - —Ex: in relation to the mind
  - —Ex: in relation to nature
  - —Ex: in relation to society
  - —Ex: in relation to culture
- On Meta-Theory
  - —Ex: in relation to methodology
  - —Ex: in relation to ontology

Each category, in its turn, can then be further sub-divided into different levels of analysis—which will be elaborated in the next two sections (and summarized in *Table 4.43*).

With this reminder in mind, I already proposed (in my previous books) my distinctive contribution to meta-theory, namely, methodology (like "theory of methodological holism," "complex theory of data analysis," "critical-dialectic theory of formal science," and "interpretivist theory of probability," for instance) and ontology (like "theory of existential dialectics" and "contrastive theory of rationality," for instance).

The next two sections are to introduce my meta-theory—especially, existential dialectics and sophisticated methodological holism, respectively—to be repeated, *often verbatim* (with only a few updated revisions), from my previous works, as this is something that I often do in each new book of mine as an introduction to some background information for the convenience of the readers.

In the process, I shall also introduce my numerous new theories (or to be exact, 72 of them in 65 books so far, as of January 05, 2014) in relation to the four categories of the mind, nature, society, and culture in the context of theory (as also summarized in *Table 4.43*).

A Unified Theory of Everything

In the end, my distinctive contribution to both meta-theory and theory serves as a foundation to unify all domains of knowledge (i.e., the social sciences, the formal sciences, the natural sciences, and the humanities) for an overarching understanding of all forms of reality (by way of some ontological principles at the meta-theoretical level and the comprehensive perspectives of the mind, nature, society, and culture at the theoretical level—as will be explained in the next two sections).

This is so, without commiting the sins of reductionism and reverse-reductionism (as often seen in many holistic approaches, with the fad of systems approach as a most recent notorious example, as already critically explained in *The Future of Post-Human Formal Science*, or *FPHFS*), since my distinctive contribution makes good use of different schools of thought without heavily favoring any of them nor trying to integrate them (as they are not necessarily compatible with each other), so as to include (or adjust for) heterogeneity, conflict, subjectivity, and complexity, for example.

In this sense, the word "unified" does not have to mean an integrative approach, in a narrow sense, nor a systems approach, in a broad one, both of which I have critically rebuked (in my previous books), while learning from them.

This overarching project (consisting of numerous theories of mine, as will be summarized in the next section), thus, constitutes my grand unifed theory of everything in the world and beyond.

The Logic of Existential Dialectics

To start, the ontology of existential dialectics can be summarized in this section—*often verbatim* (with only a few updated revisions), from my previous works, as this is something that I regularly do in each new book of mine as an introduction to some background information for the convenience of the readers.

After all, this book, like all others of mine, is in conversation with all previous ones of mine, for the final aim to converge all of them into a larger grand project on the future of intelligent life, both here on earth and there in deep space unto multiverses.

The summary can be organized in relation to (a) the conception of existential dialectics (or its ontics), (b) the syntax of existential dialectics (or its ontomethodology and ontologic), (c) the semantics of existential dialectics (or its ontosemantics), and (d) the pragmatics of existential dialectics (or its ontopragmatics), respectively hereafter—again *often verbatim*.

This means that, for those readers who had read some of my previous books, the summary below is really not much new (*almost verbatim*, with only some updated revisions).

But for those who have never read my previous books, it is a good review of them.

The Conception of Existential Dialectics (or Its Ontics)

This ontology was originally designated as "existential dialectics" in *Beyond Capitalism to Post-Capitalism* (herein abbreviated as *BCPC*), although it was already analyzed in other books of mine like the 2 volumes of *The Future of Human Civilization* (*FHC*), *The Future of Capitalism and Democracy* (*FCD*), *The Future of Post-Human Consciousness* (*FPHC*), and the 2 volumes of *Beyond Democracy to Post-Democracy* (*BDPD*).

Lest any confusion occurs, it is important to stress at the outset that the word 'existential' in "existential dialectics" has nothing to do with Existentialism, which I rebuked in *FHC*, *FCD*, and also *FPHC*. Instead, the word here merely refers to the existence of intelligent life (both primitive and advanced) in a broad sense.

Some Basic Concepts

The conception of existential dialectics (or its ontics) makes use of different concepts (like "sets," "elements," "relations," "operations," "functions," "truth values," "axioms," "postulates," and "principles"—as shown in *Table 4.3*), which are important for the understanding of any logic of ontology.

That clarified—existential dialectics, as a language of ontology, can be analyzed in different ways that a language, as an analogy, is often analyzed, namely, in relation to phonology ("the study of patterns of a language's sounds"), phonetics ("the study of the physical aspects of sounds of human language"), morphonology ("the study of the internal structure of words"), syntax ("the study of how words combine to form grammatical sentences"), semantics ("the study of the meaning of words [lexical semantics] and fixed word combinations [phraseology]"), and pragmatics ("the study of how utterances are used…literally, figuratively, or otherwise…in communicative acts"). (WK 2007)

But since my theory of existential dialectics makes use of the English language for communication here, it does not propose a new way to make the patterns of a language's sounds (as in phonology), to study the physical aspects of sounds (as in phonetics), or even to strive for a new organization of the internal structure of words for the English language (as in morphonology).

For this reason, the language of existential dialectics to be analyzed hereafter is not concerned with the phonological, phonetic, and morphonological aspects of the English language. Instead, the analysis will explore the syntax, semantics, and pragmatics of existential dialectics as a language of ontology to improve our understanding of reality.

The Ontic Dispute

But first, there is an ontic dispute to be addressed, which is related to what I called "the dilemma of ontology" as introduced earlier in *Sec. 1.4* of *The Future of Post-Human Knowledge* (*FPHK*).

A good point of departure concerns what I want to call *the ontic dispute*, on how to understand the very nature of ontology.

Traditionally, scholars in the field used to study the conception of ontology in terms of its entities (e.g., material objects, abstract numbers), properties (e.g., duration, plurality), and relationships (e.g., causation).

This is a good starting point, but it suffers from different problems. Consider three examples of bias below, which are quite common in the literature, and the recent version (of treating concepts and principles in ontology on the basis of "plurality," "dynamism," "duration," "interaction," "life," "consciousness," and "volition") by Reginald Firehammer (2005) only constitutes a latest addition to this old (impoverishing) tradition.

Firstly, the classification of entities, properties, and relationships is too material-centric, in often focusing more on the existence of material entities than on non-material ones.

Secondly, the classification is too anthropo-centric, in often favoring the factor of human agency in depicting reality.

And lastly, the classification is too confused among the three categories relating to each other, in often lumping them together without adequate consideration of their intricated interrelationships (both in relation to each and in relation to all).

A good alternative is precisely what I want to offer in this debate, that is, a new classification of ontological entities, properties, and their relationships in terms of *method*, *structure*, *process*, *agency*, and *outcome* instead. Let me explicate what these words mean below.

In relation to "method"—it concerns how ontology is to be studied. In this sense, the word 'method' here should not be confused with the different usage of the same term 'method' as a technique of investigation in the context of methodology (as already introduced in *Sec. 1.2* of *FPHK*).

In relation to "structure"—it concerns what the general patterns of entities, properties, and relationships are.

In relation to "process"—it concerns how these general patterns change over time.

In relation to "agency"—it concerns how agency affects any change of these general patterns over time (without putting, however, too much emphasis on the role of human agency).

And in relation to "outcome"—it concerns what the outputs of this change over time are.

The Syntax of Existential Dialectics (or Its Ontomethodology and Ontologic)

The syntax of existential dialectics, analogously speaking, refers here to the structure of an ontology which can be studied by way of combining ideas into complex relationships like ontological principles to depict reality in the world—and can be analyzed in seven sub-sections, namely,

(a) the selection criteria, (b) further clarifications on classification, (c) ontomethodology, (d) ontologic, (e) the principles as short cuts, (f) the principles as family resemblances, and (g) the dialectic constraints imposed by the principles, respectively hereafter.

The Selection Criteria

To start, not any pair of relationship can be chosen as an ontological principle in existential dialectics; otherwise, there could be billions of them under the sky.

Six illustrations below suffice to clarify this point.

(a) Firstly, an ontological principle cannot *overlap* with other principles, in that it is too closely related with other ones.

Of course, one cannot totally rule out any relationship between two entities, especially in a complex world of everything being linked to everything else—but the comparison here is relative, not absolute.

In this sense, the flexibility-inflexibility pair cannot be an ontological principle, since it has something closely in common with other ones (like the simpleness-complicatedness principle, for instance).

As an illustration, something which is flexible in interpretation is more likely to allow multiple viewpoints and thus makes the overall picture more complicated than otherwise (simple).

(b) Secondly, an ontological principle cannot be *redundant* in relation to other principles, in that it somehow duplicates other principles.

In this sense, for instance, the directness-indirectness pair cannot be an ontological principle, because it somehow duplicates other ontological principles (like the preciseness-vagueness principle, for instance).

As an illustration, something which is direct means to be right to the point and does not need to go to an unnecessarily long loop—so it is less vague, in being clearer about the thing in question.

(c) Thirdly, an ontological principle cannot be *derived* from other ones—as if it is a child-parent relationship.

In this sense, the convenience-inconvenience pair cannot be an ontological principle, because it can be derived from other ones (like the slowness-quickness principle, for instance).

For instance, something which is convenient already implies that it can be obtained quickly (rather than slowly)—all things considered.

(d) Fourthly, an ontological principle cannot be *trivial* (or *parochial*)—as if it is only one component competing with other ones for inclusion in a set.

In this sense, the consistency-inconsistency pair cannot be an ontological principle, because its validity is limited (or parochial), as it can be easily challenged and replaced by opposing major (not minor) theories (e.g., the correspondence theory of truth, the pragmatic theory of truth, the semantic theory of truth, the disclosedness theory of truth, and whatnot), not just in accordance to the consistency (or coherence) theory of truth (for consistency).

Thus, the consistency argument is only one among different opposing arguments in relation to the logic of what constitutes "truth."

(e) Fifthly, an ontological principle cannot be *one-sided*, that is, part of one side (category) of a dialectic pair.

In this sense, the extroversion-introversion pair cannot be an ontological principle, since it is part of the side (category) of "normality" in the normality-abnormaltiy pair (because both extroversion and introversion are two different "normal" personality traits in psychology).

In addition, the normality-abnormality pair is already related to the symmetry-asymmetry principle, the density-emptiness principle, the convention-novelty principle, and the same-difference principle (as will be explained in the section on "the principles as family resemblances" later).

(f) And sixthly, an ontological principle cannot be *misleading*, in that the two apparent opposites are not really opposites at all, upon closer examination.

In this sense, the east-west pair cannot be an ontological principle, since east and west are not really opposites, as there are also north, south, and center to form a set of five opposing elements.

By the same logic, the left-right pair cannot be an ontological principle, since left and right (width) are not really opposites, as there are also top and bottom (height), back and front (depth), and center to form a set of seven elements in four categories. In the field of visual design, this set of seven elements is known as "form" (as opposed to "function"), as addressed in *FPHARCH*. (WK 2011nn)

Although these six examples (criteria) are not exhaustive, they illustrative the selectiveness of any ontological principle to be existentially dialectic.

Further Clarifications on Classification

Lest any misunderstanding carelessly occurs, some further clarifications on classification should be made here (and summarized in *Table 4.8*).

(a) Firstly, the total number of ontological principles is unknown, to be discovered later, as our knowledge of the world becomes more advanced.

Consequently, the principles as introduced in my books are not exhaustive, with new ones being added, whenever more of them are discovered in later research.

At least, future generations can pick up where I leave off and continue the discovery.

In this sense, my work should be treated as a pioneering effort for the development of a systematic, comprehensive analysis of a new general ontology for the future of knowledge.

(b) Secondly, the principles are not rigidly classified, as they can be reclassified in a different way.

For instance, the preciseness-vagueness principle is classified under the category of "structure" but can be reclassified under the category of "method," although in so doing, it has a different meaning in the context of method.

The same logic applies to the same-difference principle under the category of "outcome," which can be reclassified under the category of "structure," although in so doing, once more, it has a different meaning in the context of structure.

(c) Thirdly, the principles are to be understood together, not that each prinicple is to be treated independently of others. After all, the principles are mutually constraining, in that they work together as a whole.

(d) Fourthly, the principles are relevant to all subject matters, but some principles are, relatively speaking of course, more useful to some subject matters than others under certain conditions—as implied in the symmetry-asymmetry principle.

(e) And fifthly, the principles constitute only two levels of analysis, this time, at the ontological and methodological levels—while other levels of analysis (from the perspectives of the mind, nature, society, and culture) are also needed, in order to understand reality in its totality (as explained in the section on "sophisticated methodological holism," so as to avoid the dual dangers of reductionism and reverse-reductionism).

Ontomethodology

With these selection criteria and clarifications in mind—the syntax of existential dialectics can be divided into two main smaller sections for analysis, since they are related to each other in the context of some fundamental principles, that is, (a) ontomethodology and (b) ontologic, to be addressed hereafter, respectively.

In the context of ontomethodology, some ontological principles are proposed here in relation to the nature of methodology, which however is tied up with the general issue of ontology in the context of knowledge.

In this sense, ontomethodology (in the context of these ontological principles in the category of method) are related with ontologic in general (in the context of ontological principles in general).

Some good instances of ontological principles involving ontomethodology include the formalness-informalness principle, the absoluteness-relativeness principle, the partiality-totality principle, the predictability-unpredictability principle, the explicability-inexplicability principle, the fiction-reality principle, and the cognitiveness-noncognitiveness principle, which will be introduced in conjunction with ontological principles in the ontologic of existential dialectics in general hereafter.

Ontologic

With this clarification of the relationship between ontomethodology and ontologic in mind (which is comparable to the relationship between methodology and ontology as described in *Sec. 1.3* of *FPHK*)—the first three principles in existential dialectics were introduced in *BCPC*. And more principles were later added to the logic in my subsequent books (like *FC*, *FAE*, and others), as explained below.

More specifically, in *BCPC*, I proposed three principles (based on previous works of mine, not just something out of the blue) for the ontological logic, namely, (a) the regression-progression principle on the direction of history, (b) the symmetry-asymmetry principle on the relationships among existents, and (c) the change-constancy principle on the alteration of things—or in a more elegant term, the dynamics of space-time (in *FC*).

In *The Future of Complexity* (*FC*), I added three more principles, on top of the three aforementioned, namely, (d) the partiality-totality principle on the relationships between parts and whole, (e) the order-chaos principle on the pattern of things, and (f) the predictability-unpredictability principle on the occurrence of things—as part of the ontological logic of existential dialectics.

In *The Future of Aesthetic Experience* (or simply *FAE*), another principle is added, namely, (g) the evolution-transformation principle (on the multiple kinds of agency).

Also in *FAE*, I then reclassified the 7 principles into four categories, namely, (i) in relation to method, for the partiality-totality principle and the predictability-unpredictability principle, (ii) in relation to process, for the change-constancy principle and the order-chaos principle, (iii) in relation to agency, for the symmetry-asymmetry principle and the evolution-transformation principle, and (iv) in relation to outcome, for the regression-progression principle.

In *The Rise of Authoritarian Liberal Democracy* (*ALD*), I added two more principles, namely, (h) the softness-hardness principle on the force of agency (which is to be classified under the category about agency) and (i) the same-difference principle on the metamorphosis of change (which is to be classified under the category about outcome).

In *The Future of Information Architecture* (*FIA*), I introduce three additional principles, that is, (j) the simpleness-complicadness principle on the interconnection among things, (k) the preciseness-vagueness principle on the refinement of things, and (l) the slowness-quickness principle on the speed of change—with the first two in relation to structure and the third in relation to process.

In *The Future of Post-Human Unconsciousness* (*FPHU*), I added three more principles, that is, (m) the openness-hidden principle on the detection of things, in relation to structure, (n) the explicability-inexplicability principle on the underlying mechanisms of things, in relation to method, and (o) the expansion-contraction principle on the growth of things, in relation to process.

In *The Future of Post-Human Knowledge* (*FPHK*), I added one more principle, that is, (p) the absoluteness-relativeness principle on the multiplicity of entities, in relation to method.

In *The Future of Post-Human Mathematical Logic* (*FPHML*), I added one more principle, that is, (q) the formalness-informalness principle on the formal requirements of systems, in relation to method.

In *The Future of Post-Human Engineering* (*FPHE*), I added one more principle, that is, (r) the theory-praxis principle on the duality of knowledge, in relation to agency.

In *The Future of Post-Human Creative Thinking* (*FPHCT*), I added one more principle, that is, (s) the convention-novelty principle on the nature of creative thinking, in relation to agency.

In *The Future of Post-Human Geometry* (*FPHG*) here, I added one more principle, that is, (t) the finiteness-transfiniteness principle on the nature of numbers, in relation to structure.

In *The Future of Post-Human Urban Planning* (*FPHUP*), I added one more principle, that is, (u) the denseness-emptiness principle on the distribution of entities in space, in relation to structure.

In *The Future of Post-Human Humor* (*FPHH*), I added one more principle, that is, (v) the seriousness-playfulness principle on the extent of seriousness, in relation to agency.

In *The Future of Post-Human Literature* (*FPHLIT*), I added one more principle, that is, (w) the fiction-reality principle on the different faces of reality, in relation to method.

In *The Future of Post-Human Chemistry* (*FPHCHEM*), I added one more principle, that is, (x) the stability-reaction principle on the feedback mechanism for further changes, in relation to outcome.

In *The Future of Post-Human Transportation* (*FPHTRAN*), I added one more principle, that is, (y) the activeness-inactiveness principle on the extent of activeness, in relation to agency.

In *The Future of Post-Human Culinary Art* (*FPHCA*), I added two more principles, that is, (z) the selfness-otherness principle on the magnitude of self-centeredness in relation to agency and (aa) the functionality-nonfunctionality principle on the presence of function in relation to outcome.

In *The Future of Post-Human Waste* (*FPHW*), I added one more principle, that is, (bb) the intentionality-nonintentionality principle on the planning for outcomes in relation to outcome.

In *Beyond Natural Resources to Post-Human Resources* (*BNR*), I added one more principle, that is, (cc) the cognitiveness-noncognitiveness principle on the understanding of meaning in relation to method.

In *The Future of Post-Human Probability* (*FPHPROB*),), I added one more principle, that is, (dd) the survivability-nonsurvivability principle on the survival of things in relation to outcome.

In *The Future of Post-Human Morphology* (*FPHMORP*),), I added two more principles, that is, (ee) the prototypicality-variation principle on the nature of duplication in relation to process, and (ff) the optimality-nonoptimality principle on the nature of highly favorable options in relation to process.

In *The Future of Post-Human Thanatology* (*FPHTHAN*,), I added one more principle, that is, (gg) the rule-exception principle on the extent of typicality in relation to structure.

In *The Future of Post-Human Phonology* (*FPHPHON*), I added one more principle, that is, (hh) the simultaneity-nonsimultaneity principle on the sequence of time in relation to process.

And in *The Future of Post-Human Accounting* (*FPHACCO*), I add one more principle, that is, (ii) the isolation-interaction principle on the interaction among entities in relation to process.

With this update, there are thirty-five principles so far in existential dialectics, that is:

- 7 principles in the category of "method"—viz., the formalness-informalness principle, the absoluteness-relativeness principle, the partiality-totality principle, the predictability-unpredictability principle, the explicability-inexplicability principle, the fiction-reality principle, and the cognitiveness-noncognitiveness principle
- 6 principles in the category of "structure"—viz., the finiteness-transfiniteness principle, the simpleness-complicatedness principle, the preciseness-vagueness principle, the openness-hiddenness principle, the denseness-emptiness principle, and the rule-exception principle
- 8 principles in the category of "process"—viz., the prototypicality-variation principle, the change-constancy principle, the slowness-quickness principle, the order-chaos principle, the expansion-contraction principle, the optimality-nonoptimality principle, the simultaneity-nonsimultaneity principle, and the isolation-interaction principle
- 8 principles in the category of "agency"—viz., the theory-praxis principle, the convention-novelty principle, the symmetry-asymmetry principle, the evolution-transformation principle, the softness-hardness principle, the seriousness-playfulness principle, the activeness-inactiveness principle, and the selfness-otherness principle
- 6 principles in the category of "outcome"—viz., the regression-progression principle, the same-difference principle, the stability-reaction principle, the functionality-nonfunctionality principle, the intentionality-nonintentionality principle, and the survivability-nonsurvivability principle

There can be more than thirty-five principles in existential dialectics, but they are what I have discovered so far (as summarized in *Table 4.4*).

The Principles as Short Cuts

Yet, these principles should be treated with caution, lest misunderstanding occurs, since they do not constitute rigid dualities (or dichotomies).

The reason is that each pair in an ontological principle consists of two opposites, which are, however, merely short cuts both for multiple variations and degrees, as well as for different interactions with multiple other entities unlike them.

In this light, each pair can end up having hundred (or even thousand, if not more) different versions, which interact with hundred (or even thousand, if not more) other entities—as summarized in *Table 4.5*.

There are two clarifications here.

Firstly, this conception of shortcuts is not mutually exclusive nor absolute, in that the opposites can come in all shapes and sizes, with different degrees. For instance, by analogy, just as there are different degrees of the two colors "white" and "black"—there are likewise different degrees of the opposites in each ontological pair, to the extent that there can be multiple entities (not only two) interacting with each other in each pair.

And secondly, this conception of shortcuts do not ignore other possible entities in interacting with the two opposites (with their different versions) in each pair. For instance, by analogy, there are not only the two colors "white" and "black" as opposites, since there are other colors too besides them like "yellow," "green," "purple," or else. The same logic can be applied to each ontological pair, in that they also interact with other entities, not with only two of them (with their different versions).

Therefore, with these two clarifications in mind—each pair in an ontological principle serves only as an abbreviation for something more complicated and, therefore, although it contains two opposites, it should not be confused as a duality (dualism).

Existential dialectics rejects any dualism (or dichotomy) as too rigid and instead allows the multiplicity of entities, to the extent that between the two opposites in each pair exist many other alternatives to choose from. They are named in that short form for aesthetic elegance only, instead of listing all possible entities between the two opposites in the title (which would be too cumbersome and sound extremely awkward).

To be dialectic is to go beyond any rigid dichotomy and transcend into something different altogether in the long haul. One may be tempted to call the dialectic logic here with a different name like existential "multilectics" (instead of "dialectics"), but this naming is incorrect (or even misleading), for the two reasons aforestated.

After all, all the possible entities between any two opposites in an ontological pair are still about the two opposites, although each ontological pair come in all shapes and sizes (with different degrees) and do not exist by themselves but also interact with other entities unlike them. The virtue of revealing an ontological pair is to show how they relate within themselves (in multiple versions) and also interact with others unlike them (also in multiple versions).

This will be clear in a later sub-section titled "the dialectic constraints imposed by the principles."

The Principles as Family Resemblances

In addition, each ontological principle is generic, with some other comparable ontological pairs to be put in the same family (like a *family resemblance*)—as summarized in *Table 4.6*.

Ludwig Wittgenstein (1953) once suggested the idea of "family resemblance" in explaining different games classified under the same family called *games*.

Why should, for instance, playing football and chess as playing "games," when it is well understood that football is not the same as chess? The answer is that, although each game is different and has different rules, many of them (though not all) share, more or less, some commonalites (e.g., scoring as necessary for winning).

And this is so, even though some games share more than some others in any given selection of critieria, and no two games are exactly identical. So, Wittgenstein's point here is that there is no essential core which is common to all games, and the best that one can look for is some characteristics which are common to many (but not all) games. (A. Biletzki 2006)

(a) Appropriate Family Resemblances

In the context of existential dialectics, I want to distinguish two types of family resemblances, and, in the absence of better words, let me call them (a) *appropriate* family resemblances and (b) *inappropriate* family resemblances.

In regard to appropriate family resemblances—good examples include some of the pairs as already explained in an earlier section on selection criteria, like the flexibility-inflexibility pair, which can be put in the family resemblance of the simpleness-complicatedness principle, although the two pairs are not exactly identical. Likewise, the directness-indirectness pair can be put in the family resemblance of the preciseness-vagueness principle, although, again, the two pairs are not exactly identical.

In this sense, which specific pair in a family should be used to designate the name of the family can be at times a bit arbitrary, but with good reason.

For illustration, in the context of method, the partiality-totality principle can take the different form like individualisticness-holisticness, just as the explicability-inexplicability principle can take the different form like underlyingness-regularness—although each two pairs are not exactly identical and have slightly different meanings and usages.

A more comprehensive listing of this family resemblance for all other principles is shown in *Table 4.6* on the syntax of existential dialectics.

In fact, this section on family resemblances overlaps with the semantics of existential dialectics, which will be introduced later in the next section (and summarized in *Table 4.11*).

(b) Inappropriate Family Resemblances

However, there are some family resemblances which are not appropriate, and let me distinugish them by calling them, in the absence of better words, (b1) *inadequate* family resemblances, (b2) *compound* family resemblances, (b3) *one-sided* family resemblances, and (b4) *vague* family resemblances (as summarized in *Table 4.10*).

(b1) In regard to "inadequate" family resemblances—there are some pairs which are not adequate, because they fail the test of selection criteria (as already pointed out in an earlier section), especially in relation to the criterion of not being *trivial* (or *parochial*)—as if it is only one component competing with other ones for inclusion in a set.

For instance, the *consistency-inconsistency* pair is inadequate, because it is only one component competing with other ones for inclusion in the larger set of the formalness-informalenss principle (like the *completeness-incompleteness* pair and the *soundness-unsoundness* pair, as discussed in my book titled *The Future of Post-Human Mathematical Logic*, or *FPHML* in short).

The same is true for the *cognition-noncognition* pair, which is related with both the completeness-incompleteness pair and the soundness-unsoundness pair in the larger set of the formalness-informalenss principle.

(b2) In regard to "compound" (or non-atomic) family resemblances—there are some pairs which are not appropriate for a different reason, because they fail the test of selection criteria (as already pointed out in an earlier section), especially in relation to the criterion of not *overlapping* with other principles, in that it is too closely related with other ones.

For instance, the *normality-abnormality* pair is compound (and thus inappropriate), because it overlaps with several other principles, like the symmetry-asymmetry principle, the density-emptiness principle, the convention-novelty principle, and the same-difference principle, for instance (as already discussed in *FPHP*).

The normality-abnormality pair overlaps with the symmetry-asymmetry principle, because being "abnormal" (say, for those with mental illnesses) is often treated not equally by others who are "normal." So, the relationship reveals the symmetry-asymmetry principle.

The normality-abnormality pair also overlaps with the density-emptiness principle, because being "abnormal" (say, for those with mental illnesses) is often not as dense in the overall (not local) distribution of a population as those who are "normal." So, the relationship reveals the density-emptiness principle.

The normality-abnormality pair also overlaps with the convention-novelty principle, because being "abnormal" (say, for those with mental illnesses) often exhibits non-conventional (non-conformist) ideas and behaviors when contrasted with those who are "normal." So, the relationship reveals the convention-novelty principle.

The normality-abnormality pair also overlaps with the same-difference principle, because being "abnormal" (say, for those with mental illnesses) shares some similarities in outcome when compared with those who are "normal" (in that both contribute to the control and regulation of human thoughts and behaviors in society, albeit in different ways). Yet, being "abnormal" also has something different from being "normal," in that abnormal thoughts and behaviors follow a different pattern when contrasted with those which are normal. So, the relationship reveals the same-difference principle.

(b2.1) There is a special case of compound family resemblances (as summarized in *Table 4.10*), which can be called, in the absence of better words, *one-sided and compound* family resemblances.

For instance, the extroversion-introversion pair is a good example of *one-sided and compound* family resemblances, because, firstly, it is "one-sided," in being part of the side (category) of "normality" in the normality-abnormality pair—and, secondly, the normality-abnormality pair is in turn "compounded," because it is related to the symmetry-asymmetry principle, the density-emptiness principle, the convention-novelty principle, and the same-difference principle, as already explained above.

(b3) In regard to "one-sided" family resemblances—there are some pairs which are not appropriate for another different reason, because they fail the test of selection criteria (as already pointed out in an earlier section), especially in relation to the criterion of not being *one-sided*, that is, part of one side (category) of a dialectic pair.

For instance, the extroversion-introversion pair cannot be an ontological principle, since it is part of the side (category) of "normality" in the normality-abnormaltiy pair (because both extroversion and introversion are two different "normal" personality traits in psychology).

By the same logic, the vision-blindness pair and the sound-silence pair cannot be ontological principles, because the pairs are part of the side (category) of "asymmetry" in the symmetry-asymmetry principle (since being able to see and hear for some individuals and being unable to see or hear for others are two contrastive abilities in the world of animals and humans, for instance).

These illustrations of indequate, compound, and one-sided family resemblances show how difficult it is to come up with an ontological principle in existential dialectics out of the blue, because many of them violate the selection criteria in one way or the other. It is all too easy to join two opposing words and simply make them into a pair, but it is all too difficult to pass the test of the selection criteria to be qualified as an ontological principle in existential dialectics.

(b4) And in regard to "vague" family resemblances—there are some pairs which are not appropriate for still another different reason, because the terms in the pairs are vague in meaning, so that they are inappropriate family resemblances if defined in some particular ways, although they can be appropriate family resemblances if defined in other particular ways.

For instance, such pairs like shortness-longness, narrowness-broadness, smallness-largeness, smallness-greatness, microness-macroness, and littleness-muchness can be understood in at least three different meanings and thus have three different possibilities. The first possibility is that they are one-sided as part of finiteness (in the finiteness-transfiniteness principle), if both opposites are defined on a finite scale. The second possibility is that they are one-sided as part of transfiniteness (in the finiteness-transfiniteness principle) if both opposites are defined on a transfinite scale. And the third possibility is that they are appropriate family resemblances in relation to the finiteness-transfiniteness principle if one opposite is defined on a finite scale and the other opposite is defined on a transfinite scale.

Other examples of inappropriate family resemblances are shown in *Table 4.10* on "The Syntax of Existential Dialectics VII: Types of Inappropriate Family Resemblances."

The Dialectic Constraints Imposed by the Principles

The principles, as they constitute the syntax of existential dialectics, are dialectic in character, such that, when they are applied, they impose dialectic constraints on how reality is to be understood.

Consider, say, the symmetry-asymmetry principle as an illustration here, in order to summarize two main characters of the dialectic constraints in question (as summarized in *Table 4.7*).

Firstly, to be dialectic here is to go beyond the narrow dichotomies (and, for that matter, any rigid multi-dimensional classificatory scheme), be they about "self" vs. "world," "freedom" vs. "unfreedom," "barbarity" vs. "civilization," "individuality" vs. "communality," and so on.

One way to do so (to go beyond) is to consider them all in terms of co-existence (without heavily favoring one over the rest). For instance, my theory of "post-civilization" (to be summarized later in the section on the pragmatics of existential dialectics) is to go beyond barbarity and civilization in terms of understanding barbarity and civilization as being co-existent.

And the same logic can be said in relation to my theories of "post-democracy," "post-capitalsm," and others (also to be introduced later in the section on the pragmatics of existential dialectics), in regard to freedom vs. unfreedom, equality vs. inequality, communality vs. individuality, spirituality vs. materiality, and so on.

But to consider them all (in the dichotomies—and, for that matter, in any rigid multi-dimensional classifactory scheme) as co-existent is not the same as to imply that the opposing categories in any classificatory scheme are all equal, since, in accordance to the symmetry-asymmetry principle (as an illustration here), if they are equal in terms of being considered as co-existent, they are asymmetric in terms of being unequal in dialectic interaction (e.g., X can be more dominant than Y in case A, or Y is more dominant than X in case B), although in some special cases, they can be relatively equal (e.g., X and Y are relatively equal in case C).

For this reason, there are different versions of "post-democracy" and "post-capitalism" in my theories. As an illustration, in version I of the theory of post-democracy, freedom is more dominant than equality, whereas in version II of the theory of post-democracy, equality is more so than freedom.

But this "X more than Y" has to be understood in the context of dialectic logic (not in symbolic logic, as conventionally understood), in that both "X" and "Y" are important in post-democracy (in the context of dialectic logic), but in an asymmetry way.

By contrast, in symbolic logic, it often favors one over the other—be it in regard to privileging freedom over equality (in Fascism), favoring freedom relatively more than equality (in Liberal Democracy), or favoring equality relatively more than freedom (in Socialist Democracy). In regard to the latter two cases (about Liberal Democracy and Socialist Democracy), the difference between dialectic logic and symbolic logic can be one in degree, not in kind—in this sense, albeit not in other senses.

The same reasoning can be said about the relatonships between individuality and communality, between spirituality and materiality, and between formal legalism and informal legalism in the different versions of my theory of "post-capitalism."

Secondly, to be dialectic is to go beyond the narrow dichotomies (and, for that matter, any rigid multi-dimensional classificatory scheme) in another way, this time, in a transcendent way, that is, in exploring other possibilities or even other issues not considered within the narrow confines of narrow dichotomies (and, for that matter, any rigid multi-dimensional classificatory scheme).

As an analogy, in this second meaning, to go beyond the narrow dichotomy of "black" vs. "white" is not just to choose both "black" and "white" (as in the first meaning) but also to explore other options (e.g., "green," "purple," "blue," etc.—and, alternatively, "shade," "line," "curve," etc.).

By the same reasoning, to go beyond "democracy" is to transcend democracy (as in version III of my theory of "post-democracy") and to explore other possibilties of lifeforms (e.g., "floating consciousness," "hyper-spatial consciousness," etc., to live beyond the narrow obsession with freedom and equality).

This dialectic character of the principles in existential dialectics has important implications for the pragmatics of existential dialectics (as will be clear shortly, in the section on the pragmatics of existential dialectics).

The Semantics of Existential Dialectics (or Its Ontosemantics)

The syntax of existential dialectics so understood in terms of ontological principles only gives us the structure of ontology in the world, in an abstract (general) sense. These principles by themselves do not tell us the specific meanings in a given context.

In order to grasp the specific meanings of the principles in a given context, it is necessary to study the semantics of existential dialectics (as summarized in *Table 4.11*).

And the analysis of the ontological principles as family resemblances in the previous section on syntax is only a starting point (and thus overlaps a bit with the semantics of existential dialectics here).

For this reason, I have often gone to great lengths in my previous books on different subjects to explain the specific meanings of the principles when applied in different contexts.

Consider two examples below for illustration, which are not exhaustive.

In *FPHST*, for instance, I used the first three principles (i.e., the change-constancy principle, the regression-progression principle, and the symmetry-asymmetry principle) to propose "the perspectival theory of space-time," for a better way to understand space and time—especially, though not exclusively, in relation to future post-human history (as summarized in *Table 1.3*, *Table 1.4*, *Table 1.5*, and *Table 1.6*).

In so doing, I had to introduce concepts and theories specific to the field of physics and other related fields (e.g., "absolute space" and "absolute time" in "classical mechanics" and "relative space-time" in "the theory of relativity").

And in *BNN*, I also exploited the three principles to propose the "transcendent" approach to the study of genes and memes as a new way to understand the interaction between nature and nurture.

In so doing, I had to explore concepts and theories in the world of evolutionary biology (e.g., "mutation," "variation," "adaptation," "selection," and "inheritance" in classical evolutionary theory) and neural biology (e.g., "chromosome," "gene," "DNA," "RNA," "protein," "neuron," "neural network," and "behavior").

Other examples, besides these two here, are given in other books of mine on different topics.

Therefore, this semantic dimension of existential dialectics is important, since, in each of my books when the ontological principles are used, each principle becomes relevant in a different meaning when contextualized within a given case study in question.

In this book, for instance, the principles are relevant in relation to a different subject matter (as summarized in the concluding chapter), and they can be shown in all shapes and sizes (since they are also to be treated as "short cuts," as indicated before), without, however, being reduced to "specific ontology" (as opposed to "general ontology"), as explained in *Table 4.9*.

In the technical jargons of semantics, these exercises perform two essential functions of translation (or interpretation), namely, (a) "intensive" interpretation (about the properties or relations of the principles in existential dialectics), and (b) "extensive" interpretation (about the classes or cases to which the properties and relations of the principles in existential dialectics are applied). (WK 2011oo)

But there are limits of translation (or interpretation), and an excellent example is that "there is a problem with the 'objectivity' of translation, as it is relative to different cultural (or sub-cultural) contexts. For instance, one of the still unresolved problems in linguistics is: 'Is there an objective gauge for the quality of translation?'" (*FPHCOM* & WK 2011pp) And the problems of "radical translation" by W. V. Quine and "radical interpretation" by Donald Davidson are another well-known examples, as already discussed in *FPHSEM*.

This is one serious problem about the logic of translation and interpretation that I already addressed in *Sec. 2.5.2* of *FPHSEM*, *Sec. 3.5.1* of *FPHCOM* (and also, for that matter, in *FHC*, in the context of hermeneutics).

The Pragmatics of Existential Dialectics (or Its Ontopragmatics)

Even the semantics of existential dialectics is not enough to understand reality, since semantics by itself does not tell us the nature of intentionality and interests in relation to the matrix of power formations and conflicting interests when existential dialectics as an ontology is applied.

Surely, one cannot completely separate the pragmatics of existential dialectics from its semantics (and for that matter, its syntax), but the point here is to give each a more elaborated analysis.

For this reason, the pragmatics of existential dialectics needs to be studied too, in special relation to the inclusion of power formation and conflicting interests in the application of the principles of existential dialectics (as summarized in *Table 4.12*).

It is thus no wonder that all my previous books have shown in different ways how and why these ontological principles and their theoretical applications can reveal the future world to come in a way that humans have never known, in a totally different battleground for competing human (and later post-human) interests by myriad groups fighting for their own versions of hegemony.

Perhaps a summary of my previous works in relation to existential dialectics and their contributions to my new theories as proposed over the years is deemed revealing here, for the further understanding of the pragmatics of existential dialectics (in close relation to the semantics of existential dialectics too).

That stressed—the following summary in four sub-sections is something that I often do in my previous books too, as a way to introduce the intellectual background of my ideas.

So, for those readers who had read my previous works, the following sub-sections serve as a reminder, as they are repeated hereafter *often verbatim* (with only some updated changes, with the rest kept intact).

These sub-sections are, namely, (a) the two-way street connecting theory with meta-theory, (b) the theoretical application of existential dialectics, (c) direct and indirect applications of existential dialectics, and (d) multiple applications of existential dialectics—to be summaried herefafter, respectively.

The Two-Way Street Connecting Theory and Meta-Theory

However, lest any misunderstanding occurs, it is important to remember that the pragmatics of existential dialectics is not a one-way street (that is, using the ontological principles for theoretical insights in praxis) but a two-way one, namely, (a) from meta-theory to theory, and (b) from theory to meta-theory.

(a) On one side of the street, the ontological principles can inspire some theoretical insights in praxis, that is, in relation to some specific fields.

(b) On the other (opposing) side of the street, however, the study of a subject matter in the specific fields in turn reveals some more hitherto unknown ontological principles to be discovered and identified.

For this reason, three new principles were added in *FC*, one in *FAE*, two in *ALD*, and three in *FIA*, on top of the original three in *BCPC*, for example (as there are more)—after some research on the specific subject matters.

Both theory and meta-theory enrich each other in all my works over the years.

The Theoretical Application of Existential Dialectics

This two-way street connecting theory and meta-theory has contributed to my original construction of many theories in numerous books of mine on numerous fields in all domains of knowledge, which can be classified in terms of specific "meta-theories" (about epistemology, methodology, logic, ontology, etc.) and specific "theories" (about the mind, nature, culture, society, and the rest—as will be further explained later in the sub-section on "the holistic organization of an inquiry" in the section about "sophisticated methodological holism").

The final aim of my numerous new specific theories (on both theory and meta-theory) is to converge all of them (as proposed in my numerous books) into a grand project about the future of intelligent life, both here on earth and in deep space.

In this sense, all these theories of mine serve as a foundation to unify all domains of knowledge for an overarching understanding of all forms of reality (by way of some ontological principles and the comprehensive perspectives of the mind, nature, society, and culture), while allowing for heterogeneity, conflict, subjectivity, and complexity, for example.

With this grand project in mind—the summary of my specific theories on meta-theory are given in this section (on "existential dialectics") and the next (on "sophisticated methodological holism"), whereas the specific theories on theory will be summarized in the following categories of (a) society, (b) culture, (c) the mind, (d) nature, and (e) the rest, as shown below, in that order.

(a) Society

The category of "society" here can be divided into four sub-categories, i.e., (a1) social organizations, (a2) social institutions, (a3) social structure, and (a4) social systems (or trends).

(a1) Social Organizations

In relation to social organizations, I proposed a new thesis on the ambivalent nature of virtual organizations in the future, known as "the theory of e-civic alienation" in *FCD* (and summarized in *Table 4.19*).

And in *FPHO*, another theory called "the combinational theory of organization" was worked out, although it was first proposed in Ch.6 of *ALD*.

(a2) Social Institutions

There are different social institutions, be they about (a2.1) politics, (a2.2) economics, (a2.3) military affairs, (a2.4) mass media, (a2.5) law, (a2.6) education, and (a.7) gambling, for instance.

(a2.1) In relation to politics, I offered different visions on (a2.1.1) political theory, (a2.1.2) comparative politics, and (a2.1.3) international relations, to cite three main ones.

(a2.1.1) In the context of political theory, my vision of a path-breaking political system in future times concerns the different forms of "post-democracy" to supersede democracy unto the post-human age.

My vision of post-democracy was called "the theory of post-democracy" (as summarized in *Table 4.22*, *Table 4.23*, and *Table 4.24*), whereas *Table 4.25* explains the distinctions among democracy, non-democracy, and post-democracy.

And *Table 4.26* stresses the multiple causes of the emergence of post-democracy, while *Table 4.27* clarifies some possible misunderstandings in regard to post-capitalism and post-democracy.

(a2.1.2) In the context of comparative politics, I also showed, in *ALD*, how and why authoritarianism can be an advanced form of liberal democracy (just as it has its own counterpart in non-liberal democracy), or what I originally called "authoritarian liberal democracy" (as summarized in *Table 4.21*), before it is to be superceded by "post-democracy."

(a2.1.3) In the context of international relations (both at the institutional and structural levels), I provided, in *BWT*, "the theory of cyclical progression of empire-building," for a better way to understand the logic of empire-building on earth and beyond—with the rise of what I originally envisioned as "hyper-empires" (like "the Chinese Union" and "the Indian Union"), "meso-empires" (like "the European Union" and "the North American Union") and "micro-empires" (like "the Middle Eastern

Union" and "the Latin American Union") in the coming "post-post-Cold War era" as a visionary illustration, as summarized in *Table 4.20*.

This is also related to the issues of "social structure" in (a3), "migration" in (a4,6), "identity" in (b11), and "geography" in (d4), for instance.

In addition, in *FPHWP*, I proposed "the multifaceted theory of war and peace" to understand war and peace, in a small sense—or aggression and pacificity, in a broad sense.

(a2.2) In relation to economics, I offered different visions on (a2.2.1) economic theory, (a2.2.2) labor economics, and (a2.2.3) applied economics.

(a2.2.1) In the context of economic theory, I offered another vision, this time to understand the future forms of economic systems, that is, the different forms of "post-capitalism" to supersede capitalism unto the post-human age.

My vision of post-capitalism was referred to (in *FCD*, *BDPD* and *BCPC*) as "the theory of post-capitalism" (as summarized in *Table 4.28*, *Table 4.29*, *Table 4.30*, *Table 4.31*, and *Table 4.32*). And *Table 4.33* clarifies the differences among capitalism, non-capitalism, and post-capitalism, whereas *Table 4.34* shows multiple causes of the emergence of post-capitalism.

This vision on "post-capitalism" constitutes the third theoretical application of existential dialectics, this time, in that there is no wealth without poverty (or the wealth/poverty dialectics).

(a2.2.2) In the context of labor economics, that is, in *BHR*, I proposed "the post-human theory of human resources" to understand quantity and quality (especially, though not solely, in the context of demographics)—which is therefore related to demographics in (a4.3).

(a2.2.3) In the context of applied economics, that is, in *FPHACCO*, I propose "the double-sided theory of accounting" to understand addition and subtraction (especially, though not solely, in the context of accounting).

(a2.3) In relation to military affairs, I also proposed "the multifaceted theory of war and peace" (in *FCD*, although it was first so labeled in *FPHK*) for a new way to understand the causes of war—and, for that matter, the rise and fall of different forms of warfare over time.

This theory was further expanded in *FPHWP*, where I suggested a new way to understand war and peace, in a small sense—or aggression and pacificity, in a broad sense.

(a2.4) In relation to mass media, I also proposed "the totalistic theory of communication" in *The Future of Post-Human Mass Media* (or in short, *FPHMM*).

(a2.5) In relation to law, I also proposed "the reconstruction theory of law" in *The Future of Post-Human Law* (or in short, *FPHLAW*). And in relation to criminality, I proposed "the reflective theory of criminality" in *FPHCRIM* for a different way to understand heroes and villains.

(a2.6) In relation to education, I proposed "the heterodox theory of education" in *The Future of Post-Human Education* (or in short, *FPHEDU*).

(a2.7) And in relation to gambling, I proposed "the detached theory of gambling" in *The Future of Post-Human Gambling* (or in short, *FPHGAM*).

(a3) Social Structure

In relation to social structure, I first proposed two additional theories, that is, "the theory of the cyclical progression of hegemony" in *FCD* (and also in *FPHC* and *BDPD*), and "the theory of the cyclical progression of empire-building" (in *BWT*, as already indicated above)—with important implications for the symmetry-asymmetry princple in existential dialectics, just to cite one example.

In addition, this is also related to the issues of "migration" in (a4,6), "identity" in (b11), and "geography" in (d4), for instance.

Then, in *BNN*, I proposed another theory, namely, "the theory of contrastive advantages" to understand the persistence of social structure by gender, race, ethnicity, or any other groupings.

(a4) Social Systems (or Trends)

In relation to social systems, there are several sub-categories for illustration, which can be about (a4.1) urbanization, (a4.2) technology, (a4.3) demography, (a4.4) the environment, (a4.5) social trends as a whole (which also include the current phenomenon of globalization), (a4.6) migration, and (a4.7) resources.

(a4.1) In relation to urbanization, I proposed "the contingent theory of urban planning" in *FPHUP* for a new way to understand density and void—which is also related to the issues of "social trends" (especially in relation to globalization) in (a4.5), "migration" in (a4.6), and "identity" in (b11).

(a4.2) In relation to technology, I worked out "the ambivalent theory of technology" in *FPHE* for a new way to understand the nature of technology.

Some other books (like *FPHMM*, *FAE*, *FCD*, and *FHC*) also deal with technology, and a good example concerns "the synthetic theory of information architecture" in *FIA* on taxonomy, network, and intelligence, in conjunction with the rise of technology.

(a4.3) In relation to demography, I proposed, first in *FHC* (and later in other books like *FCD*, *FPHC*, *FPHU*, etc.), "the theory of post-humanity," in relation to the "post-human" age at some distant point of "after-postmodernity," long after human extinction, to be eventually superseded by "post-humans" of various forms.

This theory is further elaborated in *BHR*, in which I proposed "the post-human theory of human resources" to understand quantity and quality (especially, though not solely, in the context of demographics)—which is related to economics in (a2.2).

Some good candidates of these post-human lifeforms include, for instance, "thinking robots," "thinking machines," "cyborgs," "genetically altered superior beings," "human-animal hybrids," "floating consciousness," "hyper-spatial consciousness," "unfolding unconsciousness," "the hyper-sexual body," and "the hyper-martial body." This post-human vision of mine was first originally worked out in *FHC* and further elaborated in both *FCD*, *FPHC*, *FPHU*, *FPHS*, and *FPHMA*, for instance.

Lest any confusion occurs, I need to emphasize two clarifications here about the term "post-human" as a neologism in my works.

Firstly, the word "post-human" here should *not* be confused with another term which looks similar but has a totally different meaning in the literature of postmodernism, namely, "post-humanism"—which constitutes a critique of "humanism" as tradionally understood (especially, though not exclusively, in relation to the idea of progress in science and reason in the Enlightenment project). (WK 2008)

And secondly, the word "post-human" here should also *not* be confused with a similar term which is used to champion the ideology of technology for the future co-existence between humans and cyborgs in "trans-humanism." (WK 2008a)

Instead, my term "post-human" in relation to "posthuman-ism" also rejects "transhumanism" (especially, though not exclusively, in relation to the promises of technology) and refers to something else altogether, that is, the future extinction of humans and its post-human successors in deep space and beyond unto multiverses.

Later, in *BEPE*, I went further and provided a more comprehensive critique of "transhumanism," together with other issues.

With these two clarifications in mind (as summarized in *Table 1.31*)—in the end, a most fundamental question about intelligent life now has an answer, in that, if asked, "What is the future of human civilization?"—my answer in *FCD* (89) is thus: "As addressed in Ch.7 of *FHC*, a later epoch of the age of after-postmodernity (that is, at some point further away from after-postmodernity) will begin, as what I called the 'post-human' history (with the term 'post-human' originally used in my doctoral dissertation at M.I.T., which was finished in November 1995, under the title *After Postmodernity*, still available at M.I.T. library, and was later revised and published as *FHC*). The post-human history will be such that humans are nothing in the end, other than what culture, society, and nature (with some luck) have shaped them into, to be eventually superseded by post-humans (e.g., cyborgs, thinking machines, genetically altered superior beings, and others), if humans are not destroyed long before then."

(a4.4) In relation to the environment, I also addressed the issue of sustainability, especially in *FHC*, *FCD*, and *FPHUP*. For instance, in *FPHUP*, I specifically proposed a new theory, "the contingent theory of urban planning," on the need to go beyond the contemporoary obsession with sustainability, in the context of not only the environment but also other issues.

I also proposed "the interventive-reshaping theory of geography" in *FPHGEOG*—which is related to the issues of "migration" in (a4.6), "geography" in (d4), and "geology" in (d5).

In addition, I proposed "the transfigurative theory of waste" in *FPHW*.

(a4.5) In relation to social trends as a whole, I proposed in Ch.9 of *FCD* (367-8), that "civilizational history will continue into the following cyclical progression of expansion, before it is to be superseded (solely as a high probability, since humans might be destroyed sooner either by themselves or in a gigantic natural calamity) by posthumans at some distant point in after-postmodernity (which I already discussed in *FHC*)" unto multiverses (different constellations of universes):

Local → Regional → Global →Solar →
Galactic → Clustery…→ Multiversal

The current phenomeon of "globalization" lies in the third category on the line above and was extensively analyzed in *FCD*, in close connection with "globalism" at the cultural level—in the context of "postmodernity" (which will be introduced later in the section on history).

In *BDPD*, this thesis of mine was referred to as "the theory of the cyclical progression of system integration and fragmentation," at the systemic level—in close relation to another thesis analyzed in Ch.9 of

FCD, which was called, in the absence of better words, "the multifaceted theory of war and peace" in *FPHK* (as already indicated above).

In *FC*, I also proposed "the dialectic theory of complexity" to account for the factor of uncertainty, which can have a major role to play in the world.

All this is also related to the issues of "urbanization" in (a4.1), "the environment" in (a4.4), "migration" in (a4,6), "identity" in (b11), "geography" in (d4), and "geology" in (d5).

(a4.6) In relation to migration, I proposed "the resettlement theory of geology" in *FPHGEOL* and "the theory of the cyclical progression of migration" in *FPHMIG*—which are also related to the issues of "social structure" in (a3), "urbanization" in (a4.1), "the environment" in (a4.4), "social trends" in (a4.5), "identity" in (b11), "geography" in (d4), and "geology" in (d5).

(a4.7) And in relation to resources, I proposed "the resilient theory of natural resources" in *BNR* —which is also related to the issue of "natural resources" in (d10).

(b) Culture

The category of "culture" here can also be divided into different sub-categories like (b1) aesthetics, (b2) religion, (b3) popular (mass) culture, (b4) logic, (b5) civilization as a whole, (b6) ethics, (b7) epistemology (methodology), (b8) metaphysics (ontology), (b9) ideology, (b10) language, (b11) identity, and (b12) folklore/mythology.

(b1) Aesthetics

In relation to aesthetics, I proposed "the transformative theory of aesthetic experience" in *FAE* about some great transformations of aesthetic experience in future history.

In *FPHLIT*, I further proposed "the comparative-impartial theory of literature" for a new way to understand fiction and non-fiction.

In *FPHPA*, I further proposed "the transdisciplinary theory of performing arts" for a new way to understand the body and its presence—which is also related to the issues of "popular (mass) culture" in (b3) and of "the human body" in (c1.2). And the aesthetic issue also overlaps with (d8) on architecture.

In *FPHCA*, I further proposed "the inquisitive theory of culinary art" for a new way to understand ingredients and techniques.

And in *FPHVA*, I further proposed "the ephemeral theory of the visual arts" for a new way to understand techniques and spirits.

(b2) Religion

In relation to religion, I proposed "the comparative theory of religion" in *FHC* (and also in *FCD*, although it was first so labeled in *FPHK*).

In *FPHR*, I provided a solid foundation of this theory with a better name, that is, "the comparative-substitutive theory of religion" for a new way to understand spirituality.

And in *FPHTHAN*, I suggested "the sequential theory of thanatology" for a new way to understand after-death (in the larger context of life and death)—which is also related to biology (on "death") in the category of "the mind" in (c.1.4) and to folklore/mythology (after-death) in (b.12).

(b3) Popular (Mass) Culture

In relation to popular (mass) culture, there are some sub-categories for illustration, which can be, for example, (b3.1) mass culture in general and (b3.2) sports—both of which are also related to the issues of "aesthetics" in (b1) and of "the human body" in (c1.2).

(b3.1) In relation to mass culture in general, I proposed, in *FHC*, a new way to understand mass culture in terms of some fundamental dilemmas, which is called, in the absence of better words, "the dualistic theory of mass culture" in *FPHO*.

(b3.2) In relation to sports (like chess), I proposed, in *FPHCHESS*, "the mediative-variative theory of chess" for a new way to understand tactics and strategy (in different traditions).

And in *FPHSPORT*, I proposed "the modificative theory of sports" for a new way to understand training and winning.

(b4) Logic

In relation to logic, I also worked out, in *FPHML*, "the contrastive theory of rationality" for a new way to understand rationality and knowledge (in the context of mathematical logic).

And in *FPHPROB*, I proposed "the interpretivist theory of probability" for a different way to understand objectivity and subjectivity—which is related to the issue of logic.

(b5) Civilization as a Whole

In relation to civilization as a whole, I also argued, in *FHC*, no freedom without unfreedom (as shown in *Table 4.13*) and no equality without inequality (as shown in *Table 4.14*), especially in relation to the seven dimensions of life existence first worked out in *FHC* (i.e., the technological, the everyday, the true, the holy, the sublime/beautiful, the good, and the just).

Even more interestingly, in *BDPD* and later in *BCIV*, I further revealed a theoretical refinement of existential dialectics, this time, in arguing that there is no civilization without barbarity, with human civilization to be eventually superseded by what I originally analyzed as "post-human post-civilization" (which should not be confused with "post-human civilization"), in the context of the freedom/unfreedom and equality/inequality dialectics.

In the final analysis, civilization cannot live without barbarity and has to learn to co-exist with it in ever new ways. Preposteorus as this may seem to many contemporaries—it is no more imperative to preserve civilization than necessary to destroy barbarity, and the ideal of civilization is essentially bankrupt, to be eventually superseded by "post-civilization" (as summarized in *Tables 4.35-4.43*).

This freedom/unfreedom and equality/inequality dialectics holds in pre-moderny, modernity, postmodernity, and what I orignally called "after-postmodernity" in *FHC* and *FCD*—especially, though not exclusively, in the context of the emergence of "post-humans."

There are two specific theories about "after-postmodernity" here. The first theory about "after-postmodernity" is about the trinity of pre-modernity, modernity, postmodernity, and after-postmodernity at the cultural level—and was called, in *BDPD*, "the theory of the trinity of modernity to its after-postmodern counterpart," although it was first analyzed in *FHC* and later in *FCD* and also *BCIV* (as summarized in *Table 1.34*, *Table 1.35*, *Table 1.36*, and *Table 1.37* and will be elaborated further in the later sub-section on "the rest" about future history).

The second theory about "after-postmodernity" is about pre-modernity, modernity, postmodernity, and after-postmodernity at the historical level. For instance, a conclusion in *FHC*, which is shocking indeed for many contemporaries, is none other than that "[t]he post-human history will therefore mark the end of human history as we know it and, for that matter, the end of human dominance and, practically speaking, the end of humans as well. The entire history of human civilization, from its beginning to the end, can be summarized by four words, linked by three arrows (as already discussed in *FHC*)":

Pre-Modernity → Modernity →
Post-Modernity → After-Postmodernity

In *BDPD*, this thesis of mine was specifically called "the theory of the evolution from pre-modernity to after-postmodernity," at the historical level—as shown in *Table 1.34*, *Table 1.35*, *Table 1.36*, and *Table 1.37* (and will be elaborated further in the later sub-section on "the rest" about future history).

Therefore, "[t]he end of humanity in the coming human extinction is the beginning of post-humanity. To say an untimely farewell to humanity is to foretell the future welcome of post-humanity." (P. Baofu 2002: 89)

This thesis of mine was known in *BDPD* as "the theory of post-humanity," at the systemic level—as already indicated in (a4.3) above.

(b6) Ethics

In relation to ethics, I proposed, in *BEPE*, "the theory of post-ethics" for a new way to understand morality and immorality

(b7) Epistemology (Methodology)

In relation to epistemology (methodology), I proposed "the complex theory of data analysis" in *FPHDA* for a different way to understand quantitative and qualitative research.

In addition, I already proposed "the critical-dialectic theory of formal science" in *FPHFS* and "the theory of methodological holism" in several books (like Ch.1 *FCD*, Ch.1of *FPHC*, *Sec.2.1* and *Sec.2.5* of *BCPC*, *FC*, *FPHK*, etc.) for a different way to do methodology.

In *FPHPROB*, I proposed "the interpretivist theory of probability" for a different way to understand objectivity and subjectivity.

(b8) Metaphysics (Ontology)

I proposed "the theory of existential dialectics" in several books (like *FHC*, *FCD*, *FPHC*, *BDPD*, *FC*, *FAE*, *ALD*, *FIA*, *FPHK*, etc,) and "the contrastive theory of rationality" in *FPHML* for a different way to understand ontology.

(b9) Ideology

The issue about ideology was already discussed under the categories of "social institutions" and of "civilization as a whole," especially in relation to my theories of "post-democracy," "post-capitalism," "post-civilization," and so on.

(b10) Language

I already proposed "the multilogical theory of learning" in *FPHL* for a new way to understand the nature of language structure and context.

I also worked out, in *FPHSEM*, "the interactive theory of semantics" for a new way to understand internality and externality (in the context of semantics).

In *FPHFS*, I proposed "the critical-dialectic theory of formal science," which is also related to pragmatics in regard to theory and application.

In *FPHMORP*, I proposed "the fusional theory of morphology" for a different way to understand typologies and rules.

And in *FPHPHON*, I proposed "the inclusionist theory of phonology" for a different way to understand generality and specificity.

(b11) Identity

I proposed "the theory of the cyclical progression of migration" in *FPHMIG* for a new way to understand the nature of sameness, otherness, and identity—which is also related to the issues of "social structure" in (a3), "urbanization" in (a4.1), "social trends" (especially in relation to globalization) in (a4.5), "migration" in (a4.6), religion in (b2), and ideology in (b9).

(b12) Folklore/Mythology

In relation to folklore/mythology, I proposed "the sequential theory of thanatology" in *FPHTHAN* for a different way to understand after-death—which is also related to biology (on "death") in the category of "the mind" in (c.1.4) and to religion in (b.2).

(c) The Mind

The important vision about "post-humans" (as summarized earlier) brings us to the category of "the mind" here, which, as is often understood in everyday's usage, refers to the "brain" and the "body" (together with their interactions) and has different levels of consciousness which can be both "spiritual" and non-spiritual.

Therefore, the "mind" already presupposes the "brain," "body," and "spirit" (or "soul"), so that one does not fall into the intellectual trap of the "mind-body dualism" as debunked in modern (and especially, postmodern) philosophy (which I already went to great lengths to explain in *The Future of Human Civilization*).

That said—the category of "the mind" can be divided into two sub-categories, namely, in relation to (c1) biology and (c2) psychology

(c1) Biology

In relation to biology (which also overlaps with psychology too, however), there are three concerns here, namely, (c1.1) genes and intelligence, (c1.2) the human body, (c1.3) health, and (c1.4) death.

(c1.1) In relation to genes and intelligence, I worked out "the theory of contrastive advantages" (which was originally proposed in *FCD* and further worked out in *BNN*, as summarized in *Table 1.24*), to show the nature-nurture interactions of multiple levels in action (e.g., the biological, the psychological, the structural, the systemic, the cultural, and others) for humans and post-humans on earth and beyond, to the extent that different groups, be they on the basis of race, gender, ethnicity, class, age, or else, are not equal, on average, and have not been, nor will they be, even if there can be exceptions.

(c1.2) In relation to the human body, for instance, in *FPHMA*, I suggested "the expansive-contractive theory of martial arts," in which two great future transformations of the martial body were envisioned, that is, both "with the hyper-martial body" and "without the body" (as summarized in *Table 1.32*)—although this issue can also be classified under the category of "culture."

In *FPHS*, I proposed "the theory of virtual sexuality," in which three great future transformations of the sexual body were envisioned, that is, "virtual pleasure," "the hyper-sexual body," and "the post-sexual floating mind" (as summarized in *Table 1.33*).

And in *FPHPA*, I further proposed "the transdisciplinary theory of performing arts" for a new way to understand the body and its presence, with seven great future transformations of the body in performing arts—which is also related to the issues of "aesthetics" in (b1) and "popular (mass) culture" in (b3).

(c1.3) In relation to health, for instance, in *FPHHC*, I suggested "the interconnected theory of health care" for a new way to understand mental health and physical health (in the context of mind and body)—which is also related to psychology in the category of "the mind" (as well as the issues in the categories of "culture" and "society").

(c1.4) And in relation to death, for instance, in *FPHTHAN*, I suggested "the sequential theory of thanatology" for a new way to understand death (in the larger context of life and death)—which is also related to "religion" in (b.2) and to folklore/mythology in (b.12).

(c2) Psychology

In relation to psypchology (which also overlaps with biology too, however), I also proposed some greater transformations of the mind to come in the future, especially though not exclusively in the post-human age.

These greater transformations of the mind in psychology can be further sub-divided into different categories, namely, (c2.1) consciousness, (c2.2) unconsciousness (together with subconsciousness), (c2.3) creativity, (c2.4) learning, (c2.5) aggression and pacificity, (c2.6) personality, (c2.7) humor, and (c2.8) artificial intelligence.

(c2.1) In relation to consciousness, I analyzed, in *FPHC*, the nature of consciousness in relation to the conceptual dimensions of consciousness (as shown in *Table 1.12*), the theoretical levels of consciousness (as shown in *Table 1.13*), the thematic issues of consciousness (as shown in *Table 1.14*), and the different dimensions of human existence (viz., having-ness, belonging-ness, and being-ness) in relation to consciousness (as shown in *Table 1.15*, *Table 1.16*, *Table 1.17*, and *Table 1.18*).

In *BCPC*, the nature of consciousness was further explored in relation to cognition, emotion, and behavior, which can be in consciousness and other mental states (e.g., subconsciousness and unconsciousness), in the context of existential dialectics, as shown in *Table 1.19*, *Table 1.20*, *Table 1.21*, *Table 1.22*, and *Table 1.23*.

In fact, in *FCD*, I already proposed three different theories in relation to the limits of cognition, emotion, and behavior, namely, "the theory of cognitive partiality," "the theory of emotional non-neutrality," and "the theory of behavioral alteration."

Then, I argued that the existence of human consciousness will be superseded one day with "floating consciousness" (as first proposed in *FCD*) and "hyper-spatial consciousness" (as first proposed in *FPHC*) as a climax of evolution in consciousness, after the future extinction of human consciousness (as shown below):

Primordial consciousness → Human consciousness →
Post-human consciousness (with floating consciousness and hyper-spatial consciousness as a climax in the evolution of consciousness)

In *BDPD*, these latest theses of mine were called "the theory of floating consciousness" and "the theory of hyper-spatial consciousness," both at the cosmological and psychological levels (as summarized in *Table 1.26* and *Table 1.25*).

(c2.2) In relation to unconsciousness, for instance, in *FPHU*, the unconsious mind will undergo tremendous change, to the rise of what I originally called "unfolding unconsciousess" in relation to anomalous experience, as a climax of evolution in unconsciousness, in conjunction with the conscious mind (both human now and post-human later), especially at the psychological level, as shown below:

Primordial unconsciousness → Human unconsciousness →
Post-human unconsciousness (with unfolding unconsciousness as a climax in the evolution of unconsciousness)

In *FPHU*, this thesis of mine is called "the theory of unfolding unconsciousness" in a small sense, or "the unfolding theory of anomalous experience," in a larger one, as shown in *Table 1.27* and *Table 1.28*.

(c2.3) In relation to creativity, for instance, in *FPHCT*, I also proposed "the comprehensive theory of creative thinking" to explore the role of creative thinking, in relation to technology and other factors, together with my original proposal of important creative techniques and traits (as shown in *Table 1.29*) and its double nature of desirability and undesirability (as shown in *Table 1.30*).

(c2.4) In relation to learning, for instance, in *FPHL*, I worked out "the multilogical theory of learning" for a new way to understand the nature of structure and context in language, in relation to learning and its future development in the context of the mind.

(c2.5) In relation to aggression (and pacificity), I proposed "the multifaceted theory of war and peace" in *FPHWP* (as mentioned above).

(c2.6) In relation to personality, I worked out "the contrarian theory of personality" in *FPHP* for a new way to understand normality and abnormality.

(c2.7) In relation to humor, I worked out "the metamorphic theory of humor" in *FPHH* for a new way to understand joking and laughing.

(c2.8) And in relation to artificial intelligence, I proposed "the supersession theory of computing" in *FPHCOM* for a new way to understand hardware, software, and the mind.

In addition, I also proposed "the complex theory of data analysis" in *FPHDA* for a different way to understand quantitative and qualitative research in the context of the evolution of new minds.

(d) Nature

The category of "nature" here refers to the state of nature, which can be divided into some sub-categories like (d1) physics, (d2) cosmology, (d3) biology, (d4) geography, (d5) geology, (d6) acoustics, (d7) chemistry, (d8) architecture, (d9) transportation, (d10) natural resources, and (d11) aerology (both meteorology and climatology).

But since biology in (d3) overlaps with biology in (c1)—only the rest of the categories are addressed hereafter.

(d1) Physics

In relation to physics, I also proposed "the perspectival theory of space-time" in *FPHST* (based on *FPHC*) for a new way to understand space and time, as shown in *Table 1.3*, *Table 1.4*, *Table 1.5*, and *Table 1.7*.

In *FPHG* (based on *FPHC* and *FPHST*), I proposed "the selective theory of geometry" for a new way to understand geometry, with implications for the understanding of space-time in the context of infinity, symmetry, and dimensionality for future lifeforms that our world has never known (as shown in *Table 1.6*, *Table 1.7*, *Table 1.8; Table 1.9*, *Table 1.10,* and *Table 1.11*).

And in *FC*, I proposed "the dialectic theory of complexity" for a new way to understand order and chaos in the state of nature and beyond.

(d2) Cosmology

In relation to cosmology, my theories on physics (as indicated above) already have important implications for understanding the nature and the future of the universe (which was also addressed in *FHC*, among other works aforcited).

I also proposed "the theory of hyper-spatial consciousness" in *FPHC* (and also in *FPHG*) for a new way to understand the future evolution of the mind to enter into different dimensions of space-time that our current world has never known.

In addition, in *BCOS*, I proposed "the theory of post-cosmology" for a new way to understand the beginnings and ends of the cosmos and beyond.

(d4) Geography

In relation to geography, I proposed "the theory of the geopower of nature" in *ALD* for a different way to understand the relationships between geography and other entities (like the mind, culture, and society).

I further refined this issue of geography by proposing "the interventive-reshaping theory of geography" in *FPHGEOG*—which is related to the issues of "the environment" in (a4.4), "migration" in (a4.6), and "geology" in (d5).

(d5) Geology

In relation to geology, I proposed "the resettlement theory of geology" in *FPHGEOL* for a different way to understand the relationships between statics and dynamics—which is related to the issues of "the environment" in (a4.4), "migration" in (a4.6), and "geography" in (d4).

(d6) Acoustics

In relation to acoustics, I proposed "the multilateral theory of acoustics" in *FPHA* for a different way to understand the relationships between sound and silence.

(d7) Chemistry

In relation to chemistry, I proposed "the creational theory of chemistry" in *FPHCHEM* for a different way to understand substances and their changes.

(d8) Architecture

In relation to architecture, I proposed "the transcendent theory of architecture" in *FPHARCH* for a different way to understand form and function.

(d9) Transportation

In relation to transportation, I proposed "the panoramic theory of transportation" in *FPHTRAN* for a different way to understand networks and operations.

(d10) Natural Resources

In relation to natural resources, I proposed "the resilient theory of natural resources" in *BNR*—which is also related to the issue of "resources" in (a4.7).

(d11) Aerology (Meteorology and Climatology)

And in relation to aerology (both meteorology and climatology), I proposed "the constructivist theory of aerology" in *FPHAERO* for a different way to understand predictability and nonpredictability.

(e) The Rest

The category of "the rest" here refers to what cannot be strictly classified in any of the previous categories (i.e., society, culture, the mind, and nature). A good illustration of "the rest" refers to (e1) history.

(e1) History

The category of history can be divided into some sub-categories like (e1.1) empirical history and (e.1.2) meta-history.

(e1.1) In relation to empirical history, I proposed "the theory of the evolution from pre-modernity to after-postmodernity" (as first worked out in *FHC* and later in *FCD* and *FPHC*), which overlaps—as a reminder from (b5)—with the issue concerning "civilization as a whole" in (b5).

For instance, in both *FCD* and *FPHC*, I worked out the structure of "post-human civilization" in terms of the trinity of after-postmodernity (i.e., "free-spirited after-postmodernity," "post-capitalist after-postmodernity," and "hegemonic after-postmodernity").

Both conceptually and theoretically, the trinity of after-postmodernity is a sequential extension of the trinity of modernity (i.e., "free-spirited modernity," "capitalist modernity,"and "hegemonic modernity") and the trinity of postmodernity (i.e., "free-spirited postmodernity," "capitalist postmodernity," and "hegemonic postmodernity") as first proposed in *FHC*.

And the trinity of pre-modernity (i.e., "pre-free-spirited pre-modernity," "pre-capitalist pre-modernity" and "hegemonic pre-modernity") was later conceived in *BCIV* to complete the historical set from pre-modernity to after-postmodernity.

In *BDPD*, this thesis about the trinity of pre-modernity, modernity, postmodernity, and after-postmodernity was collectively known as "the theory of the trinity of modernity to its after-postmodern counterpart," at the cultural level—as already so mentioned in (b5) earlier, and summarized in *Table 1.34*, *Table 1.35*, *Table 1.36*, and *Table 1.37*.

At the structural level, all these trinities are subject to the existential constraints (e.g., the freedom/unfreedom and equality/inequality dialectics in the context of "the cyclical progression of hegemony"), be the historical epoch in pre-modernity, modernity, postmodernity, or after-postmodernity in future times. Each of the historical epochs has its ever new ways of coming to terms with the ever new (different) mixtures of freedom/unfreedom and equality/inequality.

This is importantly so, not because, as is falsely assumed in conventional wisdom, one certain way is superior (or better) than another in terms of achieving more freedom and less unfreedom, or more equality with less inequality.

On the contrary, indeed, in each of the historical epochs, each increase of unfreedom greets each freedom achieved, just as each increase of inequality welcomes each equality achieved, albeit in ever new (different) ways. In *BDPD*, this thesis of mine was labeled as "the theory of the cyclical progression of hegemony," at the structural level, though it was first analyzed in *FCD*.

In *BDPD*, more theoretical applications of existential dialectics were further examined, in relation to five main features, in the context of the duality of oppression, namely, (a) that each freedom/equality achieved is also each unfreedom/inequality created, (b) that the subsequent oppressiveness is dualistic, both by the Same against the Others and itself and by the Others against the Same and themselves, (c) that both oppression and self-oppression can be achieved by way of downgrading differences (between the Same and the Others) and of accentuating them, (d) that the relationships are relatively asymmetric among them but relatively symmetric within them, even when the Same can be relatively asymmetric towards itself in self-oppression, and the Others can be likewise towards themselves, and (e) that symmetry and asymmetry change over time, with ever new players, new causes, and new forms, be the locality here on Earth or in deep space unto multiverses—as summarized in *Table.4.13*, *Table 4.14*, *Table 4.15*, and *Table 4.16*.

The same logic also holds both in relation to wealth and poverty (as addressed in *BCPC* and summarized in *Table 4.17* on the wealth/poverty dialectics) and in relation to civilization and barbarity (as addressed in *BCIV* and summarized in *Table 4.18* on the civilization/barbarity dialectics).

In *BDPD*, this thesis on existential dialectics was labeled as "the theory of existential dialectics."

(e.1.2) And in relation to meta-history, I proposed "the multifold theory of history" in *FPHHIST* for a different way to understand universality and relativity (together with other debates).

Direct and Indirect Applications of Existential Dialectics

A different way to appreciate the usefulness of existential dialectics is by way of the analysis of its (a) direct and (b) indirect applications.

(a) In direct applications, on the one hand, the logic of existential dialectics can shed some theoretical insights on diverse phenomena in the world, and good instances are the usage of the principles of existential dialectics for the theoretical insights on the freedom/unfreedom dialectics, the equality/inequality dialectics, and the wealth/poverty dialectics—as introduced earlier in (b5) and (e) in the previous sub-section.

For instance, my books like *FPHST, BNN, FAE, FC, FIA, FPHU, FPHK, FPHML, FPHE, FPHMM, FPHCT, FPHG, FPHUP, FPHL, FPHO, FPHMA, FPHS, FPHLAW, FPHFS, FPHWP, FPHP, BEPE, BCOS, FPHGEOL, FPHCHESS, FPHR, FPHEDU, FPHH, FPHA, FPHGAM, FPHCOM, FPHLIT, FPHCHEM, FPHDA, FPHGEOG, FPHPA, FPHSEM, FPHARCH, FPHCRIM, FPHHIST, FPHMIG, FPHTRAN, FPHCA, FPHW, BNR, FPHSPORT, FPHPROB, FPHMORP, FPHHC, FPHTHAN, BHR, FPHVA, FPHAERO, FPHPHON,* and *FPHACCO* use the principles to reveal some theoretical insights on the perspectives of space and time (as in *FPHST*), of nature and nurture (as in *BNN*), of beauty and ugliness (as in *FAE*), of order and chaos (as in *FC*), of taxonomy and network (as in *FIA*), of the unconscious mind (as in *FPHU*), of the nature of knowledge (as in *FPHK*), of mathematical logic (as in *FPHML*), of engineering (as in *FPHE*), of mass media (as in *FPHMM*), of creative thinking (as in *FPHCT*), of geometry (as in *FPHG*), of urban planning (as in *FPHUP*), of language (as in *FPHL*), of organization (as in *FPHO*), of martial arts (as in *FPHMA*), of sexuality (as in *FPHS*), of law (as in *FPHLAW*), of formal science (as in *FPHFS*), of aggression and pacificity (as in *FPHWP*), of normality and abnormality (as in *FPHP*). of morality and immorality (as in *BEPE*), of the beginnings and ends of the universe (as in *BOCS*), of statics and dynamics (as in *FPHGEOL*), of tactics and strategy (as in *FPHCHESS*), of secularness and sacredness (as in *FPHR*), of teaching and learning (as in *FPHEDU*), of joking and laughing (as in *FPHH*), of sound and silence (as in *FPHA*), of risk and caution (as in *FPHGAM*), of hardware and software (as in *FPHCOM*), of fiction and non-fiction (as in *FPHLIT*), of substances and their changes (as in *FPHCHEM*), of quantitative and qualitative research (as in *FPHDA*), of environments and their interactions (as in *FPHGEOG*), of the body and its presence (as in *FPHPA*), of internality and externality (as in *FPHSEM*). of form and function (as in *FPHARCH*), of heroes and villains (as in *FPHCRIM*), of universality and relativity (as in *FPHHIST*), of Sameness and Otherness (as in *FPHMIG*), of networks and operations (as in *FPHTRAN*), of ingredients and techniques (as in *FPHCA*), of uselessness and usefulness (as in *FPHW*), of diversity and discontinuity (as in *BNR*), of training and winning (as in *FPHSPORT*), of objectivity and subjectivity (as in *FPHPROB*), of typologies and rules (as in *FPHMORP*), of mental health and physical health (as in *FPHHC*), of death and after-death (as in *FPHTHAN*), of quantity and quality (as in *BHR*), of techniques and spirits (as in *FPHVA*), of predictability and nonpredictability (as in *FPHAERO*), of generality and specificity (as in *FPHPHON*), and of addition and subtraction (as in *FPHACCO*).

(b) In indirect applications, on the other hand, the theoretical insights can further be used to reveal other phenomena directly from them (viz., the theoretical insights) and therefore indirectly from the principles themselves.

A good illustration is of course the use of the theoretical insights on the freedom/unfreedom and equality/inequality dialectics for the understanding of the civilization/barbarity dialectics.

This distinction between direct and indirect applications may sound a bit academic, since even in indirect applications, the phenomena under study can still be directly related back to the principles themselves.

In the previous example, as an illustration, the civilization/barbarity dialectics can be directly related to the principles of existential dialectics without the intermediate role of the freedom/unfreedom and equality/inequality dialectics.

Multiple Levels of Application

There is another issue to be clarified, however. In other words, the theoretical insights can be applied to multiple levels of analysis—even though, in a given example, it may refer to one level or a few only.

For instance, in the example concerning the freedom/unfreedom dialectics, it can be used at the structural level (e.g., in relation to the theory of the cyclical progression of hegemony), but it can also be exploited for other levels (e.g., the theory of post-capitalism at the institutional level).

All these levels of application should not be misleadingly construed, as I stress this before, as a one-way street (that is, to use the ontological principles for theoretical application at multiple levels of specific analysis) but a two-way one, in which theoretical insights in praxis, when studied in more specific contexts, can in turn refine the nature of existential dialectics (for example, with the addition of new principles).

Sophisticated Methodological Holism

The summary of my original meta-theory on methodology is provided below, again, to be repeated *often verbatim* from my previous works—as this is something I regularly do in each new book to introduce my past works, which are in conversation with this current one.

This means that, for those readers who have not read any of my previous books, the summary below is helpful.

That clarified—this original contribution of mine to the study of methodology takes the form of an approach known as "the theory of methodological holism" or "methodological holism" in short, as already worked out in *FPHC* (2004).

I have made good use of this methodological approach of mine for all of my previous works.

The Confusion of Holism with Systems Approach

Yet, it is imperative to stress at the outset, as I often emphasized this in all my previous books, that my approach of "methodological holism" does not oppose or exclude "methodological individualism" (as some readers may be tempted to assume, as is conventionally understood) but actually includes it.

The confusion here has to do with equating holism with systems approach in general (be it about systems theory, chaos theory, complexity theory, or else), which I debunked, while learning something from it, as already elaborated in *The Future of Complexity* (2007)—and *The Future of Post-Human Formal Science* (2010) is another addition to this critique of the intellectual fad of systems approach.

In other words, my meta-theory of holism, or sophisticated methodological holism in short, makes a distinction between doing holistic work and doing systems work, for two reasons.

The first reason is that holism, in my theory, does not imply systems approach, since it rejects systems approach because of the myriad problems that the literature on systems approach has been unable to resolve, while learning something from it.

And the second reason is that my theory is related to the other theory of mine, that is, existential dialectics, which does not exclude methodological individualism (but includes it, because it simply goes beyond both reductionism and reverse-reductionism in a dialectic way).

For these reasons (and others too, as summarized in *Table 4.1*), my version of methodological holism is *sophisticated*—not *vulgar* as sometimes used by inapt scholars employing the same term.

The Ontological Constraints

Sophisticated methodological holism is subject to some ontological constraints, and some good examples include "the formalness-informalness principle," "the absoluteness-relativeness principle," "the partiality-totality principle," "the predictability-unpredictability principle," "the explicability-inexplicability principle," and "the fiction-reallity principle" in existential dialectics.

Of course, there can be more than six ontological principles in relation to method, but the examples here are sufficient to illustrate the point in question.

With this caveat in mind—the first two were already analyzed in previous books of mine like *FC* and *FAE*, whereas the third one was addressed in *FPHU*, the fourth and fifth ones were accounted for in *FPHK* and *FPHML*, respectively, and the last one was addressed in *FPHLIT*.

For now, it suffices to show that sophisticted methodological holism, because of these ontological constraints on methodology, targets two major sins of methodology, namely, what I call, in the absence of better words, (a) *reductionism* and (b) *reverse-reductionism*—both of which come in all shapes and sizes, to be summarized below (and also shown in *Table 4.2*).

Against the Varieties of Reductionism

There are four versions of reductionism to be summarized here, which sophisticated methodological holism rejects, namely, (a1) conceptual, (a2) theoretical, (a3) methodological, and (a4) ontological forms of reductionism.

(a1) In conceptual reductionism, a good case in point concerns myriad dualities like mind vs. body, self vs. world, democracy vs non-democracy, and the like (as already addressed in *FHC*, *FPHC*, and *BDPD*, for instance).

(a2) In theoretical reductionism, an illuminating case study is best exemplified by what I originally called "the foundation fallacy" in *FPHST*, in any attempt to naively understand space-time from the physical perspective as the foundation and, consequently, to dangerously dismiss (or belittle) other perspectives.

In *FAE*, I elaborated further these versions of reductionism in the literature on aesthetics (e.g., form vs. content, representation vs. expression, critics vs. artists, and externalism vs. internalism).

In *FIA*, I revealed other forms of reductionism in the literature on information architecture (e.g., the constructivist argument and the representational argument).

In *FPHU*, I showed the persistent legacy of reductionism, this time, in the literature on anomalous experience (e.g., the obsession with physics, chemistry, and biology for explaining anomalous experience).

In *FPHE*, I examined another case of reductionism in action, in the context of engineering (e.g., technical constraints vs. normative constraints).

In *FPHMM*, I elaborated one more version of reductionism, in relation to the three domains of communication (e.g., the competing views on sending, connecting, and receiving).

In *FPHCT*, I explored another version of reductionism, in relation to invention and innovation (e.g., the bio-psychological argument vs. the socio-cultural arguments).

In *FPHG*, I identified another version of reductionism, in relation to infinity, symmetry, and dimensionality (e.g., the Euclidean argument vs. the non-Euclidean arguments).

In *FPHUP*, I analyzed another version of reductionism, in relation to density and void (e.g., the engineering argument and the ecology argument).

In *PFHL*, I examined another version of reductionism, in relation to structure and context (e.g., the structuralist argument and the contextualist argument).

In *PFHO*, I explored another version of reductionism, in relation to communcation, decision-making, and leadership (e.g., the rational-system argument vs. the natural-system argument vs. the open-system argument).

In *PFHMA*, I revealed another version of reductionism, in relation to the martial body and spirit (e.g., the spiritual argument vs. the materialist argument vs. the defensive argument).

In *PFHS*, I analyzed another version of reductionism, in relation to the sexual body and spirit (e.g., the naturalist argument vs. the constructivist argument).

In *PFHLAW*, I examined another version of reductionism, in relation to law (e.g., the necessity argument vs. the contengency argument).

In *FPHWP*, I show another version of reductionism, in relation to war and peace (e.g., the aggressivist argument vs. the pacifist argument).

In *BEPE*, I analyzed another version of reductionism, in relation to morality and immorality (e.g., the objectivist argument vs. the non-objectivist argument vs. the skeptical argument).

In *BCOS*, I examined another version of reductionism, in relation to the contested beginnings and speculative ends of the universe (e.g., the scientific argument vs. the religious argument vs. the esoteric argument vs. the metaphysical argument).

In *FPHP*, I showed another version of reductionism, in relation to normality and abnormality (e.g., the natural argument vs. the social argument vs. the cultural argument vs. the mental argument).

In *FPHGEOL*, I scrutinized another version of reductionism, in relation to statics and dynamics (e.g., the catastrophe argument vs. the uniformity argument vs. the revision argument).

In *FPHCHESS*, I showed another version of reductionism, in relation to tactics and strategy (e.g., the natural argument vs. the social argument vs. the cultural argument vs. the mental argument).

In *FPHR*, I revealed another version of reductionism, in relation to secularness and sacredness (e.g., the critical argument vs. the skeptical argument vs. the theist argument).

In *FPHEDU*, I examined another version of reductionism, in relation to teaching and learning (e.g., the teacher-centered argument vs. the student-centered argument vs. the balanced argument).

In *FPHH*, I explored another version of reductionism, in relation to joking and laughing (e.g., the natural argument vs. the social argument vs. the cultural argument vs. the mental argument).

In *FPHA*, I revealed another version of reductionism, in relation to sound and silence (e.g., the natural argument vs. the social argument vs. the cultural argument vs. the mental argument).

In *FPHGAM*, I revealed another version of reductionism, in relation to risk and caution (e.g., the natural argument vs. the social argument vs. the cultural argument vs. the mental argument).

In *FPHCOM*, I showed another version of reductionism, in relation to hardware and software (e.g., the natural argument vs. the social argument vs. the cultural argument vs. the mental argument).

In *FPHLIT*, I proposed another version of reductionism, in relation to fiction and non-fiction (e.g., the natural argument vs. the social argument vs. the cultural argument vs. the mental argument).

In *FPHCHEM*, I proposed another version of reductionism, in relation to substances and their changes (e.g., the natural argument vs. the social argument vs. the cultural argument vs. the mental argument).

In *FPHDA*, I proposed another version of reductionism, in relation to quantitative and qualitative research (e.g., the natural argument vs. the social argument vs. the cultural argument vs. the mental argument).

In *FPHGEOG*, I proposed another version of reductionism, in relation to environments and their interactions (e.g., the natural argument vs. the social argument vs. the cultural argument vs. the mental argument).

In *FPHPA*, I proposed another version of reductionism, in relation to the body and its presence (e.g., the natural argument vs. the social argument vs. the cultural argument vs. the mental argument).

In *FPHSEM*, I proposed another version of reductionism, in relation to internality and externality (e.g., the natural argument vs. the social argument vs. the cultural argument vs. the mental argument).

In *FPHARCH*, I proposed another version of reductionism, in relation to form and function (e.g., the natural argument vs. the social argument vs. the cultural argument vs. the mental argument).

In *FPHCRIM*, I proposed another version of reductionism, in relation to heroes and villains (e.g., the natural argument vs. the social argument vs. the cultural argument vs. the mental argument).

In *FPHHIST*, I proposed another version of reductionism, in relation to universality and relativity (e.g., the natural argument vs. the social argument vs. the cultural argument vs. the mental argument).

In *FPHMIG*, I proposed another version of reductionism, in relation to Sameness and Otherness (e.g., the natural argument vs. the social argument vs. the cultural argument vs. the mental argument).

In *FPHTRAN*, I proposed another version of reductionism, in relation to networks and operations (e.g., the natural argument vs. the social argument vs. the cultural argument vs. the mental argument).

In *FPHCA*, I proposed another version of reductionism, in relation to ingredients and techniques (e.g., the natural argument vs. the social argument vs. the cultural argument vs. the mental argument).

In *FPHW*, I proposed another version of reductionism, in relation to uselessness and usefulness (e.g., the natural argument vs. the social argument vs. the cultural argument vs. the mental argument).

In *BNR*, I proposed another version of reductionism, in relation to diversity and discontinuity (e.g., the natural argument vs. the social argument vs. the cultural argument vs. the mental argument).

In *FPHSPORT*, I proposed another version of reductionism, in relation to training and winning (e.g., the natural argument vs. the social argument vs. the cultural argument vs. the mental argument).

In *FPHPROB*, I proposed another version of reductionism, in relation to objectivity and subjectivity (e.g., the natural argument vs. the social argument vs. the cultural argument vs. the mental argument).

In *FPHMORP*, I proposed another version of reductionism, in relation to typologies and rules (e.g., the natural argument vs. the social argument vs. the cultural argument vs. the mental argument).

In *FPHHC*, I proposed another version of reductionism, in relation to mental health and physical health (e.g., the natural argument vs. the social argument vs. the cultural argument vs. the mental argument).

In *FPHTHAN*, I proposed another version of reductionism, in relation to death and after-death (e.g., the natural argument vs. the social argument vs. the cultural argument vs. the mental argument).

In *BHR*, I proposed another version of reductionism, in relation to quantity and quality (e.g., the natural argument vs. the social argument vs. the cultural argument vs. the mental argument).

In *FPHVA*, I proposed another version of reductionism, in relation to techniques and spirits (e.g., the natural argument vs. the social argument vs. the cultural argument vs. the mental argument).

In *FPHAERO*, I proposed another version of reductionism, in relation to predictability and nonpredictability (e.g., the natural argument vs. the social argument vs. the cultural argument vs. the mental argument).

In *FPHPHON*, I proposed another version of reductionism, in relation to generality and specificity (e.g., the natural argument vs. the social argument vs. the cultural argument vs. the mental argument).

And in *FPHACCO*, I propose another version of reductionism, in relation to addition and subtraction (e.g., the natural argument vs. the social argument vs. the cultural argument vs. the mental argument).

(a3) In methodological reductionism, a good illustration can be the debate between different versions of qualitative and quantitative methods (as already analyzed in *FC* and also *FHC*).

In *FPHML*, I examined similar reductionism, this time, in the literature on mathematical logic (e.g., the obsession with consistency, soundness, and completeness).

And in *FPHFS*, I explored the problems of reductionism in the context of formal science (e.g., the analytical argument, as opposed to the synthetic argument).

(a4) In ontological reductionism, an excellent instance involves another debate, this time between emergentism and reductionism in complexity theory (as addressed in *FC*) and also in psychology (as elaborated in *FPHC*, in the context of Being and Belonging).

But to target against the varieties of reductionism constitutes only a side of the same coin, and the other side concerns the varieties of reverse-reductionism.

Against the Varieties of Reverse-Reductionism

The other side of the same coin is the reverse version of reductionism, which is what I want to call, in the absence of better words, *reverse-reductionism.*

My sophisticated version of methodological holism targets against the varieties of reverse-reductionism (just as it also rejects the ones of reductionism).

Perhaps nothing expresses better the popularity of reverse-reductionim than the "anything-goes" mentality in postmodernism of our time, as shown in the following four versions.

(b1) In conceptual reverse-reductionism, any concept of "art" (e.g., fine art, applied art, outsider art, junk art) is welcome in postmodernism (as already analyzed in Ch.4 of *FHC*—and also in *FAE*).

(b2) In theoretical reverse-reductionism, a variety of art and literary theories co-exist. Take the case of literary studies, as there are now Literary Structuralism, Marxist Literary Criticism, New Criticism, Phenomenology, Hermeneutics, Language-Game Literary Criticism, Feminist Literary Criticism, Reception Theory, Reader Response Criticism, Poststructuralism, Semiotics, Pyschoanalytic Literary Criticism, just to cite some well-known ones, with no one being said to be better than any others (as detailedly analyzed in Ch.4 of *FHC*). (S. Raman 1997)

In *BNN*, I even introduced "the compromise fallacy" as another good illustration of theoretical reverse-reductionism, in misleadingly treating both genetic and environmental approaches as equally valid.

(b3) In methodological reverse-reductionism, multiple methodologies are deemed as acceptable in postmodernism (e.g., doing art without praxis, doing art with praxis, and doing art by sublation), as analyzed in Ch.4 of *FHC*.

And in *FPHFS*, I exposed the problems of reverse-reductionism in the context of formal science in relation to systems theory (e.g., the synthetic argument).

(b4) In ontological reverse-reductionism, no privileged ontology is allowed, and the door is open practically for anything in postmodernism (e.g., the equal status of the ontology of Being vs. that of Becoming, as already addressed in Ch.4 of *FHC*—and also in *FPHC*).

In *FAE*, I also introduced another version of reverse-reductionism, that is, "the pluralist fallacy," in the context of understanding aesthetic experience, for instance—although this fallacy has been committed not exclusively in relation to the ontological level (but also at the conceptual, theoretical, and methodological ones).

These dual dangers against reverse-reductionism (in this sub-section) and reductionism (in the previous sub-section) are something that sophisticated methodological holism rejects. And in *FPHFS*, I addressed these two problems further.

This point about the dual dangers is important enough, since many scholars often favor one against the other, but it is vital to target both.

The Holistic Organization of an Inquiry

With these dual dangers against reductionism and reverse-reductionism in mind—sophisticated methodological holism suggests that an inquiry of any given phenomenon is more complete, if treated in the context of a comprehensive analysis at all relevant levels, which challengingly encompass all the fields of human knowledge, ranging from the natural sciences and formal sciences through the social sciences to the humanities.

This section is something that I had already stressed in all my previous books and repeat (*often verbatim*) hereafter. But, for those readers who had read my previous books, this serves as a reminder.

With this reminder in mind—there are multiple ways to engage in a holistic inquiry with all relevent levels of analysis. Over the years, I have proposed different ways to fulfill this holistic methodological requirement, as repeated in my previous books.

Hereafter is a summary of four major ways, namely, (a) by discipline, (b) by domain, (c) by subject, and (d) by meta-analysis—all of which fulfill the holistic requirement of sophisticated methodological holism on a given issue.

(a) In a holistic organization by discipline, a good illustration is the multiple levels of analysis, namely, (a1) the micro-physical, (a2) the chemical, (a3) the biological, (a4) the psychological, (a5) the organizational, (a6) the institutional, (a7) the structural, (a8) the systemic, (a9) the cultural, (a10) the macro-physical (cosmological), and (a11) other relevant levels, which are either a combination of all these levels or the practical applications with a combination of them.

This holistic organization by discipline is used in *FCD* and *BCPC*, for instance.

(b) In a holistic organization by domain, some good candidates include the classification of different perspectives of analysis in relation to (b1) nature, (b2) the mind, (b3) society, and (b4) culture, as already worked out in *BCIV*, *FPHST*, *FC*, *FAE*, *FPHU*, *FPHK*, *FPHE*, and *FPHFS*, just to cite some examples.

Culture in (b4), in this re-classification, is the same as culture in (a9) and can be further divided into different sub-categories, with good examples like (a9i) epistemology (methodology), (a9ii) aesthetics, (a9iii) ethics, (a9iv) metaphysics (ontology), and (a9v) religion. For clarification, epistemology (methodology) in (a9i) and metaphysics (ontology) in (a9iv) overlap with the category of "meta-theory."

Society in (b3) includes the organizational in (a5), the institutional in (a6), the structural in (a7), and the systemic in (a8).

The mind (b2) has more to do with the chemical in (a2), the biological in (a3), and the psychological in (a4), although it can overlap with (a1), (b3) and (b4), for instance.

It should be stressed that the "mind" here also includes the categories like the "brain," "body," and the "spirit" (or the "soul")—whenever needed in a given context, so that one does not fall into the intellectual trap of the "mind-body dualism" in the history of modern (and especially, postmodern) philosophy, which I already went to great lengths to explain in *The Future of Human Civilization*. After all, in everyday's usage, the "mind" already implies the "brain" and the "body" (together with their interactions) and has different levels of consciousness which can be both "spiritual" and non-spiritual.

And nature in (b1) refers to the micro-physical in (a1) and the macro-physical (cosmological) in (a10), although it can also overlap with (a2), (a3) and (a4).

(c) In a holistic organization by subject, some excellent examples concern the analysis of consciousness in *FPHC* (i.e., on Having, Belonging, and Being); of civilization in *FHC* (i.e., the True, the Holy, the Everyday, the Technological, the Beautiful/Sublime, the Good, and the Just); of international politics in *BWT* (i.e., Hyper-Empires, Meso-Empires, and Micro-Empires); of nature and nurture in *BNN* (i.e., genes and memes); of information architecture in *FIA* (i.e., taxonomy and network); of mathematical logic in *FPHML* (i.e., consistency, soundness, and completeness); of communication in *FPHMM* (i.e., sending, connecting, and receiving); of creative thinking in *FPHCT* (i.e., invention and innovation); of geometry in *FPHG* (i.e., infinity, symmetry, and dimensionality); of urban planning in *FPHUP* (i.e., density and void), of language in *FPHL* (i.e., structure and context), of organization in *FPHO* (i.e., communication, decision-making, and leadership), of martial arts in *FPHMA* (i.e., the martial body and the martial spirit), of sexuality in *FPHS* (i.e., the sexual body and the sexual spirit), of law in *FPHLAW* (i.e., necessity and contingency), of war and peace in *FPHWP* (i.e., aggression and pacificity), of personality in *FPHP* (i.e., normality and abnormality), of ethics in *BEPE* (i.e., morality and immorality), of cosmology in *BCOS* (i.e., the beginnings and ends of the universe), of geology in *FPHGEOL* (i.e., statics and dynamics), of chess in *FPHCHESS* (i.e., tactics and strategy), of religion in *FPHR* (i.e., secularness and sacredness), of education in *FPHEDU* (i.e., teaching and learning), of humor in *FPHH* (i.e., laughing and joking), of acoustics in *FPHA* (i.e., sound and silence), of gambling in *FPHGAM* (i.e., risk and caution), of computing in *FPHCOM* (i.e., hardware and software), of literature in *FPHLIT* (i.e., fiction and non-fiction), of chemistry in *FPHCHEM* (i.e., substances and their changes), of data analysis in *FPHDA* (i.e., quantitative and qualitative research), of geography in *FPHGEOG* (i.e., environments and their interactions), of performing arts in *FPHPA* (i.e., the body and its presence), of semantics in *FPHSEM* (i.e., internality and externality), of architecture in *FPHARCH* (i.e., form and function), of criminality in *FPHCRIM* (i.e., heroes and villains), of history in *FPHHIST* (i.e., universality and relativity), of migration in *FPHMIG* (i.e., Sameness and Otherness), of transportation in *FPHTRAN* (i.e., networks and operations), of culinary art in *FPHCA* (i.e., ingredients and techniques), of waste in *FPHW* (i.e., uselessness and usefulness), of natural resources in *BNR* (i.e., diversity and discontinuity), of sports in *FPHSPORT* (i.e., training and winning), of probability in *FPHPROB* (i.e., objectivity and subjectivity), of morphology in *FPHMORP* (i.e., typologies and rules), of health care in *FPHHC* (i.e., mental health and physical health), of thanatology in *FPHTHAN* (i.e., death and after-death), of human resources in *BHR* (i.e., quantity and quality), of the visual arts in *FPHVA* (i.e., techniques and spirits), of aerology in *FPHAERO* (i.e., predictability and nonpredictability), of phonology in *FPHPHON* (i.e., generality and specificity), and of accounting in *FPHACCO* (i.e., addition and subtraction)—by incorporating all the relevent levels of analysis as cited above in each.

(d) And in a holistic organization by meta-analysis, two good cases in point involve the works on post-capitalism in *BCPC* and on authoritarian liberal democracy in *ALD*, with the classification of analysis in theory and meta-theory.

The essential point to remember here, as I thus stress in all my previous books, is that the multiple levels of analysis can be reorganized in many different ways, insofar as none of the levels (if relevant to an inquiry in question) is ignored or dismissed, to avoid the danger of reductionism (and for that matter, the one of reverse-reductionism, in the opposite direction).

Three Clarifications

Three clarifications are needed here, however, to avoid any hasty misunderstanding.

Firstly, the entities in each classification are not mutually exclusive. For instance, in the holisitc organization by domain, nature in (b1) can also be linked to the chemical in (a2), the biological in (a3), and the systemic in (a8). And the mind (b3) can alternatively be related to the micro-physical in (a1) and the chemical in (a2), for example.

Secondly, there is also the important factor of luck (or uncertainty in conventional usage), but it is already implied or allowed in each of the organizations of an inquiry—especially in relation to the predictability-unpredictability principle and the order-chaos principle in existential dialectics. This topic about uncertainty was already addressed in *FHC*—and more extensively, in *FC*, in the context of order and chaos (and later, to an extent, in *FPHK*, *FPHFS*, and *FPHGAM*, for example).

And lastly, the comparison in a classification is not absolute, but relative, as there are often some shades of gray, not exactly black or white (figuratively speaking).

Some further clarifications and qualifications of sophisticated methodological holism are shown in *Table 4.2*.

Some Distinctive Usefulness

In the end, sophisticated methodological holism—when applied, especially though not exclusively, as illustrated in my numerous works—can enrich the understanding of reality in some distinctive ways. Here are four examples (as revealed in each of my books).

Firstly, it provides a comprehensive analysis of a subject matter, from which much can be learned about reality, both in relation to the perspectives of the mind, nature, society, and culture.

Secondly, it offers a new classification of the subject matter.

Thirdly, it suggests some visions of the future in relation to the subject matter in question.

And fourthly, it proposes some insights on meta-theory (e.g., methodology and ontology) in general—with the clear understanding, however, of the dilemma of specific vs. general ontology (as shown in *Table 4.9* on the syntax of existential dialectics in the context of the dilemma of ontology).

For this very reason, all of these ways are important, without reducing one into the analysis of another. (*FPHCT*)

Chapter Outline

With this summary of my two original meta-theories (on methodology and ontology, *almost verbatim* from my previous works) in mind—this book is organized in four main parts, namely, (a) Part I on the introduction to this book, (b) Part II on addition, (c) Part III on subtraction, and (d) Part IV on the conclusion of the book.

The book is thus organized in four chapters.

Chapter One titled "Introduction—The Invention of Accounting" begins with a quarrel about an invention, provides a definition and history of accounting, and then introduces addition and subtraction in accounting, as well as the diverse aspects of accounting, together with the theoretical debate on accounting, my double-sided theory of accounting, the relationship between theory and meta-theory, existential dialectics, sophisticated methodological holism, and some further clarifications.

Chapter Two titled "Addition and its Two Faces" investigates the nature of addition in accounting from the four perspectives of the mind, nature, society, and culture, with a constructive evaluation of each.

Chapter Three titled "Subtraction and its Dual Attributes" explores the nature of subtraction in accounting from the four perspectives of the mind, nature, society, and culture, with a deconstructive assessment of each.

Chapter Four titled "Conclusion—The Future of Accounting" summarizes the analysis in the book in the dialectic context of addition and subtraction in accounting (together with the need to transcend them) with thirty-six major theses, namely, (a) the first thesis: the formalness-informalness principle, (b) the second thesis: the absoluteness-relativeness principle, (c) the third thesis: the partiality-totality principle, (d) the fourth thesis: the predictability-unpredictability principle, (e) the fifth thesis: the explicability-inexplicability principle, (f) the sixth thesis: the fiction-reality principle, (g) the seventh thesis: the cognitiveness-noncognitiveness principle, (h) the eighth thesis: the finiteness-transfiniteness principle, (i) the ninth thesis: the preciseness-vagueness principle, (j) the tenth thesis: the simpleness-complicatedness principle, (k) the eleventh thesis: the openness-hiddenness principle, (l) the twelfth thesis: the denseness-emptiness principle, (m) the thirteenth thesis: the rule-exception principle, (n) the fourteenth thesis: the prototypicality-variation principle, (o) the fifteenth thesis: the change-constancy principle, (p) the sixteenth thesis: the order-chaos principle, (q) the seventeenth thesis: the slowness-quickness principle, (r) the eighteenth thesis: the expansion-contraction principle, (s) the nineteenth thesis: the optimality-nonoptimality principle, (t) the twentieth thesis: the simultaneity-nonsimultaneity principle, (u) the twenty-first thesis: the isolation-interaction principle, (v) the twenty-second thesis: the theory-praxis principle, (w) the twenty-third thesis: the convention-novelty principle, (x) the twenty-fourth thesis: the evolution-transformation principle, (y) the twenty-fifth thesis: the symmetry-asymmetry principle, (z) the twenty-sixth thesis: the softness-hardness principle, (aa) the twenty-seventh thesis: the seriousness-playfulness principle, (bb) the twenty-eighth thesis: the activeness-inactiveness principle, (cc) the twenty-ninth thesis: the selfness-otherness principle, (dd) the thirtieth thesis: the regression-progression principle, (ee) the thirty-first thesis: the sameness-difference principle, (ff) the thirty-second thesis: the stability-reaction principle, (gg) the thirty-third thesis: the functionality-nonfunctionality principle, (hh) the thirty-fourth thesis: the intentionality-nonintentionality principle, (ii) the thirty-fifth thesis: the survivability-non-survivability principle, and (jj) the thirty-sixth thesis: the post-human double-sidedness—to be elaborated in the rest of the book.

This seminal project will fundamentally change the way that we think about addition and subtraction in accounting (together with other debates as will be discussed in the rest of the book), from the combined perspectives of the mind, nature, society, and culture, with enormous implications for the human future and what I originally called its "post-human" fate.

Some Clarifications

However, some additional (last) clarifications are needed.

These clarifications, lest any potential misunderstanding by the readers occurs, refer to (a) the conversation with my previous books, (b) the illustrative selection of case studies and examples, (c) detailed analysis versus overall analysis, (d) three distinctive features of using quotations, (e) two difficulties with understanding, (f) repetition, (g) chapter organization to counter conventional wisdom, (h) the political correctness of mainstream publishing, (i) specific context and wider implication, and (j) the use of neologisms.

These clarifications are something that I also regularly repeat in my previous books too, so these points are repeated hereafter, *often verbatim*—albeit in a different context this time.

In Conversation with My Previous Books

Firstly, my theory in this book is constructed on the theoretical foundation of my previous books (i.e., *FHC, FCD, FPHC, BDPD, BCPC, BCIV, FPHST, BNN, BWT, FC, FAE, ALD, FIA, FPHU, FPHK, FPHML, FPHE, FPHMM, FPHCT, FPHG, FPHUP, FPHL, FPHO, FPHMA, FPHS, FPHLAW, FPHFS, FPHWP, FPHP, BCOS, BEPE, FPHGEOL, FPHCHESS, FPHR, FPHEDU, FPHH, FPHA, FPHGAM, FPHCOM, FPHLIT, FPHCHEM, FPHDA, FPHGEOG, FPHPA, FPHSEM, FPHARCH, FPHCRIM, FPHHIST, FPHMIG, FPHTRAN, FPHCA, FPHW, BNR, FPHSPORT, FPHPROB, FPHMORP, FPHHC, FPHTHAN, BHR, FPHVA, FPHAERO*, and *FPHPHON*—as explicated in "The List of Abbreviations").

For this reason, many of my books have some familiar words in their titles like "The Future of Post-Human ____" or "Beyond ____to ____," as well as some tables for background information—as they constitute a series of books with the final aim to converge all of my original theories on numerous fields into a unified theory of everything (as already summarized in the section on theory and meta-theory in Chapter One).

Just consider the following two illustrations, namely, (a) in relation to other related theoretical debates and (b) in relation to different visions as worked out in my previous books.

(a) The first illustration is that the theoretical debate here also involves a few other related theoretical issues.

These other related theoretical issues have to do with some deeply contested theoretical debates, which, however, I had already gone to great lengths to analyze in my numerous (previous) books, as summarized below. It should also be stressed, lest misunderstanding occurs, that each of the theoretical debates (as listed below) does not exhaust the issues covered in the books, as each book covers more issues than what is listed below (for illustration only).

- On Economics (and Business)
  - –Ex: *The Future of Capitalism and Democracy*
  - –Ex: *Beyond Capitalism to Post-Capitalism*
  - –Ex: *Beyond Human Resources to Post-Human Resources*
  - –Ex: 2 volumes, *The Future of Human Civilization*
- On Mathematics
  - –Ex: *The Future of Post-Human Mathematical Logic*
  - –Ex: *The Future of Post-Human Data-Analysis*
  - –Ex: *The Future of Post-Human Geometry*

–Ex: *The Future of Post-Human Probability*
–Ex: *The Future of Complexity*
- On Information
  –Ex: *The Future of Information Architecture*
  –Ex: *The Future of Post-Human Formal Science*
  –Ex: *The Future of Post-Human Mass Media*
- On Classification
  –Ex: *The Future of Information Architecture*
  –Ex: *The Future of Post-Human Morphology*
- On Knowledge
  –Ex: *The Future of Post-Human Knowledge*
  –Ex: 2 volumes, *The Future of Human Civilization*
- On Organization
  –Ex: *The Future of Post-Human Organization*
- On Computing
  –Ex: *The Future of Post-Human Computing*
- On Criminality
  –Ex: *The Future of Post-Human Criminality*
- On Law
  –Ex: *The Future of Post-Human Law*
- On Ethics
  –Ex: *Beyond Ethics to Post-Ethics*
  –Ex: 2 volumes, *The Future of Human Civilization*
- On Nature and Nurture
  –Ex: *Beyond Nature and Nurture*
  –Ex: *The Future of Post-Human Health Care*
- On Resources
  –Ex: *Beyond Human Resources to Post-Human Resources*
  –Ex: *Beyond Natural Resources to Post-Human Resources*
  –Ex: *The Future of Post-Human Waste*
- On Communication
  –Ex: *The Future of Post-Human Mass Media*
- On Technology
  –Ex: *The Future of Post-Human Engineering*
  –Ex: *The Future of Post-Human Transportation*
  –Ex: *The Future of Post-Human Communication*
- On Complexity
  –Ex: *The Future of Complexity*
  –Ex: *The Future of Post-Human Formal Science*
- On Space and Time
  –Ex: *The Future of Post-Human Space-Time*

–Ex: *Beyond Cosmology to Post-Cosmology*

- On Post-Humans
  - –Ex: *Beyond Human Resources to Post-Human Resources*
  - –Ex: *Beyond Civilization to Post-Civilization*
  - –Ex: 2 volumes, *The Future of Human Civilization*
  - –Ex: *The Future of Capitalism and Democracy*
  - –Ex: *The Future of Post-Human Consciousness*
  - –Ex: *The Future of Post-Human Unconsciousness*
- On the Body
  - –Ex: *The Future of Post-Human Martial Arts*
  - –Ex: *The Future of Post-Human Performing Arts*
  - –Ex: *The Future of Post-Human Sexuality*
- On the Mind
  - –Ex: *The Future of Post-Human Consciousness*
  - –Ex: *The Future of Post-Human Unconsciousness*
  - –Ex: *The Future of Post-Human Personality*
  - –Ex: *The Future of Post-Human Computing*
  - –Ex: *The Future of Post-Human Data Analysis*
  - –Ex: *The Future of Post-Human Humor*
  - –Ex: *The Future of Post-Human Creative Thinking*
  - –Ex: *Beyond Nature and Nurture*
- On Memory
  - –Ex: *The Future of Post-Human Data Analysis*
  - –Ex: *The Future of Post-Human Semantics*
  - –Ex: *The Future of Post-Human Consciousness*
- On Learning
  - –Ex: *The Future of Post-Human Language*
  - –Ex: *The Future of Post-Human Education*
- On Cognition
  - –Ex: *The Future of Post-Human Unconsciousness*
  - –Ex: *The Future of Post-Human Consciousness*
  - –Ex: *The Future of Capitalism and Democracy*
  - –Ex: *The Future of Post-Human Creative Thinking*
  - –Ex: *The Future of Post-Human Personality*
- On Intuition
  - –Ex: *The Future of Post-Human Unconsciousness*
  - –Ex: *The Future of Post-Human Consciousness*
  - –Ex: *Beyond Capitalism to Post-Capitalism*
  - –Ex: *The Future of Capitalism and Democracy*
- On Emotion
  - –Ex: *The Future of Post-Human Unconsciousness*

–Ex: *The Future of Post-Human Consciousness*
–Ex: *The Future of Aesthetic Experience*
–Ex: *Beyond Capitalism to Post-Capitalism*
–Ex: *The Future of Capitalism and Democracy*
–Ex: *The Future of Post-Human Personality*

- On Behavior
  –Ex: *The Future of Post-Human Language*
  –Ex: *The Future of Post-Human Unconsciousness*
  –Ex: *The Future of Post-Human Consciousness*
  –Ex: *Beyond Capitalism to Post-Capitalism*
  –Ex: *The Future of Capitalism and Democracy*
  –Ex: *The Future of Post-Human Sexuality*
  –Ex: *The Future of Post-Human Personality*
  –Ex: *The Future of Post-Human Martial Arts*
- On Spirituality
  –Ex: 2 volumes, *The Future of Human Civilization*
  –Ex: *The Future of Capitalism and Democracy*
  –Ex: *The Future of Post-Human Religion*
  –Ex: *Beyond Civilization to Post-Civilization*
  –Ex: *The Future of Post-Human Sexuality*
  –Ex: *The Future of Post-Human Martial Arts*
  –Ex: *The Future of Post-Human Unconsciousness*
  –Ex: *The Future of Post-Human Consciousness*
  –Ex: *Beyond Ethics to Post-Ethics*
- On History
  –Ex: *The Future of Post-Human History*
- On Civilization
  –Ex: *Beyond Civilization to Post-Civilization*
  –Ex: 2 volumes, *The Future of Human Civilization*
- On Freedom
  –Ex: 2 volumes, *The Future of Human Civilization*
  –Ex: *The Future of Capitalism and Democracy*
  –Ex: *Beyond Capitalism to Post-Capitalism*
  –Ex: 2 volumes, *Beyond Democracy to Post-Democracy*
  –Ex: *Beyond Civilization to Post-Civilization*
- On Being and Belonging
  –Ex: *The Future of Post-Human Consciousness*
  –Ex: 2 volumes, *The Future of Human Civilization*
- On Different Dimensions of Human Existence
  –Ex: 2 volumes, *The Future of Human Civilization*
- On Modernity, Post-Modernity, and After-Postmodernity

–Ex: 2 volumes, *The Future of Human Civilization*
–Ex: *Beyond Civilization to Post-Civilization*

- On Pre-Modernity
  –Ex: *Beyond Civilization to Post-Civilization*

Therefore, those who are interested in these other theoretical issues (and debates) which are related to the current context can consult these other books of mine for more details.

(b) And the second illustration is that one of the major theses of this book concerns the future, be it human and later post-human, which is closely linked with the different visions that I had worked out in my previous books, in relation to the mind, nature, society, and culture.

Of course, for the convenience of the readers, a summary of some of these visions will be provided whenever necessary, as shown in the two sections on meta-theory above (*often verbatim*) and in different tables at the end of this chapter and the concluding chapter, for instance.

Yet, I still expect the readers to read my previous books directly for more details—as this book is written in conversation with my previous ones.

Case Studies and Examples

Secondly, the case studies and examples in the chapters are not exhaustive but solely illustrative. Nor are they necessarily mutually exclusive, as they can be reclassified in a different way. And exceptions are allowed.

One is tempted to assume, however, that, since the case studies and their examples are not exhaustive, more case studies and examples are needed. But the problem here is that just adding more of them does not necessarily change the conclusions to be drawn and may even lead to redundancy.

After all, the case studies and examples used in this book are chosen with care and deemed sufficient for the purpose at hand, even though they are solely illustrative.

Detailed Analysis versus Overall Synthesis

Thirdly, there is a distinction between detailed analysis and overall synthesis, in that the former is, especially though not exclusively, more for the specialized viewpoint of a specific discipline, whereas the latter is, especially though not exclusively, more for the broad horizon of an interdisciplinary approach—relatively speaking, of course.

Or, by analogy, there is a distinction between depicting a "tree" (with tiny details) in a "forest" and viewing the entire "forest" itself (with the broader horizon).

This book, like all previous ones of mine, tries to strike a balance between the two forms of understanding, although this does not imply that the two forms should be exactly equally offered, in an exact 50-50 balance.

Instead, the book, more often than not, does not allow the tiny details to obscure the larger picture in a given case study, because the entire book is inter-disciplinary in nature and tries to look at the entire "forest," not being bogged down with the detailed analysis of an individual "tree" in the forest, by analogy.

In this sense, although detailed analysis can be provided whenever needed for a given case study, the important point to remember is the overall horizon with its broadness of scope.

Three Distinctive Features of Using Quotations

Fourthly, the way that quotes are used in this book (and others of mine) has three distinctive features, as explained below.

The first distinctive feature is that, at the end of each quotation, sometimes there are more than one reference—in which case the first reference is for the original citation, and the rest of the references are for the convenience of the reader for more information about the issue(s) under quotation.

The second distinctive feature is that quotations are used as often as possible, so as to let others speak for themselves, without the nuisance of putting my own words into their mouths. Even more importantly, this allows me to use the original text as an evidence for critical analysis at the end of each section.

And the third distinctive feature is that all these quotations are used for my own purposes, in that I often creatively (or originally) re-organize them for a new way of analyzing the issue(s) in question.

Two Difficulties with Understanding

Fifthly, to understand this book (or any other book of mine) is difficult for two main reasons.

In regard to the first difficulty, this book, just like many other books of mine, is in conceptual and theoretical conversation with other books of mine (65 books, as of January 05, 2014), so this requires a vast background information on numerous subject matters.

And in regard to the second difficulty, this book also analyzes the subject matter in a way very different from the expectation of a conventional specialist reader, since it examines reality outside the specialist's interest (or outside his or her "comfort zone"), from the four broad perspectives of the mind, nature, society, and culture, such that it covers all domains of knowledge (in the social sciences, the natural sciences, the formal sciences, and the humanities). In other words, those who are truly interested in my project for a unified theory of everything would find many books of mine highly stimulating, but those who are specialists in a subject matter (like physics, geography, chess, and so on, in a strictly specialist sense) will not find them suitable to their specialist taste (or expectation).

Not surprisingly, a conventional specialist reader would find these two difficulties (with understanding) almost practically impossible to overcome, as they almost require the mind of a polymath. Even then, the best advice is that one should still try one's best to understand it.

Repetition

Sixthly, many of my books have something in common, and the reason is to show "the unified theory of everything" in all domains of knowledge.

For this reason, there is certain repetition in all books (especially in regard to the background information, together with the tables, in the introductory chapter and the concluding chapter). Therefore, these books are all titled "The Future of Post-Human ____" (or at times "Beyond ____ to ____") and constitute the "post-human" series.

Chapter Organization to Counter Conventional Wisdom

Seventhly, all of my books aim to deconstruct conventional wisdom and to replace it with new thinking, and the chapters in my books are often organized in two distinctive ways for this purpose

The first distinctive way is to show the polarized thinking in conventional wisdom by devoting two main chapters on opposites and then deconstructing them later. For example, in *Beyond Ethics to Post-Ethics* (2011), Chapter Two first introduces the conventional wisdom for morality but then shows that morality has its negative side often unsaid—and Chapter Three does just the opposite, that is, first introduces the conventional wisdom against immorality but then shows that immorality has its bright side often unspoken of.

And the second distinctive way is to show the opposites (embedded in conventional wisdom) in a dialectic way. For instance, in the above example, the task here is to show how and why the analysis of morality (in Chapter Two) is inadequate without the inclusion of its immorality, just as the examination of immorality (in Chapter Three) is insufficient without the consideration of its morality—to the extent that there is no morality without immorality, and vice versa.

But there are two qualifications.

The first qualification is that the opposites do not have to be always mutually exclusive in the conventional meaning (like morality vs. immorality in *BEPE* and heroes vs. villains in *FPHCRIM*)—but are simply disproportional in focus in other cases (in that some thinkers focus more on one than on the other), and good examples are networks vs. operations (in *FPHTRAN*), ingredients vs. techniques (in *FPHCA*), and form vs. function (in *FPHARCH*). In other words, there are multiple meanings of the opposites.

And the second qualification is that some books (like the 2 volumes of *The Future of Human Civilization*, the 2 volumes of *Beyond Democracy to Post-Democracy*, and *The Future of Capitalism and Democracy*) are very lengthy and analyze many issues in more than two main chapters. Yet, there is a general pattern of introducing and then deconstructing conventional wisdom in my works.

The Political Correctness of Mainstream Publishing

Eighthly, all my books are politically incorrect—and consequently, they are not suitable for publication by major mainstream publishers, for three main reasons.

The first reason is that major mainstream publishers (in the U.S. as a prime example) have their own ideological agenda. For instance, any book which is perceived, either rightly or wrongly, as too "critical" to the U.S. (or to the West) is most unlikely to be accepted for publication by a top publishing house in the U.S. (as a prime example).

The second reason is that major mainstream publishers have their own methodological bias. For instance, any book which is truly interdisciplinary (with an analysis of almost everything, with no focus on a specific discipline or a few) is most unlikely to be accepted for publication by a top publishing house in the U.S. (as a prime example).

And the third reason is that major mainstream publishers have their own preference for visions. For instance, any book which envisions the future in the most distant horizon (with no practical relevance to the present) is most unlikely to be accepted for publication by a top publishing house in the U.S. (as a prime example).

These three reasons (or a combination of them) then add up to their argument that there is no "profitable market" for a book of such political incorrectness in a capitalist society (like the U.S. as a prime example).

Specific Context and Wider Implication

Ninthly, each book proposes a new way to treat a subject matter in a specific context, but its implication is wider in scope.

For example, in the book titled *The Future of Post-Human Waste* (2012), the subject matter was treated in the new way of "uselessness" and "usefulness," within the specific context of waste—but the implication here is wider in scope, since the understanding of "uselessness" and "usefulness" can indeed be applied to other subject matters outside the concern with waste (e.g., education, strategy, communication, economics, philosophy, etc.).

This wider implication can be said about all my other books.

The Use of Neologisms

And finally, I use different neologisms in my books, mostly to introduce my original concepts and theories, and good instances include those here or elsewhere in my previous books (e.g., "the double-sided theory of accounting," "post-democracy," "hyper-spatial consciousness," "post-capitalism," and whatnot). Needless to say, they are used here solely for our current intellectual convenience, as they will be renamed differently in different ways in future history.

As I thus wrote in *FCD* (508-9), "all these terms 'post-capitalism,' 'post-democracy'...and other ones as introduced in...[the] project (e.g.,...'posthuman elitists,' and 'posthuman counter-elitists,' just to cite a few of them) are more for our current intellectual convenience than to the liking of future humans and post-humans, who will surely invent more tasteful neologisms to call their own eras, entities, and everything else, for that matter. But the didactic point here is to use the terms to foretell what the future might be like, not that its eras and entities must be called so exactly and permanently. After all, William Shakespeare (1995: Act II, Scene II, Line 47) well said long ago: 'What is in a name? That which we call a rose by any other name would smell as sweet.'"

As I had also stressed time and again before, each of the neologisms can be re-written as a different "X," only to be re-named differently by the powers that be in different eras of future history.

With these clarifications for thought in mind—I now proceed to Chapter Two on addition and its two faces.

Table 1.1. Diverse Aspects of Accounting

- **Addition and Subtraction in Accounting**
  - —Addition
    - Ex: debit (in asset)
    - Ex: credit (in liability)
    - Ex: quality audit (to increase effectiveness)
  - —Subtraction
    - Ex: credit (in expense)
    - Ex: debit (in income)
    - Ex: lean accounting (to decrease waste)
- **Main Distinctions of Issues**
  - —Ex: debit vs. credit
  - —Ex: a single point in time vs. a period over time
  - —Ex: purchasing power accounting vs. historical cost accounting
  - —Ex: accounting vs. bookkeeping

Notes: The categories are solely illustrative (not exhaustive), and the comparison is relative (not absolute), nor are they necessarily mutually exclusive. And some can be easily re-classified elsewhere. As generalities, they allow exceptions.
Source: From *Sec. 1.3* & *Sec. 1.4* of *FPHACCO*. See book for citations.

Table 1.2. Nine Great Future Transformations of Accounting (Part I)

- **In Relation to the Mind**
  —*New Goals*
  - This search search for "new goals" will give future advanced humans and later their successors, namely, the post-humans, different motivations to deal with addition and subtraction in a way that current humans cannot—in relation to my previous visions on the future "post-human" supersession of the human mind like "interconnected health care," "supersession computing," "metamorphic humor," "contrarian personality," "hyper-martial body," "hyper-sexual body," "hyper-spatial consciousness," "thinking machines," "human-animal hybrids," "unfolding unconsciousness," "floating consciousness," "thinking robots," "genetically altered superior beings," "cyborgs," etc.

  —*New Means*
  - This quest for "new means" will give future advanced humans and later their successors, namely, the post-humans, the capacity to find novel options to achieve the goals in regard to addition and subtraction in a way that current humans cannot.

- **In Relation to Nature**
  —*New Comparisons*
  - This search for "new comparisons" will give future advanced humans and later their successors, namely, the post-humans, the ability to learn from animals, plants, and the universe, through comparison and contrast, about addition and subtraction in a way that current humans cannot.

(continued on next page)

Table 1.2. Nine Great Future Transformations of Accounting (Part II)

- **In Relation to Nature (*cont'd*)**
  - —*New Integrations*
    - This search for "new integrations" will give future advanced humans and later their successors, namely, the post-humans, the ability to live and interact with nature for addition and subtraction in a way that current humans cannot. This is especially so, in the context of the great future transformations of geology (in *FPHGEOL*), of geography (in *FPHGEOG*), of migration (in *FPHMIG*), and of transportation (in *FPHTRAN*)—in relation to my previous visions on the future "post-human" adventures into different dimensions of space-time unto multiverses like "the alteration of space-time" (*Table 1.3* & *Table 1.4*), "multiverses" (*Table 1.7* & *Table 1.8*), "hyperspace" (*Table 1.9* & *Table 1.25*), "time travel" (*Table 1.10* & *Table 1.11*), and others.

      This extension is consistent with my other theory, namely, "the theory of the cyclical progression of system integration and fragmentation" in *FCD* (2002)—at the social-systemic level (as already clarified in *Sec. 1.8* and shown below):

      Local → Regional → Global →Solar →
      Galactic → Clustery...→ Multiversal

(continued on next page)

Table 1.2. Nine Great Future Transformations of Accounting (Part III)

- **In Relation to Society**
  - —*New Praxes*
    - This search for "new praxes" will give future advanced humans and later their successors, namely, the post-humans, the ability to engage in practice (like activism, politics, personal experience, etc.) for the future of addition and subtraction in a way that current humans cannot.
  - —*New Technologies*
    - This search for "new technologies" will give future advanced humans and later their successors, namely, the post-humans, the ability to come up with different techniques, materials, and experiments for the study of addition and subtraction in a way that current humans cannot.
  - —*New Stratifications*
    - This emergence of "new stratifications" will lead future advanced humans and later their successors, namely, the post-humans, to face new inequalities in the context of addition and subtraction in a way that current humans cannot.

- **In Relation to Culture**
  - —*New Sub-Fields*
    - This search for "new sub-fields" will give future advanced humans and later their successors, namely, the post-humans, the ability to explore new specialties for accounting in a way that current humans cannot.
  - —*New Interpretations*
    - This search for "new interpretations" will give future advanced humans and later their successors, namely, the post-humans, the ability to come up with different evaluations and views about accounting in a way that current humans cannot.

Notes: The categories and examples are solely illustrative (not exhaustive), and the comparison is relative (not absolute), nor are they necessarily mutually exclusive. And some can be easily re-classified elsewhere. As generalities, they allow exceptions.
Source: From *Ch.4* of *FPHACCO*. See text for citations.

Table 1.3. The Theoretical Debate on Space-Time (Part I)

- **Isaac Newton's Absolutist (Substantivist) Theory of Space-Time**
  —space and time are independent from each other. The structure of space-time is $E3 \times E1$ (with the structure of space, *P*, as the set of spatial locations in a three-dimensional Euclidean space, *E3*, and the structure of time as the set of temporal moments, *T*, in the one-dimensional real time, *E1*).
  —space and time are also independent from the effects of matter and events. The existence of space and time is possible even in a world absent of matter (and, for that matter, even in a world absent of events), as if they were material objects but with their total unchangingness thorough time.

- **Albert Einstein's Relativist Theory of Space-Time**
  —space and time are interchangeable (not absolute), just as matter and energy are equivalent (not independent) with the famous equation, $E = mc^2$ (as in the special theory of relativity in 1905).
  —space-time and matter-energy are also relative in a grand union (as in the general theory of relativity in 1915). Thus, each pair affects the other pair, as "matter 'tells' spacetime how to curve [in a non-Euclidean geometry] and curved spacetime 'tells' matter how to behave….Space contracts near mass and dilates away from it. Time dilates near mass and contracts away from it. Clocks positioned farther away from the mass of the earth run faster than clocks closer to the earth." (L. Shlain 1991: 328-330)

(continued on next page)

Table 1.3. The Theoretical Debate on Space-Time (Part II)

- **Peter Baofu's Perspectival Theory of Space-Time**
  —space and time can be understood from multiple perspectives, be they in relation to culture, society, nature, and the mind, with each perspective revealing something about the nature of space-time and simultaneously delimiting its view. This is subject to "the regression-progression principle" in existential dialectics.
  —each perspective of space and time exists in society and culture with good reasons, with some being more successful and hegemonic (dominant) than others. This is subject to "the symmetry-asymmetry principle" in existential dialectics.
  —space and time will not last, to be eventually superseded (altered) by post-humans in different forms (e.g.,stretching/shrinking space-time, engineering more dimensions of space-time, and manipulating multiverses), be they here in this universe or in multiverses. Thus, even the physical existence of space-time cannot last forever, with ever more transformations in the process. This is subject to "the change-constancy principle" in existential dialectics.
  —the conventional wisdom (especially by physicists) of treating the physical perspective of space and time as the foundation of all other perspectives (of space and time) and of regarding them as much less important is a form of reductionism,committing what I call *the foundation fallacy*, in misleadingly dismissing the multiple perspectives of space and time in relation to culture, society, nature, and the mind.

Notes: The examples in each category are solely illustrative (not exhaustive), and the comparison is relative (not absolute), nor are they necessarily mutually exclusive. Some can be easily re-classified elsewhere. As generalities, they allow exceptions.
Source: A summary of *Sec.1.1*, *Sec.1.2*, and *Sec.1.3* of *FPHST*—and, for that matter, the rest of *FPHST*

Table 1.4. Main Reasons for Altering Space-Time

- **The Need to Make New Energy-Matter**
  —Ex: manipulating molecular bonds for new materials
  —Ex: creating nanotechnologies on the atomic scale
  —Ex: engineering the atomic nucleus
  —Ex: restructuring most elementary particles
  —Ex: inventing new forms of matter and energy

- **The Need to Create New Space-Time**
  —Ex: creating "warp drive" (as in science fiction) for space travel
  —Ex: creating "pocket universes"

- **The Need to Conquer the Cosmos unto Multiverses**
  —Ex: spreading floating consciousness and hyper-spatial consciousness, besides other forms that humans have never known, in the cosmos and beyond unto multiverses for ultimate conquest

Notes: The examples in each category are solely illustrative (not exhaustive) nor necessarily mutually exclusive, and the comparison is relative (not absolute). As generalities, they allow exceptions. Also, it should be stressed that the three reasons are all related, in that they all contribute to the evolution of intelligent life in the cosmos unto multiverses in the most distant future beyond our current knowledge.
Sources: A summary of *Sec.6.2* of *FPHST*. See also *FHC*, *FCD*, *FPHC*, *FC*, *FPHU*, and *FPHG*, for example.

Table 1.5. The Technological Frontiers of the Micro-World

- **Type I-Minus**
  —Ex: building structures and mining

- **Type II-Minus**
  —Ex: playing with the genetic makeups of living things

- **Type III-Minus**
  —Ex: manipulating molecular bonds for new materials

- **Type IV-Minus**
  —Ex: creating nanotechnologies on the atomic scale

- **Type V-Minus**
  —Ex: engineering the atomic nucleus

- **Type VI-Minus**
  —Ex: restructuring most elementary particles

- **Type Ω-Minus**
  —Ex: altering the structure of space-time

Notes: As already indicated in *Sec.4.4.2.2* of *FPHC*, the problem with this micro-classification (from Barrow's work) is that the civilization types (with the exception of Type Ω-Minus, for example) are not quite distinct, since many of them can be achieved more or less in a civilization, to the extent that Type II-minus and Type III-minus, just to cite two plausible types, can be historically contemporaneous, relatively speaking, unlike the vast historical distance between, say, Type O and Type I (or Type I and Type II) civilizations. In other words, the micro-classification here is not very useful to understand civilization types but is revealing to see the technological frontiers of the micro-world.
Sources: A reconstruction from J.Barrow (1998:133), as originally shown in *Table 4.7* of *FPHC*. See *FPHC* for more info.

Table 1.6. Finity, Transfinity, and Infinity (Part I)

- **Infinity**
  - —The term "infinity," which is "symbolically represented with ∞," derives from the Latin *infinitas* (or "unboundedness"), to refer to "several distinct concepts—usually linked to the idea of 'without end'—which arise in philosophy, mathematics, and theology." (WK 2008c)
  - —In the context of mathematics, "infinity" is often "treated as if it were a number (i.e., it counts or measures things: 'an infinite number of terms') but it is a different type of 'number' from the real numbers. Infinity is related to limits,...large cardinals,...projective geometry, extended real numbers and the absolute Infinite," for instance. (WK 2008c)

- **Transfinity**
  - —To avoid confusion, my usage of the word "transfinite" (as used here in the ontological principle) is not the same as the one which was originally coined by Georg Cantor to refer to "cardinal numbers or ordinal numbers that are larger than all finite numbers, yet not necessarily absolutely infinite" (in the mind of God), as opposed to "relative" infinity (in the mind of Man). (WK 2008d) In other words, for Cantor, there are two kinds of infinity, "absolute" and "relative"—with "relative" infinity or "transfinity" (in the mind of Man) being dependent on "absolute" infinity (in the mind of God) for its existence.
  - —By contrast, my usage of the word "transfinite" here differs radically from the one used by Cantor (and other mathematicians) and is more limited, in light of the problems confronting any attempt to understand the idea of infinity, be it by intuition, imagination, and conception (as already shown in the prevous three sub-sections).
  - —With this clarification in mind, I allow numbers which can be many times larger—or smaller, for that matter—than the finite things that we encounter in daily life, but they do not have to be related to the idea of infinity at all (which may not exist).

(continued on next page)

Table 1.6. Finity, Transfinity, and Infinity (Part II)

- **Transfinity** ***(cont'd)***
  —Of course, there may be some *borderline* cases, in which it is not clear whether the number in question is transfinite (in my usage) or simply a mathematical convenience. A good example of a borderline case is the Planck unit of length for "the smallest space possibly measured in nature," which is "less than billionths of trillionths of trillionths of an inch" (or something like 1.6×10^{-35} meters). (P. Baofu 2006a; N. McAleer 1987: 219; WKV 2008; D. Corbett 2008)
  —The ideas concerning symmetry and dimensions in the examples above were addressed in Chapter Three and Chapter Four of *FPHG*. For now, it suffices to further add that it is mind-boggling to even conceive or imagine extreme numbers like the Planck unit of length (or others). Does it really exist at all? Or is it merely a mathematical convenience to describe a physical situation that the human mind does not understand properly?
  —In any event, "transfinity" (in my usage) can be used as a further extension (of number) to the finite numbers in daily life—albeit with some borderline cases in mind.

- **Finity**
  —It refers to numbers which are bounded (that is, with an end), especially (though not exclusively) in relation to things in everyday life.

Source: From *Sec. 1.4* and *Sec. 2.2.3* of *FPHG*

Table 1.7. Theoretical Speculations of Multiverses

- **"Baby Universes" (Ex: Andre Linde and others)**
  —Ex: In a flat universe theory, "even if our part of it eventually collapses,...some spots in the cosmos would suddenly start inflating on their own, creating brand-new 'baby universes.'" (P. Baofu 2000: 623)

- **"Parallel Universes" (Ex: Stephen Hawking and others)**
  —Ex: In quantum cosmology, there allows the existence of infinite numbers of parallel universes, with tunneling among them. (M. Kaku 1994: 256) Hawking later revised his views on this.

- **"Pocket Universes" (Ex: Alan Guth)**
  —Ex: "As the pocket universes live out their lives and recollapse or dwindle away, new universes are generated to take their place....While life in our pocket universe will presumably die out, life in the universe as a whole will thrive for eternity." (A. Guth 1997: 248; P. Baofu 2002: 482)

- **"Brane Worlds" (Ex: Warren Siegel, Lisa Randall, and others)**
  —Ex: Our universe is stuck on a membrane of space-time embedded in a larger cosmos, with different brane worlds connecting and/or colliding with each other.

Notes: These examples are solely illustrative (not exhaustive), and some of the items can be reclassified somewhere else. Nor are they always mutually exclusive. Since they are generalities, exceptions are expected.
Source: From *Table 4.8* of *FPHC*

Table 1.8. The Confusion between "Many Worlds" and "Multiverse" (Part I)

- **First Problem**

—The first problem in this second confusion concerns a debate between the "many-worlds" interpretation of quantum mechanics by Hugh Everett (1957) and others, and the Copenhagen interpretation by Niels Bohr and Werner Heisenberg on the effects of quantum states after interacting with an external environment (e.g., by a measuring observer).

—On the one hand, the Copenhagan interpretation of quantum mechanics is more indeterministic in its concern with the probabilities of the outcomes of a wave function after interacting with an external environment (e.g., by a measuring observer using light to detect, for example, "an electron in a particular region around the nucleus at a particular time." (WK 2008e) Each wave function for the state of a particle refers to "a mathematical representation used to calculate the probability for it to be found in a location, or state of motion." (WK 2008f) And the "wavefunction collapse" means that "the act of measurement causes the calculated set of probabilities to 'collapse' to the value defined by the measurement." (WK 2008f)

—On the other hand, the Everett's interrpetation is more deterministic, in rejecting "the objective reality of wavefunction collapse" and "instead explaining the subjective appearance of wavefunction collapse with the mechanism of quantum decoherence." (WK 2008g) The words "quantum decoherence" here refers to "the mechanism by which quantum systems interact with their environments to exhibit probabilistically additive behavior," in such a way that "the quantum nature of the system is simply 'leaked' into the environment." (WK 2008g) For example, "[a]s a result of an interaction, the wave functions of the system and the measuring device become entangled with each other. Decoherence happens when different portions of the system's wavefunction become entangled *in different ways* [italic added] with the measuring device...." (WK 2008g) Thus, for Everett, the appearance of "wavefunction collapse" can be explained by "quantum decoherence" for quantum states to branch out in different independent worlds after interacting with an external environment.

(continued on next page)

Table 1.8. The Confusion between "Many Worlds" and "Multiverse" (Part II)

- **First Problem *(cont'd)***
  —Yet, a major criticism against Everett's interpretation is that these different independent worlds in Everett's sense "will never be accessible to us," because of the interference by the measuring observer in an external environment. (WK 2008g)
  —For illustration, "once a measurement is done, the measured system becomes entangled with both the physicist who measured it and a huge number of other particles, some of which are photons flying away towards the other end of the universe; in order to prove that the wave function did not collapse one would have to bring all these particles back and measure them again, together with the system that was measured originally. This is completely impractical, but even if one can theoretically do this, it would destroy any evidence that the original measurement took place (including the physicist's memory)." (WK 2008g)
  —But then, the critics ask, If these independent worlds in Everett's sense are inaccessible to us, how do we know much of anything really about them?

- **Second Problem**
  —What is relevant at the quantum mechanical level may not necessarily apply to the world at the physicsl level.
  —In fact, the persistent inability among contemporoary physicists to unifiy the theory of relativity for large objects at the physical level and quantum mechanics for tiny objects at the quantum mechanical level should warn us about the danger of reductionism, by reducing the understanding of different universes ("parallel universes") at the physical level from a generalization of "many worlds" at the quantum mechanical level.

- **Third Problem**
  —There is no empirical evidence to suggest that there are infinite (or extremely many) possible worlds in the cosmos as the many-worlds model by Everett would have us to believe.

Source: From *Sec.4.4.1* of *FPHG*

Table 1.9. Hyperspace and Its Challenge (Part I)

- **First Problem**
  —Hyperspace for intra- and inter-universal travel is assumed to be possible, on the basis of four main assumptions; although there may be others, these are the often cited ones.
  —The first assumption concerns the existence of "entry" and "exit" points in hyperspace, which is easier said than done to enter, go through, and then exit hyperspace.
  —For instance, using hyperspace "requires complex calculations...of single or multiple hyperspace jumps and the control of the jump....Peturbations such as those experienced by ship in space from the gravitional field around an object such as a planet or even a star are exacerbated in hyperspatial travel, since mass in real space distorts hyperspace in an equal measure. 'Jumping' near to a gravitational mass is likely to make resulting exit from hyperspace to be highly uncertain, with the level of improbability *i* increasing with the square of the distance to the nearest gravitional 'well.'" (WK 2008h)
  —Besides, "[o]nly spaceships equipped with a special force field can enter hyperspace, because exposure to hyperspace even for short period of time is hazardous to unprotected humans." (WK 2008h)

- **Second Problem**
  —The second assumption concerns the faster-than-light speed. It is well known that Einstein argued for the speed of light as the ultimate limit, but nowadays phyiscists do allow the violation of this limit under certain conditions (like the hypothetical existence of tachyons, as will be explained further in a later sub-section on time travel).
  —Even then, there is a challenge to explain in science fiction "why ships can travel faster than light in hyperspace," and two good arguments are that "hyperspace may be smaller than real space and therefore a star ship's propulsion seems to be greatly multiplied, or else the speed of light in hyperspace is not a barrier as it is in real space. Whatever the reasoning, the general effect is that ships traveling in hyperspace seem to have broken the speed of light, appearing at their destinations much more quickly and without the shift in time that the Theory of Relativity would suggest." (WK 2008h)

(continued on next page)

Table 1.9. Hyperspace and Its Challenge (Part II)

- **Third Problem**
  —The third assumption concerns the speed of time (to be measured, say, by a clock), but this is elaborated in a different section on the related issue of "time travel" (as shown in a different table on time travel).

- **Fourth Problem**
  —The fourth assumption concerns spatial shortcut, as it is possible, because of the curvature of space, in that "the manifold of ordinary three-dimensional space is curved in four or more 'higher' spacial dimensions (a 'hyperspace' in the geometric sense...). This curvature causes certain widely separated points in three-dimensional space to nonetheless be 'adjacent' to each other four-dimensionally. Creating an aperture in 4D space (a wormhole) between these locations can allow instantaneous transit between the two locations; a common comparison is that of a folded piece of paper, where a hole punched through two folded sections is more direct than a line drawn between them on the sheet. This idea probably arose out of certain popular descriptions of General Relativity and/or Riemannian manifolds, and may be the original form from which later concepts of hyperspace arose." (WK 2008h)
  —But this does not mean that hyperpsace is always safe. In fact, "in some science fiction, the danger of hyperspace travel is due to the chance that the route through hyperspace may take a ship too close to a celestial body with a large gravitational field, such as a star. In such scenarios, if a starship passes too close to a large gravitational field while in hyperspace, the ship is forcibly pulled out of hyperspace and reverts to normal space. Therefore, certain hyperspace 'routes' may be mapped out that are safe, not passing too close to stars or other dangers." (WJ 2008h)

Source: From *Sec.4.4.1* of *FPHG*

Table 1.10. The Problems of Time Travel into the Future (Part I)

- **First Problem**

—The first problem concerns "time dilation under the Theory of Special Relativity" (which separates space-time from mass-energy as special cases). Suppose you are on a moving train and an observer outside the train is watching you along the tracks. As your train moves along at the speed of light (or something close to it), "time, as measured by your watch, ticked along at a slower pace than time measured by the observer. Not only that, distance changed, too. For the observer, a one-foot ruler whizzing by on the train would have measured less than a foot." (NO 2000)

—In fact, you may not even notice that this slower clicking of the clock; in fact, everything looks normal to you inside the train as it was before: "The weird thing is that, for you on the train, time wouldn't seem to be moving slower and your ruler wouldn't be shorter—all would appear normal." (NO 2008) By contrast, you instead think that "time on the rest of the Earth would appear to be ticking along slower and its rulers would be shorter." (NO 2000) So, both of you insist that the other clock (not his) is clicking more slowly—assuming, of course, an inertial frame of reference.

—Now, change the story a bit, in that, this time, instead of an "inertial reference frame," acceleration is allowed, and you are on your way to board a spaceship. As an illustration, suppose "[y]ou board a spaceship and take off for deep space. The ship approaches the speed of light. Time for you seems to pass as it always has. It takes you about five seconds to tie your shoe. But to an observer on Earth (assuming he or she could watch you), you are moving at a snail's pace. It takes hours to tie your shoe." (NO 2000) In any event, "you continue on your journey. You slow down, stop, and accelerate back to Earth. You arrive home. You have aged two years during your flight. Two hundred years have passed on Earth. You have successfully travelled forward through time." (NO 2000) In this case, both the observer (if he is still alive) and you agree that you have aged less.

(continued on next page)

Table 1.10. The Problems of Time Travel into the Future (Part II)

- **First Problem *(cont'd)***
  —In other words, when you are back to Earth, a lot of things have changed since, and you may find rather strange to be in a new environment which you hardly recognize, when compared with what you used to think as your city, or the Earth (two years ago, well, in accordance to your time of measurement). This means that you are in the future (two hundred years later), even though you are still relatively young (only two years older). But, can you go back? Well, no. The reason is that "[a]ccording to relativity, you can only move through time in one direction." (NO 2000)

- **Second Problem**
  —The second problem concerns "time dilation under the Theory of General Relativity" (which unifies space-time with mass-energy in a general theory), the same result of time dilation occurs.
  —For instance, when "one clock is deeper in a gravity well than the other," a result is that "the clock deeper in the well" will "tick…more slowly; this effect must be taken into account when calibrating the clocks on the satellites of the Global Positioning System, and it could lead to significant differences in rates of aging for observers at different distances from a black hole." (WK 2008i)
  —More precisely, by way of another example, "it has been calculated that, under general relativity, a person could travel forward in time at a rate four times that of distant observers by residing inside a spherical shell with a diameter of 5 meters and the mass of Jupiter. For such a person, every one second of their 'personal' time would correspond to four seconds for distant observers. Of course, squeezing the mass of a large planet into such a structure is not expected to be within our technological capabilities in the near future." (WK 2008i; R. Gott 2002)
  —The point here, however, is that time travel into the future, in this sense, is allowed, theoretically speaking, but in a very limited way, from the theory of relativity. But one can also ask, What if the theory of relativity were wrong? Well, in that case, there would be a need to find a better theory to explain time travel into the future.

Source: From *Sec.4.4.2* of *FPHG*

Table 1.11. The Problems of Time Travel into the Past (Part I)

- **First Problem**
  - —The first problem concerns the faster-than-light-speed, since, in accordance to the theory of special relativity, when an object goes faster than the speed of light, something weird can happen, in that, instead of having the clock clicking more slowly, it is moving backward, in relation to some inertial frame of reference—which then raisies the issue of whether time travel into the past is possible at all.
  - —But the problem here is about the violation of the law of casuality. In everyday language, an excellent illustration of causality violation is that, suppose you travel back into the past and then kill your parents, this then means that you could not have been born. But this is not true, since you are alive.
  - —Yet, "in the case of a hypothetical signal moving faster than light, there would always be some frames in which the signal was received before it was sent, so that the signal could be said to have moved backwards in time." The violation of casuality exists in this case.
  - —The first part of the problem is that, in accordance to the theory of special relativity, "it would take an infinite amount of energy to accelerate a slower-than-light object to the speed of light"—let alone the energy needed to propel an object to go faster than the speed of light. (WK 2008i) Although some suggest the possibility of "negative energy," it remains to be seen to what extent this is true.
  - —And the second part of the problem is that, "although relativity does not forbid the theoretical possibility of tachyons which move faster than light at all times, when analyzed using quantum field theory it seems that it would not actually be possible to use them to transmit information faster than light and there is no evidence for their existence." (WK 2008i; S. Chase 1993)

(continued on next page)

Table 1.11. The Problems of Time Travel into the Past (Part II)

- **Second Problem**
  - —The second problem concerns cosmic string and black holes, since it is the theory of general relativity which "extends the special theory to cover gravity, illustrating it in terms of curvature in spacetime caused by mass-energy and the flow of momentum." (WK 2008i)
  - —On the other hand, "[g]eneral relativity describes the universe under a system of field equations...and there exist solutions to these equations that permit what are called 'closed time-like curves,' and hence time travel into the past....The first of these was proposed by Kurt Gödel, a solution known as the Gödel metric, but his (and many others') example requires the universe to have physical characteristics that it does not appear to have." (WK 2008i; K. Thorne 1994)
  - —But then, the problem now becomes: "Whether general relativity forbids closed time-like curves for all realistic conditions is unknown." (WK 2008i)

- **Third Problem**
  - —The third problem concerns wormholes and related ones, and "[w]ormholes are a hypothetical warped spacetime which are also permitted by the Einstein field equations of general relativity, although it would be impossible to travel through a wormhole unless it was what is known as a traversable wormhole." (WK 2008i; M. Visser 1995) In other words, "a wormhole is a hypothetical topological feature of spacetime that is fundamentally a 'shortcut' through space and time. Spacetime can be viewed as a 2D surface, and when 'folded' over, a wormhole bridge can be formed. A wormhole has at least two mouths which are connected to a single throat or tube. If the wormhole is traversable, matter can 'travel' from one mouth to the other by passing through the throat." (WK 2008j)
  - —But the problem here is three-fold, as shown in what follows.

(continued on next page)

Table 1.11. The Problems of Time Travel into the Past (Part III)

- **Third Problem** ***(cont'd)***
  —The first part of the problem is that "there is no observational evidence for wormholes." (WK 2008j)
  —The second part of the problem is that it is not clear how exactly wormholes (and black holes, for that matter) could allow time travel into the past at all.
  —And the third part of the problem is that if using wormholes (and black holes, for that matter) to travel into the past requires something to go faster than light too, then this raises another issue concerning the two problems as previously indicated in the sub-section on the faster-than-light-speed.

- **Fourth Problem**
  —The fourth problem is that some physicists have argued that, even should time travel into the past be possible, by way of these three methods, the person still would not return to the same past history that he exactly experienced before, but in a different one. (WK 2008i)
  —In other words, "strictly speaking," time travel into the past is not "really" possible, in light of these criticisms, so there is no need to worry that one could travel into the past and kill one's parent, as a way to violate the law of casuality.

Source: From *Sec.4.4.2* of *FPHG*

Table 1.12. The Conceptual Dimensions of Consciousness (and Other Mental States)

- **On Heredity and Time**
  —Heredity and the Environment
  —The Past and the Present

- **On Layers of Mental States and Abnormality**
  —Consciousness, Unconsciousness, and Preconsciousness
  —Normality and Paranormality

- **On Organicity and Motivation**
  —Mechanicity and Organicity
  —Primary Motivations and Multiple Motivations

- **On Other (Mostly Epistemic) Considerations**
  —Synthesis and Analysis
  —Situation and the Subject
  —Process and Outcome
  —Reasoning and Other Modes of Thinking
  —Meta-Conceptual Nominalism and Realism

Notes:: These examples are solely illustrative (not exhaustive), and some of the items can be reclassified somewhere else. Nor are they always mutually exclusive. Since they are generalities, exceptions are expected.
Sources: From *FPHC*. A re-construction, but with my own contribution, originally from G.Lindzey & C.Hall, *Introduction to Theories of Personality* (NY: John Wiley & Sons, 1985).

Table 1.13. The Theoretical Levels of Consciousness (and Other Mental States) (Part I)

- **At the Micro-Physical Theoretical Level**
  - —Quantum-Mechanics
  - —Electromagnetism

- **At the Chemical Theoretical Level**
  - —Biochemistry

- **At the Biological Theoretical Level**
  - —Evolutionary Biology
  - —Neuroscience
  - —Artificial Intelligence

- **At the Psychological Theoretical Level**
  - —Psychodynamic Psychology
    - Psychoanalytical Psychology
    - Analytical Psychology
    - Socially Oriented Psychology
  - —Experimental (Behavioral) Psychology
    - Operant Reinforcement Theory
    - Stimulus-Response Theory
    - Social Learning Theory
  - —Cognitive (Gestalt) Psychology
    - Humanist Psychology
    - Existential Psychology
    - Field Theory
  - —Psychometric Psychology
    - Idiographics
    - Constitutional Psychology
  - —Social Psychology
    - Symbolic Interactive Theory
    - Social Exchange Theory

(continued on next page)

Table 1.13. The Theoretical Levels of Consciousness (and Other Mental States) (Part II)

- **At the Organizational Theoretical Level**
  - —Managerial-Bureaucratic Theory
  - —Oligarchic Theory
  - —Network Theory

- **At the Institutional Theoretical Level**
  - —Functionalist Theory
  - —Anomic Theory

- **At the Structural Theoretical Level**
  - —Conflict Theory
    - Marxian Theory
    - Critical Theory
    - Weberian Theory
  - —Games Theory (in Formal Theory)
  - —Feminist Theory
    - Feminist Concerns
    - Feminist Hopes

- **At the Systemic Theoretical Level**
  - —Equilibrium Theory
  - —System Theory
  - —Chaos Theory

- **At the Cosmological Theoretical Level**
  - —Superluminal Model
  - —The Theory of Floating Consciousness

(continued on next page)

Table 1.13. The Theoretical Levels of Consciousness (and Other Mental States) (Part III)

- **At the Cultural Theoretical Level**
  - —Substantive Theories
    - Structuralist Theory
    - Post-Structuralist Theory (in Postmodernism)
  - —Meta-Theories
    - Epistemic Objectivism vs. Epistemic Historicism
    - Epistemic Subjectivism vs. Epistemic Non-Subjectivism
      - –Phenomenology
      - –Ethnomethodology
      - –Hermeneutics
    - Epistemic Relativism vs. Epistemic Absolutism
    - Epistemic Reductionism vs. Epistemic Emergencism

- **At Other Levels**
  - —Structuration Theory
  - —Reflexive Socioanalysis

Notes: These examples are solely illustrative (not exhaustive), and some of the items can be reclassified somewhere else. Nor are they always mutually exclusive. Since they are generalities, exceptions are expected.
Source: From many different sources as indicated in *FPHC*

Table 1.14. The Thematic Issues of Consciousness (and Other Mental States)

- **The Factor of History**
  - —Pre-Modernity
  - —Modernity
  - —Postmodernity
  - —After-Postmodernity

- **The Importance of Needs**
  - —Having (e.g., the everyday, the technological)
  - —Belonging (e.g., the just and the good)
  - —Being (e.g., the true, the holy, the sublime/beautiful, and the good)

Notes: These examples are solely illustrative (not exhaustive). Nor are they always mutually exclusive. Since they are generalities, exceptions are expected. Also, some categories overlap with others; for instance, the good is also related with being-ness, just as the everyday can be connected with belong-ness.
Sources: From *FPHC*, based on *FHC* and *FCD*

Table 1.15. Having, Belonging, and Being in Consciousness

- **Having (e.g., About the Everyday and Technological)**
  - —Ex: Physiological needs (e.g., thirst, hunger, sex)
  - —Ex: Safety (e.g., stability, freedom from fear and anxiety)
- **Belonging (e.g., About the Good and Just)**
  - —Ex: Social interaction
  - —Ex: Friendship, acquaintance
  - —Ex: Love, family
  - —Ex: Self-respect, respect from others
- **Being (e.g., About the True, Holy, Beautiful/Sublime, and Good)**
  - —Ex: Understanding of reality about self and world
  - —Ex: Spiritual quest for holiness, beauty, sublimity, and goodness

Notes: These examples are solely illustrative (not exhaustive), and some of the items can be reclassified somewhere else. Nor are they always mutually exclusive. Since they are generalities, exceptions are expected.
Sources: From *FPHC*, based on *FHC* and *FCD*

Table 1.16. The Having-Ness of Consciousness (and Other Mental States)

- **At the Micro-Physical Level**
  —Ex: The finer physics of nutrition

- **At the Chemical Level**
  —Ex: Food guide pyramid

- **At the Bio-Psychological Level**
  —Ex: The hierarchy of needs

- **At the Institutional Level**
  —Ex: The production of commodity fetishism

- **At the Organizational Level**
  —Ex: Mass standardization (and customization); bureaucratic rationality

- **At the Structural Level**
  —Ex: The divide between the haves and the have-nots

- **At the Cultural Level**
  —Ex: The bondage of greater expectations

- **At the Systemic Level**
  —Ex: The technological transformation of poverty with transvaluations

- **At the Cosmological Level**
  —Ex: Floating consciousness in deep space without the human physiological needs

Notes: These examples are solely illustrative (not exhaustive), and some of the items can be reclassified somewhere else. Nor are they always mutually exclusive. Since they are generalities, exceptions are expected.
Sources: From *FPHC*. See also *FCD* and *FHC*.

Table 1.17. The Belonging-Ness of Consciousness (and Other Mental States)

- **At the Micro-Physical Level**
  —Ex: Statistical physics and social networks
- **At the Chemical Level**
  —Ex: Olfaction and social relations
- **At the Bio-Psychological Level**
  —Ex: Dual human nature on relationships
- **At the Institutional Level**
  —Ex: Private property and social cleavage
- **At the Organizational Level**
  —Ex: Legal formalism and iron cage
- **At the Structural Level**
  —Ex: Class conflicts and social alienation
- **At the Cultural Level**
  —Ex: Ruthless society and the culture of law
- **At the Systemic Level**
  —Ex: Human extinction and post-human relations
- **At the Cosmological Level**
  —Ex: The transcendence of equality in the cosmos and beyond

Notes: These examples are solely illustrative (not exhaustive), and some of the items can be reclassified somewhere else. Nor are they always mutually exclusive. Since they are generalities, exceptions are expected.
Sources: From *FPHC*. See also *FCD* and *FHC*.

Table 1.18. The Being-Ness of Consciousness (and Other Mental States) (Part I)

- **At the Micro-Physical Level**
  —Ex: Space-time and matter-energy in classical mechanics, quantum mechanics, and the theory of relativity

- **At the Chemical Level**
  —Ex: Drugs and well being

- **At the Biological Level**
  —Ex: Neuroscience and mystical experiences

- **At the Psychological Level**
  —Ex: Reductionism and emergencism

- **At the Institutional Level**
  —Ex: The proliferation of cults and sects, and the corporate-art industry

- **At the Organizational Level**
  —Ex: The organization of ideas in the scheme of things

- **At the Structural Level**
  —Ex: Disciplinary power and control in the order of things

- **At the Systemic Level**
  —Ex: Mutualism and ecological psychology

- **At the Cultural Level**
  —Ex: The recurrence of competing spirits, and floating consciousness as a climax of evolution

(continued on next page)

Table 1.18. The Being-Ness of Consciousness (and Other Mental States) (Part II)

- **At the Cosmological Level**
  —Ex: The search for elsewherewhen, and the evolution of other consciousnesses (e.g., the hyper-spatial consciousness)

Notes: These examples are solely illustrative (not exhaustive), and some of the items can be reclassified somewhere else. Nor are they always mutually exclusive. Since they are generalities, exceptions are expected.
Sources: From *FPHC*. See also *FCD* and *FHC*.

Table 1.19. Cognitive Partiality in Different Mental States

- **The Conscious**
  - —*Biased*
    - Ex: Self-overratingness
    - Ex: Self-modesty
    - Ex: Self-fulfilling prophesy
  - —*Shallow*
    - Ex: Attribution error
    - Ex: Heuristics
    - Ex: Illusory thinking
  - —*Short-term*
    - Ex: Deindividuation
  - —*Materialistic*
    - Ex: Hierarchy of needs

- **The Subconscious**
  - Ex: mental reconstruction
  - Ex: stereotyping

- **The Unconscious**
  - Ex: automatic processing
  - Ex: cognitive intuition

Notes: These examples are solely illustrative (not exhaustive), and some of the items can be reclassified somewhere else. Nor are they always mutually exclusive. Since they are generalities, exceptions are expected. And the comparison is relative, not absolute.
Source: From *BCPC*

Table 1.20. Emotional Non-Neutrality and Behavioral Alteration in Different Mental States

- **The Emotional**
  - —*Conscious*
    - Ex: Feelings
  - —*Subconscious*
    - Ex: Moods
  - —*Unconscious*
    - Ex: Empathy
    - Ex: Instincts

- **The Behavioral**
  - —*Conscious*
    - Ex: Foot-in-the-door
    - Ex: Role playing
    - Ex: Saying-is-believing
  - —*Subconscious*
    - Ex: Discrimination
  - —*Unconscious*
    - Ex: Non-Verbal Communication

Notes: These examples are solely illustrative (not exhaustive), and some of the items can be reclassified somewhere else. Nor are they always mutually exclusive. Since they are generalities, exceptions are expected. And the comparison is relative, not absolute.
Source: From *BCPC*

Table 1.21. The Limits of Intuition in Unconsciousness

- **Cognitive Intuition (or Intuitive Thinking)**
  - —*Features*
    - Ex: mental schemas
    - Ex: expertise
    - Ex: blindsight
  - —*Problems*
    - Ex: inaccurate
    - Ex: non-explanatory

- **Emotional Intuition**
  - —*Features*
    - Ex: empathy
    - Ex: instincts
  - —*Problems*
    - Ex: over-generalized
    - Ex: non-explanatory

- **Behavioral Intuition**
  - —*Features*
    - Ex: prosemics
    - Ex: kinesics
  - —*Problems*
    - Ex: unreliable
    - Ex: non-explanatory

Notes: These examples are solely illustrative (not exhaustive), and some of the items can be reclassified somewhere else. Nor are they always mutually exclusive. Since they are generalities, exceptions are expected. And the comparison is relative, not absolute.
Source: From *BCPC*

Table 1.22. The Wealth/Poverty Dialectics in Different Mental States: The Case of Cognition

- **The Conscious**
  - —*Biased*
    - Ex: If richer in self-serving bias, then poorer in understanding others
  - —*Shallow*
    - Ex: If richer in heuristic thinking, then poorer in comprehending reality
  - —*Short-term*
    - Ex: If richer in short-term tendency, then poorer in long-term planning
  - —*Materialistic*
    - Ex: If richer in physiological obsession, then poorer in spiritual enlightenment

- **The Subconscious**
  - Ex: If richer in mental reconstruction, then poorer in accuracy of recall
  - Ex: If richer in stereotyping, then poorer in understanding groups

- **The Unconscious**
  - Ex: If richer in automatic processing, then poorer in depth of analysis
  - Ex: If richer in cognitive intuition, then poorer in reliability of judgments

Notes: These examples are solely illustrative (not exhaustive), and some of the items can be reclassified somewhere else. Nor are they always mutually exclusive. Since they are generalities, exceptions are expected. And the comparison is relative, not absolute.
Source: From *BCPC*

Table 1.23. The Wealth/Poverty Dialectics in Different Mental States: The Case of Emotion and Behavior

- **Emotion**
  - —*Conscious*
    - Ex: If richer in relying on familiarity and similarity for social attraction, then poorer in having diverse social relations
  - —*Subconscious*
    - Ex: If richer in moodiness, then poorer in sober thinking
  - —*Unconscious*
    - Ex: If richer in empathy, then poorer in task-oriented efficiency
    - Ex: If richer in instincts, then poorer in critical thinking

- **Behavior**
  - —*Conscious*
    - Ex: If richer in foot-in-the-door technique, then poorer in instant Gratification
    - Ex: If richer in role playing, then poorer in identity stability
  - —*Subconscious*
    - Ex: If richer in discriminating, then poorer in social harmony
  - —*Unconscious*
    - Ex: If richer in kinesics and proxemics, then poorer in verbal skills, relatively speaking.

Notes: These examples are solely illustrative (not exhaustive), and some of the items can be reclassified somewhere else. Nor are they always mutually exclusive. Since they are generalities, exceptions are expected. And the comparison is relative, not absolute.
Source: From *BCPC*

Table 1.24. The Theoretical Debate on Nature vs. Nurture (Part I)

• **The Environmental Approach**

—*Thesis*: It focuses, relatively speaking, more on the environment (culture and society) in explaining the achievement gap among individuals and for that matter, countries or regions, when contrasted with the natural factors.

—*Discourse*: Examples include Jose Ortega y Gasset ("Man has no nature; what he has is history"), Ashley Montagu ("Man is man because he has no instincts, because everything he is and has become he has learned from his culture, from the man-made part of the environment, from other human beings"), Stephen Jay Gould ("[The] brain [is] capable of a full range of behaviors and predisposed to none"), and Jesse Jackson (who blames white racism for the failure of blacks to close the achievement gap between whites and blacks in America). The works on dependency theory in international political economy (with a Marxian influence) and on the Protestant work ethic (by Max Weber) also point to this environmental direction.

• **The Genetic Approach**

—*Thesis*: It focuses instead, relatively speaking again, on hereditory factors (e.g., genes and evolution) in explaining the achievement gap among individuals and for that matter, countries or regions, when contrasted with the envrionmental factors.

—*Discourse*: Examples are Hans Eysenck and William Sheldon (in constitutional psychology), Konrad Lorenz (in his work on innate aggressive human nature), Gary Marcus (on the complexities of human thought by a tiny number of genes) and Robert Plomin, together with Michael Owen and Peter McGuffin (on the genetic basis of complex human behaviors).

(continued on next page)

Table 1.24. The Theoretical Debate on Nature vs. Nurture (Part II)

- **The Compromise Approach**
  —*Thesis*: It seeks the middle-of-the-road argument in regard to nature and nurture and regards all differences among individuals and groups as the result of the mixture of both nature and nurture, more or less equally.
  —*Discourse*: Examples include C. Murray and R. Herrnstein ("It seems highly likely to us that both genes and environment have something to do with this issue") and Dan Dennett ("Surely 'everyone knows' that the nature-nurture debate was resolved long ago, and neither side wins since everything – is – a – mixture – of – both – and - it's – all – very -complicated....")

- **The Transcendent Approach**
  —*Thesis*: It goes beyond both nature and nurture (without, however, committing the compromise fallacy) in showing their closely intertwined interactions in producing the behavioral differences as often seen in individual human endeavors on the micro scale, and for that matter, in country (or regional) endeavors on the macro one—in the context of my five theses, namely, (a) the compromise fallacy, (b) no oppression without self-oppression, (c) no success without failure, (d) the factor of randomness, and (e) the post-human vision, to be elaborated in Chapter Six.
  —*Discourse*: Peter Baofu proposed this approach on the basis of his "theory of contrastive advantages" (as an original theoretical contribution to the debate, which was first proposed and analyzed in *The Future of Capitalism and Democracy*). In the end, the human genes will not last, to be eventually superseded by post-human life forms, so the debate between genes and memes has obscured something profoundly important about the future that the world has never known. And the debate is also misleading and faulty in its dichotomy.

Source: From *BNN*. See the book for citations and details.

Table 1.25. Physical Challenges to Hyper-Spatial Consciousness

- **The Understanding of a Higher-Dimensional World of Space-Time**
  —Ex: 4 for traditional aspects of space-time (e.g., length, width, breadth and time) plus 6 more new dimensions in theory of hyper-space, with profound implications for practical applications to new forms of consciousness.

- **The Mastering of Dark Matter and Dark Energy**
  —Ex: "ordinary matter" (e.g., atoms, molecules) as a mere 4.4% of the universe, with 23% made of "cold dark matter" and the rest (about 73%) of mysterious "dark energy," with fundamental significance to questions about the limit of the speed of energy (or info), the availability of energy for use, and the nature of space-time, just to cite some examples.

- **The Exploration of Multiverses**
  —Ex: theoretical speculation of other universes (e.g., "baby universes," "gateways" in black holes, "wave function of the universe," "many worlds," "brane worlds"), with potentially seminary discoveries of different physical laws in relation to matter-energy and space-time, and vital differences to the future of post-human conquest of other universes (for the emergence of new forms of consciousness).

Notes: These examples are solely illustrative (not exhaustive), and some of the items can be reclassified somewhere else. Nor are they always mutually exclusive. Since they are generalities, exceptions are expected. The point here is to give a rough picture of the evolution of consciousness to the hyper-spatial consciousness and others totally unknown to current earthlings. As a note of clarification, it makes no difference to my argument as to whether or not the hyper-spatial consciousness may emerge before, during, and after floating consciousness.
Source: From *Table 4.5* of *FPHC*

Table 1.26. The Theory of Floating Consciousness (Part I)

- **At the Micro-Physical Level**
  —Ex: intelligent life without the human physical-chemical system
- **At the Chemical Level**
  —Ex: space radiation and toxins
- **At the Bio-Psychological Level**
  —Ex: exo-biological evolution in deep space
  —Ex: genetic engineering of new beings
- **At the Institutional Level**
  —Ex: post-capitalism
  —Ex: post-democracy
- **At the Organizational Level**
  —Ex: less legal-formalistic routines
- **At the Structural Level**
  —Ex: alien forms of violence
- **At the Cultural Level**
  —Ex: transcending freedom
  —Ex: transcending equality

(continued on next page)

Table 1.26. The Theory of Floating Consciousness (Part II)

- **At the Cosmological Level**
  —Ex: parallel universes
  —Ex: pocket universes

- **At the Systemic Level**
  —Ex: space habitats (in zero-gravity)

Notes: Each example draws from the works of different scholars in the field. For instance, at the cosmological level, the idea of parallel universes is from the theoretical speculation in quantum cosmology by Stephen Hawking and others, while the one of pocket universes comes from the theoretical work of Allan Guth at MIT. And at the institutional level, I proposed post-capitalism and post-democracy in *FCD* (and later, from *BDPD* and *BCPC*). In addition, the examples are solely illustrative (not exhaustive), and some of the items can be reclassified somewhere else. Nor are they always mutually exclusive. Since they are generalities, exceptions are expected.
Sources: From *FPHC*—and, originally, from *FCD*

Table 1.27. The Potential of Unfolding Unconsciousness

- **Superior Senses**
  - —*Through space*
    - Ex: clair-sentience (in feeling)
    - Ex: clair-voyance (in seeing)
    - Ex: clair-audience (in hearing)
    - Ex: clair-austance (in tasting)
    - Ex: clair-alience (in smelling)
    - Ex: clair-cognizance (in knowing)
  - —*Through time*
    - Ex: pre-cognition (in knowing), pre-sentiment (in feeling)
    - Ex: retro-cognition (in knowing), retro-sentiment (in feeling)

- **Superior Contacts**
  - —*With living minds*
    - Ex: telepathy
  - —*With the dead*
    - Ex: mediumship
  - —*With other-living entities*
    - Ex: telekinesis

Notes: The categories and examples are solely illustrative (not exhaustive), and the comparison is relative (not absolute), nor are they necessarily mutually exclusive. And some can be easily re-classified elsewhere. As generalities, they allow exceptions.
Source: A summary of Ch.5 of *FPHU*

Table 1.28. The Future Exploration of Unfolding Unconsciousness

- **In the Context of Nature**
  - —Ex: the higher-dimensional space-time in hyperspace theory
  - —Ex: the particle-wave interactions in quantum mechanics
  - —Ex: the study of brain waves in electromagnetic theory

- **In the Context of the Mind**
  - —Ex: the unconscious fantasies in psychoanalysis
  - —Ex: the mind-altering drugs in biochemistry
  - —Ex: the manipulation of neural activity in neurobiology

- **In the Context of Culture**
  - —Ex: the correlation between anomalous experience and moral/religious/aesthetic interests in culture studies
  - —Ex: the role of myths and language in epistemology

- **In the Context of Society**
  - —Ex: different organizational agendas in anomalous research
  - —Ex: the scholarly divide on anomalous study
  - —Ex: interests and propaganda in institutioanl research
  - —Ex: the transformative power of technology in research on social systems
  - —Ex: the natural and non-natural selection of other worlds in exo-biology
  - —Ex: the role of the post-human conscious mind in qualitative demography

Notes: The examples in each category are solely illustrative (not exhaustive), and the comparison is relative (not absolute), nor are they necessarily mutually exclusive. And some can be easily re-classified elsewhere. As generalities, they allow exceptions.
Source: A summary of Chs.2-5 of *FPHU*, in conjunction with the rest of the book

Table 1.29. Creative Techniques and Traits

- **Creative Techniques**
  - —*Comprehensiveness*
    - Ex: Broadness of Scope
    - Ex: In-Depth Analysis
    - Ex: Multifaceted Taxonomy
    - Ex: Numerous Theories
  - —*Arrangement*
    - Ex: Separation (or Subtraction)
    - Ex: Combination (or Addition)
    - Ex: Replacement
    - Ex: Accommodation
    - Ex: Reversal
    - Ex: Stretching
  - —*Serendipity*
    - Ex: Play
    - Ex: Chance

- **Creative Traits**
  - —*Openness / Risk*
    - Ex: Ability to welcome new ideas or to do new things, with risk
  - —*Discipline*
    - Ex: Ability to sacrifice whatever it takes to succeed
  - —*Resilience / Confidence*
    - Ex: Ability to take failures to mind, not to heart
  - —*Trust*
    - Ex: Ability to cooperate with others for team work, if necessary

Notes: The examples in the categories are solely illustrative (not exhaustive), and the comparison is relative (not absolute), nor are they necessarily mutually exclusive. And some can be easily re-classified elsewhere. As generalities, they allow exceptions.
Source: From *Sec.4.2* of *FPHCT*. For more details, also read the rest of the book.

Table 1.30. The Desirability of Creativity, and Its Dark Sides (Part I)

- **In Relation to Instrumental Rationality**
  - *—Desirability*
    - Ex: For the sake of efficiency and effectiveness in this capitalist era of our time.
  - *—Undesirability*
    - Ex: What Max Weber (1930) called the "iron cage" of capitalism, where live the "sensualists without heart, specialists without spirit"—or something which the Frankfurt School has forcefully asked us to be freed from.
- **In Relation to Substantive Rationality**
  - *—Desirability*
    - Ex: For the sake of God, the King, Motherland, or other comparable ideals.
  - *—Undesirability*
    - Ex: Domination and oppression of various forms—something which the moderns had struggled to free themselves from in the first place, since the dawn of modernity.
- **In Relation to Autonomous Rationality**
  - *—Desirability*
    - Ex: For the sake of itself, the autonomy of creative endeavor.
  - *—Undesirability*
    - Ex: The decadence of its own degeneration (e.g., the donwsides of postmodernism in our postmodern times, or what Nietzsche once called the unbearable "unreality" and "falseness" of the autonomous artist's "innermost existence"—and, in other cases, the painful suffering from different mental illnesses for those highly creative individuals).

(continued on next page)

Table 1.30. The Desirability of Creativity, and Its Dark Sides (Part II)

Notes: The examples in the categories are solely illustrative (not exhaustive), and the comparison is relative (not absolute), nor are they necessarily mutually exclusive. And some can be easily re-classified elsewhere. As generalities, they allow exceptions.
Also, I already went to great lenghts to explain the desirability and its dark sides of all these three rationalities (although there may be others too) in the two volumes of *FHC*—and, for that matter, *FCD*.
Source: From *Sec.4.9* of *FPHCT*. For more details, also read the rest of the book.

Table 1.31. Posthuman-Ism, Post-Humanism, and Trans-Humanism

- **Post-Humanism**
  —The neologism "post-human" used in my books should *not* be confused with another term which looks similar but has a totally different meaning in the literature of postmodernism, namely, "post-humanism"—which constitutes a critique of "humanism" as tradionally understood (especially, though not exclusively, in relation to the idea of progress in science and reason in the Enlightenment project). (WK 2008)
  —My works reject the project of "postmodernism" and propose the future world of what I orignally called "after-postmodernity" in *FHC* and *FCD*, for instance.

- **Trans-Humanism**
  —Also, the neologism "post-human" used in my books should *not* be confused with a similar term which is used to champion the ideology of technology for the fturue co-existence between humans and cyborgs in "trans-humanism." (WK 2008a)
  —Instead, my term "post-human" in relation to "posthuman-ism" also rejects "trans-humanism" (especially, though not exclusively, in relation to the promises of technology) and refers to something else altogether, that is, the future extinction of humans and its post-human successors in deep space and beyond unto multiverses.
  —Chapter Two of my book on ethics (*BEPE*) provides a more comprehensive critique of trans-humanism.

- **Posthuman-Ism**
  —With these two clarifications in mind, the "post-humans" (as envisioned in my books) can take different forms, and I proposed, in my different books, some of them, such as "unfolding unconsciousness," "floating consciousness," "hyper-spatial consciousness," "thinking machines," "thinking robots," "genetically altered superior beings," "cyborgs," and others (as already analyzed in my previous works).
  —They will evolve in conjunction with other visions of mine in relation to nature, society, and culture.

Sources: From *Sec.1.6* of *FPHE* (and also *BEPE*). For more details, also read the rest of the books (and other books of mine).

Table 1.32. Three Great Future Transformations of the Martial Body (Part I)

- **Virtual Battle**
  —Ex: virtual fight
  —Ex: virtual memory

- **The Hyper-Martial Body**
  —*In relation to the physical dimension*
  - Ex: striking (e.g., punching, kicking, trapping, acupressure-striking, and so on in an amazing way)
  - Ex: grapping (e.g., throwing, pinning, joint-locking, and so on in an amazing way)
  - Ex: running and walking (e.g., on different surfaces)
  - Ex: jumping (e.g., through walls, trees, etc.)
  - Ex: flying (e.g., through space)
  - Ex: enduring pain (e.g., absorbing punches, kicks, and so on in an amazing way)
  - Ex: using extremely advanced hi-tech weapons
  - Ex: healing (e.g., with special energy, etc.)
  - Ex: entering (e.g., into different dimensions of space)
  - Ex: sensing (e.g., in extraordinary ways through space and time, etc.)
  - Ex: wearing hi-tech suits for defensive and offensive purposes (e.g., like a hi-tech armor)

  —*In relation to the mental dimension*
  - Ex: the ability to relax the martial body in a way that current humans cannot.
  - Ex: the ability to concentrate the martial mind in a way that current humans are not accustomed to.
  - Ex: the ability to breathe and be timely in preparing the martial body for fighting in a way that current humans are not good at.

- **The Post-Martial Floating Mind**
  —Ex: the mind without the human body
  —Ex: the quest for spirituality beyond martial arts

(continued on next page)

Table 1.32. Three Great Future Transformations of the Martial Body (Part II)

Notes: The examples in the categories are solely illustrative (not exhaustive), and the comparison is relative (not absolute), nor are they necessarily mutually exclusive. And some can be easily re-classified elsewhere. As generalities, they allow exceptions.
Source: From *Sec. 2.5.1* of *FPHMA*. See text for citations.

Table 1.33. Three Great Future Transformations of the Sexual Body

- **Virtual Pleasure**
  —Ex: virtual Sex
  —Ex: virtual Recall

- **The Hyper-Sexual Body**
  —*In relation to the physical dimension*
  - Ex: the ability to perform differenet sexual positions without the various physical limitations that humans currently have.
  - Ex: the ability to produce unusual amount of sexual energy to endure in sexual acts in a way that humans currently cannot.
  - Ex: the ability to engage in multiple sexual experiences comparable to (but different from) non-human sexuality (e.g., hermaphrodite, monoecious, sex-switching, and many other types) that current humans cannot.

  —*In relation to the mental dimension*
  - Ex: the ability to relax the sexual body in a way that current humans are not accustomed to.
  - Ex: the ability to concentrate the sexual mind in a way that current humans are not good at.
  - Ex: the ability to breathe and be patient in preparing the sexual body for sexuality in a way that current humans are not able to.

- **The Post-Sexual Floating Mind**
  —Ex: the mind without the human body
  —Ex: the quest for spirituality beyond sexuality

Notes: The examples in the categories are solely illustrative (not exhaustive), and the comparison is relative (not absolute), nor are they necessarily mutually exclusive. And some can be easily re-classified elsewhere. As generalities, they allow exceptions.
Source: From *Sec. 3.4.2* of *FPHS*. See text for more info.

Table 1.34. The Trinity of Pre-Modernity

- **Pre-Free-Spirited Pre-Modernity (Pre-Modernism) and Its Internal Split**
  —Competing worldviews and values both within and between linear centric (e.g., Islamic, Christian, Judaic, Imperial Roman) and cyclical-centric (e.g., Confucian, Taoist, Hindu, and Buddhist) orientations
  —Compare modernism with pre-modernism here in relation to the seven dimensions of human existence like the true and the holy (e.g., different versions of epistemic dogmas and religious superstitions), the everyday and the technological (e.g.,different versions of non-technophilism and non-consumerism), the beautiful/sublime (e.g., different versions of aesthetic non-autonomy), and the good and the just (e.g., different versions of moral particularism).

- **Pre-Capitalist Pre-Modernity (Pre-Modernization) and Its Own Discontents**
  —Competing versions of societal arrangements (e.g., feudalism, monarchism, and the holy order)

- **Hegemonic Pre-Modernity and Its Countervailing Forces**
  —Different power centers and their enemies (e.g., the Roman Empire and the "barbarian hordes," the "Holy Crusades" and the Muslims, the Middle Kingdom and the invading tribes, different social castes in India, and warring Greek city-states)

Notes: The examples in each category are solely illustrative (not exhaustive) nor necessarily mutually exclusive, and the comparison is relative (not absolute). As generalities, they allow exceptions. Also, it does not matter what the "base" era is in the analysis of any trinity. And in the present context, the "base" era is modernity (for instance, with its "free-spirited modernity" and the other two parts). So, for pre-modernity, the trinity takes the form of, say, "pre-free-spirited pre-modernity," together with the other two parts.
Sources: From Ch.2 of *BCIV* and also the 2 volumes of *FHC*

Table 1.35. The Trinity of Modernity (Part I)

- **Free-Spirited Modernity (Modernism) and Its Internal Split**

  —*On the True and the Holy*
  - The freedom from the dogmas of the past to the better understanding of, and union with, the world and self (Ch.3 of *FHC*)
  - Alternative discourses: about the true (e.g., anti-science discourses) and the holy (non-mainstream theologies) (Ch.3 of *FHC*)

  —*On the Technological and the Everyday*
  - The freedom from life harshness to the higher technophilic, consumeristic lifeform (Ch.2 of *FHC*)
  - Alternative discourses: about the everyday (e.g., transcendental mindsets) and the technological (e.g., Arcadianism) (Ch.2 of *FHC*)

  —*On the Good and the Just*
  - The freedom from the theo-aristocratic tyranny to the moral universality for a just society (Ch.5 of *FHC*)
  - Alternative discourses: about the just (e.g., Communism, Anarchism) and the good (e.g., Nazism/Fascism, and Zarathustrianism) (Chs.5-6 of *FHC*)

  —*On the Beautiful and the Sublime*
  - The freedom from the external distortion of aesthetic pleasure to the boundless infinity of totality in artistic autonomy (Ch.4 of *FHC*)
  - Alternative discourses: about the beautiful/sublime (e.g., kitsch and historical avant-gardism) (Ch.4 of *FHC*)

- **Capitalist Modernity (Modernization) and Its Own Discontents**

  —*During the Industrial Revolution*
  - Ex: Marx on the institution of inequality (Ch.1 of *FHC*)

  —*During the Modern Rational-Instrumental Epoch*
  - Ex: Weber on the politics of soft liberal institutions (Ch.5 of *FHC*)

  —*During the Great Depression*
  - Ex: Keynes on the myth of the free market (Chs.1-3 of *FHC*)

  —*During the Cold War*
  - Ex: Lasch on the narcissistic culture industry (Chs.2-3 of *FHC*)

(continued on next page)

Table 1.35. The Trinity of Modernity (Part II)

- **Hegemonic Modernity and Its Countervailing Forces**
  —*The Legacies of Colonialism and Imperialism*
  - Ex: European colonization of most of the modern world (Ch.1 of *FHC*)

  —*The Struggle for Decolonialization*
  - Ex: The countervailing forces of resentment, rechantment, and regionalism (Chs.1 & 6 of *FHC*)

Notes: The examples in each category are solely illustrative (not exhaustive) nor necessarily mutually exclusive, and the comparison is relative (not absolute). As generalities, they allow exceptions.
Sources: From the 2 volumes of *FHC*—and also from *FCD*

Table 1.36. The Trinity of Post-Modernity

- **Free-Spirited Postmodernity (Postmodernism) and Its Internal Split**
  —*On the True and the Holy*
    - Postmodern performative turn for knowing and its enemies (Ch.3 of *FHC*)
    - Postmodern comparative theology and its opponents (Ch.3 of *FHC*)

  —*On the Technological and the Everyday*
    - Postmodern corporate technological mindset and its adversaries (Ch.2 of *FHC*)
    - Postmodern postmaterialism and its critics (Ch.2 of *FHC*)

  —*On the Good and the Just*
    - Postmodern politics of difference and its foes (Ch.5 of *FHC*)

  —*On the Beautiful and the Sublime*
    - Postmodern deconstruction and its dissenters (Ch.4 of *FHC*)

- **Capitalist Postmodernity (Postmodernization) and Its Own Discontents**
  —*During the Post-Cold War and Beyond*
    - Ex: post-Fordism and its shortcomings (Ch.6 of *FHC*; Chs.6-7 of *FCD*)

- **Hegemonic Postmodernity and Its Countervailing Forces**
  —*The Debate on the Global Village*
    - Ex: uni-civilizationalism vs. multi-civilizationalism (Ch.6 of *FHC*)

  —*The Resistance Movement*
    - Ex: rechantment and the politics of civilizational claims (e.g., Islamic, Confucian and other ethos in relation to the Same) (Ch.6 of *FHC*; Ch.10 of *FCD*)
    - Ex: resentment and the politics of resurgence (e.g., the rising Chinese superpower, the growing EU, and other players in relation to the U.S. and her allies) (Ch.6 of *FHC*; Ch.8 of *FCD*)
    - Ex: regionalism and the politics of inequality (e.g., trans- or inter-national blocs, the North-South divide, NGO's) (Ch.6 of *FHC*; Ch.5 of *FCD*)

Notes: The examples in each category are solely illustrative (not exhaustive) nor necessarily mutually exclusive, and the comparison is relative (not absolute). As generalities, they allow exceptions.
Sources: From *FCD* and the 2 volumes of *FHC*

Table 1.37. The Trinity of After-Postmodernity

- **Free-Spirited After-Postmodernity (After-Postmodernism) and Its Internal Split**
  —The discourse of naked contingency (Ch.10 of *FCD*; Ch.4 of *FPHC*)

- **Post-Capitalist After-Postmodernity (After-Postmodernization) and Its Own Discontents**
  —Different versions of post-capitalism and post-democracy, and their enemies (Ch.10 of *FCD*; Chs.3-4 of *FPHC*)

- **Hegemonic After-Postmodernity and Its Countervailing Forces**
  —The Cyclical Progression of Hegemony in Multiverses (Chs.9-10 of *FCD*; Ch.4 of *FPHC*)

Notes: The examples in each category are solely illustrative (not exhaustive) nor necessarily mutually exclusive, and the comparison is relative (not absolute). As generalities, they allow exceptions.
Sources: From *FCD* and also *FHC*

• Part Two •

Addition

CHAPTER 2
ADDITION AND ITS TWO FACES

> Life does not proceed by the association and addition of elements, but by dissociation and division."
> —Henri Bergson (THI 2013)

The Use of Addition

The study of accounting involves both addition and subtraction—as already indicated in *Sec. 1.3*.

The relationship between addition and subtraction (in the context of accounting) is dialectic, since there is no addition without subtraction, and vice versa—to be eventually transcended altogether later.

Yet, a scholarly danger here is to reduce the analysis of one to that of the other; thus, this book is to devote a whole chapter on each, with the understanding, of course, that the two are related.

This dialectic relationship between addition and subtraction (in the context of accounting) can be revealed by showing how and why the analysis of addition (in Chapter Two) is inadequate without the inclusion of subtraction, just as the examination of subtraction (in Chapter Three) is insufficient without the consideration of addition.

The issues concerning addition in the study of accounting are to be addressed in this chapter (Chapter Two), whereas the issues concerning subtraction in the study of accounting will be analyzed in the next chapter (Chapter Three) instead.

With this dialectic treatment in mind—a good way to examine addition and its two faces in the study of accounting is by way of the evaluation of the extent to which it is in fact both possible (or impossible) and desirable (or undesirable).

This can be done by way of a comprehensive analysis of addition in accounting from the four perspectives of the mind, nature, society, and culture (in accordance to my sophisticated methodological holism, as explained in *Sec. 1.9*).

In other words, this chapter is organized in four main sections, namely, in relation to (2.2) addition and the mind, (2.3) addition and nature, (2.4) addition and society, and (2.5) addition and culture—to be addressed hereafter, in that order (and summarized in *Table 3.1*).

Accounting and the Mind

Addition, when examined from the perspective of the mind, can illuminate its possibility (or impossibility) and desirability (or undesirability), and this can be shown by way of two case studies, namely, (2.2.1) addition, deception, and creative accounting, and (2.2.2) addition, reputation, and intangible assets—to be addressed in what follows, respectively.

Addition, Deception, and Creative Accounting

A good way to evaluate the possibility (or impossibility) and desirability (or undesirability) of addition from the perspective of the mind is to explore addition, deception, and creative accounting.

An argument in the literature is that people use "creative accounting" to complicate (or add) things (which are not really necessary) so as to deceive others for private gains.

Characteristics of Creative Accounting

The term "creative accounting," according to the critic David Ehrenstein, "was first used in 1968 in the film *The Producers* by Mel Brooks" and is also known as "earnings management," as both are "euphemisms referring to accounting practices that may follow the letter of the rules of standard accounting practices, but certainly deviate from the spirit of those rules. They are characterized by excessive complication and the use of novel ways of characterizing income, assets, or liabilities and the intent to influence readers towards the interpretations desired by the authors." (WK 2013q)

In general, "creative accounting" refers to "systematic misrepresentation of the true income and assets of corporations or other organizations. 'Creative accounting' is at the root of a number of accounting scandals, and many proposals for accounting reform—usually centering on an updated analysis of capital and factors of production that would correctly reflect how value is added." (WK 2013q)

For example, in the situation involving "earnings management," creative accounting occurs when "managers use judgment in financial reporting and in structuring transactions to alter financial reports to either mislead some stakeholders about the underlying economic performance of a company or influence contractual outcomes that depend on reported accounting numbers." (WK 2013q; P. Healy 1999)

Often, "earnings management...involves the artificial increase (or decrease) of revenues, profits, or earnings per share figures through aggressive accounting tactics. Aggressive earnings management is a form of fraud and differs from reporting error." (WK 2013q)

A good illustration is when "management wishing to show earnings at a certain level or following a certain pattern seek loopholes in financial reporting standards that allow them to adjust the numbers as far as is practicable to achieve their desired aim or to satisfy projections by financial analysts. These adjustments amount to fraudulent financial reporting when they fall 'outside the bounds of acceptable accounting practice.'" (WK 2013q)

A historical case in point involves "the systemic over-reporting of corporate income which came to light in 2002," which "was the granting of stock options as part of executive compensation packages. Since stock prices reflect earning reports, stock options could be most profitably exercised when income is exaggerated, and the stock can be sold at an inflated profit." (WK 2013q)

With this historical example in mind, there are four main forms of earnings management as shown below (and summarized in *Table 2.2*): (WK 2013q)

- Ex: "unsuitable revenue recognition"
- Ex: "inappropriate accruals and estimates of liabilities"
- Ex: "excessive provisions and generous reserve accounting"
- Ex: "intentional minor breaches of financial reporting requirements that aggregate to a material breach"

One of the most notable activists against earnings management is "Abraham Briloff (professor emeritus of CUNY Baruch) who for years wrote a column for Barron's that constantly analyzed breaches of ethics and audit professionalism among CPA firms. His book is called *Unaccountable Accounting*. The profession, in turn, was not kind to Dr.

Briloff but much of what he advocated has been forced on the industry in the wake of the Enron scandal." (WK 2013q; D. Owsen 1999)

Motivations of Creative Accounting

Creative accountants can cheat for different reasons; for instance, "drivers for such behaviour include market expectations, personal realisation of a bonus, and maintenance of position within a market sector." (WK 2013q)

In general, some main motivations can be identified, as shown below (and summarized in *Table 2.2*): (WK 2013q)

- Ex: "personal incentives"
- Ex: "bonus-related pay"
- Ex: "benefits from shares and share options"
- Ex: "job security"
- Ex: "personal satisfaction"
- Ex: "cover-up fraud"
- Ex: "tax management"
- Ex: "management buyouts"

Of course, the list above is illustrative, not exhaustive.

Cognitive Psychology to Detect Creative Accounting

But creative accounting can be better detected with the help of cognitive psychology. (G. Cohen 2013)

In the early days of psychology as a discipline, "when Freud, Jung, Adler and contemporary colleagues first formulated psychological theories and published their findings, each school of thought took a different approach. Most could be classified as 'behavioral,' and practitioners used external influences—mothers, fathers, spouses—to explain why patients behave in certain ways. The cognitive model didn't come along until 1967, when Ulric Neisser proposed a psychology type focusing on the mind's internal processes as they impact behavior. The word 'cognition' means 'knowing' and suggests insights into perception, language, memory and thinking. While some critics find cognitive psychology lacking in validity, accountants adopting cognitive methods when dealing with clients have to come to rely upon it." (G. Cohen 2013)

In the past few decades, there have been several areas where cognitive psychology is applied to the field of creative accounting, and hereafter are five main examples (as summarized in *Table 2.2*).

(a) Decision Making

The first area has to do with "decision making." (G. Cohen 2013)

For instance, "among the behaviors an accountant can assess when entering into a relationship with a client is the theory and practice of decision making, which she likely learned during human subject experimentation in undergraduate school. Accounting students learn to identify characteristics in decision makers that can help them spot personality types most likely to commit fraud or 'cook the books.' Students develop case studies focused on problem-solving skills and ferret out information not disclosed during typical client-accountant communications. Understanding the way a client approaches decision making related to corporate earnings and expenditures is at the heart of the practice of cognitive psychology." (G. Cohen 2013)

(b) Numbers and Behaviors

The second area has to do with "numbers and behaviors." (G. Cohen 2013)

For instance, "when accountants and auditors enter into relationships with clients, an undercurrent develops between the financial practitioner and client that involves dependency, trust and control. Initial interactions, based solely on fiscal matters, become 'subject to emotions, pressures, fears and cognitive limitations,' according to Northeastern University Professor Arnie Wright. The evolution of the relationship between accountant and client morphs from number cruncher to counselor, but if the accountant has no grounding in cognitive psychology, the relationship that develops may not be as successful." (G. Cohen 2013)

(c) Complexity of Audits

The third area has to do with the "complexity of audits." (G. Cohen 2013)

For instance, "when an auditor is hired to verify financial records, part of her job is looking for financial irregularities. To probe possible cases of fraud or figure manipulation, auditors can be trained in cognitive psychology so that their approach to examining books and forming independent conclusions produces a more successful audit. Audits require accountants to balance human behavior with financial transactions, and many of the newest theories and practices pairing audits as they relate to psychological approaches are regularly published in the 'Journal of Accounting Research.' Standard research techniques used during these audits are observation, theory development, background collection and hypothesis testing that often changes the accountant's role from figure collector and analyst to independent investigative researcher, probing beyond the books." (G. Cohen 2013)

(d) Personalities of Folks

The fourth area has to do with the "personalities of folks." (G. Cohen 2013)

For instance, "understanding the personalities of folks for whom accountants do tax returns is another aspect of cognitive psychology education that's important, even for those who do nothing but prepare tax returns. Knowing a client's personality can help an accountant do his job more efficiently. Greg Crabtree, writing for AccountingToday.com, profiles 'types' of taxpayers so he has a better understanding of their motives. From 'No Taxes, No Way Nate,' who avoids paying taxes to 'Accidentally Profitable Alex,' who's so business-involved, he forgets to file, Crabtree stresses the importance of learning as much as possible about the psychology of tax customers. After all, if you can't figure out what's driving your clients' motives, you may not be able to get them on the road to financial health." (G. Cohen 2013)

(e) Investor Mentality

And the fifth area has to do with the "investor mentality." (G. Cohen 2013)

For instance, "one interesting demographic recognized within the cognitive accounting psychology field focuses on the behaviors of investors. High-risk takers often give their accountants a particularly difficult time because they tend to be overly confident, may lack self-control and often have a propensity for 'memory loss' following financial failures. Additionally, people falling into this category tend to be reticent about disclosing the information an accountant needs to do her job, so the more training in cognitive psychology an accountant receives, the better able she is to do her job and manage her client as well." (G. Cohen 2013)

Problems with the Argument on Deception and Creative Accounting

However, to the critics, this argument on addition, deception, and creative accounting has some major problems. Consider, for illustration, a few examples below (as summarized in *Table 2.2*).

Firstly, one main criticism of of the argument on addition, deception, and creative accounting is that the problem with accounting scandals is not wholly the fault of creative accountants.

For instance, to the critics, although "earnings management is seen as a pressing issue in current accounting practice," one needs to understand that "part of the difficulty lies in the accepted recognition that there is no such thing as a single 'right' earnings figure," so this tempts "creative accountants" to offer alternative interpretations of earnings figures for private gains. (WK 2013q)

Secondly, another main criticism of the argument on addition, deception, and creative accounting is that the nature of business cycles in the free market makes it easier for creative accountants to cheat.

For instance, as the critics are right to point out, "conformance to acceptable accounting practices is a matter of personal integrity. Aggressive earnings management becomes more probable when a company is affected by a downturn in business." (WK 2013q)

Thirdly, still another main criticism of the argument on addition, deception, and creative accounting is that the identification of "fraud" from "error" is difficult in practice.

For instance, to the critics, "whilst it is comparatively easy for an auditor to detect error, earnings management can involve sophisticated fraud that is covert and much harder to detect. The requirement for management to assert that the accounts have been prepared properly offers no protection where those managers have already entered into conscious deceit and fraud." (WK 2013q)

Fourthly, an additional main criticism of the argument on addition, deception, and creative accounting is that the analysis of "the mind's internal processes" in cognitive psychology to detect fraud is easier said than done.

For instance, "some critics find cognitive psychology lacking in validity," because it is a daunting task to prove "the mind's internal processes," so the intellectual disagreement between the "behavioral psychologists" for the analysis of "external influences" and the "cognitive psychologists" for "internal processes" still remains today.

Fifthly, another additional main criticism of the argument on addition, deception, and creative accounting is that extensive lobbying efforts in the business world can hinder accounting reform to prevent fraud.

For instance, to the critics, "not only do most businesses raise capital based on numbers derived from current standards, there are extensive lobbying efforts by the accounting industry to keep those standards roughly as they are: complex, loopholed, and unable to be applied or audited easily....Heads of the U.S. Securities and Exchange Commission since the 1980s have consistently complained that this lobbying makes it impossible for them to apply meaningful reform, even in the wake of accounting scandals, e.g. that which felled Arthur Andersen in 2002." (WK 2013x)

And sixthly, still another additional main criticism of the argument on addition, deception, and creative accounting is that the role of the government can be part of the problem.

For instance, "one barrier to accounting reform are governments themselves. They have the authority to determine what are accepted accounting principles, while using questionable accounting practices themselves. Governments, for example, pay off operating costs with longer-term debt and thus overstate budgetary surpluses or conceal operating deficits. This is not unlike the allegedly fraudulent practices of some corporations." (WK 2013x)

Beyond Creative Accounting

These criticisms of the argument on addition, deception, and creative accounting should not be misunderstood as a total rejection of its validity but serve to show us the two opposing sides of the debate, such that the possibility (or impossibility) and desirability (or undesirability) of addition (from the perspective of the mind with addition, deception, and creative accounting as a case study here) are not to the extent that the respective defenders would like us to believe.

Moreover, the analysis of addition, deception, and creative accounting is to shed some light on the ontological principles in existential dialectics, and good examples include the formalness-informalness principle, the absoluteness-relativeness principle, the partiality-totality principle, the predictability-unpredictability principle, the explicability-inexplicability principle, the fiction-reality principle, the cognitiveness-noncognitiveness principle, the finiteness-transfiniteness principle, the preciseness-vagueness principle, the simpleness-complicatedness principle, the openness-hiddenness principle, the denseness-emptiness principle, the rule-exception principle, the prototypicality-variation principle, the change-constancy principle, the order-chaos principle, the slowness-quickness principle, the expansion-contraction principle, the optimality-nonoptimality principle, the simultaneity-nonsimultaneity principle, the isolation-interaction principle, the theory-praxis principle, the convention-novelty principle, the evolution-transformation principle, the symmetry-asymmetry principle, the softness-hardness principle, the seriousness-playfulness principle, the activeness-inactivess principle, the selfness-otherness principle, the regression-progression principle, the same-difference principle, the stability-reaction principle, the functionality-nonfunctionality principle, the intentionality-nonintentionality principle. and the survivability-non-survivability principle.

For instance, in relation to the formalness-informalness principle, if there is formalness (e.g., the formal logical requirement of soundness and completeness in a system of ideas, as in the argument on deception and creative accounting), there is informalness (e.g., the non-formal existence of unsoundness and incompleteness in the argument on deception and creative accounting, as shown in the criticisms that the problem with accounting scandals is not necessarily the fault of creative accountants, that the nature of business cycles in the free market makes it easier for creative accountants to cheat, that the identification of "fraud" from "error" is difficult in practice, that the analysis of "the mind's internal processes" in cognitive psychology to detect fraud is easier said than done, that extensive lobbying efforts in the business world can hinder accounting reform to prevent fraud, and that the role of the government can be part of the problem). And the reverse direction also holds true.

In relation to the absoluteness-relativeness principle, if there is absoluteness (e.g., the absolute view on "external influences" by the advocates of "behavioral psychology"), there is relativeness (e.g., what is right for the advocates of "behavioral psychology" in regard to "external influences" is not necessarily so for "cognitive psychologists," since the latter focuses on "internal processes" instead). And the reverse direction also holds true.

In relation to the partiality-totality principle, if there is partiality (e.g., the partial view on "external influences" by the advocates of "behavioral psychology"), there is totality (e.g., the more holistic view on psychology in regard to deception, such that the whole is not the sum of its parts, that is, it cannot be reduced to the sum of opposing views by "behavioral psychologists," "cognitive psychologists," etc., since there will emerge new views in the future not yet known today, just as there are already alternative views nowadays which do not side exclusively with any of them, as is the analysis here). And the reverse direction also holds true.

In relation to the predictability-unpredictability principle, if there is predictability (e.g., the predictable tendency of "high-risk takers" to "be overly confident,...lack self-control and often have a propensity for 'memory loss' following financial failures," according to the study on cognitive psychology and creative accounting aforecited), there is unpredictability (e.g., the more difficult task to predict exactly the extent to which a particular "high-risk taker" may end up actually committing fraud on a particular future occasion). And the reverse direction also holds true.

In relation to the explicability-inexplicability principle, if there is explicability (e.g., the explanation, by "behavioral psychologists," of deception, on the basis of "external influences"), there is inexplicability (e.g., the lack of sufficient explanation, by "behavioral psychologists," of why "external influences" are necessarily useful to explain deception, since, to "cognitive psychologists," "internal processes" in the mind are more relevant). And the reverse direction also holds true.

In relation to the fiction-reality principle, if there is fiction (e.g., the fictional aspect of information, insofar as decision makers "cook the books" for private gains, such as the information in question is fraudulent), there is reality (e.g., the realistic aspect of information, insofar as decision makers open their books for auditing with no errors found). And the reverse direction also holds true.

In relation to the cognitiveness-noncognitiveness principle, if there is cognitiveness (e.g., a view of things on the basis of reason and evidences, as shown in the finding that governments can be asked to carry out "accounting reforms"), there is noncognitiveness (e.g., a view of things on the basis of non-cognitive factors like envy, jealousy, power, nationality, race, gender, age, class, greed, lust, status, faith, anger, sadness, joy, fear, wish, etc.—as shown in the politics of corruption, when "one barrier to accounting reform are governments themselves. They have the authority to determine what are accepted accounting principles, while using questionable accounting practices themselves"). And the reverse direction also holds true.

In relation to the finiteness-transfiniteness principle, if there is finiteness (e.g., the finite number of the categories of people who commit fraud in business, like "creative accountants," "earnings managers," etc.), there is transfiniteness (e.g., the transfinite amount of all the money that has ever been stolen or taken away due to all the "fraudulent" activities in human history). And the reverse direction also holds true.

In relation to the preciseness-vagueness principle, if there is preciseness (e.g., the precise identification of 5 areas where cognitive psychology is applied to the field of creative accounting in the article aforecited), there is vagueness (e.g., the vagueness in the identification of the 5 areas where cognitive psychology is applied to the field of creative accounting in the article aforecited, since it is not clear why there must be only 5, not 6, 7, 8, 9, and so on). And the reverse direction also holds true.

In relation to the simpleness-complicatedness principle, if there is simpleness (e.g., the relatively simple analysis of the "fault" of creative accountants in the article aforecited), there is complicatedness (e.g., the relatively more complicated analysis of the "fault" of creative accountants, by challenging the claims and assumptions, as shown in the problems pointed out by the critics as cited). And the reverse direction also holds true.

In relation to the openness-hiddenness principle, if there is openness (e.g., the open availability of "financial reporting" and "structuring transactions" to be used by creative accountants), there is hiddenness (e.g., the hidden bias in the use of "financial reporting" and "structuring transactions" by creative accountants, because "managers" can then misuse them "to alter financial reports to either mislead some stakeholders about the underlying economic performance of a company or influence contractual outcomes that depend on reported accounting numbers"). And the reverse direction also holds true.

In relation to the denseness-emptiness principle, if there is denseness (e.g., the relatively denser concentration of concerns with finding "errors" by auditors), there is emptiness (e.g., the relatively less dense, or more empty, concentration of concerns with "fraudulent" reports by creative accountants). And the reverse direction also holds true.

In relation to the rule-exception principle, if there are rules (e.g., the usual generalization, procedure, habit, or tendency that people normally "conform…to acceptable accounting practices" due to the pressure of "personal integrity"), there are exceptions (e.g., a case to which a rule does not apply, such as the exceptional case when "aggressive earnings management becomes more probable when a company is affected by a downturn in business," contrary to the rule that people normally "conform…to acceptable accounting practices" due to the pressure of "personal integrity"). And the reverse direction also holds true.

In relation to the prototypicality-variation principle, if there is prototypicality (e.g., the prototype, like the example about the original usage of the term "creative accounting" in 1968 in the film *The Producers* by Mel Brooks), there is variation (e.g., different variants of the protoptye, as shown in many studies since the original usage of the term "creative accounting" in 1968 in the film *The Producers* by Mel Brooks—like the later ones by Abraham Briloff in the 1970s, the application of cognitive psychology to the field of creative accounting, etc., though each with its own variation, contribution, and disagreement). And the reverse direction also holds true.

In relation to the change-constancy principle, if there is change (e.g., the ever new ways to study fraudulent activities, as shown in the application of cognitive psychology to the field of creative accounting), there is constancy (e.g., the ever constant existence of different problems with the ever new ways to study fraudulent activities, as shown in the problems with the application of cognitive psychology to the field of creative accounting pointed out by the critics). And the reverse direction also holds true.

In relation to the order-chaos principle, if there is order (e.g., the more or less orderly view on acounting, if looked from the sole vantage point of "acceptable accounting practices" as "a matter of personal integrity"), there is chaos (e.g., the more or less chaotic view on accounting, if looked from the multiple, conflicting vantage points of both the advocates of creative accounting and the critics, especially in regard to the problems aforecited, such that they do not add up to much of anything coherent). And the reverse direction also holds true.

In relation to the slowness-quickness principle, if there is slowness (e.g., the relatively slower speed for "an auditor to detect…sophisticated fraud that is covert and much harder to detect"), there is quickness (e.g., the relatively quicker speed for "an auditor to detect error"). And the reverse direction also holds true.

In relation to the expansion-contraction principle, if there is expansion (e.g., the relatively more developed ability to analysis "external influences" among "behavioral psychologists"), there is contraction (e.g., the relatively less developed ability to analyze "internal processes" among "behavioral psychologists"). And the reverse direction also holds true.

In relation to the optimality-nonoptimality principle, if there is optimality (e.g., the search for highly favorable options, like the quest for systematic "accounting reform" by the International Accounting Standards Board), there is nonoptimality (e.g., the existence of non-optimal alternatives to optimality, like the argument for more realistic studies by the critics, who pointed out different problems with "accounting reform" aforediscussed). (WK 2013x) And the reverse direction also holds true.

In relation to the simultaneity-nonsimultaneity principle, if there is simultaneity (e.g., the occurrence of different things more or less at the same time, such as the case about governments which "have the authority to determine what are accepted accounting principles, while using questionable accounting practices themselves"), there is nonsimultaneity (e.g., the occurrence of something after something else, such as the term "Creative Accounting" which was first used in 1968, before Abraham Briloff wrote against it in the 1970s). And the reverse direction also holds true.

In relation to the isolation-interaction principle, if there is isolation (e.g., the state of a situation that is separate from others, like a creative accountant before being audited by an auditor), there is interaction (e.g., the influence of entities on one another, like the interaction between a creative accountant and an auditor in the process of an audit). And the reverse direction also holds true.

In relation to the theory-praxis principle, if there is theory (e.g., the theoretical construction of professional ethics against creative accounting by Abraham Briloff), there is praxis (e.g., what Abraham Briloff "advocated has been forced on the industry in the wake of the Enron scandal"). (WK 2013q) And the reverse direction also holds true.

In relation to the convention-novelty principle, if there is convention (e.g., the conventional wisdom about the role of governments "to determine what are accepted accounting principles"), there is novelty (e.g., the alternative novel challenge to the conventional wisdom about the role of governments "to determine what are accepted accounting principles"—by the new idea that "governments "us[e] questionable accounting practices themselves"). And the reverse direction also holds true.

In relation to the evolution-transformation principle, if there is evolution (e.g., the natural evolution of humans to count things in the state of nature in a primitive way), there is transformation (e.g., the technical transformation of human ability to count by the invention of "creative accounting"). And the reverse direction also holds true.

In relation to the symmetry-asymmetry principle, if there is symmetry (e.g., the co-existence of governments and ordinary citizens), there is asymmetry (e.g., governments "have the authority to determine what are accepted accounting principles"—but ordinary citizens do not have this authority). And the reverse direction also holds true.

In relation to the softness-hardness principle, if there is softness (e.g., the defense of the application of "cognitive psychology" to accounting by the advocates aforecited), there is hardness (e.g., the critique of the application of "cognitive psychology" to accounting, as shown in the problems pointed out by the critics aforecited). And the reverse direction also holds true.

In relation to the seriousness-playfulness principle, if there is seriousness (e.g., the serious business of those who try to alter financial reports for creative accounting), there is playfulness (e.g., the playful part of those who try to alter financial reports for creative accounting, when they play around with different ways over the decades, like "unsuitable revenue recognition," "inappropriate accruals and estimates of liabilities," "excessive provisions," "generous reserve accounting," "intentional minor breaches of financial reporting requirements that aggregate to a material breach," etc.). And the reverse direction also holds true.

In relation to the activeness-inactivess principle, if there is activeness (e.g., the relative activeness of those executives in Enron in the 1990s who engaged in earnings management), there is inactiveness (e.g., the relative inactiveness of those executives in Enron after the Enron scandal, when they were sentenced to prison, so that they could no longer engage in creative accounting). (WK 2013y) And the reverse direction also holds true.

In relation to the selfness-otherness principle, if there is selfness (e.g., the tendency to be mostly centered on one's self or group, as shown in those accountants who engage in earnings management for private gains), there is otherness (e.g., the tendency to be mostly centered on others, as shown in the activism of Abraham Briloff, "who for years wrote a column for Barron's that constantly analyzed breaches of ethics and audit professionalism among CPA firms" and "much of what he advocated has been forced on the industry in the wake of the Enron scandal," as an advancement in the field of accounting reform). And the reverse direction also holds true.

In relation to the regression-progression principle, if there is regression (e.g., the regression made by the fight against creative accounting, as shown in the problems pointed out by the critics), there is progression (e.g., the progress made by the fight against creative accounting, as shown in the contribution to accounting reform aforecited). And the reverse direction also holds true.

In relation to the same-difference principle, if there is similarity in outcome (e.g., the contribution to the molding and control of beliefs, values, and behaviors, regardless of whether this be done by way of creative accounting, or by way of accounting reform), there is difference in outcome (e.g., the contribution to the molding and control of beliefs, values, and behaviors by way of creative accounting for a relatively more amoral lifeworld—but the contribution to the molding and control of beliefs, values, and behaviors by way of accounting reform for a relatively more moral lifeworld). And the reverse direction also holds true.

In relation to the stability-reaction principle, if there is stability (e.g., the relatively more stable condition of "breaches of ethics and audit professionalism among CPA firms" in the older days), there is reaction (e.g., "breaches of ethics and audit professionalism among CPA firms" in the older days then led to further changes later on, when Abraham Briloff "for years wrote a column for Barron's that constantly analyzed breaches of ethics and audit professionalism among CPA firms" and "much of what he advocated has been forced on the industry in the wake of the Enron scandal"). And the reverse direction also holds true.

In relation to the functionality-nonfunctionality principle, if there is functionality (e.g., the presence of action for which something or someone is fitted, as shown in the role of governments "to determine what are accepted accounting principles," which serves the function of maintaining professional ethics in the business world), there is nonfunctionality (e.g., the relative lack of action for which something or someone is fitted, as shown in the problems with the role of governments "to determine what are accepted accounting principles," since they also use "questionable accounting practices themselves," such that they are not as functional as the advocates would like us to believe). And the reverse direction also holds true.

In relation to the intentionality-nonintentionality principle, if there is intentionality (e.g., the planning or design of something for a certain outcome, as shown in the original intention of Ulric Neisser who, in 1967, "proposed a psychology type focusing on the mind's internal processes as they impact behavior," that is, "cognitive psychology," so as to counter the influence of "behavioral psychology"), there is nonintentionality (e.g., the relative absence of planning or design of something for a certain outcome, such that the outcome can even be counter-intentional in being the opposite of what was originally planned, as shown in the rejection of "cognitive psychology" by "some critics" nowadays who "find cognitive psychology lacking in validity," because it is a daunting task to prove "the mind's internal processes," so "behavioral psychologists" for the analysis of "external influences" still remains influential in our time—contrary to the original intention of Neisser). And the reverse direction also holds true.

And in relation to the survivability-nonsurvivability principle, if there is survivability (e.g., the relative survival, nowadays, of the accounting reform against creative accounting), there is non-survivability (e.g., the relative non-survival, nowadays, of those firms like Enron, Worldcom, etc., because of the accounting scandals). And the reverse direction also holds true.

Addition, Reputation, and Intangible Assets

Another way to evaluate the possibility (or impossibility) and desirability (or undesirability) of addition from the perspective of the mind is to explore addition, reputation, and intangible assets.

An argument in the literature is that goodwill in accounting can be managed so as to increase reputation capital (as an intangible asset) for the benefit of all stakeholders concerned.

Goodwill and Intangible Assets

The word "goodwill" refers to the situation in accounting "when one company acquires another, but pays more than the fair market value of the net assets (total assets – total liabilities). It is classified as an intangible asset on the balance sheet"; therefore, "management is responsible to value goodwill every year end to determine if an impairment is required. If the fair market value goes below historical cost (what goodwill was purchased for), an impairment must be recorded to bring it down to its fair market value." (WK 2013z)

(a) Measurement of Goodwill

To measure goodwill, "the purchase price is deducted from the fair market value of identifiable assets and liabilities of the company acquired. For instance, if company A acquired 100% of company B, a goodwill occurs. In order to calculate goodwill, it is necessary to have a list of all of company B's assets and liabilities at fair market value," as shown below: (WK 2013z)

	Fair market value
Accounts Receivable	$10
Inventory	$5
Accounts payable	$6
Total Net assets	= 10 + 5 - 6
	= $9

In other words, "in order to acquire company B, company A paid $20. Hence, goodwill would be $11 ($20 - $9). The journal entry in the books of company A to record the acquisition of company B would be" like the following: (WK 2013z)

DR Goodwill	$11
DR Accounts Receivable	$10
DR Inventory	$5
CR Accounts Payable	$6
CR Cash	$20

Of course, this is just an example for illustration.

(b) Modern Meaning of Goodwill

In verbal explanation, for example, "a software company may have net assets (consisting primarily of miscellaneous equipment, and assuming no debt) valued at $1 million, but the company's overall value (including brand, customers, and intellectual capital) is valued at $10 million. Anybody buying that company would book $10 million in total assets acquired, comprising $1 million physical assets and $9 million in goodwill. In a private company, goodwill has no predetermined value prior to the acquisition; its magnitude depends on the two other variables *by definition*. A publicly traded company, by contrast, is subject to a constant process of market valuation, so goodwill will always be apparent." (WK 2013z)

And "while a business can invest to increase its reputation, by advertising or assuring that its products are of high quality, such expenses cannot be capitalized and added to goodwill, which is technically an intangible asset. Goodwill and intangible assets are usually listed as separate items on a company's balance sheet." (WK 2013z)

Goodwill, and the Psychology of Reputation

The analysis of goodwill is essentially psychological, because of its connection to the psychology of "reputation," which refers to an "opinion" about "a social entity (a person, a social group, an organization) ,...typically a result of social evaluation on a set of criteria. It is important in business, education, online communities, and many other fields." (WK 2013aa)

(a) A Cognitive View of Reputation

But "reputation" should not be confused with "image."

On the one hand, "image and reputation...are social in two senses: they concern properties of another agent (the target's presumed attitude toward socially desirable behavior), and they may be shared by a multitude of agents." (WK 2013aa)

On the other hand, "the two notions operate at different levels. Image is a belief, namely, an evaluation. Reputation is a meta-belief, i.e., a belief about others' evaluations of the target with regard to a socially desirable behavior. To better understand the difference between image and reputation, the mental decisions based upon them must be analyzed at the following three levels": (WK 2013aa)

- "Epistemic"—refers to a situation when you "accept the beliefs that form either a given image or acknowledge a given reputation. This implies a believed evaluation gives rise to one's direct evaluation. Suppose I know the friend I mostly admire has a good opinion of Mr. Berlusconi. However puzzled I may be by this dissonance-inducing news, I may be convinced because of my friendship to accept this evaluation and share it."

- "Pragmatic–Strategic"—refers to a situation when you "use image to decide whether and how to interact with the target. Once I have my own opinion (perhaps resulting from acceptance of others' evaluations) about a target, I will use it to make decisions about my future actions concerning that target. Perhaps, I may abstain from participating in political activity against Mr. Berlusconi."
- "Memetic"— refers to a situation when you "transmit my (or others') evaluative beliefs about a given target to others. Whether or not I act in conformity with a propagating evaluation, I may decide to spread the news to others."

But then what are causes and consequences of reputation?

(b) Causes and Consequences of Reputation

In regard to the causes—especially in business, the key lies "in stakeholder experiences. Stakeholder experiences relate to a company's day-to-day business operations, its branding and marketing and 'noise' in the system, such as the media and word of mouth. Further causes of reputation may include the perceived innovativeness of a company, the customers' expectations, the (perceived) quality of the company's goods and services and the subsequent customer satisfaction, all of which differ according to the respective customers' cultural background." (WK 2013aa; C. Falkenreck 2011)

Thus, "reputation is seen to reside in the beliefs that stakeholders hold about a company (the cognitive element)and the feelings that stakeholders have about a company (the affective element). While the cognitive element of reputation can reflect the uniqueness of a company or of products in term characteristics such as brand attributes (whether an organisation is delivering high quality products, is international, friendly etc.), the affective element is always evaluative. In other words, it gives an indication of whether stakeholders like, admire or trust a company and its attributes. A unique and distinctive cognitive evaluation of a company only has value if this results in a positive affective evaluation and positive consequences of reputation." (WK 2013aa)

And in regard to the consequences—the key lies "in the behaviors (supportive or otherwise) that stakeholders demonstrate towards a company. Behaviors such as advocacy, commitment and cooperation are key positive outcomes of a positive reputation." (WK 2013aa)

(c) Reputation Capital and Management

And reputation can be regarded as a form of capital and thus can be managed.

In regard to reputation management—"many businesses have public relations departments dedicated to managing their reputation. In addition, many public relations firms describe their expertise in terms of reputation management. The public relations industry is growing due to the demand for companies to build corporate credibility and hence reputation. Incidents which damage a company's reputation for honesty or safety may cause serious damage to finances. For example, in 1999 Coca-Cola lost $60 million (by its own estimate) after schoolchildren reported suffering from symptoms like headaches, nausea and shivering after drinking its products at the Wayback Machine." (WK 2013aa)

And in regard to reputation capital—"according to Joachim Klewes and Robert Wreschniok, reputation can be managed, accumulated and traded in for trust, legitimisation of a position of power and social recognition, a premium price for goods and services offered, higher customer loyalty, a stronger willingness among shareholders to hold on to shares in times of crisis, or a stronger readiness to invest in the company's stock. Therefore, reputation is one of the most valuable forms of 'capital' of a company." (WK 2013aa; C. Falkenreck 2011)

As Klewes and Wreschniok aptly put it, "delivering functional and social expectations of the public on the one hand and manage to build a unique identity on the other hand creates trust and this trust builds the informal framework of a company. This framework provides 'return in cooperation' and produces reputation capital. A positive reputation will secure a company or organisation long-term competitive advantages. The higher the Reputation Capital, the less the costs for supervising and exercising control." (WK 2013aa)

Problems with the Argument on Reputation and Intangible Assets

However, to the critics, this argument on addition, reputation, and intangible assets has some major problems. Consider, for illustration, a few examples below.

Firstly, one main criticism of the argument on addition, reputation, and intangible assets is that the calculation of "reputation" is not objective.

For example, goodwill is psycholgically tied to the "opinion," "gossip," etc. about an entity (like a company), as discussed earlier;

furthermore, “a company's reputation is...influenced by culture, as nationalities differ with regard to how valued specific aspects of the company's brand identity are in the respective national culture (e.g. environmental concerns or work ethics) as well as with regard to popular cultural dimensions (e.g. Hofstede).” (WK 2013aa)

Secondly, another main criticism of the argument on addition, reputation, and intangible assets is that there are competing methods to account for goodwill over the decades.

For instance, “previously, companies could structure many acquisition transactions to determine the choice between two accounting methods to record a business combination: [either] purchase accounting or pooling-of-interests accounting. Pooling-of-interests method combined the book value of assets and liabilities of the two companies to create the new balance sheet of the combined companies. It therefore did not distinguish between who is buying whom. It also did not record the price the acquiring company had to pay for the acquisition. Since 2001, U.S. Generally Accepted Accounting Principles (FAS 141) no longer allows the pooling-of-interests method.” (WK 2013z)

Thirdly, still another main criticism of the argument on addition, reputation, and intangible assets is that there are conflicting models of reputation.

For instance, “according to Kevin Money and Carola Hillenbrand, there are many different and often conflicting models of reputation. Terminology such as reputation, branding, image and identity is often used interchangeably, or to distinguish differences between related constructs.” (WK 2013aa)

Fourthly, an additional main criticism of the argument on addition, reputation, and intangible assets is that there are different approaches to reputation management.

For instance, there are competing approaches to reputation management, be they focusing on the “shareholders” or the “stakeholders.” As an illustration, “the stakeholder theory says corporations should be run for the benefit of all 'stakeholders,' not just the shareholders. Stakeholders of a company include any individual or group that can influence or is influenced by a company's practices. The stakeholders of a company can be suppliers, consumers, employees, shareholders, financial community, government, and media. Companies must properly manage the relationships between stakeholder groups and they must consider the interest(s) of each stakeholder group carefully.” (WK 2013aa)

Fifthly, another additional main criticism of the argument on addition, reputation, and intangible assets is that the obsession with reputation can be a psychological problem instead.

For instance, to the critics, "concern over reputation is sometimes considered a human fault, exaggerated in importance due to the fragile nature of the human ego. William Shakespeare provides the following insights from *Othello*": "Reputation is an idle and most false imposition; oft got without merit, and lost without deserving: you have lost no reputation at all, unless you repute yourself such a loser." (WK 2013aa)

And sixthly, still another additional main criticism of the argument on addition, reputation, and intangible assets is that reputation building is easier said than done.

For instance, to the critics, "reputation recovery is the long and arduous path to rebuilding equity in a company's good name," and "the 'stumble rate' among companies has risen exponentially over the past five years. In fact, 79% of the world's most admired companies have lost their number one positions in industries in that time period. Companies which were once heralded as invincible no longer are." (WK 2013aa; L. Gaines-Ross 2008)

Beyond Reputation and Intangible Assets

These criticisms of the argument on addition, reputation, and intangible assets do not mean a total rejection of its usefulness but serve to show us the two opposing sides of the debate, such that the possibility (or impossibility) and desirability (or undesirability) of addition (from the perspective of the mind with addition, reputation, and intangible assets as a case study here) are not to the extent that the respective defenders would like us to believe.

In fact, the analysis of addition, reputation, and intangible assets can teach us a valuable lesson about the ontological principles in existential dialectics, and good examples include the formalness-informalness principle, the absoluteness-relativeness principle, the partiality-totality principle, the predictability-unpredictability principle, the explicability-inexplicability principle, the fiction-reality principle, the cognitiveness-noncognitiveness principle, the finiteness-transfiniteness principle, the preciseness-vagueness principle, the simpleness-complicatedness principle, the openness-hiddenness principle, the denseness-emptiness principle, the rule-exception principle, the prototypicality-variation principle, the change-constancy principle, the order-chaos principle, the slowness-quickness principle, the expansion-contraction principle, the optimality-nonoptimality principle, the simultaneity-nonsimultaneity principle, the isolation-interaction principle, the theory-praxis principle, the convention-novelty principle, the evolution-transformation principle, the symmetry-asymmetry principle, the softness-hardness principle, the seriousness-playfulness principle, the activeness-inactivess principle, the selfness-otherness principle, the regression-progression principle, the same-difference principle, the stability-reaction principle, the functionality-nonfunctionality principle, the intentionality-nonintentionality principle. and the survivability-non-survivability principle.

For instance, in relation to the formalness-informalness principle, if there is formalness (e.g., the formal logical requirement of soundness, completeness, and consistency in a system of ideas, as in the argument on reputation capital and management), there is informalness (e.g., the non-formal existence of unsoundness, incompleteness, and inconsistency in the argument on reputation capital and management, as shown in the criticisms that the calculation of "reputation" is not objective, that there are competing methods to account for goodwill over the decades, that there are conflicting models of reputation, that there are different approaches to reputation management, that the obsession with reputation can be a psychological problem instead, and that reputation building is easier said than done). And the reverse direction also holds true.

In relation to the absoluteness-relativeness principle, if there is absoluteness (e.g., the absolute view on "reputation capital and management" by the advocates in business), there is relativeness (e.g., what is right for the advocates in regard to "reputation capital and management" in business is not necessarily so for the critics who view the obsession with reputation as a psychological problem instead, that is, as "a human fault, exaggerated in importance due to the fragile nature of the human ego"). And the reverse direction also holds true.

In relation to the partiality-totality principle, if there is partiality (e.g., the partial view on "reputation" in accordance to a specific "model"), there is totality (e.g., the more holistic view on "reputation," such that the whole is not the sum of its parts, that is, it cannot be reduced to the sum of "often conflicting models of reputation," according to Kevin Money and Carola Hillenbrand, since there will emerge new views in the future not yet known today, just as there are already alternative views nowadays which do not side exclusively with any of them, as is the analysis here). And the reverse direction also holds true.

In relation to the predictability-unpredictability principle, if there is predictability (e.g., the predictable tendency of firms nowadays to pay attention to "reputation capital and management"), there is unpredictability (e.g., the more difficult task to predict exactly the extent to which a particular firm will be sucessful in building "reputation" on a particular future occasion, since reputation building is easier said than done, as not everyone can make it and "reputation recovery is the long and arduous path to rebuilding equity in a company's good name"). And the reverse direction also holds true.

In relation to the explicability-inexplicability principle, if there is explicability (e.g., the explanation, by those advocates of "reputation capital and management," of its desirability on the basis of "the benefit" to all "stakeholders"), there is inexplicability (e.g., the lack of sufficient explanation, by those advocates of "reputation capital and management," of why it is necessarily desirable, since, to the critics, the obsession with reputation is a psychological problem instead, "a human fault, exaggerated in importance due to the fragile nature of the human ego"). And the reverse direction also holds true.

In relation to the fiction-reality principle, if there is fiction (e.g., the fictional aspect of "reputation," insofar as the calculation of "reputation" is not objective, since "goodwill" in business is psycholgically tied to the "opinion," "gossip," etc. about an entity like a company), there is reality (e.g., the realistic aspect of "reputation," insofar as "nationalities differ with regard to how valued specific aspects of the company's brand identity are in the respective national culture"). And the reverse direction also holds true.

In relation to the cognitiveness-noncognitiveness principle, if there is cognitiveness (e.g., a view of things on the basis of reason and evidences, as shown in the finding that "some individuals and organizations hire reputation management companies to attempt to hide truthful but unflattering information about themselves"), there is noncognitiveness (e.g., a view of things on the basis of non-cognitive factors like envy, jealousy, power, nationality, race, gender, age, class, greed, lust, status, faith, anger, sadness, joy, fear, wish, etc.—as shown in the psychological pathology of the obsession with reputation, as "a human fault, exaggerated in importance due to the fragile nature of the human ego," regardless of whether the person in question really loses his reputation in reality or not). (WK 2013aa) And the reverse direction also holds true.

In relation to the finiteness-transfiniteness principle, if there is finiteness (e.g., the finite number of fields in which the concern with reputation is important, like "business, education, online communities," etc.), there is transfiniteness (e.g., the transfinite number of all the individual instances that have ever occurred in human hisory, where the concern with reputation becomes important "in business, education, online communities, and many other fields"). (WK 2013aa) And the reverse direction also holds true.

In relation to the preciseness-vagueness principle, if there is preciseness (e.g., the precise identification of 3 levels of analyzing "the difference between image and reputation," in the article aforecited), there is vagueness (e.g., the vagueness in the identification of the 3 levels of analyzing "the difference between image and reputation" in the article aforecited, since it is not clear why there must be only 3, not 4, 5, 6, 7, and so on). And the reverse direction also holds true.

In relation to the simpleness-complicatedness principle, if there is simpleness (e.g., the relatively simple analysis of goodwill in accounting by the advocates, especially in relation to its psychological connectin to reputation), there is complicatedness (e.g., the relatively more complicated analysis of "goodwill," especially in relation to its psychological connectin to reputation, by challenging the claims and assumptions, as shown in the problems pointed out by the critics as cited). And the reverse direction also holds true.

In relation to the openness-hiddenness principle, if there is openness (e.g., the open exploration of the "benefit" of reputation capital and management for a company), there is hiddenness (e.g., the hidden bias in the exploration of the "benefit" of reputation capital and management for a company, since it can become a psychological pathology instead, "exaggerated in importance due to the fragile nature of the human ego,"). And the reverse direction also holds true.

In relation to the denseness-emptiness principle, if there is denseness (e.g., the relatively denser concentration of concerns with "reputation capital and management" among those businesses with "public relations department" nowadays), there is emptiness (e.g., the relatively less dense, or more empty, concentration of concerns with "reputation" as a psychological pathology, "exaggerated in importance due to the fragile nature of the human ego," among those businesses with "public relations department" nowadays). And the reverse direction also holds true.

In relation to the rule-exception principle, if there are rules (e.g., the usual generalization, procedure, habit, or tendency that assets can be amortized), there are exceptions (e.g., a case to which a rule does not apply, such as the exceptional case that, "according to International Financial Reporting Standards, goodwill is never amortized" since 2001). (WK 2013z) And the reverse direction also holds true.

In relation to the prototypicality-variation principle, if there is prototypicality (e.g., the prototype, like the example about the original U.S. Generally Accepted Accounting Principles [FAS 141] on goodwill), there is variation (e.g., different variants of the protoptye, as shown in many studies since the original U.S. Generally Accepted Accounting Principles [FAS 141] on goodwill—like the later ones such as "U.S. GAAP [FAS 142]," "International Financial Reporting Standards," etc., though each with its own variation, contribution, and disagreement). (WK 2013z) And the reverse direction also holds true.

In relation to the change-constancy principle, if there is change (e.g., the ever new ways to study accounting, as shown in the work on goodwill), there is constancy (e.g., the ever constant existence of different problems with the ever new ways to study accounting, as shown in the problems with the work on goodwill pointed out by the critics). And the reverse direction also holds true.

In relation to the order-chaos principle, if there is order (e.g., the more or less orderly view on "reputation," if looked from the sole vantage point of a specific "model" of reputation by its advocates), there is chaos (e.g., the more or less chaotic view on "reputation," if looked from the multiple "often conflicting models of reputation," according to Kevin Money and Carola Hillenbrand, such that they do not add up to much of anything coherent, especially when compounded with different problems aforediscussed). And the reverse direction also holds true.

In relation to the slowness-quickness principle, if there is slowness (e.g., the relatively slower speed of "reputation recovery," which is "long and arduous" in our time), there is quickness (e.g., the relatively quicker speed of "reputation loss" in our time, since "the 'stumble rate' among companies has risen exponentially over the past five years" and "79% of the world's most admired companies have lost their number one positions in industries in that time period"). (WK 2013aa) And the reverse direction also holds true.

In relation to the expansion-contraction principle, if there is expansion (e.g., the relative spread of "purchase accounting" nowadays), there is contraction (e.g., the relative decline of "pooling-of-interests accounting" nowadays). And the reverse direction also holds true.

In relation to the optimality-nonoptimality principle, if there is optimality (e.g., the search for highly favorable options, like the quest for a comprehensive account for "reputation building," like "the stakeholder theory," which says "corporations should be run for the benefit of all 'stakeholders,' not just the shareholders"), there is nonoptimality (e.g., the existence of non-optimal alternatives to optimality, like the argument for more realistic studies by the critics, who pointed out different problems with reputation capital and management aforediscussed). And the reverse direction also holds true.

In relation to the simultaneity-nonsimultaneity principle, if there is simultaneity (e.g., the occurrence of different things more or less at the same time, such as the simultaneous co-existence of both "cognitive" and "affective" elements of "reputation"), there is nonsimultaneity (e.g., the occurrence of something after something else, such as the "causes" of reputation and its subsequent "consequences"). (WK 2013aa) And the reverse direction also holds true.

In relation to the isolation-interaction principle, if there is isolation (e.g., the state of a situation that is separate from others, such as "the good reputation of a company and its existings products" before any "reputation transfer" to "new markets and new products"), there is interaction (e.g., the influence of entities on one another, such as the "reputation transfer" of "the good reputation of a company and its existings products" to "new markets and new products" elsewhere). (WK 2013aa) And the reverse direction also holds true.

In relation to the theory-praxis principle, if there is theory (e.g., the theoretical construction of the 3 levels of "reputation-based decisions" in the field of psychology), there is praxis (e.g., the practical application of the 3 levels of "reputation-based decisions" in the field of psychology to the field of accounting, like "reputation management"). And the reverse direction also holds true.

In relation to the convention-novelty principle, if there is convention (e.g., the conventional wisdom about "reputation capital and management"), there is novelty (e.g., the alternative novel challenge to the conventional wisdom about "reputation capital and management"—by the new idea that the obsession with reputation can be a psychological problem instead, "exaggerated in importance due to the fragile nature of the human ego"). And the reverse direction also holds true.

In relation to the evolution-transformation principle, if there is evolution (e.g., the natural evolution of humans to hold certain opinions about others in the state of nature), there is transformation (e.g., the technical transformation of human ability to to hold certain opinions about others by the invention of "reputation management"). And the reverse direction also holds true.

In relation to the symmetry-asymmetry principle, if there is symmetry (e.g., the co-existence of different methods in the field of accounting, like "purchase accounting" and "pooling-of-interests accounting"), there is asymmetry (e.g., in practice, "since 2001, U.S. Generally Accepted Accounting Principles [FAS 141] no longer allows the pooling-of-interests method," but the "purchase accounting" is allowed). And the reverse direction also holds true.

In relation to the softness-hardness principle, if there is softness (e.g., the defense of "reputation management" by the advocates aforecited), there is hardness (e.g., the critique of "reputation management," as shown in the problems pointed out by the critics aforecited). And the reverse direction also holds true.

In relation to the seriousness-playfulness principle, if there is seriousness (e.g., the serious business of those who try to understand the nature of "reputation"), there is playfulness (e.g., the playful part of those who try to understand "reputation," when they play around with different possible levels of analysis over the last decades, like the "epistemic," the "pragmatic–strategic," the "memetic," etc.). And the reverse direction also holds true.

In relation to the activeness-inactivess principle, if there is activeness (e.g., the relative activeness of "purchase accounting" nowadays, which is allowed to use, under the U.S. Generally Accepted Accounting Principles), there is inactiveness (e.g., the relative inactiveness of the "pooling-of-interests method," which is no longer allowed nowadays, due to the U.S. Generally Accepted Accounting Principles [FAS 141] since 2001). And the reverse direction also holds true.

In relation to the selfness-otherness principle, if there is selfness (e.g., the tendency to be mostly centered on one's self or group, as shown by those who are obsessed with their "reputation"), there is otherness (e.g., the tendency to be mostly centered on others, as shown in the academic analysis of both sides of the debate on "reputation capital and management," as an advancement in the field of accounting for humanity). And the reverse direction also holds true.

In relation to the regression-progression principle, if there is regression (e.g., the regression made by "reputation management," as shown in the problems pointed out by the critics), there is progression (e.g., the progress made by "reputation management," as shown in the contribution to the "transfer" of "the good reputation of a company and its existings products to new markets and new products" aforecited). (WK 2013aa) And the reverse direction also holds true.

In relation to the same-difference principle, if there is similarity in outcome (e.g., the contribution to the molding and control of beliefs, values, and behaviors, regardless of whether this be done by way of the "stakeholder theory," or by way of the "shareholder theory," on building reputation), there is difference in outcome (e.g., the contribution to the molding and control of beliefs, values, and behaviors by way of the "stakeholder theory" on building reputation for a relatively broader view of management—but the contribution to the molding and control of beliefs, values, and behaviors by way of the "shareholder theory" on building reputation for a relatively more narrow view of management). And the reverse direction also holds true.

In relation to the stability-reaction principle, if there is stability (e.g., the relatively more stable condition of using "pooling-of-interests accounting" in the older days), there is reaction (e.g., the use of "pooling-of-interests accounting" in the older days then led to further changes later on, when, "since 2001, U.S. Generally Accepted Accounting Principles no longer allows the pooling-of-interests method"). (WK 2013z) And the reverse direction also holds true.

In relation to the functionality-nonfunctionality principle, if there is functionality (e.g., the presence of action for which something or someone is fitted, as shown in "reputation management," which serves the function of increasing the "benefit" of "stakeholders," for instance), there is nonfunctionality (e.g., the relative lack of action for which something or someone is fitted, as shown in different problems with "reputation management" aforecited, such that it is not as functional as the advocates would like us to believe). And the reverse direction also holds true.

In relation to the intentionality-nonintentionality principle, if there is intentionality (e.g., the planning or design of something for a certain outcome, as shown in the original intention of the advocates of the "stakeholder theory" to build reputation for the benefit of all concerned), there is nonintentionality (e.g., the relative absence of planning or design of something for a certain outcome, such that the outcome can even be counter-intentional in being the opposite of what was originally planned, as shown in the critique that the obsession with reputation is a psychological problem instead, "a human fault, exaggerated in importance due to the fragile nature of the human ego"—contrary to the original intention of the advocates of the "stakeholder theory" to build reputation for the benefit of all concerned). And the reverse direction also holds true.

And in relation to the survivability-nonsurvivability principle, if there is survivability (e.g., the relative survival, nowadays, of "purchasing accounting"), there is non-survivability (e.g., the relative non-survival, nowadays, of "pooling-of-interests" accounting). And the reverse direction also holds true.

Accounting and Nature

Addition, when examined from the perspective of nature, can show us in a controversial way its possibility (or impossibility) and desirability (or undesirability), and this can be shown by way of two case studies, namely,

(2.3.1) addition, eco-management, and audit scheme , and (2.3.2) addition, space-time, and accounting—to be addressed in what follows, respectively.

Addition, Eco-Management, and Audit Scheme

A good way to evaluate the possibility (or impossibility) and desirability (or undesirability) of addition from the perspective of nature is to explore addition, eco-management, and audit scheme.

An argument in the literature is that a voluntary environmental management instrument like an "audit scheme" for "eco-management" can be used to increase "environmental performance." (WK 2013)

Introduction to EMAS

A good example is "the Eco-Management and Audit Scheme (EMAS)...developed in 1993 by the European Commission. It enables organizations to assess, manage and continuously improve their environmental performance. The scheme is globally applicable and open to all types of private and public organizations. In order to register with EMAS, organisations must meet the requirements of the EU EMAS-Regulation. Currently, more than 4,600 organisations and more than 7,900 sites are EMAS registered." (WK 2013r)

In fact, "the EMAS Regulation includes the environmental management system requirements of the international standard for environmental management, ISO 14001, and additional requirements for EMAS registered organisations such as employee engagement, ensuring legal compliance or the publication of an environmental statement. Because of its additional requirements, EMAS is known as the premium instrument for environmental management." (WK 2013r; GER 2011)

Development of EMAS

In 1993, "the first EMAS Regulation (EMAS I) was adopted...and became operational in 1995. It was originally restricted to companies in industrial sectors." (WK 2013r)

In 2001, "with the first revision of the EMAS Regulation...(EMAS II), the scheme opened to all economic sectors including public and private services. In addition, EMAS II was strengthened by the integration of the environmental management requirements of ISO 14001; by adopting a new EMAS logo to signal engagement to stakeholders; and by considering more strongly indirect effects such as those related to financial services or administrative and planning decisions." (WK 2013r)

And on January 11, 2010, "the latest revision of EMAS came into effect...(EMAS III). With the introduction of EMAS III, the scheme is globally applicable and no longer limited to EU Member States. With EMAS III the EU also introduced obligatory Key Performance Indicators (KPI) in order to harmonize reporting on environmental performance. The number of EMAS registered organizations increased from 2,140 in 1997 to 4,659 in 2011." (WK 2013r)

Implementation of EMAS

The implementation of EMAS can be a challenging task, For instance, "in order to register with EMAS an organization must comply with the following implementation steps (Article 4 of the EMAS-Regulation)," as summarized in *Table 2.3*: (WK 2013r)

- "Environmental Review"—"Initial comprehensive analysis of the organization's activities, products and services and their environmental impact; cataloguing applicable environmental law, etc."
- "Environmental Policy"—"definition of the organization's overarching environmental objectives; commitment to continuous improvement of environmental performance."
- "Environmental Programme"—"description of measures, responsibilities and means to achieve environmental objectives and targets."
- "Environmental Management System"—"part of an organization's management entailing structure, planning activities, responsibilities, practices, procedures, processes and resources for developing, implementing, achieving, reviewing and maintaining the environmental policy and managing the environmental aspects."
- "Environmental Audit"—"systematic, documented, periodic and objective evaluation of the organization's environmental performance, management system and processes designed to protect the environment, conducted by internal auditor(s)."
- "Environmental Statement"—"comprehensive, regular reports to the public on the organization's structure and activities; environmental policy and management system; environmental aspects and impacts; environmental programme, objectives and targets; environmental performance and compliance with applicable environmental law etc."

- "Verification and Registration"—"The steps above must be verified by an accredited/licensed environmental verifier; the validated environmental statement needs to be sent to the EMAS Competent Body (exists in each EU country) for registration and made publicly available before an organisation can use the EMAS logo."

Of course, these examples are illustrative, not exhaustive.

Benefits of EMAS

And the "benefits" of EMAS are not difficult to identify, as shown below (and summarized in *Table 2.3*): (WK 2013r)

- "Environmental and Financial Performance through a Systematic Framework"—"e.g. increased resource and energy efficiency," etc.
- "Risk and Opportunity Management"—"e.g., legal compliance, regulatory relief."
- "Credibility, Transparency and Reputation"—"e.g., environmental statement, key performance indicators, verification and validation through independent environmental verifiers."
- "Employee Empowerment and Motivation"—"e.g., improved involvement of staff, higher awareness, often leading to innovations."

Again, these examples are illustrative, not exhaustive.

Problems with Eco-management and Audit Scheme

However, to the critics, this argument on addition, eco-management, and audit scheme has some major problems. Consider, for illustration, a few examples below (as summarized in *Table 2.3*).

Firstly, one main criticism of the argument on addition, eco-management, and audit scheme is that the required measurement to implement EMAS is not practical.

For instance, to the critics, "given the heterogeneity of registered organisations" (with "more than 4,600 organisations and more than 7,900 sites"), "calculating average benefits and costs of EMAS is nearly unfeasible." (WK 2013r)

Secondly, another main criticism of the argument on addition, eco-management, and audit scheme is that the benefits of EMAs vary from country to country and from organization to organization.

For instance, to the critics, "the financial benefits of an EMAS registered organisation can differ per country. For example, some countries offer substantial reductions of waste fees, lower permitting costs, faster licensing procedures, etc."—but others do not. (WK 2013r)

Thirdly, still another main criticism of the argument on addition, eco-management, and audit scheme is that the implementation is not mandatory.

For instance, "although EMAS is an official EU Regulation, it is binding only for organizations which voluntarily decide to implement the scheme." (WK 2013r) And "a common misconception is that ISO 14001 certification automatically implies legal compliance. Certification under ISO 14001 does not directly reflect compliance with any legal requirements, although ISO 14001 demands the organization to evaluate its compliance with legal requirements." (WK 2013s & 2013bb)

Fourthly, an additional main criticism of the argument on addition, eco-management, and audit scheme is that there are costs of EMAS.

For instance, there are different costs of EMAS, and three main examples are (a) "fixed costs: validation/verification fees, registration fees, integrating EMAS logo into corporate design," (b) "external costs: consultancy expertise to support implementation and reporting," and (c) "internal costs: personnel and technical resources needed for implementing, administering and reporting." (WK 2013r)

And fifthly, another additional main criticism of the argument on addition, eco-management, and audit scheme is that there is no consensus on the audit tools to be used for EMAS.

For instance, "there is a long standing debate among environmental audit professionals on the value of large, highly detailed and prescriptive protocols (i.e., that can, in theory, be completed by an auditor with little or no technical experience) versus more flexible protocols that rely on the expertise and knowledge of experienced auditors and source documents (regulations, permits, etc.) directly." (WK 2013bb)

Beyond Eco-management and Audit Scheme

These criticisms of the argument on addition, eco-management, and audit scheme should not be taken as a total rejection of its usefulness but serve to show us the two opposing sides of the debate, such that the possibility (or impossibility) and desirability (or undesirability) of addition (from the perspective of nature with addition, eco-management, and audit scheme as a case study here) are not to the extent that the respective defenders would like us to believe.

Furthermore, the analysis of addition, eco-management, and audit scheme can shed some light on the ontological principles in existential dialectics, and good examples include the formalness-informalness principle, the absoluteness-relativeness principle, the partiality-totality principle, the predictability-unpredictability principle, the explicability-inexplicability principle, the fiction-reality principle, the cognitiveness-noncognitiveness principle, the finiteness-transfiniteness principle, the preciseness-vagueness principle, the simpleness-complicatedness principle, the openness-hiddenness principle, the denseness-emptiness principle, the rule-exception principle, the prototypicality-variation principle, the change-constancy principle, the order-chaos principle, the slowness-quickness principle, the expansion-contraction principle, the optimality-nonoptimality principle, the simultaneity-nonsimultaneity principle, the isolation-interaction principle, the theory-praxis principle, the convention-novelty principle, the evolution-transformation principle, the symmetry-asymmetry principle, the softness-hardness principle, the seriousness-playfulness principle, the activeness-inactivess principle, the selfness-otherness principle, the regression-progression principle, the same-difference principle, the stability-reaction principle, the functionality-nonfunctionality principle, the intentionality-nonintentionality principle, and the survivability-nonsurvivability principle.

For instance, in relation to the formalness-informalness principle, if there is formalness (e.g., the formal logical requirement of soundness, completeness, and consistency in a system of ideas, as in the argument on EMAS), there is informalness (e.g., the non-formal existence of unsoundness, incompleteness, and inconsistency in the argument on EMAS, as shown in the criticisms that the required measurement to implement EMAS is not practical, that the benefits of EMAs vary from country to country and from organization to organization, that the implementation is not mandatory, that there are costs of EMAS, and that there is no consensus on the audit tools to be used for EMAS). And the reverse direction also holds true.

In relation to the absoluteness-relativeness principle, if there is absoluteness (e.g., the absolute view on the implementation of EMAS by those organizations which "voluntarily decide to implement the scheme"), there is relativeness (e.g., what is right for those organizations which "voluntarily decide to implement the scheme" is not necessarily so for other rganizations which do not voluntarily decide to implement the scheme," because "a common misconception is that ISO 14001 certification automatically implies legal compliance"). And the reverse direction also holds true.

In relation to the partiality-totality principle, if there is partiality (e.g., the partial view on audit tools by the advocates of "large, highly detailed and prescriptive protocols"), there is totality (e.g., the more holistic view on audit tools, such that the whole is not the sum of its parts, that is, it cannot be reduced to the sum of opposing views by the advocates of "large, highly detailed and prescriptive protocols," the other advocates of "more flexible protocols," etc., since there will emerge new views in the future not yet known today, just as there are already alternative views nowadays which do not side exclusively with any of them, as is the analysis here). And the reverse direction also holds true.

In relation to the predictability-unpredictability principle, if there is predictability (e.g., the predictable tendency that "the benefits of EMAs vary from country to country and from organization to organization"), there is unpredictability (e.g., the more difficult task to predict exactly the extent to which, or the specific way in which, a particular country or a particular organization will be benefited from EMAS on a particular future occasion). And the reverse direction also holds true.

In relation to the explicability-inexplicability principle, if there is explicability (e.g., the explanation, by those advocates of "large, highly detailed and prescriptive protocols," of their desirability on the basis of "easier review by other parties"), there is inexplicability (e.g., the lack of sufficient explanation, by those advocates of of "large, highly detailed and prescriptive protocols," of why they are necessarily desirable, since, to the critics, "there is a long standing debate among environmental audit professionals on the value of large, highly detailed and prescriptive protocols...versus more flexible protocols"). And the reverse direction also holds true.

In relation to the fiction-reality principle, if there is fiction (e.g., the fictional aspect of ISO 14001 certification, insofar as there is "a common misconception... that ISO 14001 certification automatically implies legal compliance"), there is reality (e.g., the realistic aspect of ISO 14001 certification, insofar as "ISO 14001 demands the organization to evaluate its compliance with legal requirements"). And the reverse direction also holds true.

In relation to the cognitiveness-noncognitiveness principle, if there is cognitiveness (e.g., a view of things on the basis of reason and evidences, as shown in the finding that there are different "audit tools" available for EMAS implementation), there is noncognitiveness (e.g., a view of things on the basis of non-cognitive factors like envy, jealousy, power, nationality, race, gender, age, class, greed, lust, status, faith, anger, sadness, joy, fear, wish, etc.—as shown in the politics of "preference," when some auditors prefer "large, highly detailed and prescriptive protocols," in spite of other alternative protocols available too, like the "more flexible protocols that rely on the expertise and knowledge of experienced auditors and source documents [regulations, permits, etc.] directly"). And the reverse direction also holds true.

In relation to the finiteness-transfiniteness principle, if there is finiteness (e.g., the finite number of the categories of the environmental impact by organizations to be identified for environmental audit, like "activities," "products," "services," etc.), there is transfiniteness (e.g., the transfinite number of all the individual "activities," "products," "services," etc. that have ever produced in human history and have enviromental impact). (WK 2013bb) And the reverse direction also holds true.

In relation to the preciseness-vagueness principle, if there is preciseness (e.g., the precise identification of 4 benefits of EMAS in the article aforecited), there is vagueness (e.g., the vagueness in the identification of the 4 benefits of EMAS in the article aforecited, since it is not clear why there must be only 4, not 5, 6, 7, 8, and so on). And the reverse direction also holds true.

In relation to the simpleness-complicatedness principle, if there is simpleness (e.g., the relatively simple analysis of EMAS by the advocates), there is complicatedness (e.g., the relatively more complicated analysis of EMAS, by challenging the claims and assumptions, as shown in the problems pointed out by the critics as cited). And the reverse direction also holds true.

In relation to the openness-hiddenness principle, if there is openness (e.g., the open exploration of using "large, highly detailed and prescriptive protocols" as an audit tool), there is hiddenness (e.g., the hidden bias in using "large, highly detailed and prescriptive protocols" as an audit tool, since, to the critics, "there is a long standing debate among environmental audit professionals on the value of large, highly detailed and prescriptive protocols…versus more flexible protocols"). And the reverse direction also holds true.

In relation to the denseness-emptiness principle, if there is denseness (e.g., the relatively denser concentration of concerns with "easier reviews" by those advocates of using "large, highly detailed and prescriptive protocols" as an audit tool), there is emptiness (e.g., the relatively less dense, or more empty, concentration of concerns with "more flexible protocols" by those advocates of using "large, highly detailed and prescriptive protocols" as an audit tool). And the reverse direction also holds true.

In relation to the rule-exception principle, if there are rules (e.g., the usual generalization, procedure, habit, or tendency that there are benefits and costs of EMAS), there are exceptions (e.g., a case to which a rule does not apply, such as the exceptional cass that "some countries offer substantial reductions of waste fees, lower permitting costs, faster licensing procedures, etc."). And the reverse direction also holds true.

In relation to the prototypicality-variation principle, if there is prototypicality (e.g., the prototype, like the example about the original EMAS Regulation [EMAS I] in 1993), there is variation (e.g., different variants of the protoptye, as shown in many studies since the original EMAS Regulation [EMAS I] in 1993—like the later ones such as EMAS II in 2001, EMAS III in 2010, etc., though each with its own variation, contribution, and disagreement). And the reverse direction also holds true.

In relation to the change-constancy principle, if there is change (e.g., the ever new ways to study accounting, as shown in the research on environmental auditing like EMAS), there is constancy (e.g., the ever constant existence of different problems with the ever new ways to study accounting, as shown in the problems with the research on environmental auditing like EMAS pointed out by the critics). And the reverse direction also holds true.

In relation to the order-chaos principle, if there is order (e.g., the more or less orderly view on EMAS, if looked from the sole vantage point of of its official regulations), there is chaos (e.g., the more or less chaotic view on EMAS, if looked from the multiple, conflicting vantage points of both the advocates and the critics, especially in regard to the problems aforecited, such that they do not add up to much of anything coherent). And the reverse direction also holds true.

In relation to the slowness-quickness principle, if there is slowness (e.g., the relatively slower speed of auditing with the use of "more flexible protocols that rely on the expertise and knowledge of experienced auditors and source documents [regulations, permits, etc.] directly" for more technical reviews), there is quickness (e.g., the relatively quicker speed of auditing with the use of "large, highly detailed and prescriptive protocols" for "easier reviews"). And the reverse direction also holds true.

In relation to the expansion-contraction principle, if there is expansion (e.g., the relative spread of the influence of EMAS III nowadays), there is contraction (e.g., the relative decline of the influence of EMAS I nowadays). And the reverse direction also holds true.

In relation to the optimality-nonoptimality principle, if there is optimality (e.g., the search for highly favorable options, like the quest for the global eco-management with the introduction of EMAS III in 2010), there is nonoptimality (e.g., the existence of non-optimal alternatives to optimality, like the argument for more realistic studies by the critics, who pointed out different problems with EMAS aforecited). And the reverse direction also holds true.

In relation to the simultaneity-nonsimultaneity principle, if there is simultaneity (e.g., the occurrence of different things more or less at the same time, such as the simultaneous co-existence of organizaitons which implement EMAS and those which do not), there is nonsimultaneity (e.g., the occurrence of something after something else, such as the development of EMAS III in 2010 after that of EMAS II in 2001). And the reverse direction also holds true.

In relation to the isolation-interaction principle, if there is isolation (e.g., the state of a situation that is separate from others, such as the separation of organizations in the non-industrial sector from the jurisdiction of EMAS in 1995), there is interaction (e.g., the influence of entities on one another, such as the diverse relationships between different organizations in the non-industrial sector and EMAS in 2010). And the reverse direction also holds true.

In relation to the theory-praxis principle, if there is theory (e.g., the theoretical construction of "voluntary environmental management instrument" in the interdisciplinary field of environmental studies and accounting), there is praxis (e.g., the practical application of "voluntary environmental management instrument" in the interdisciplinary field of environmental studies and accounting to the political realm, like the use of EMAS by the European Commission). And the reverse direction also holds true.

In relation to the convention-novelty principle, if there is convention (e.g., the conventional wisdom about "the value of large, highly detailed and prescriptive protocols" as an audit tool for "easier review"), there is novelty (e.g., the alternative novel challenge to the conventional wisdom about "the value of large, highly detailed and prescriptive protocols" as an audit tool" for "easier review"—by the alternative idea of a "more flexible protocols that rely on the expertise and knowledge of experienced auditors and source documents [regulations, permits, etc.] directly" for technical review). And the reverse direction also holds true.

In relation to the evolution-transformation principle, if there is evolution (e.g., the natural evolution of humans to deal with the environment in the state of nature in a primitive way), there is transformation (e.g., the technical transformation of human ability to deal with the environment by the invention of EMAS). And the reverse direction also holds true.

In relation to the symmetry-asymmetry principle, if there is symmetry (e.g., the co-existence of different entities, like the European Commission, companies, etc.), there is asymmetry (e.g., the European Commission has the authority to set up EMAS for companies to comply—but companies do not have the authority to set up EMAS for the European Commission to comply). And the reverse direction also holds true.

In relation to the softness-hardness principle, if there is softness (e.g., the defense of EMAS by the advocates aforecited), there is hardness (e.g., the critique of EMAS, as shown in the problems pointed out by the critics aforecited). And the reverse direction also holds true.

In relation to the seriousness-playfulness principle, if there is seriousness (e.g., the serious business of those who try to understand the nature of eco-management and audit scheme), there is playfulness (e.g., the playful part of those who try to understand the nature of eco-management and audit scheme, when they play around over the last few decades with different implementation steps for organizations to follow, as shown in the 7 examples aforecited). And the reverse direction also holds true.

In relation to the activeness-inactivess principle, if there is activeness (e.g., the relative activeness of EMAS III nowadays), there is inactiveness (e.g., the relative inactiveness of EMAS I nowadays). And the reverse direction also holds true.

In relation to the selfness-otherness principle, if there is selfness (e.g., the tendency to be mostly centered on one's self or group, as shown in the implementation of EMAS by some organizations for the "benefits" as cited before), there is otherness (e.g., the tendency to be mostly centered on others, as shown in the academic analysis of both sides of the debate on EMAS, as an advancement in the knowledge of eco-management and audit scheme for humanity). And the reverse direction also holds true.

In relation to the regression-progression principle, if there is regression (e.g., the regression made by EMAS, as shown in the problems pointed out by the critics), there is progression (e.g., the progress made by EMAS, as shown in the contribution to a global scheme for eco-management and audit scheme aforecited). And the reverse direction also holds true.

In relation to the same-difference principle, if there is similarity in outcome (e.g., the contribution to the molding and control of beliefs, values, and behaviors, regardless of whether this be done by way of "highly detailed and prescriptive protocols," or by way of "more flexible protocols"), there is difference in outcome (e.g., the contribution to the molding and control of beliefs, values, and behaviors by way of "highly detailed and prescriptive protocols" for a relatively more novice worldview—but the contribution to the molding and control of beliefs, values, and behaviors by way of "more flexible protocols" for a relatively more experienced worldview). And the reverse direction also holds true.

In relation to the stability-reaction principle, if there is stability (e.g., the relatively more stable condition of EMAS I in the 1990s), there is reaction (e.g., EMAS I in the 1990s then led to further changes later on, when nowadays, there is EMAS III). And the reverse direction also holds true.

In relation to the functionality-nonfunctionality principle, if there is functionality (e.g., the presence of action for which something or someone is fitted, as shown in the creation of EMAS, which serves the function of providing a way to do eco-management and audit scheme), there is nonfunctionality (e.g., the relative lack of action for which something or someone is fitted, as shown in the problems with EMAS, such that it is not as functional as the advocates would like us to believe). And the reverse direction also holds true.

In relation to the intentionality-nonintentionality principle, if there is intentionality (e.g., the planning or design of something for a certain outcome, as shown in the original intention of those to design "highly detailed and prescriptive protocols" for "easier review," so as to become institutionalized as the preferred audit tool), there is nonintentionality (e.g., the relative absence of planning or design of something for a certain outcome, such that the outcome can even be counter-intentional in being the opposite of what was originally planned, as shown in the alternative proposal nowadays to use "more flexible protocols that rely on the expertise and knowledge of experienced auditors and source documents [regulations, permits, etc.] directly"—contrary to the original intention of those who design to design "highly detailed and prescriptive protocols" for "easier review," so as to become institutionalized as the preferred audit tool). And the reverse direction also holds true.

And in relation to the survivability-nonsurvivability principle, if there is survivability (e.g., the relative survival, nowadays, of EMAS III), there is non-survivability (e.g., the relative non-survival, nowadays, of EMAS I). And the reverse direction also holds true.

Addition, Space-Time, and Accounting

Another good way to evaluate the possibility (or impossibility) and desirability (or undesirability) of addition from the perspective of nature is to explore addition, space-time, and accounting.

An argument in the literature is that the modern income accounting constructs reality, in regard to some conceptions of space-time (e.g., "demarcated space" "constant time value of money," etc.), so as to deal with the growing (or "increasing") business world in our time, as suggested by Sadao Takatera (2000) and Norio Sawabe.

Evolution of Accounting Conventions

Sadao Takatera (2000) and Norio Sawabe argued that the evolutionary process of modern accounting conventions in the last 5 centuries can be divided into two periods, namely, (a) "the pre-going concern" and (b) "the going concern," as explained below (and summarized in *Table 2.4*).

(a) The Pre-Going Concern Period

In the older days of the "pre-going concern period"—"early venture (limited term) business typically employed total income accounting, by which total income from establishment till liquidation was calculated. Total income accounting had no concept of accounting periods. There was no need for the concept to take place in accounting calculation, since total income was calculated over a material lifetime of the venture business which was given exogenously to the accounting system. The accounting system was based on the actual amount of cash obtained and paid (receipts and expenditures) within the total lifetime of the venture business concerned." (S. Takatera 2000)

According to D. Scott (1931), "in the field of merchandising in which accounting first developed, the scope and function of accounts were limited, in the beginning, to recording decisions rendered by the market....Accounting was, thus, entirely subordinated to the market." (S. Takatera 2000)

As a comparison, "the embryonic stage of a modern corporation with going concern status carried properties similar to those of early venture business, though they introduced artificially partitioned accounting periods for periodic income calculation. This 'equitemporal accounting' was not yet based on accrual but on contemplated cash flows. For instance, some going concern enterprises included the holding gain and loss on operating assets unsold at the end of each accounting period in profit and loss accounts by assuming as if those assets had been sold at the marketable value at the time concerned. In this case, the marketable value of the unsold stock could be regarded as 'uncollected sales.' In terms of the autonomy of the accounting calculative logic, it was still subordinated to the logic of the market, i.e. the market price fluctuation was immediately reflected in the periodic income." (S. Takatera 2000)

(b) The Going Concern Period

Later on, in the "going concern period" starting from the 19th century—"in England and Scotland, modern corporations emerged that developed periodic income accounting that fit a larger going concern. From this stage on, the internal calculative logic of the accounting system has been structurally separated from the external logic of the market....The accounting system that begun to embrace its own internal calculative logic is now called the accrual accounting system. The accrual accounting system allocates income earning cash inflows and outflows as periodic revenues and expenditures. The allocation process was dictated by its internally specified logic (generally termed as accrual-basis), independent from the external market conditions. It is worth noting that the income smoothing accounting practice was extensively developed about the same period when accrual concept emerged." (S. Takatera 2000; M. Mathews 1991)

This "periodic income is calculated as a basis for income distribution because periodic income is accepted as periodic income by one person because he or she knows that periodic income is accepted as periodic income by others. In this way, periodic income becomes a de facto common ground for the interactions within and without organizations. The self-reproducing process is not eternal, however. The sustainablity of the process depends both on the availability of resources, and on the rigidity and immunity of the internal logic that provides the rationale and the technologies of its existence. Understanding the idiosyncratic nature of accounting's internal logic is a prerequisite for answering these questions. because the core of the reproduction process is the accrual process incorporated in income accounting." (S. Takatera 2000)

Assumptions on Time in Accrual Accounting Conventions

There are two important assumptions on time in the accrual accounting system (as summarized in *Table 2.4*).

(a) Assumption 1: The Constant Value of Money

The first assumption on time has to do with "the constant value of money" (or simply "the indifferentiating external transaction time"). (S. Takatera 2000)

In other words, "when it comes to valuation of money in the accrual accounting system, the 'constant value of money' assumption is what comes to mind for accountants. It is true that the purchasing power of money must be assumed constant throughout accounting periods, however fictional, in order for the consistency of accounting calculation to hold, unless we introduce some sort of inflation accounting or the like. In the case of accounting for changing monetary value, it is simply that we replace one fiction of monetary value, i.e. constant value of money, by another, e.g. inflating or deflating by general price index." (S. Takatera 2000)

In this way, "when value of money is assumed constant, we have no base to differentiate transactions that are identical except for the temporal difference. Therefore, we term the assumption an assumption of 'indifferentiating external transaction time'....Since the assumption of the constant value of money is defined in terms of the purchasing power of money, it should be understood within the context of transactions that involve money and goods and services purchased by money. This assumption is directly related to the material basis of economic transactions. Transactions that do not satisfy this criterion are, strictly speaking, irrelevant to the application of the assumption. This is critical for our argument that the accrual accounting system is capable of creating internal space independent of external time flow." (S. Takatera 2000)

(b) Assumption 2: The Constant Time Value of Money

And the second assumption on time has to do with "the constant time value of money" (or "the emptying-out of internal transaction time)." (S. Takatera 2000)

In other words, "by artificially partitioning continuous external transactions into accounting periods of a going concern, current accounting practices calculate periodic income as a 'ganz unentbehrliche Fiktion (completely indispensable fiction)'....The resulting periodic income from this particular type of accounting practice would not be temporally comparable and consistent unless the temporal uniqueness of external transactions are ignored." (S. Takatera 2000)

For instance, "the accounting logic of matching expenses with revenues...is sometimes referred to as the accrual process. For the accounting logic of periodic allocation to be consistent, we need to assume that the time value of money is negligible. By the time value of money, we mean the expression of changes in the value of money in different time periods measured by the present monetary scale." (S. Takatera 2000)

In this way, "in order to differentiate this 'constant time value of money' assumption from the constant value of money assumption, we call the constant time value of money assumption the 'emptying-out of internal transaction time,' since this assumption decontextualizes the value of money in terms of time differences." (S. Takatera 2000)

Assumptions on Space in Accrual Accounting Conventions

There are likewise two important assumptions on space in the accrual accounting system (as summarized in *Table 2.4*).

(a) Assumption 1: Primary Enclosing as a Gatekeeper

The first assumption on space has to do with "accounting as a gatekeeper: primary enclosing." (S. Takatera 2000)

As Takatera (2000) and Sawabe wrote, "one of the parallel conditions for an organization to coordinate economic activities is to recognize itself as an independent economic entity. It is here that accounting plays a vital role. Accounting serves to construct the self-image of the organization that has a distinctive territory of its own. The status of an organization as an independent economic entity is an artificially created image that is contestable if we take different views from those of transaction cost economics. For instance, those who view the economy as a nexus of contracts cannot take the ontological existence of a firm as taken for granted." (S. Takatera 2000)

Thus, "as long as the current accounting system maintains its legitimacy, organizations are likely to keep their status as elementary units in the business world. A basic function of the accounting system in this framework is that of the gatekeeper of organizations. All the records of external transactions are systematically collected even by a primitive accounting system. Those external transaction records are equivalent to gate records for the organization, entry and exit records of items confirm the boundary of the organization marked in financial terms. Accordingly, we can observe that the financial boundary of the organization is maintained by the accounting system. In this sense, the accounting system constructs islands of conscious power in the ocean of unconsciousness. This basic constructive role of the accounting system for organizations is what we call 'the primary enclosing.'" (S. Takatera 2000)

(b) Assumption 2: Secondary Enclosing as Demarcated Space

And the second assumption on space has to do with "demarcated accounting space: secondary enclosing." (S. Takatera 2000)

The concern here is about "the valuation process of business activities within organizations. Valuation is here defined in relation to expectation: taking a certain set of expectations in order for the actor to position him or herself in an uncertain world....In the market system, expectations of each participant might vary, but a unique valuation of a certain commodity is obtained as a price at a time. If all the market participants held the same future expectations, we would be in a situation where no market clearing price was obtained. If all participants expect that the price will move in one direction, either up or down, there will be no one who wants to sell or buy. Heterogeneity of expectations in the market is a necessary condition for the well-functioning of the price mechanism. The price mechanism (ideally) produces a unique value for the item concerned as a system's outcome, i.e. one price for a good or service at a given time and space as a consequence of unconscious cooperation between heterogeneous individuals. In this sense, diversity of expectations is transformed into a single universal value by the price mechanism. The current market price is the dominant value in the market at any one time." (S. Takatera 2000)

Valuation in terms of expectation is "an integral and inseparable part of the accounting process, even though it takes a form of allocation on the surface of the accrual accounting system. The valuation process is intertwined with the way in which time and space is constructed in the accounting system," which can be understood in terms of two subsystems: "the external transaction" and "the internal transaction" accounting systems. (S. Takatera 2000)

In the "external transaction accounting system"—"goods or services that are transacted in the external market are recognized as receipts or expenditures measured by the amount of accompanied cash flows. Those receipts or expenditures are not bound by any accounting period as far as they are in the domain off the external transaction accounting system." (S. Takatera 2000)

But in the "internal transaction accounting system"—revenues and expenses that are successively translated from those receipts and expenditures are allocated to particular accounting periods. This allocation process is not dependent upon the external transaction timing since the assumption of indifferentiating external transaction time is held, but it is carried out in a way that it represents the fiction of the accrual process. An outcome of the allocation is the creation of an internal transaction accounting space insulated from the external market. We call this whole process of creating internal transaction accounting space 'the secondary enclosing.'" (S. Takatera 2000)

Problems with the Study of Space-Time in Accounting

However, to the critics, this study of addition, space-time, and accounting has some major problems. Consider, for illustration, a few examples below (as summarized in *Table 2.4*).

Firstly, one main criticism of the study of addition, space-time, and accounting is that Takatera and Sawabe used a writing sytle which is too obstruse (and flowery) to be empirically useful for clear analysis.

For example, consider these obtruse (and flowery) sentences used by Takatera and Sawabe, like "the accounting system constructs islands of conscious power in the ocean of unconsciousnes," "the dictatorship of 'money' as a carrier of time is withheld in the 'accrual process,'" etc. (S. Takatera 2000: 794-795)

Secondly, another main criticism of the study of addition, space-time, and accounting is that the claim by Takatera and Sawabe about the social construction of space-time in modern accrual accounting conventions is exaggeratedly misleading.

For instance, the replacement of the "constant value of money" by a more historical accounting (e.g., "inflating or deflating by general price index") is not just another fiction (as the two authors falsely claimed) but can be more realistic, by contrast. The contentious debate in accounting between "purchasing power accounting" and "historical cost accounting" (as discussed in *Sec. 1.4*) is a good case in point.

Thirdly, still another main criticism of the study of addition, space-time, and accounting is that the claims in the article are often argumentatively relative to the theories that Takatera and Sawabe personally preferred.

For instance, Takatera and Sawabe one-sidedly used one approach to economics (i.e., the "transaction cost economics") against another (i.e., the "accrual accounting system" under attack by the two authors, without questioning the institutional approach itself), when they thus wrote: "The status of an organization as an independent economic entity is an artificially created image that is contestable if we take different views from those of transaction cost economics." (S. Takatera 2000: 793)

Fourthly, an additional main criticism of the study of addition, space-time, and accounting is that Takatera and Sawabe confused "chaos theory" with "complexity theory," which then raises the thorny question about their scholarly qualification to use the two fields for the analysis of accounting.

For instance, Takatera and Sawabe wrote that "chaos theory, sometimes more generally referred to as complexity theory, has become widely acknowledged," so the two authors then used them interchangeably to attack the acrrual accounting conventions. (S. Takatera 2000: 796) But this is incorrect, because chaos theory was constructed in mathematics about "small differences in initial conditions (such as those due to rounding errors in numerical computation)," which can "yield widely diverging outcomes for such dynamical systems, rendering long-term prediction impossible in general"—whereas "complexity theory" is more about the study of "complex systems" in relation to such phenomena like "emergence," "self-organization," etc. (WK 2013cc & 2013dd)

And fifthly, another additional main criticism of the study of addition, space-time, and accounting is that Takatera and Sawabe are polemically driven by the discourse on "chaos" to analyze accounting in their idiosyncratic way (with the subsequent bias).

For instance, Takatera and Sawabe claim that "accounting processes are complex phenomena," and it is important to "account...order out of chaos," in that "the accrual accounting system can be interpreted as a complex system in which a noise-induced order of smoothed income series is generated." (S. Takatera 2000: 796-797) In so doing, Takatera and Sawabe twisted the historical analysis of the accrual accounting conventions from the perspective of "accounting order out of chaos"—but one of the major problems with "complexity theory" is how exactly the phenomenon of "emergence" is to be explained, as already critically discussed in my book *The Future of Complexity* (2007).

Beyond Space-Time in Accounting

These criticisms of the study of addition, space-time, and accounting do not mean, however, a total rejection of its usefulness but serve to show us the two opposing sides of the debate, such that the possibility (or impossibility) and desirability (or undesirability) of addition (from the perspective of nature with addition, space-time, and accounting as a case study here) are not to the extent that the respective defenders would like us to believe.

Besides, the analysis of addition, space-time, and accounting is didactic about the ontological principles in existential dialectics, and good examples include the formalness-informalness principle, the absoluteness-relativeness principle, the partiality-totality principle, the predictability-unpredictability principle, the explicability-inexplicability principle, the fiction-reality principle, the cognitiveness-noncognitiveness principle, the finiteness-transfiniteness principle, the preciseness-vagueness principle, the simpleness-complicatedness principle, the openness-hiddenness principle, the denseness-emptiness principle, the rule-exception principle, the prototypicality-variation principle, the change-constancy principle, the order-chaos principle, the slowness-quickness principle, the expansion-contraction principle, the optimality-nonoptimality principle, the simultaneity-nonsimultaneity principle, the isolation-interaction principle, the theory-praxis principle, the convention-novelty principle, the evolution-transformation principle, the symmetry-asymmetry principle, the softness-hardness principle, the seriousness-playfulness principle, the activeness-inactivess principle, the selfness-otherness principle, the regression-progression principle, the same-difference principle, the stability-reaction principle, the functionality-nonfunctionality principle, the intentionality-nonintentionality principle. and the survivability-non-survivability principle.

For instance, in relation to the formalness-informalness principle, if there is formalness (e.g., the formal logical requirement of soundness and completeness in a system of ideas, as in the argument on space-time in accounting), there is informalness (e.g., the non-formal existence of unsoundness and incompleteness in the argument on space-time in accounting, as shown in the criticisms that Takatera and Sawabe used a writing sytle which is too obstruse and flowery to be empirically useful for clear analysis, that the claim by Takatera and Sawabe about the social construction of space-time in modern accrual accounting conventions is exaggeratedly misleading, that the claims in the article are often argumentatively relative to the theories that Takatera and Sawabe personally preferred, that Takatera and Sawabe confused "chaos theory" with "complexity theory" and this thus raises the thorny question about their scholarly qualification to use the two fields for the analysis of accounting, and that Takatera and Sawabe are polemically driven by the discourse on "chaos" to analyze accounting in their idiosyncratic way with the subsequent bias). And the reverse direction also holds true.

In relation to the absoluteness-relativeness principle, if there is absoluteness (e.g., the absolute view on "the status of an organization as an independent economic entity" by the advocates of the "accrual accounting system"), there is relativeness (e.g., what is right for the advocates in regard to "the status of an organization as an independent economic entity" is not necessarily so for the critics in "transaction cost economics," since "the status of an organization as an independent economic entity is an artificially created image that is contestable if we take different views from those of transaction cost economics"). And the reverse direction also holds true.

In relation to the partiality-totality principle, if there is partiality (e.g., the partial view on the evolutionary process of modern accounting conventions in the last 5 centuries by Takatera and Sawabe in terms of two periods, namely, "the pre-going concern" and "the going concern"), there is totality (e.g., the more holistic view on the evolutionary process of modern accounting conventions in the last 5 centuries, such that the whole is not the sum of its parts, that is, it cannot be reduced to the sum of opposing views by Takatera and Sawabe, and their critics, since there will emerge new views in the future not yet known today, just as there are already alternative views nowadays which do not side exclusively with any of them, as is the analysis here). And the reverse direction also holds true.

In relation to the predictability-unpredictability principle, if there is predictability (e.g., the predictable tendency of the supporters of Takatera and Sawabe to argue about the evolutionary process of modern accounting conventions in the last 5 centuries in terms of two periods, namely, "the pre-going concern" and "the going concern"), there is unpredictability (e.g., the more difficult task to predict exactly the extent to which the argument by Takatera and Sawabe about the evolutionary process of modern accounting conventions in the last 5 centuries in terms of two periods will be well accepted in a particular future era). And the reverse direction also holds true.

In relation to the explicability-inexplicability principle, if there is explicability (e.g., the explanation, by those advocates of the "accrual accounting system," of "the status of an organization as an independent economic entity" on the basis of its assumptions on space aforecited), there is inexplicability (e.g., the lack of sufficient explanation, by those advocates of the "accrual accounting system," of why "the status of an organization as an independent economic entity" on the basis of its assumptions on space is necessarily plausible, since, to the critics, "the status of an organization as an independent economic entity is an artificially created image that is contestable if we take different views from those of transaction cost economics"). And the reverse direction also holds true.

In relation to the fiction-reality principle, if there is fiction (e.g., the fictional aspect of "money" in accounting, insofar as it is treated from the perspective of the "constant value of money," which is a "fiction," according to Takatera and Sawabe), there is reality (e.g., the realistic aspect of "money" in accounting,, insofar as it can be understood from the perspective of a more historical accounting, e.g., "inflating or deflating by general price index"). And the reverse direction also holds true.

In relation to the cognitiveness-noncognitiveness principle, if there is cognitiveness (e.g., a view of things on the basis of reason and evidences, as shown in the finding about the "accrual accounting system" in modern times), there is noncognitiveness (e.g., a view of things on the basis of non-cognitive factors like envy, jealousy, power, nationality, race, gender, age, class, greed, lust, status, faith, anger, sadness, joy, fear, wish, etc.—as shown in the power of "institutional norms," in that "periodic income is calculated as a basis for income distribution because periodic income is accepted as periodic income by one person because he or she knows that periodic income is accepted as periodic income by others," according to Takatera and Sawabe). (S. Takatera 2000: 791) And the reverse direction also holds true.

In relation to the finiteness-transfiniteness principle, if there is finiteness (e.g., the finite number of the categories of entities in an accounting system, like "receipts and expenditures"), there is transfiniteness (e.g., the transfinite number of all the individual instances of "receipts and expenditures" that have ever existed in history, regardless of whether they followed the accounting conventions in the older days or nowadays). And the reverse direction also holds true.

In relation to the preciseness-vagueness principle, if there is preciseness (e.g., the precise identification of 2 assumptions on time and space in the article aforecited), there is vagueness (e.g., the vagueness in the identification of the 2 assumptions on time and space in the article aforecited, since it is not clear why there must be only 2, not 1, 3, 4, 5, 6, and so on). And the reverse direction also holds true.

In relation to the simpleness-complicatedness principle, if there is simpleness (e.g., the relatively simple analysis of space-time in accounting by Takatera and Sawabe), there is complicatedness (e.g., the relatively more complicated analysis of space-time in accounting, by challenging the claims and assumptions, as shown in the problems pointed out by the critics as cited). And the reverse direction also holds true.

In relation to the openness-hiddenness principle, if there is openness (e.g., the open exploration of using "transaction cost economics" to question the "accrual accounting system" in regard to "the status of an organization as an independent economic entity"), there is hiddenness (e.g., the hidden bias in the use of "transaction cost economics" to question the "accrual accounting system" in regard to "the status of an organization as an independent economic entity," since Takatera and Sawabe did not question the institutional approach itself)). And the reverse direction also holds true.

In relation to the denseness-emptiness principle, if there is denseness (e.g., the relatively denser concentration of concerns with "accounting income" to be "determined completely by the market" in the "pre-going concern" period), there is emptiness (e.g., the relatively less dense, or more empty, concentration of concerns with "accounting income" to be "determined completely by the market" in the "going concern" period). And the reverse direction also holds true.

In relation to the rule-exception principle, if there are rules (e.g., the usual generalization, procedure, habit, or tendency that accounting conventions can be understood in terms of two main periods, namely, "pre-going concern" and "going-concern"), there are exceptions (e.g., a case to which a rule does not apply, such as the exceptional case that the two periods of accounting conventins do not apply to historical periods before the previous 5 centuries, since Takatera and Sawabe did not address them). And the reverse direction also holds true.

In relation to the prototypicality-variation principle, if there is prototypicality (e.g., the prototype, like the example about the original works by D. R. Scott and Thorstein Veblen on institutional economics), there is variation (e.g., different variants of the protoptye, as shown in many studies since the original works by D. R. Scott and Thorstein Veblen on institutional economics—like the later ones by Takatera and Sawabe, though with their own variation, contribution, and disagreement). And the reverse direction also holds true.

In relation to the change-constancy principle, if there is change (e.g., the ever new ways to study accounting, as shown in the work on space-time in accounting), there is constancy (e.g., the ever constant existence of different problems with the ever new ways to study accounting, as shown in the problems with te work on space-time in accounting pointed out by the critics). And the reverse direction also holds true.

In relation to the order-chaos principle, if there is order (e.g., the more or less orderly view on accounting, if looked from the sole vantage point of "accounting order out of chaos" by Takatera and Sawabe), there is chaos (e.g., the more or less chaotic view on accounting, if looked from the vantage point of "noise-induced" chaos, when "the micro level noise is sometimes amplified in the system to the extent that it qualitatively alters the behavior of the whole system when chaos apparently dominates at elementary level interactions," according to Takatera and Sawabe). And the reverse direction also holds true.

In relation to the slowness-quickness principle, if there is slowness (e.g., the relatively slower speed of "the value of money" to change throughout accounting periods if "the purchasing power of money" is assumed to be "constant" or more or less so), there is quickness (e.g., the relatively quicker speed of "the value of money" to change throughout accounting periods if "the purchasing power of money" is not assumed to be constant and can change rapidly, in response to the "volatile market prices" in "exernal transactions"). (S. Takatera 2000: 791-792) And the reverse direction also holds true.

In relation to the expansion-contraction principle, if there is expansion (e.g., the relative spread of the "going-concern" in the accrual accounting system), there is contraction (e.g., the relative decline of the "pre-going concern" in the accrual accounting system). And the reverse direction also holds true.

In relation to the optimality-nonoptimality principle, if there is optimality (e.g., the search for highly favorable options, like the quest for "certainty" in the accrual accounting system, when it "systematically exclude[s] uncertainties from the accrual process" and "the accrual accounting system deals with uncertain phenomena in a deterministic manner," in light of its assumptions on space and time aforediscussed), there is nonoptimality (e.g., the existence of non-optimal alternatives to optimality, like the argument for more realistic studies by the critics like Takatera and Sawabe, who pointed out the problems with the "fictional" assumptions" on space and time in the accrual accounting system). (S. Takatera 2000: 795) And the reverse direction also holds true.

In relation to the simultaneity-nonsimultaneity principle, if there is simultaneity (e.g., the occurrence of different things more or less at the same time, such as the simultaneous co-existence of two sub-systems of the accrual accounting system, that is, "external transaction" and "internal transaction"), there is nonsimultaneity (e.g., the occurrence of something after something else, such as the "pre-going concern" period before the "going concern" period). And the reverse direction also holds true.

In relation to the isolation-interaction principle, if there is isolation (e.g., the state of a situation that is separate from others, such as the isolation or independence of the "demarcated space" in the "going concern" period, away from "external transactions"), there is interaction (e.g., the influence of entities on one another, such as the close connection between the function and scope of accounts and external transactions in the "pre-going concern" period). And the reverse direction also holds true.

In relation to the theory-praxis principle, if there is theory (e.g., the theoretical construction of "chaos theory" in mathematics), there is praxis (e.g., the practical application of chaos theory" in mathematics to the field of accounting, as shown in the work by Takatera and Sawabe). And the reverse direction also holds true.

In relation to the convention-novelty principle, if there is convention (e.g., the conventional wisdom about "the status of an organization as an independent economic entity" in the "accrual income accounting"), there is novelty (e.g., the alternative novel challenge to the conventional wisdom about "the status of an organization as an independent economic entity" in the "accrual income accounting"—by the new idea that the independent status is fictional on the basis of its assumptions on space and time, as pointed out by Takatera and Sawabe). And the reverse direction also holds true.

In relation to the evolution-transformation principle, if there is evolution (e.g., the natural evolution of humans to experience space and time in the state of nature), there is transformation (e.g., the technical transformation of human ability to experience space and time by the invention of different assumptions on space and time in the modern accrual accounting system). And the reverse direction also holds true.

In relation to the symmetry-asymmetry principle, if there is symmetry (e.g., the co-existence of "certain" and "uncertain" phenomena in the world), there is asymmetry (e.g., the accrual accounting system favors "certainty" and "systematically exclude[s] uncertainties from the accrual process," so "the accrual accounting system deals with uncertain phenomena in a deterministic manner," in a misleadingly "fictional" way, according to Takatera and Sawabe). And the reverse direction also holds true.

In relation to the softness-hardness principle, if there is softness (e.g., the defense of "the status of an organization as an independent economic entity" by those for the "accrual accounting system"), there is hardness (e.g., the critique of "the status of an organization as an independent economic entity" by those in "transaction cost economics"). And the reverse direction also holds true.

In relation to the seriousness-playfulness principle, if there is seriousness (e.g., the serious business of those who try to analyze the new accounting system known as the "accrual accounting system"), there is playfulness (e.g., the playful part of those who who try to analyze the new accounting system known as the "accrual accounting system," when they play around with different assumptions over the last few centuries, like the 2 assumptions on time and the 2 assumptions on space aforediscussed). And the reverse direction also holds true.

In relation to the activeness-inactivess principle, if there is activeness (e.g., the relative activeness of those who use the accrual accounting system nowadays), there is inactiveness (e.g., the relative inactiveness of those who still want to do accounting in the old way, when "the scope and function of accounts were limited…to recording decisions rendered by the market" during the "pre-going concern" period, because these accountants are not in much demand anymore). And the reverse direction also holds true.

In relation to the selfness-otherness principle, if there is selfness (e.g., the tendency to be mostly centered on one's self or group, as shown by those individuals in the older days who acted solely as the "results" of their "subjective judgments" and interests in the market, prior to the rise of modern accounting conventions), there is otherness (e.g., the tendency to be mostly centered on others, as shown in the academic analysis of both sides of the debate on the evolution of accounting conventions in history, as an advancement in the knowledge of space-time in income accounting for humanity). (S. Takatera 2000: 790) And the reverse direction also holds true.

In relation to the regression-progression principle, if there is regression (e.g., the regression made by the work on space-time in income accounting, as shown in the problems pointed out by the critics), there is progression (e.g., the progress made by the work on space-time in income accounting, as shown in the contribution to the understanding of the evolution of accounting conventions aforecited). And the reverse direction also holds true.

In relation to the same-difference principle, if there is similarity in outcome (e.g., the contribution to the molding and control of beliefs, values, and behaviors, regardless of whether this be done by way of using the accrual accounting system, or by way of using chaos theory), there is difference in outcome (e.g., the contribution to the molding and control of beliefs, values, and behaviors by way of using the accrual accounting system for a relatively more deterministic worldview—but the contribution to the molding and control of beliefs, values, and behaviors by way of using chaos theory for a relatively more uncertain worldview). And the reverse direction also holds true.

In relation to the stability-reaction principle, if there is stability (e.g., the relatively more stable condition of the "pre-going concern" in the older days), there is reaction (e.g., the "pre-going concern" in the older days then led to further changes later on, when it was superseded by the "going concern" in modern times). And the reverse direction also holds true.

In relation to the functionality-nonfunctionality principle, if there is functionality (e.g., the presence of action for which something or someone is fitted, as shown in the creation of the "accrual accounting system," which serves the function of providing an "internal calculative logic" of accounting which is "structurally separated from the external logic of the market"), there is nonfunctionality (e.g., the relative lack of action for which something or someone is fitted, as shown in the problems with the "accrual accounting system" in regard to its fictional assumptions, such that it is not as functional as the advocates would like us to believe, to be further accounted for by an alternative model known as "accounting order out of chaos" by Takatera and Sawabe). And the reverse direction also holds true.

In relation to the intentionality-nonintentionality principle, if there is intentionality (e.g., the planning or design of something for a certain outcome, as shown in the original intention of those in the older days who acted solely as the "results" of their "subjective judgments" and interests, prior to the rise of modern accounting conventions), there is nonintentionality (e.g., the relative absence of planning or design of something for a certain outcome, such that the outcome can even be counter-intentional in being the opposite of what was originally planned, as shown in the rise of modern accounnting conventions, which involved "individual actions that are often interpreted as results of the subjective judgments of individual actors" but "these norms are achieved without any explicit human intentions or interventions or design," with the "paradox of individual actions resulting unintentionally in the formation of social institutions," according to Takatera and Sawabe—contrary to the original intention of those in the older days who acted solely as the "results" of their "subjective judgments" and interests). And the reverse direction also holds true.

And in relation to the survivability-nonsurvivability principle, if there is survivability (e.g., the relative survival, nowadays, of the "going concern"), there is non-survivability (e.g., the relative non-survival, nowadays, of the "pre-going concern"). And the reverse direction also holds true.

Accounting and Society

Addition, when examined from the perspective of society, can reveal to us some powerful features in relation to its possibility (or impossibility) and desirability (or undesirability), and this can be shown by way of two case studies, namely, (2.4.1) addition, financial performance, and auditing, and (2.4.2) addition, government revenue, and the accounting of VAT—to be addressed hereafter.

Addition, Financial Performance, and Auditing

A good way to evaluate the possibility (or impossibility) and desirability (or undesirability) of addition from the perspective of society is to explore addition, financial performance, and auditing.

An argument in the literature is that an "audit" is important for a business to increase financial performance.

Auditing in Accounting

Historically, "the word 'auditing' has been derived from Latin word 'audire' which means 'to hear.' According to Dicksee, traditionally auditing can be understood as an examination of accounting records undertaken with a view to establishing whether they completely reflect the transactions correctly for the related purpose." (WK 2013t)

With this definition in mind, "auditing is a vital part of accounting. Traditionally, audits were mainly associated with gaining information about financial systems and the financial records of a company or a business. Financial audits are performed to ascertain the validity and reliability of information, as well as to provide an assessment of a system's internal control." (WK 2013t)

There are different types of auditing. For example, "in the case of financial audits, a set of financial statements are said to be true and fair when they are free of material misstatements—a concept influenced by both quantitative (numerical) and qualitative factors." (WK 2013t) And in the case of "cost accounting," there is the "process for verifying the cost of manufacturing or producing of any article, on the basis of accounts measuring the use of material, labor or other items of cost. In simple words, the term *cost audit* means a systematic and accurate verification of the cost accounts and records, and checking for adherence to the cost accounting objectives. According to the Institute of Cost and Management Accountants of Pakistan, a cost audit is 'an examination of cost accounting records and verification of facts to ascertain that the cost of the product has been arrived at, in accordance with principles of cost accounting.'" (WK 2013t)

In any event, "an audit must adhere to generally accepted standards established by governing bodies. These standards assure third parties or external users that they can rely upon the auditor's opinion on the fairness of financial statements, or other subjects on which the auditor expresses an opinion." (WK 2013t)

Three Aspects of Auditing

Auditing, in general, can be analyzed in terms of three major aspects, in what follows.

(a) Integrated Audits

The first aspect of auditing has to do with "integrated audits."

For instance, in the U.S., "audits of publicly traded companies are governed by rules laid down by the Public Company Accounting Oversight Board (PCAOB), which was established by Section 404 of the

Sarbanes-Oxley Act of 2002. Such an audit is called an integrated audit, where auditors, in addition to an opinion on the financial statements, must also express an opinion on the effectiveness of a company's internal control over financial reporting, in accordance with PCAOB Auditing Standard No. 5." (WK 2013t)

In addition, "there are…new types of integrated auditing becoming available that use unified compliance material….Due to the increasing number of regulations and need for operational transparency, organizations are adopting risk-based audits that can cover multiple regulations and standards from a single audit event. This is a very new but necessary approach in some sectors to ensure that all the necessary governance requirements can be met without duplicating effort from both audit and audit hosting resources." (WK 2013t)

(b) Assessments

The second aspect of auditing has to do with "assessments."

After all, "the purpose of an assessment is to measure something or calculate a value for it. Although the process of producing an assessment may involve an audit by an independent professional, its purpose is to provide a measurement rather than to express an opinion about the fairness of statements or quality of performance." (WK 2013t)

In general, "audits should always be an independent evaluation that will include some degree of quantitative and qualitative analysis whereas an assessment implies a less independent and more consultative approach. The outcome of the assessment should relate to the norms that were set for the task, product or event." (WK 2013t)

(c) Auditors

And the third aspect of auditing has to do with "auditors."

For illustration, consider the following five types of "auditors": (WK 2013t)

- "External Auditor/Statutory Auditor"—"is an independent firm engaged by the client subject to the audit, to express an opinion on whether the company's financial statements are free of material misstatements, whether due to fraud or error. For publicly traded companies, external auditors may also be required to express an opinion over the effectiveness of internal controls over financial reporting. External auditors may also be engaged to perform other agreed-upon procedures, related or unrelated to financial statements. Most importantly, external auditors, though engaged and paid by the company being audited, are regarded as independent auditors."

- "Cost Auditor/Statutory Cost Auditor"—"is an independent firm engaged by the client subject to the Cost audit, to express an opinion on whether the company's Cost statements and Cost Sheet are free of material misstatements, whether due to fraud or error. For publicly traded companies, external auditors may also be required to express an opinion over the effectiveness of internal controls over Cost reporting. These are Specialized Person called Cost Accountants in India & CMA globally either Cost & management Accountant or Certified management Accountants."
- "Internal Auditors"—"are employed by the organizations they audit. They work for government agencies (federal, state and local); for publicly traded companies; and for non-profit companies across all industries. The internationally recognised standard setting body for the profession is the Institute of Internal Auditors–IIA (www.theiia.org)....While internal auditors are not independent of the companies that employ them, independence and objectivity are a cornerstone of the IIA professional standards; and are discussed at length in the standards and the supporting practice guides and practice advisories. Professional internal auditors are mandated by the IIA standards to be independent of the business activities they audit. This independence and objectivity are achieved through the organizational placement and reporting lines of the internal audit department. Internal auditors of publicly traded companies in the United States are required to report functionally to the board of directors directly, or a sub-committee of the board of directors (typically the audit committee), and not to management except for administrative purposes."
- "Consultant Auditors"—"are external personnel contracted by the firm to perform an audit following the firm's auditing standards. This differs from the external auditor, who follows their own auditing standards. The level of independence is therefore somewhere between the internal auditor and the external auditor. The consultant auditor may work independently, or as part of the audit team that includes internal auditors. Consultant auditors are used when the firm lacks sufficient expertise to audit certain areas, or simply for staff augmentation when staff are not available."
- "Quality Auditors"—"may be consultants or employed by the organization."

Of course, these examples are illustrative, not exhaustive.

Case Study: Internal Audit

A more elaboration of auditing can best be provided with a case study, say, on "internal audit."

(a) Brief History

Historically, "the Internal Auditing profession evolved steadily with the progress of management science after World War II. It is conceptually similar in many ways to financial auditing by public accounting firms, quality assurance and banking compliance activities. While some of the audit technique underlying internal auditing is derived from management consulting and public accounting professions, the theory of internal auditing was conceived primarily by Lawrence Sawyer (1911-2002), often referred to as 'the father of modern internal auditing'; and the current philosophy, theory and practice of modern internal auditing as defined by the International Professional Practices Framework (IPPF) of the Institute of Internal Auditors owes much to Sawyer's vision." (WK 2013ff; S. Ramamoorti 2003)

Later, in 2002, "with the implementation in the United States of the Sarbanes-Oxley Act of 2002, the profession's exposure and value was enhanced, as many internal auditors possessed the skills required to help companies meet the requirements of the law." (WK 2013ff)

(b) Organizational Independence

Internal auditors are required to hold some kind of "independence" and "objectivity," although "internal auditors are not independent of the companies that employ them," since "independence and objectivity are a cornerstone of the IIA professional standards; and are discussed at length in the standards and the supporting practice guides and practice advisories. Professional internal auditors are mandated by the IIA standards to be independent of the business activities they audit. This independence and objectivity are achieved through the organizational placement and reporting lines of the internal audit department. Internal auditors of publicly traded companies in the United States are required to report functionally to the board of directors directly, or a sub-committee of the board of directors (typically the audit committee), and not to management except for administrative purposes." (WK 2013ff)

The reason is that "the required organizational independence from management enables unrestricted evaluation of management activities and personnel and allows internal auditors to perform their role effectively. Although internal auditors are part of company management and paid by the company, the primary customer of internal audit activity is the entity charged with oversight of management's activities. This is typically the

Audit Committee, a sub-committee of the Board of Directors. To provide independence, most Chief Audit Executives report to the Chairperson of the Audit Committee and can only be replaced with the concurrence of that individual." (WK 2013ff)

(c) Roles

In general, internal auditors have 3 major roles, as explained below.

(c1) Role in Internal Control

The first role of "internal auditors" concerns "internal control."

In regard to "internal control"—"under the COSO Framework, internal control is broadly defined as a process, effected by an entity's board of directors, management, and other personnel, designed to provide reasonable assurance regarding the achievement of the following core objectives for which all businesses strive": (WK 2013ff)

- "effectiveness and efficiency of operations"
- "reliability of financial and management reporting"
- "compliance with laws and regulations"
- "safeguarding of assets"

And "management is responsible for internal control, which comprises five critical components: the control environment; risk assessment; risk focused control activities; information and communication; and monitoring activities. Managers establish policies, processes, and practices in these five components of management control to help the organization achieve the four specific objectives listed above. Internal auditors perform audits to evaluate whether the five components of management control are present and operating effectively, and if not, provide recommendations for improvement." (WK 2013ff)

(c2) Role in Risk Management

The second role of "internal auditors" concerns "risk management."

In regard to "risk management"—it refers to "the process by which an organization identifies, analyzes, responds, gathers information about, and monitors strategic risks that could actually or potentially impact the organization's ability to achieve its mission and objectives." (WK 2013ff)

For instance, "under the COSO enterprise risk management (ERM) Framework, an organization's strategy, operations, reporting, and compliance objectives all have associated strategic business risks—the negative outcomes resulting from internal and external events that inhibit the organization's ability to achieve its objectives. Management assesses risk as part of the ordinary course of business activities such as strategic planning, marketing planning, capital planning, budgeting, hedging, incentive payout structure, credit/lending practices, mergers and acquisitions, strategic partnerships, legislative changes, conducting business abroad, etc. Sarbanes-Oxley regulations require extensive risk assessment of financial reporting processes. Corporate legal counsel often prepares comprehensive assessments of the current and potential litigation a company faces. Internal auditors may evaluate each of these activities, or focus on the overarching process used to manage risks entity-wide. For example, internal auditors can advise management regarding the reporting of forward-looking operating measures to the Board, to help identify emerging risks; or internal auditors can evaluate and report on whether the board and other stakeholders can have reasonable assurance the organization's management team has implemented an effective enterprise risk management program." (WK 2013ff)

(c3) Role in Corporate Governance

And the third role of "internal auditors" concerns "corporate governance."

In regard to "corporate governance"—it refers to "the policies, processes and structures used by the organization's leadership to direct activities, achieve objectives, and protect the interests of diverse stakeholder groups in a manner consistent with ethical standards. The internal auditor is often considered one of the 'four pillars' of corporate governance, the other pillars being the Board of Directors, management, and the external auditor." (WK 2013ff)

And "a primary focus area of internal auditing as it relates to corporate governance is helping the Audit Committee of the Board of Directors (or equivalent) perform its responsibilities effectively. This may include reporting critical management control issues, suggesting questions or topics for the Audit Committee's meeting agendas, and coordinating with the external auditor and management to ensure the Committee receives effective information. In recent years, the IIA has advocated more formal evaluation of Corporate governance, particularly in the areas of board oversight of enterprise risk, corporate ethics, and fraud." (WK 2013ff)

Problems with the Argument on Financial Performance and Auditing

However, to the critics, this argument on addition, financial performance, and auditing has some major problems. Consider, for illustration, a few examples below.

Firstly, one main criticism of the argument on addition, financial performance, and auditing is that an audit cannot guarantee a financial statement to be "free from material error."

For instance, as the critics are quick to point out, "due to constraints, an audit seeks to provide only reasonable assurance that the statements are free from material error. Hence, statistical sampling is often adopted in audits." (WK 2013t)

Secondly, another main criticism of the argument on addition, financial performance, and auditing is that even the conventional standard of "reasonable assurance" is no longer sufficient for some people.

For instance, to the critics, "recently, the argument that auditing should go beyond just true and fair is gaining momentum. And the US Public Company Accounting Oversight Board has come out with a concept release on the same." (WK 2013t; F. McKenna 2011)

Thirdly, still another main criticism of the argument on addition, financial performance, and auditing is that auditing does not necessarily prevent fraudulent activities.

For instance, the case of "Peregrine Systems, Inc." is instructive, since "in 2003, the U.S. Securities and Exchange Commission charged Peregrine with 'massive financial fraud' for the purposes of inflating the company's revenue and stock price. Peregrine, without admitting or denying the allegations of the complaint, agreed to a partial settlement." (WK 2013ee)

Fourthly, an additional main criticism of the argument on addition, financial performance, and auditing is that the history of auditing is not necessarily consistent in its implementation of certain standards and goals.

For instance, "with the implementation in the United States of the Sarbanes-Oxley Act of 2002, the profession's exposure and value was enhanced, as many internal auditors possessed the skills required to help companies meet the requirements of the law. However, the focus by internal audit departments of publicly traded companies on SOX related financial policy and procedures derailed progress made by the profession in the late 20th century toward Larry Sawyer's vision for internal audit. Beginning in about 2010, the IIA once again began advocating for the broader role internal auditing should play in the corporate arena, in keeping with the IPPF's philosophy." (WK 2013ff)

And fifthly, another additional main criticism of the argument on addition, financial performance, and auditing is that the evaluation of the success or failure of audit functions itself can be difficult to measure.

For instance, "internal audit functions are primarily evaluated based on the quality of counsel and information provided to the Audit Committee and top management. However, this is primarily qualitative and therefore difficult to measure." (WK 2013ff)

Beyond Financial Performance and Auditing

These criticisms of the argument on addition, financial performance, and auditing are not to mean a total rejection of its usefulness but serve to show us the two opposing sides of the debate, such that the possibility (or impossibility) and desirability (or undesirability) of addition (from the perspective of society with addition, financial performance, and auditing as a case study here) are not to the extent that the respective defenders would like us to believe.

Furthermore, the analysis of addition, financial performance, and auditing reveals the insights of the ontological principles in existential dialectics, and good examples include the formalness-informalness principle, the absoluteness-relativeness principle, the partiality-totality principle, the predictability-unpredictability principle, the explicability-inexplicability principle, the fiction-reality principle, the cognitiveness-noncognitiveness principle, the finiteness-transfiniteness principle, the preciseness-vagueness principle, the simpleness-complicatedness principle, the openness-hiddenness principle, the denseness-emptiness principle, the rule-exception principle, the prototypicality-variation principle, the change-constancy principle, the order-chaos principle, the slowness-quickness principle, the expansion-contraction principle, the optimality-nonoptimality principle, the simultaneity-nonsimultaneity principle, the isolation-interaction principle, the theory-praxis principle, the convention-novelty principle, the evolution-transformation principle, the symmetry-asymmetry principle, the softness-hardness principle, the seriousness-playfulness principle, the activeness-inactivess principle, the selfness-otherness principle, the regression-progression principle, the same-difference principle, the stability-reaction principle, the functionality-nonfunctionality principle, the intentionality-nonintentionality principle, and the survivability-nonsurvivability principle.

For instance, in relation to the formalness-informalness principle, if there is formalness (e.g., the formal logical requirement of soundness, completeness, and consistency in a system of ideas, as in the argument on auditing), there is informalness (e.g., the non-formal existence of unsoundness, incompleteness, and inconsistency in the argument on auditing, as shown in the criticisms that an audit cannot guarantee a financial statement to be "free from material error," that even the conventional standard of "reasonable assurance" is no longer sufficient for some people, that auditing does not necessarily prevent fraudulent activities, that the history of auditing is not necessarily consistent in its implementation of certain standards and goals, and that the evaluation of the success or failure of audit functions itself can be difficult to measure). And the reverse direction also holds true.

In relation to the absoluteness-relativeness principle, if there is absoluteness (e.g., the absolute view on "reasonable assurance" by the advocates), there is relativeness (e.g., what is right for the advocates in regard to "reasonable assurance" is not necessarily so for the critics, since "recently, the argument that auditing should go beyond just true and fair is gaining momentum" and "the US Public Company Accounting Oversight Board has come out with a concept release on the same"). And the reverse direction also holds true.

In relation to the partiality-totality principle, if there is partiality (e.g., the partial view on "reasonable assurance" by the advocates), there is totality (e.g., the more holistic view on "reasonable assurance," such that the whole is not the sum of its parts, that is, it cannot be reduced to the sum of opposing views by its advocates, its critics like the US Public Company Accounting Oversight Board, etc., since there will emerge new views in the future not yet known today, just as there are already alternative views nowadays which do not side exclusively with any of them, as is the analysis here). And the reverse direction also holds true.

In relation to the predictability-unpredictability principle, if there is predictability (e.g., the predictable tendency that, in the U.S., "an audit must adhere to generally accepted standards established by governing bodies" like "the US Public Company Accounting Oversight Board," etc.), there is unpredictability (e.g., the more difficult task to predict exactly the extent to which an audit of a particular firm can prevent fraudulent activities on a particular future occassion). And the reverse direction also holds true.

In relation to the explicability-inexplicability principle, if there is explicability (e.g., the explanation, by those advocates of the Sarbanes-Oxley Act of 2002, of its desirability on the basis that "the profession's exposure and value was enhanced, as many internal auditors possessed the skills required to help companies meet the requirements of the law"), there is inexplicability (e.g., the lack of sufficient explanation, by those advocates of the Sarbanes-Oxley Act of 2002 of why it is necessarily desirable, since, to the critics, "the focus by internal audit departments of publicly traded companies on SOX related financial policy and procedures derailed progress made by the profession in the late 20th century toward Larry Sawyer's vision for internal audit" in a broader sense). And the reverse direction also holds true.

In relation to the fiction-reality principle, if there is fiction (e.g., the fictional aspect of "reasonable assurance," insofar as the standard is no longer deemed sufficient for some people, contrary to a conventional claim, since "recently, the argument that auditing should go beyond just true and fair is gaining momentum" and "the US Public Company Accounting Oversight Board has come out with a concept release on the same"), there is reality (e.g., the realistic aspect of "reasonable assurance," insofar as, "due to constraints, an audit seeks to provide only reasonable assurance that the statements are free from material error" and "hence, statistical sampling is often adopted in audits"). And the reverse direction also holds true.

In relation to the cognitiveness-noncognitiveness principle, if there is cognitiveness (e.g., a view of things on the basis of reason and evidences, as shown in the finding that "an audit seeks to provide only reasonable assurance that the statements are free from material error"), there is noncognitiveness (e.g., a view of things on the basis of non-cognitive factors like envy, jealousy, power, nationality, race, gender, age, class, greed, lust, status, faith, anger, sadness, joy, fear, wish, etc.—as shown in the "political correctness" of the requirement of "reasonable assurance," in spite of the fact that the standard is no longer deemed sufficient for some people, since "recently, the argument that auditing should go beyond just true and fair is gaining momentum"). And the reverse direction also holds true.

In relation to the finiteness-transfiniteness principle, if there is finiteness (e.g., the finite number of the types of auditors, like the 5 examples aforedescribed), there is transfiniteness (e.g., the transfinite number of all the individual items that have ever been examiend by all auditors in human history, regardless of which types they belong to). And the reverse direction also holds true.

In relation to the preciseness-vagueness principle, if there is preciseness (e.g., the precise identification of 5 types of auditors in the article aforecited), there is vagueness (e.g., the vagueness in the identification of the 5 types of auditors in the article aforecited, since it is not clear why there must be only 5, not 6, 7, 8, 9, and so on). And the reverse direction also holds true.

In relation to the simpleness-complicatedness principle, if there is simpleness (e.g., the relatively simple analysis of auditing by the advocates), there is complicatedness (e.g., the relatively more complicated analysis of auditing, by challenging the claims and assumptions, as shown in the problems pointed out by the critics as cited). And the reverse direction also holds true.

In relation to the openness-hiddenness principle, if there is openness (e.g., the open "implementation in the United States of the Sarbanes-Oxley

Act of 2002" so as to enhance "the profession's exposure and value" in regard to auditing), there is hiddenness (e.g., the hidden bias in the implementation of the Sarbanes-Oxley Act of 2002, since, to the critics, "the focus by internal audit departments of publicly traded companies on SOX related financial policy and procedures derailed progress made by the profession in the late 20th century toward Larry Sawyer's vision for internal audit"). And the reverse direction also holds true.

In relation to the denseness-emptiness principle, if there is denseness (e.g., the relatively denser concentration of concerns with "reasonable assurance" by its advocates), there is emptiness (e.g., the relatively less dense, or more empty, concentration of concerns with "reasonable assurance" by the critics like the US Public Company Accounting Oversight Board, who favor "the argument that auditing should go beyond just true and fair is gaining momentum"). And the reverse direction also holds true.

In relation to the rule-exception principle, if there are rules (e.g., the usual generalization, procedure, habit, or tendency that "as a general rule, audits should always be an independent evaluation that will include some degree of quantitative and qualitative analysis"), there are exceptions (e.g., a case to which a rule does not apply, such as the exceptional case about "an assessment," which implies "a less independent and more consultative approach," since "the outcome of the assessment should relate to the norms that were set for the task, product or event"). (WK 2013t) And the reverse direction also holds true.

In relation to the prototypicality-variation principle, if there is prototypicality (e.g., the prototype, like the example about the original theory of internal auditing by Lawrence Sawyer), there is variation (e.g., different variants of the protoptye, as shown in many studies since the original theory of internal auditing by Lawrence Sawyer —like the later ones such as "the Sarbanes-Oxley Act of 2002," the revision by the IIA in 2010, etc., though each with its own variation, contribution, and disagreement). And the reverse direction also holds true.

In relation to the change-constancy principle, if there is change (e.g., the ever new ways to study accounting, as shown in the work on "internal auditing"), there is constancy (e.g., the ever constant existence of different problems with the ever new ways to study accounting, as shown in the problems with the work on "internal auditing" pointed out by the critics). And the reverse direction also holds true.

In relation to the order-chaos principle, if there is order (e.g., the more or less orderly view on the history of internal auditing, if looked from the sole vantage point of the implementation of "the Sarbanes-Oxley Act of

2002" at the time), there is chaos (e.g., the more or less chaotic view on the history of internal auditing, if looked from the multiple, conflicting vantage points as revealed in the inconsistent history of the implementation of internal auditing aforedescribed). And the reverse direction also holds true.

In relation to the slowness-quickness principle, if there is slowness (e.g., the relatively slower speed of internal auditing within an organization to complete, if it is very broad in scope, not only covering "an organization's governance, risk management and management controls" but also "conducting proactive fraud audits to identify potentially fraudulent acts; participating in fraud investigations under the direction of fraud investigation professionals, and conducting post investigation fraud audits to identify control breakdowns and establish financial loss"), there is quickness (e.g., the relatively quicker speed of internal auditing within an organization to complete, if it is narrow in scope and covers only "an organization's governance, risk management and management controls" but nothing else). (WK 2013ff) And the reverse direction also holds true.

In relation to the expansion-contraction principle, if there is expansion (e.g., the relative increase of independence if done by an "external auditor"), there is contraction (e.g., the relative decrease of independence if done by "internal auditor"). And the reverse direction also holds true.

In relation to the optimality-nonoptimality principle, if there is optimality (e.g., the search for highly favorable options, like the quest for the comprehensive "integrated audit," where "auditors, in addition to an opinion on the financial statements, must also express an opinion on the effectiveness of a company's internal control over financial reporting, in accordance with PCAOB Auditing Standard No. 5"), there is nonoptimality (e.g., the existence of non-optimal alternatives to optimality, like the argument for more realistic studies by the critics, who pointed out different problems with auditing aforecited). And the reverse direction also holds true.

In relation to the simultaneity-nonsimultaneity principle, if there is simultaneity (e.g., the occurrence of different things more or less at the same time, such as the simultaneous co-existence of different audits nowadays, like "financial audits," "information systems audits," etc.), there is nonsimultaneity (e.g., the occurrence of something after something else, such as the historical development of "financial audits" before that of "information systems audits" in our time). (WK 2013t) And the reverse direction also holds true.

In relation to the isolation-interaction principle, if there is isolation (e.g., the state of a situation that is separate from others, such as the formal separation of the role of "internal auditing" from "corporate governance" in the older days, when "internal auditing activity as it relates to corporate governance has in the past been generally informal, accomplished"), there is interaction (e.g., the influence of entities on one another, such as the more "formal" interaction between "internal auditing" and "corporate governance" nowadays, since "the IIA has advocated more formal evaluation of Corporate governance, particularly in the areas of board oversight of enterprise risk, corporate ethics, and fraud"). And the reverse direction also holds true.

In relation to the theory-praxis principle, if there is theory (e.g., the theoretical construction of "internal auditing" by Lawrence Sawyer), there is praxis (e.g., the practical application of "internal auditing" by "the IIA," which, in 2010, "began advocating for the broader role internal auditing should play in the corporate arena, in keeping with the IPPF's philosophy" envisioned by Sawyer). And the reverse direction also holds true.

In relation to the convention-novelty principle, if there is convention (e.g., the conventional wisdom that, "due to constraints, an audit seeks to provide only reasonable assurance that the statements are free from material error"), there is novelty (e.g., the alternative novel challenge to the conventional wisdom that, "due to constraints, an audit seeks to provide only reasonable assurance that the statements are free from material error"—by the new idea that the standard is no longer deemed sufficient for some people, since "recently, the argument that auditing should go beyond just true and fair is gaining momentum"). And the reverse direction also holds true.

In relation to the evolution-transformation principle, if there is evolution (e.g., the natural evolution of humans to "see" or "hear" things in the state of nature in a primitive way), there is transformation (e.g., the technical transformation of human ability to "see" or "hear" things by the invention of "internal auditing," with the word "audit" from "Latin word 'audire' which means 'to hear'"). And the reverse direction also holds true.

In relation to the symmetry-asymmetry principle, if there is symmetry (e.g., the co-existence of different auditors, like "internal auditors," "external auditors," etc.), there is asymmetry (e.g., "external auditors" have more "independence" or the power of distancing than "internal auditors"). And the reverse direction also holds true.

In relation to the softness-hardness principle, if there is softness (e.g., the defense of "internal auditing" by the advocates like Sawyer aforecited), there is hardness (e.g., the critique of "internal auditing," as shown in the problems pointed out by the critics aforecited). And the reverse direction also holds true.

In relation to the seriousness-playfulness principle, if there is seriousness (e.g., the serious business of those who try to find out the appropriate "role" of auditing in "internal control"), there is playfulness (e.g., the playful part of those who try to find out the appropriate "role" of auditing in "internal control," when they play around over the decades with possible "core objectives" like the 4 ones aforecited). And the reverse direction also holds true.

In relation to the activeness-inactivess principle, if there is activeness (e.g., the relative activeness of those auditors when it is "broad" in scope, that is, not only covering "an organization's governance, risk management and management controls" but also "conducting proactive fraud audits to identify potentially fraudulent acts; participating in fraud investigations under the direction of fraud investigation professionals, and conducting post investigation fraud audits to identify control breakdowns and establish financial loss"), there is inactiveness (e.g., the relative inactiveness of those auditors, when they are "narrow" or "minimal" in scope, like covering a small specific task like "security audit" only). (WK 2013t) And the reverse direction also holds true.

In relation to the selfness-otherness principle, if there is selfness (e.g., the tendency to be mostly centered on one's self or group, as shown by those who commit "fraudulent activities" and are discovered when audited by an auditor), there is otherness (e.g., the tendency to be mostly centered on others, as shown in the academic analysis of both sides of the debate on "internal auditing, " as an advancement in the knowledge of accounting for humanity). And the reverse direction also holds true.

In relation to the regression-progression principle, if there is regression (e.g., the regression made by "internal auditing," as shown in the problems pointed out by the critics), there is progression (e.g., the progress made by "internal auditing," as shown in the contribution to the improvement of the financial performance of businesses aforecited). And the reverse direction also holds true.

In relation to the same-difference principle, if there is similarity in outcome (e.g., the contribution to the molding and control of beliefs, values, and behaviors, regardless of whether this be done by way of appealing to "reasonable assurance," or by way of going beyond just "reasonable assurance"), there is difference in outcome (e.g., the contribution to the molding and control of beliefs, values, and behaviors by way of appealing to "reasonable assurance" for a relatively looser professional standard in the business lifeworld—but the contribution to the molding and control of beliefs, values, and behaviors by way of going beyond just "reasonable assurance" for a relatively stricter professional standard in the business lifeworld). And the reverse direction also holds true.

In relation to the stability-reaction principle, if there is stability (e.g., the relatively more stable condition of the Sarbanes-Oxley Act of 2002 at the time), there is reaction (e.g., the Sarbanes-Oxley Act of 2002 then led to further changes later on, when in 2010, there was the revised interpretation by the IIA). And the reverse direction also holds true.

In relation to the functionality-nonfunctionality principle, if there is functionality (e.g., the presence of action for which something or someone is fitted, as shown in the appeal to "reasonable assurance," which serves the function of providing a realistic standard of professional ethics under the condition of "constraints"), there is nonfunctionality (e.g., the relative lack of action for which something or someone is fitted, as shown in the problems with the appeal to "reasonable assurance," since, to the critics, the conventional standard of "reasonable assurance" is no longer sufficient for some people since "recently, the argument that auditing should go beyond just true and fair is gaining momentum," such that it is no longer as functional as the advocates would like us to believe). And the reverse direction also holds true.

In relation to the intentionality-nonintentionality principle, if there is intentionality (e.g., the planning or design of something for a certain outcome, as shown in the original intention of those advocating for "the implementation in the United States of the Sarbanes-Oxley Act [SOX] of 2002" to enhance "the profession's exposure and value"), there is nonintentionality (e.g., the relative absence of planning or design of something for a certain outcome, such that the outcome can even be counter-intentional in being the opposite of what was originally planned, as shown in the setback when "the focus by internal audit departments of publicly traded companies on SOX related financial policy and procedures derailed progress made by the profession in the late 20th century toward Larry Sawyer's vision for internal audit"—contrary to the original intention of those advocating for "the implementation in the United States of the Sarbanes-Oxley Act of 2002" to enhance "the profession's exposure and value"). And the reverse direction also holds true.

And in relation to the survivability-nonsurvivability principle, if there is survivability (e.g., the relative survival, nowadays, of "the broader role" of "internal auditing" advocated by the IIA since 2010), there is non-survivability (e.g., the relative non-survival, nowadays, of the more narrow focus of internal auditing under the Sarbanes-Oxley Act of 2002, before it was reinterpreted by the IIA in 2010). And the reverse direction also holds true.

Addition, Government Revenue, and the Accounting of VAT

Another good way to evaluate the possibility (or impossibility) and desirability (or undesirability) of addition from the perspective of society is to explore addition, government revenue, and the accounting of VAT.

An argument in the literature is that, from an accounting point of view, "valued-added tax" (VAT) is a "tax only on the value added to a product, material, or service,...by this stage of its manufacture or distribution," so as "to generate tax revenues to the government similar to the corporate income tax or the personal income tax." (WK 2013gg)

Brief History of VAT

Historically, "Maurice Lauré, Joint Director of the France Tax Authority, the Direction Générale des Impôts, was first to introduce VAT on 10 April 1954, although German industrialist Dr. Wilhelm von Siemens proposed the concept in 1918. Initially directed at large businesses, it was extended over time to include all business sectors. In France, it is the most important source of state finance, accounting for nearly 50% of state revenues." (WK 2013gg; MIN 2009)

The difference between consumers and sellers here is that "personal end-consumers of products and services cannot recover VAT on purchases, but businesses are able to recover VAT (input tax) on the products and services that they buy in order to produce further goods or services that will be sold to yet another business in the supply chain or directly to a final consumer. In this way, the total tax levied at each stage in the economic chain of supply is a constant fraction of the value added by a business to its products, and most of the cost of collecting the tax is borne by business, rather than by the state." (WK 2013gg)

The reason for the creation of VAT was "in part because they create stronger incentives to collect than a sales tax does. Both types of consumption tax create an incentive by end consumers to avoid or evade the tax, but the sales tax offers the buyer a mechanism to avoid or evade the tax—persuade the seller that the buyer is not really an end consumer, and therefore the seller is not legally required to collect it. Therefore, the burden of determining whether the buyer's motivation is to consume or re-sell is on the seller, and the seller has no direct economic incentive to collect it. The VAT approach gives sellers a direct financial stake in collecting the tax and eliminates a decision needing to be made by the seller about whether the buyer is or is not an end consumer." (WK 2013gg)

VAT vs. Sales Tax

The main difference between VAT and sale tax is that "value added tax avoids the cascade effect of sales tax by taxing only the value added at each stage of production. For this reason, throughout the world, VAT has been gaining favor over traditional sales taxes. In principle, VAT applies to all provisions of goods and services. VAT is assessed and collected on the value of goods or services that have been provided every time there is a transaction (sale/purchase). The seller charges VAT to the buyer, and the seller pays this VAT to the government. If, however, the purchaser is not an end user, but the goods or services purchased are costs to its business, the tax it has paid for such purchases can be deducted from the tax it charges to its customers. The government only receives the difference; in other words, it is paid tax on the gross margin of each transaction, by each participant in the sales chain." (WK 2013gg)

At least in theory, "sales tax is normally charged on end users (consumers). The VAT mechanism means that the end-user tax is the same as it would be with a sales tax. The main difference is the extra accounting required by those in the middle of the supply chain; this disadvantage of VAT is balanced by application of the same tax to each member of the production chain regardless of its position in it and the position of its customers, reducing the effort required to check and certify their status. When the VAT system has few, if any, exemptions such as with GST in New Zealand, payment of VAT is even simpler." (WK 2013gg)

The advantage of VAT is all the more obvious when, "if sales taxes are high enough, people start engaging in widespread tax evading activity (like buying over the Internet, pretending to be a business, buying at wholesale, buying products through an employer etc.). On the other hand, total VAT rates can rise above 10% without widespread evasion because of the novel collection mechanism." (WK 2013gg)

Implementation of VAT

The implementation of VAT depends on two factors in accounting, that is, (a) the method of collection and (b) the timing of collection, as explained below.

(a) The Method of Collection

The method of collection (in accounting) can be "accounts-based" or "invoice-based." (WK 2013gg)

On the one hand, "under the invoice method of collection, each seller charges VAT rate on his output and passes the buyer a special invoice that indicates the amount of tax charged. Buyers who are subject to VAT on their own sales (output tax), consider the tax on the purchase invoices as input tax and can deduct the sum from their own VAT liability. The difference between output tax and input tax is paid to the government (or a refund is claimed, in the case of negative liability)." (WK 2013gg)

On the other hand, "under the accounts based method, no such specific invoices are used. Instead, the tax is calculated on the value added, measured as a difference between revenues and allowable purchases. Most countries today use the invoice method, the only exception being Japan, which uses the accounts method." (WK 2013gg)

(b) The Timing of Collection

And the timing of collection (for accounting) can be either "accrual" or "cash based." (WK 2013gg)

On the one hand, "cash basis accounting is a very simple form of accounting. When a payment is received for the sale of goods or services, a deposit is made, and the revenue is recorded as of the date of the receipt of funds—no matter when the sale had been made. Cheques are written when funds are available to pay bills, and the expense is recorded as of the cheque date—regardless of when the expense had been incurred. The primary focus is on the amount of cash in the bank, and the secondary focus is on making sure all bills are paid. Little effort is made to match revenues to the time period in which they are earned, or to match expenses to the time period in which they are incurred." (WK 2013gg; S. Snyder 2008)

On the other hand, "accrual basis accounting matches revenues to the time period in which they are earned and matches expenses to the time period in which they are incurred. While it is more complex than cash basis accounting, it provides much more information about your business. The accrual basis allows you to track receivables (amounts due from customers on credit sales) and payables (amounts due to vendors on credit purchases). The accrual basis allows you to match revenues to the expenses incurred in earning them, giving you more meaningful financial reports." (WK 2013gg)

Problems with the Accounting of VAT

However, to the critics, this study on addition, government revenue, and the accounting of VAT has some major problems. Consider, for illustration, a few examples below.

Firstly, one main criticism of the study on addition, government revenue, and the accounting of VAT is that there is a trade-off between the national government and sub-national governments in regard to the revenue collected through VAT.

For example, in practice, "in many developing countries such as India, sales tax/VAT are key revenue sources as high unemployment and low per capita income render other income sources inadequate. However, there is strong opposition to this by many sub-national governments as it leads to an overall reduction in the revenue they collect as well as of some autonomy." (WK 2013gg)

Secondly, another main criticism of the study on addition, government revenue, and the accounting of VAT is that the introduction of VAT can lead to a new kind of fraud.

For instance, although "total VAT rates can rise above 10% without widespread evasion because of the novel collection mechanism"—there is a different problem, in that, "because of its particular mechanism of collection, VAT becomes quite easily the target of specific frauds like carousel fraud, which can be very expensive in terms of loss of tax incomes for states." (WK 2013gg) And the term "carousel fraud" (also known as "missing trader fraud") refers to "the theft of Value Added Tax (VAT) from a government by organised crime gangs who exploit the way VAT is treated within multi-jurisdictional trading where the movement of goods between jurisdictions is VAT-free. This allows the fraudster (person who commits fraud) to charge VAT on the sale of goods, and then instead of paying this over to the government's collection authority, to simply abscond, taking the VAT with him." (WK 2013hh)

Thirdly, still another main criticism of the study on addition, government revenue, and the accounting of VAT is that it is not necessarily "efficient" for the economy.

For instance, "a VAT, like most taxes, distorts what would have happened without it. Because the price for someone rises, the quantity of goods traded decreases. Correspondingly, some people are worse off by more than the government is made better off by tax income. That is, more is lost due to supply and demand shifts than is gained in tax. This is known as a deadweight loss. If the income lost by the economy is greater than the government's income; the tax is inefficient. It must be noted that a VAT and a Non-VAT has the same implications on the microeconomic model." (WK 2013gg) In addition,

Fourthly, an additional main criticism of the study on addition, government revenue, and the accounting of VAT is that the implementation of VAT varies from country to country and is subject to different loopholes.

For instance, in the EU, "the VAT Directive (prior to 1 January 2007 referred to as the Sixth VAT Directive)" allows "certain goods and services to be exempt from VAT (for example, postal services, medical care, lending, insurance, betting), and certain other goods and services to be exempt from VAT but subject to the ability of an EU member state to opt to charge VAT on those supplies (such as land and certain financial services)." (WK 2013gg)

Fifthly, another additional main criticism of the study on addition, government revenue, and the accounting of VAT is that there is no general consensus on the right rate of VAT.

For instance, different countries adopt different VAT rates, as shown by 25.5% in Iceland but only 4.5% in Andorra. (WK 2013gg)

Sixthly, still another additional main criticism of the study on addition, government revenue, and the accounting of VAT is that there is also no consensus on the need for VAT.

For instance, the U.S. is VAT-free, but Germany is not (with a VAT rate of 19%). (WK 2013gg) The reason is that, in the U.S., "many politicians and economists...consider VAT taxation on US goods and VAT rebates for goods from other countries to be unfair practice. E.g. the American Manufacturing Trade Action Coalition claims that any rebates or special taxes on imported goods should not be allowed by the rules of the World Trade Organisation. AMTAC claims that so-called 'border tax disadvantage' is the greatest contributing factor to the $5.8 trillion US current account deficit for the decade of the 2000s, and estimated this disadvantage to US producers and service providers to be $518 billion in 2008 alone." (WK 2013gg)

Seventhly, one additional main criticism of the study on addition, government revenue, and the accounting of VAT is that it can be a "regressive tax."

For instance, "the 'value added tax' has been criticized as the burden of it falls on personal end-consumers of products. Some critics consider it to be a regressive tax, meaning that the poor pay more, as a percentage of their income, than the rich." (WK 2013gg)

And eighthly, one last main criticism of the study on addition, government revenue, and the accounting of VAT is that it can be costly and thus lead to "higher prices."

For instance, "under VAT, however, manufacturers and wholesale companies also have to hire accountants and incur accounting expenses to handle the additional paperwork required for collecting VAT, increasing overhead costs that in turn get incorporated into the cost of the item, possibly creating a cascading effect of higher prices throughout the chain of production. Manufacturers and wholesalers have a choice of retaining less profits overall, or passing on the additional cost to their customers in the form of increased prices." (WK 2013gg)

Beyond the Accounting of VAT

These criticisms of the study on addition, government revenue, and the accounting of VAT do not imply a total rejection of its usefulness but serve to show us the two opposing sides of the debate, such that the possibility (or impossibility) and desirability (or undesirability) of addition (from the perspective of society with addition, government revenue, and the accounting of VAT as a case study here) are not to the extent that the respective defenders would like us to believe.

In fact, the analysis of addition, government revenue, and the accounting of VAT can illuminate the usefulness of the ontological principles in existential dialectics, and good examples include the formalness-informalness principle, the absoluteness-relativeness principle, the partiality-totality principle, the predictability-unpredictability principle, the explicability-inexplicability principle, the fiction-reality principle, the cognitiveness-noncognitiveness principle, the finiteness-transfiniteness principle, the preciseness-vagueness principle, the simpleness-complicatedness principle, the openness-hiddenness principle, the denseness-emptiness principle, the rule-exception principle, the prototypicality-variation principle, the change-constancy principle, the order-chaos principle, the slowness-quickness principle, the expansion-contraction principle, the optimality-nonoptimality principle, the simultaneity-nonsimultaneity principle, the isolation-interaction principle, the theory-praxis principle, the convention-novelty principle, the evolution-transformation principle, the symmetry-asymmetry principle, the softness-hardness principle, the seriousness-playfulness principle, the activeness-inactivess principle, the selfness-otherness principle, the regression-progression principle, the same-difference principle, the stability-reaction principle, the functionality-nonfunctionality principle, the intentionality-nonintentionality principle, and the survivability-nonsurvivability principle.

For instance, in relation to the formalness-informalness principle, if there is formalness (e.g., the formal logical requirement of soundness and completeness in a system of ideas, as in the argument on VAT), there is informalness (e.g., the non-formal existence of unsoundness and incompleteness in the argument on VAT, as shown in the criticisms that there is a trade-off between the national government and sub-national governments in regard to the revenue collected through VAT, that the introduction of VAT can lead to a new kind of fraud, that it is not necessarily "efficient" for the economy, that the implementation of VAT varies from country to country and is subject to different loopholes, that there is no general consensus on the right rate of VAT, that there is also no consensus on the need for VAT, that it can be a "regressive tax," and that it can be costly and thus lead to "higher prices"). And the reverse direction also holds true.

In relation to the absoluteness-relativeness principle, if there is absoluteness (e.g., the absolute view on VAT by the advocates like some national governments for its implementation), there is relativeness (e.g., what is right for the advocates in regard to the implementation of VAT is not necessarily so for the critics who "consider it to be a regressive tax, meaning that the poor pay more, as a percentage of their income, than the rich"). And the reverse direction also holds true.

In relation to the partiality-totality principle, if there is partiality (e.g., the partial view on VAT by the advocates like some national governments for its implementation), there is totality (e.g., the more holistic view on VAT, such that the whole is not the sum of its parts, that is, it cannot be reduced to the sum of those opposing views by the advocates, the critics, etc. as discussed earlier, since there will emerge new views in the future not yet known today, just as there are already alternative views nowadays which do not side exclusively with any of them, as is the analysis here). And the reverse direction also holds true.

In relation to the predictability-unpredictability principle, if there is predictability (e.g., the predictable tendency that Hong Kong or Libya is VAT-free in the foreseeable future), there is unpredictability (e.g., the more difficult task to predict exactly the extent to which Hong Kong or Libya will be VAT-free in a particular distant future era). (WK 2013gg) And the reverse direction also holds true.

In relation to the explicability-inexplicability principle, if there is explicability (e.g., the explanation, by those advocates of VAT, of its desirability on the basis of the increase of "government revenue"), there is inexplicability (e.g., the lack of sufficient explanation, by those advocates of VAT, of why it is necessarily desirable, since, to the critics, "because the price for someone rises, the quantity of goods traded decreases," so "correspondingly, some people are worse off by more than the government is made better off by tax income"). And the reverse direction also holds true.

In relation to the fiction-reality principle, if there is fiction (e.g., the fictional aspect of VAT in regard to "government revenue," insofar as, "in many developing countries such as India,...there is strong opposition to this [VAT] by many sub-national governments as it leads to an overall reduction in the revenue they collect as well as of some autonomy"), there is reality (e.g., the realistic aspect of VAT, insofar as it can increase "national" government revenue). And the reverse direction also holds true.

In relation to the cognitiveness-noncognitiveness principle, if there is cognitiveness (e.g., a view of things on the basis of reason and evidences, as shown in the finding that EU countries support VAT), there is noncognitiveness (e.g., a view of things on the basis of non-cognitive factors like envy, jealousy, power, nationality, race, gender, age, class, greed, lust, status, faith, anger, sadness, joy, fear, wish, etc.—as shown in the "politics" of supporting VAT in EU countries, in spite of the various criticisms against "loophole," "fraud, "regressive tax," "costliness," "inefficiency," etc. as discussed earlier). And the reverse direction also holds true.

In relation to the finiteness-transfiniteness principle, if there is finiteness (e.g., the finite number of the categories of taxation, like "sales tax," "VAT," etc.), there is transfiniteness (e.g., the transfinite number of all the individuals who have ever taxed in history, regardless of which categories of taxation the taxes belong to). And the reverse direction also holds true.

In relation to the preciseness-vagueness principle, if there is preciseness (e.g., the precise identification of 2 types of the method of collection and 2 types of the timing of collection in the article aforecited), there is vagueness (e.g., the vagueness in the identification of the 2 types of the method of collection and 2 types of the timing of collection in the article aforecited, since it is not clear why there must be only 2, not 3, 4, 5, 6, and so on). And the reverse direction also holds true.

In relation to the simpleness-complicatedness principle, if there is simpleness (e.g., the relatively simple analysis of VAT by the advocates), there is complicatedness (e.g., the relatively more complicated analysis of VAT, by challenging the claims and assumptions, as shown in the problems pointed out by the critics as cited). And the reverse direction also holds true.

In relation to the openness-hiddenness principle, if there is openness (e.g., the open exploration of using VAT to increase government revenue), there is hiddenness (e.g., the hidden bias in using VAT to increase government revenue, because of different problems aforedescribed, like "regressive tax," "inefficiency," etc.). And the reverse direction also holds true.

In relation to the denseness-emptiness principle, if there is denseness (e.g., the relatively denser concentration of concerns with the rates of VAT to be implemented by those in EU countries), there is emptiness (e.g., the relatively less dense, or more empty, concentration of concerns with the rates of VAT to be implemented by those in the U.S.). And the reverse direction also holds true.

In relation to the rule-exception principle, if there are rules (e.g., the usual generalization, procedure, habit, or tendency that "most countries today use the invoice method"), there are exceptions (e.g., a case to which a rule does not apply, such as the exceptional case of "Japan, which uses the accounts method" instead). (WK 2013gg) And the reverse direction also holds true.

In relation to the prototypicality-variation principle, if there is prototypicality (e.g., the prototype, like the example about the original concept of VAT by Wilhelm von Siemens in 1918), there is variation (e.g., different variants of the protoptye, as shown in many works since the original concept of VAT by Wilhelm von Siemens in 1918—like the later ones by Maurice Lauré in 1954, the EU in 2006, etc., though each with its own variation, contribution, and disagreement). And the reverse direction also holds true.

In relation to the change-constancy principle, if there is change (e.g., the ever new ways to study accounting, as shown in the implementation of VAT aforediscussed), there is constancy (e.g., the ever constant existence of different problems with the ever new ways to study accounting, as shown in the problems with the implementation of VAT pointed out by the critics). And the reverse direction also holds true.

In relation to the order-chaos principle, if there is order (e.g., the more or less orderly view on VAT, if looked from the sole vantage point of the advocates), there is chaos (e.g., the more or less chaotic view on VAT, if looked from the multiple, conflicting vantage points of both the advocates and the critics, especially in regard to the problems aforecited, such that they do not add up to much of anything coherent). And the reverse direction also holds true.

In relation to the slowness-quickness principle, if there is slowness (e.g., the relatively slower speed for "carousel fraud" to occur, if VAT is not implemented at all or only trivially implemented), there is quickness (e.g., the relatively quicker speed for "carousel fraud" to occur, if it is to be implemented, "because of its particular mechanism of collection, VAT becomes quite easily the target of specific frauds like carousel fraud"). And the reverse direction also holds true.

In relation to the expansion-contraction principle, if there is expansion (e.g., the relative increase of "national government" revenue in India due to VAT), there is contraction (e.g., the relative decrease of "many sub-national government" revenues in India due to VAT). And the reverse direction also holds true.

In relation to the optimality-nonoptimality principle, if there is optimality (e.g., the search for highly favorable options, like the quest for the more efficient way of implementing VAT, such that "the tax revenue is used for productive spending or has positive externalities'—in other words, governments may do more than simply consume the tax income"), there is nonoptimality (e.g., the existence of non-optimal alternatives to optimality, like the argument for more realistic studies by the critics, who pointed out different problems with VAT as discussed earlier). (WK 2013gg) And the reverse direction also holds true.

In relation to the simultaneity-nonsimultaneity principle, if there is simultaneity (e.g., the occurrence of different things more or less at the same time, such as the simultaneous co-existence of VAT-free countries like the U.S. and countries with VAT like those in the EU), there is nonsimultaneity (e.g., the occurrence of something after something else, such as the use of VAT in Australia in 2000, after Canada introduced its own version of VAT in 1991). (WK 2013gg) And the reverse direction also holds true.

In relation to the isolation-interaction principle, if there is isolation (e.g., the state of a situation that is separate from others, such as the immunity or separation of goods and services from any VAT in Australia prior to 2000), there is interaction (e.g., the influence of entities on one another, such as the influence of VAT on goods and services in Australia after 2000). And the reverse direction also holds true.

In relation to the theory-praxis principle, if there is theory (e.g., the theoretical construction of "VAT" by the German industrialist Dr. Wilhelm von Siemens in 1918), there is praxis (e.g., the practical application of "VAT" to the EU countries, which implemented it in the 2000s). And the reverse direction also holds true.

In relation to the convention-novelty principle, if there is convention (e.g., the conventional wisdom about using "VAT" to "avoid…the cascade effect of sales tax"), there is novelty (e.g., the alternative novel challenge to the conventional wisdom about using "VAT" to "avoid…the cascade effect of sales tax"—by the new idea that the implementation of VAT can lead to "carousel fraud"). (WK 2013gg) And the reverse direction also holds true.

In relation to the evolution-transformation principle, if there is evolution (e.g., the natural evolution of humans to consume things in the state of nature in a primitive way), there is transformation (e.g., the technical transformation of human ability to consume by the invention of a consumption tax known as "VAT"). And the reverse direction also holds true.

In relation to the symmetry-asymmetry principle, if there is symmetry (e.g., the co-existence of different entites like the government and the consumers), there is asymmetry (e.g., the government can tax the consumers with VAT—but the consumers cannot tax the government). And the reverse direction also holds true.

In relation to the softness-hardness principle, if there is softness (e.g., the defense of VAT by the advocates like Wilhelm von Siemens), there is hardness (e.g., the critique of VAT, as shown in the problems pointed out by the critics aforecited). And the reverse direction also holds true.

In relation to the seriousness-playfulness principle, if there is seriousness (e.g., the serious business of those who try to implement the VAT), there is playfulness (e.g., the playful part of those who try to implement the VAT, when they play around with different options and possibilities over the decades, like the 2 types of the method of collection and the 2 types of the timing of collection). And the reverse direction also holds true.

In relation to the activeness-inactivess principle, if there is activeness (e.g., the relative activeness of those governments in countries with VAT to implement it), there is inactiveness (e.g., the relative inactiveness of those "missing traders" in "carousel fraud" to comply with the implementation of VAT). And the reverse direction also holds true.

In relation to the selfness-otherness principle, if there is selfness (e.g., the tendency to be mostly centered on one's self or group, as shown by those who commit "carousel fraud" for their personal gains), there is otherness (e.g., the tendency to be mostly centered on others, as shown in the academic analysis of both sides of the debate on VAT, as an advancement in the knowledge of taxation and accounting for humanity). And the reverse direction also holds true.

In relation to the regression-progression principle, if there is regression (e.g., the regression made by VAT, as shown in the problems pointed out by the critics), there is progression (e.g., the progress made by VAT, as shown in the contribution to the increase of government revenue aforecited). And the reverse direction also holds true.

In relation to the same-difference principle, if there is similarity in outcome (e.g., the contribution to the molding and control of beliefs, values, and behaviors, regardless of whether this be done by way of VAT, or by way of carousel fraud), there is difference in outcome (e.g., the contribution to the molding and control of beliefs, values, and behaviors by way of VAT for a relatively more legalistic lifeworld—but the contribution to the molding and control of beliefs, values, and behaviors by way of carousel fraud for a relatively more underground lifeworld). And the reverse direction also holds true.

In relation to the stability-reaction principle, if there is stability (e.g., the relatively more stable condition of the spread of VAT in modern times), there is reaction (e.g., the spread of VAT in modern times then led to further changes later on, when nowadays, there is "carousel fraud"). And the reverse direction also holds true.

In relation to the functionality-nonfunctionality principle, if there is functionality (e.g., the presence of action for which something or someone is fitted, as shown in the creation of VAT, which serves the function of increasing government revenue), there is nonfunctionality (e.g., the relative lack of action for which something or someone is fitted, as shown in the different problems with the implementation of VAT, such that it is not as functional as the advocates would like us to believe). And the reverse direction also holds true.

In relation to the intentionality-nonintentionality principle, if there is intentionality (e.g., the planning or design of something for a certain outcome, as shown in the original intention of the German industrialist Dr. Wilhelm von Siemens to propose VAT as a way to increase government revenue), there is nonintentionality (e.g., the relative absence of planning or design of something for a certain outcome, such that the outcome can even be counter-intentional in being the opposite of what was originally planned, as shown in the different problems with the implementation of VAT, like the decrease of sub-national government revenues, tax evasion via "carousel fraud," etc.—contrary to the original intention of the German industrialist Dr. Wilhelm von Siemens to propose VAT as a way to increase government revenue). And the reverse direction also holds true.

And in relation to the survivability-nonsurvivability principle, if there is survivability (e.g., the relative survival, nowadays, of VAT), there is non-survivability (e.g., the relative non-survival, nowadays, of the older way to implement VAT without addressing the "regressive" impact on the poor, since "to maintain the progressive nature of total taxes on individuals, countries implementing VAT have reduced income tax on lower income-earners as well as instituted direct transfer payments to lower-income groups, resulting in lower tax burdens on the poor"). (WK 2013gg; C. Min 2004) And the reverse direction also holds true.

Accounting and Culture

Addition, when examined from the perspective of culture, can cast an illuminating light on its possibility (or impossibility) and desirability (or undesirability), and this can be shown by way of two case studies, namely, (2.5.1) addition, values, and accounting principles, and (2.5.2) addition, legal traditions, and accounting standards—to be addressed hereafter, in that order.

Additon, Values, and Accounting Principles

An excellent case study here to understand addition from the perspective of culture has to do with addition, values, and accounting principles.

An argument in the literature is that certain accounting principles which have developed over the last few hundreds of years govern the norms of doing accounting, with its hidden values, so as to facilitate the growing (or increasing) business transactions in the modern world.

Debate on Accounting Princples for Values

For instance, the "Accounting for Value" website published an interesting article titled "Encouraging a Wider Discussion on the Nature of Accounting Principles," in which the opening question is: "Are the basic accounting principles that have developed over hundreds of years still fit for purpose?" (ACCO 2013)

This debate on values in accounting principles is important, because "the principles upon which accounting and reporting are based have a massive influence on the world we live in, on relative prices and on investment flows. The relationships are becoming ever more critical between business and society, and between financial markets and inequality." (ACCO 2013)

In fact, "over the last couple of decades more and more voices have raised concerns and proposed solutions. Recent examples are Michael Porter's thinking on *Shared Value* [2011], Tim Jackson's on *Prosperity without Growth* [2009], and the International Integrated Reporting Council's focus on corporate reporting [2001] that demonstrates the links with the social, environmental and economic context within which companies operate." (ACCO 2013)

Yet, in spite of all these, "there is one area that seems to have slipped under the radar when solutions are being suggested. An area that has profound implications for the way our world works. Yet, from Marx onwards, it hasn't been part of the debate. Why not? Mainly because, as Monty Python's lion tamer reminds us, it is very, very boring. And so it has been left to the experts, our modern day temple priests—the accountancy profession." (ACCO 2013)

The conventional wisdom is that "financial accounting is generally thought of (when at all) as a technique: here are some transactions and here's how we account for them. But take away our internationally accepted approach to measuring the financial performance of organisations, from multinationals to unincorporated charities, and what do you think would happen?" (ACCO 2013)

In reality, "accounting principles did not arrive on tablets of stone, nor were they discovered by an eminent scientist. Most of us will never read a set of accounts—but they are the ghost in a machine on which our lives depend. All the products and services we use are the result of investment decisions that were based on information arising from these principles. This is what makes them so powerful and important. Accounting is one of the great social innovations, an approach that allows us to make decisions based on a reasonable (and audited) application of a set of principles." (ACCO 2013)

But "we should remember that they are principles rather than truths. Different principles could work just as well as the existing ones. But they would result in different prices, patterns of investment, products and services. Which raises a question: are we sure that this complex accounting structure built on a 900-year-old framework is still fit for purpose?" (ACCO 2013)

Nowadays, "there is increasing pressure for businesses to provide new information, particularly relating to sustainability, primarily in corporate responsibility reports. Company directors are required to have regard for the impact of their decisions on employees and customers, but also the community and the environment. This trend reflects society's growing concern that the organisations providing us with goods and services should not be doing so at a greater cost to society in the long term." (ACCO 2013)

Therefore, "this has led to some progress, but is it time to go back to first principles and look at accounting itself? Are the basic accounting principles that have developed over hundreds of years still fit for purpose? After all, though most of us think that accounting is about money, it is firstly about people carrying out activities and preparing an account to those who have an interest in those activities." (ACCO 2013)

Three Accounting Principles

For illustration, there are three accounting principles to be identified and addressed, namely, (a) the the principle of the accounting entity, (b) the principle of the unit of measure and (c) the principle of the historic cost convention," as summarized in *Table 2.5*. (ACCO 2013)

(a) The Principle of Accounting Entity

The first accounting principle to be considered here concerns the "accounting entity," which means that "it is separate and distinct from its owners and other firms in order that the entity can provide information to current and future investors and creditors, particularly in relation to the financial value that has been created and could be distributed to owners." (ACCO 2013)

Indeed, "this is the basis for deciding what should be included in an account. Investors may want to know about potential dividends and the overall financial position but they may also want to know that the business isn't creating value at a cost to others. Certainly society as a whole may want to know this. The decisions about what should be included could be based, as is currently the case, on known and probable liabilities." (ACCO 2013)

But "the accounts could be expanded to consider the effects of the company's activities on all those whose access to resources is affected. Not necessarily to recognize a legal liability to all those affected but to make provision for the effect, to be accountable for them. Uncertainty about the extent to which issues can be attributed to a company's activities shouldn't be a barrier to inclusion." (ACCO 2013)

For example, "in some situations such issues are already being considered. For example, mining companies are now required to provide for the costs of closing mining operations that may not be incurred for many years: costs relating to environmental cleanup and possibly social decline." (ACCO 2013)

(b) The Principle of the Unit of Measure

The second accounting principle to be considered here concerns the "unit of measure," which means that "transactions should be included using a standard unit of measure and that the monetary unit is used." (ACCO 2013)

And "the reliance on monetary units in the unit of measure principle may mean that we need to improve valuation—coming up with values for broader issues that have now been identified. Uncertainty of valuation shouldn't act as a barrier to including the value of issues that should be included." (ACCO 2013)

But there remains another challenge, in that "in the social sciences, there are no standard units of measurement," although "the theory and practice of measurement is studied in psychometrics and the theory of conjoint measurement," as well as others. (WK 2013ii)

(c) The Principle of Historical Cost

And the third accounting principle to be considered here concerns "historical cost," which means that "costs incurred in the past…is the basis for how items are valued." (ACCO 2013)

This reliance on historical cost has some "advantages," as shown below: (WK 2013n)

- "Historical cost accounts are straightforward to produce."
- "Historical cost accounts do not record gains until they are realized."
- "Historical cost accounts are still used in most accounting systems."

The popularity of the principle remains so, even though in 1989, the "International Accounting Standards Board" proposed "the Capital Maintenance in Units of Constant Purchasing Power model" as "an alternative basic accounting model to the traditional Historical Cost Accounting model"—as already distinguished earlier in *Sec. 1.4.* (WK 2013n)

Problems with the Study of Values and Accounting Principles

However, to the critics, the argument on addition, values, and accounting principles has some major problems. Consider, for illustration, a few examples below (as summarized in *Table 2.5*).

Firstly, one main criticism of the argument on addition, values, and accounting principles is that there are opposing sides in the debate on "the principle of historical cost."

For instance, on the one hand, the defenders of the principle of historical cost point to its implementation in the last few hundred years (aforediscussed). On the other hand, the opponents of the principle of historical cost point to the problem that "the historical costs convention…has always been difficult since significant areas of financial accounting are not based on historical costs. Many require an assessment of cost (or value) based on a forecast of the future where costs have not yet been incurred. Think of the provision for bad debts, fair value and depreciation." (ACCO 2013)

Secondly, another main criticism of the argument on addition, values, and accounting principles is that there are opposing sides in the debate on "the principle of the unit of measure."

For instance, on the one hand, the defenders of the principle of the unit of measure point to its implementation in the last few hundred years (as discussed earlier). On the other hand, the opponents of the principle of the unit of measure point to the problem that, when "transactions should be included using a standard unit of measure and that the monetary unit is used,…accounts will be restricted to those transactions that can be expressed using monetary units"—but "there are other transactions which cannot be so easily expressed using monetary units, and good examples include "carbon trading," etc., because of "the socially constructed character of carbon conversion factors or of the accountants' work practice which cannot implement abstract accounting schemes into reality." (ACCO 2013; WK 2013s; L. Lohmann 2009; I. Lippert 2012; D. MacKenzie 2009) In addition, how can, say, the costs and benefits of "spiritual salvation" be monetized?

Thirdly, still another main criticism of the argument on addition, values, and accounting principles is that there are opposing sides in the debate on "the principle of the accounting entity."

For instance, there are competing models to deal with the "accounting entity" of a firm, as shown in the debate between one model focusing on the "shareholders" and the other focusing on the "stakeholders." As an illustration, "the stakeholder theory says corporations should be run for the benefit of all 'stakeholders,' not just the shareholders. Stakeholders of a company include any individual or group that can influence or is influenced by a company's practices. The stakeholders of a company can be suppliers, consumers, employees, shareholders, financial community, government, and media. Companies must properly manage the relationships between stakeholder groups and they must consider the interest(s) of each stakeholder group carefully." (WK 2013aa)

Fourthly, an additional main criticism of the argument on addition, values, and accounting principles is that any change to the principles is not costless, even with benefits.

For instance, to the critics, "if there were a broader definition of what should be included in the accounts, this would affect company profitability. It would also result in changes in strategy and business models as businesses sought to maintain profitability." (ACCO 2013)

And fifthly, another additional main criticism of the argument on addition, values, and accounting principles is that each principle, be it the old one (as defended by its advocates) or any new one (as suggested by the opponents of the old one), has both advantages and disadvantages.

For instance, the principle of the historical cost has some advantages (as shown in the 3 examples aforecited), but there are also some disadvantages, as shown below: (WK 2013n)

- "Historical cost accounts give no indication of current values of the assets of a business."
- "Historical cost accounts do not record the opportunity costs of the use of older assets, particularly property which may be recorded at a value based on costs incurred many years ago."
- "Historical cost accounts do not measure the loss of value of monetary assets as a result of inflation."

Surely, there can be other examples, so the ones above are illustrative, not exhaustive. But an important implication to remember here is that, should the "Constant Purchasing Power model" as "an alternative basic accounting model to the traditional Historical Cost Accounting model" be chosen instead, then it would have both the advantages (that the older model does not have) and the disadvantages (that the older model does not have) too.

Beyond Accounting Principles

These criticisms of the argument on accounting principles and values should not be misconstrued as a total rejection of its usefulness but serve to show us the two opposing sides of the debate, such that the possibility (or impossibility) and desirability (or undesirability) of addition (from the perspective of culture with addition, values, and accounting principles as a case study here) are not to the extent that the respective defenders would like us to believe.

Furthermore, the analysis of addition, values, and accounting principles can cast a bright light on the ontological principles in existential dialectics, and good examples include the formalness-informalness principle, the absoluteness-relativeness principle, the partiality-totality principle, the predictability-unpredictability principle, the explicability-inexplicability principle, the fiction-reality principle, the cognitiveness-noncognitiveness principle, the finiteness-transfiniteness principle, the preciseness-vagueness principle, the simpleness-complicatedness principle, the openness-hiddenness principle, the denseness-emptiness principle, the rule-exception principle, the prototypicality-variation principle, the change-constancy principle, the order-chaos principle, the slowness-quickness principle, the expansion-contraction principle, the optimality-nonoptimality principle, the simultaneity-nonsimultaneity principle, the isolation-interaction principle, the theory-praxis principle, the convention-novelty principle, the evolution-transformation principle, the symmetry-asymmetry principle, the softness-hardness principle, the seriousness-playfulness principle, the activeness-inactivess principle, the selfness-otherness principle, the regression-progression principle, the same-difference principle, the stability-reaction principle, the functionality-nonfunctionality principle, the intentionality-nonintentionality principle, and the survivability-nonsurvivability principle.

For instance, in relation to the formalness-informalness principle, if there is formalness (e.g., the formal logical requirement of soundness and completeness in a system of ideas, as in the argument on accounting principles), there is informalness (e.g., the non-formal existence of unsoundness and incompleteness in the argument on accounting principles, as shown in the criticisms that there are opposing sides in the debate on "the principle of historical cost," that there are opposing sides in the debate on "the principle of the unit of measure," that there are opposing sides in the debate on "the principle of the accounting entity," that any change to the principles is not costless [even with benefits], and that each principle [be it the old one as defended by its advocates or any new one as suggested by the opponents of the old one] has both advantages and disadvantages). And the reverse direction also holds true.

In relation to the absoluteness-relativeness principle, if there is absoluteness (e.g., the absolute view on the fundamental change of the 3 accounting principles by the advocates like Jeremy Nicholls of the SROI Network), there is relativeness (e.g., what is right for the advocates like Jeremy Nicholls of the SROI Network in regard to the fundamental change of the 3 accounting principles is not necessarily so for the defenders of the principles). (ACCO 2013) And the reverse direction also holds true.

In relation to the partiality-totality principle, if there is partiality (e.g., the partial view on the fundamental change of the 3 accounting principles by the advocates like Jeremy Nicholls), there is totality (e.g., the more holistic view on the fundamental change of the 3 accounting principles, such that the whole is not the sum of its parts, that is, it cannot be reduced to the sum of opposing views by the defenders, the advocates, etc., as discussed earlier, since there will emerge new views in the future not yet known today, just as there are already alternative views nowadays which do not side exclusively with any of them, as is the analysis here). And the reverse direction also holds true.

In relation to the predictability-unpredictability principle, if there is predictability (e.g., the predictable tendency that the advocates like Jeremy Nicholls favor the fundamental change of the 3 accounting principles in the foreseeable future), there is unpredictability (e.g., the more difficult task to predict exactly in which particular ways any change of the accounting principles will be in a particular future era). And the reverse direction also holds true.

In relation to the explicability-inexplicability principle, if there is explicability (e.g., the explanation, by those advocates of the principle of the historical cost, of its desirability on the basis of its advantages), there is inexplicability (e.g., the lack of sufficient explanation, by those advocates of the principle of the historical cost, of why it is necessarily desirable, since, to the critics, there are disadvantages, such that it should be replaced by the "Constant Purchasing Power model" instead). And the reverse direction also holds true.

In relation to the fiction-reality principle, if there is fiction (e.g., the fictional aspect of the principle of the historical cost, insofar as "historical cost accounts do not measure the loss of value of monetary assets as a result of inflation"), there is reality (e.g., the realistic aspect of the principle of the historical cost, insofar as "historical cost accounts do not record gains until they are realized"). And the reverse direction also holds true.

In relation to the cognitiveness-noncognitiveness principle, if there is cognitiveness (e.g., a view of things on the basis of reason and evidences, as shown in the finding that there is a debate on the desirability of the 3 accounting principles), there is noncognitiveness (e.g., a view of things on the basis of non-cognitive factors like envy, jealousy, power, nationality, race, gender, age, class, greed, lust, status, faith, anger, sadness, joy, fear, wish, etc.—as shown in the subjective factor of "boredom," such that some people find the discussion of the 3 accounting principles "very, very boring," in spite of the fact that "the principles upon which accounting and reporting are based have a massive influence on the world we live in"). And the reverse direction also holds true.

In relation to the finiteness-transfiniteness principle, if there is finiteness (e.g., the finite number of the principles in accounting, like the 3 main ones aforediscussed), there is transfiniteness (e.g., the transfinite number of all the individual instances of business transactions that have ever been affected by the 3 accounting principles in human history). And the reverse direction also holds true.

In relation to the preciseness-vagueness principle, if there is preciseness (e.g., the precise identification of 3 accounting principles in the article aforecited), there is vagueness (e.g., the vagueness in the identification of the 3 accounting principles in the article aforecited, since it is not clear why there must be only 3, not 4, 5, 6, 7, and so on). And the reverse direction also holds true.

In relation to the simpleness-complicatedness principle, if there is simpleness (e.g., the relatively simple analysis of the 3 accounting principles by the advocates), there is complicatedness (e.g., the relatively more complicated analysis of the 3 accounting principles, by challenging the claims and assumptions, as shown in the problems pointed out by the critics as cited). And the reverse direction also holds true.

In relation to the openness-hiddenness principle, if there is openness (e.g., the open exploration of using the 3 principles for accounting in the last few hundred years), there is hiddenness (e.g., the hidden bias in the use of the 3 principles for accounting in the last few hundred years, as shown in the different problems pointed out by the critics aforecited). And the reverse direction also holds true.

In relation to the denseness-emptiness principle, if there is denseness (e.g., the relatively denser concentration of concerns with "historical cost" by those in the traditional Historical Cost Accounting model), there is emptiness (e.g., the relatively less dense, or more empty, concentration of concerns with "the loss of value of monetary assets as a result of inflation" by those in the traditional Historical Cost Accounting model). And the reverse direction also holds true.

In relation to the rule-exception principle, if there are rules (e.g., the usual generalization, procedure, habit, or tendency that the issue of accounting princples is "very, very boring" to many ordinary people), there are exceptions (e.g., a case to which a rule does not apply, such as the exceptional case that the issue of accounting princples "has been left to the experts, our modern day temple priests—the accountancy profession"). And the reverse direction also holds true.

In relation to the prototypicality-variation principle, if there is prototypicality (e.g., the prototype, like the example about the original "accounting structure" built over the last few hundred years), there is variation (e.g., different variants of the protoptye, as shown in many studies since the original "accounting structure" built over the last few hundred years—like the later ones such as the accounting reform suggested by Jeremy Nicholls of the SROI Network, the "Constant Purchasing Power model" proposed by the "International Accounting Standards Board," etc., though each with its own variation, contribution, and disagreement). And the reverse direction also holds true.

In relation to the change-constancy principle, if there is change (e.g., the ever new ways to study accounting, as shown in the work on accounting principles), there is constancy (e.g., the ever constant existence of different problems with the ever new ways to study accounting, as shown in the problems with the work on accounting principles pointed out by the critics). And the reverse direction also holds true.

In relation to the order-chaos principle, if there is order (e.g., the more or less orderly view on accounting princples, if looked from the sole vantage point of the advocates of the 3 main ones aforediscussed), there is chaos (e.g., the more or less chaotic view on accounting princples, if looked from the multiple, conflicting vantage points of both the advocates and the critics, especially in regard to the different problems aforecited, such that they do not add up to much of anything coherent). And the reverse direction also holds true.

In relation to the slowness-quickness principle, if there is slowness (e.g., the relatively slower speed for businesses to conduct transactions nowadays, in the absence of any accounting conventions like the 3 accounting principles), there is quickness (e.g., the relatively quicker speed for businesses to conduct transactions nowadays, with the help of accounting conventions like the 3 accounting principles). And the reverse direction also holds true.

In relation to the expansion-contraction principle, if there is expansion (e.g., the relatively more developed ability to deal with "those transactions that can be expressed using monetary units" by those favoring "the principle of the unit of measure"), there is contraction (e.g., the relatively less developed ability to deal with "other transactions which cannot be so easily expressed using monetary units, and good examples include "carbon trading," the costs and benefits of "religious salvation," etc. among those favoring "the principle of the unit of measure"). And the reverse direction also holds true.

In relation to the optimality-nonoptimality principle, if there is optimality (e.g., the search for highly favorable options, like the quest for the comprehensive reform of accounting principles, such as the proposal by Jeremy Nicholls of the SROI Network), there is nonoptimality (e.g., the existence of non-optimal alternatives to optimality, like the argument for more realistic studies by the critics, who pointed out different problems with accounting reform as viewed by different sides in the debate). And the reverse direction also holds true.

In relation to the simultaneity-nonsimultaneity principle, if there is simultaneity (e.g., the occurrence of different things more or less at the same time, such as the simultaneous co-existence of the "Historical Cost model" and the "Constant Purchasing Power model" in our time), there is nonsimultaneity (e.g., the occurrence of something after something else, such as the creation of the "Constant Purchasing Power model" by the "International Accounting Standards Board" in 1989 after the establishment of the "Historical Cost model" several hundred years ago). And the reverse direction also holds true.

In relation to the isolation-interaction principle, if there is isolation (e.g., the state of a situation that is separate from others, such as the isolation of the "Historical Cost model" from any accounting reform, prior to the creation of the "Constant Purchasing Power model" by the "International Accounting Standards Board"), there is interaction (e.g., the influence of entities on one another, such as the influence of the "Constant Purchasing Power model" on accounting, after it was proposed by the "International Accounting Standards Board"). And the reverse direction also holds true.

In relation to the theory-praxis principle, if there is theory (e.g., the theoretical construction of "purchasing power" in the field of economics), there is praxis (e.g., the practical application of "purchasing power" in the field of economics to the field of accounting, like the "Constant Purchasing Power model" by "International Accounting Standards Board"). And the reverse direction also holds true.

In relation to the convention-novelty principle, if there is convention (e.g., the conventional wisdom about the 3 accounting principles), there is novelty (e.g., the alternative novel challenge to the conventional wisdom about the 3 accounting principles—by the new idea that they are out-dated and need to be reformed, as discussed earlier). And the reverse direction also holds true.

In relation to the evolution-transformation principle, if there is evolution (e.g., the natural evolution of humans to measure things in the state of nature in a primitive way), there is transformation (e.g., the technical transformation of human ability to measure things by the invention of accounting principles like "the unit of measure"). And the reverse direction also holds true.

In relation to the symmetry-asymmetry principle, if there is symmetry (e.g., the co-existence of different accounting principles, like the "Constant Purchasing Power model" and the "Historical Cost model"), there is asymmetry (e.g., the "Historical Cost model" is "still used in most accounting systems"—but the "Constant Purchasing Power model" is not widely used yet). And the reverse direction also holds true.

In relation to the softness-hardness principle, if there is softness (e.g., the defense of the 3 accounting principles by the advocates aforecited), there is hardness (e.g., the critique of the 3 accounting principles, as shown in the problems pointed out by the critics aforecited). And the reverse direction also holds true.

In relation to the seriousness-playfulness principle, if there is seriousness (e.g., the serious business of those who try to evaluate the "historical cost principle"), there is playfulness (e.g., the playful part of those who try to evaluate the "historical cost principle," when they play around with different possible advantages and disadvantages over the centuries, like the 3 advantages and 3 disadvantages aforecited). And the reverse direction also holds true.

In relation to the activeness-inactivess principle, if there is activeness (e.g., the relative more activeness of the "Historical Cost model," because it is "still used in most accounting systems"), there is inactiveness (e.g., the relative inactiveness of the "Constant Purchasing Power model," because it is still not used in most accounting systems). And the reverse direction also holds true.

In relation to the selfness-otherness principle, if there is selfness (e.g., the tendency to be mostly centered on one's self or group, as shown in the defense of the continued use of the "Historical Cost model" in most accounting systems," in spite of the problems aforecited), there is otherness (e.g., the tendency to be mostly centered on others, as shown in the academic analysis of both sides of the debate on accounting principles, as an advancement in the knowledge of accounting for humanity). And the reverse direction also holds true.

In relation to the regression-progression principle, if there is regression (e.g., the regression made by the 3 accounting principles, as shown in the problems pointed out by the critics), there is progression (e.g., the progress made by the 3 accounting principles, as shown in the contribution to the facilitation of business transactions aforecited). And the reverse direction also holds true.

In relation to the same-difference principle, if there is similarity in outcome (e.g., the contribution to the molding and control of beliefs, values, and behaviors, regardless of whether this be done by way of the continuation of the existing 3 accounting principles, or by way of the proposal of new accounting principles as discussed earlier), there is difference in outcome (e.g., the contribution to the molding and control of beliefs, values, and behaviors by way of the continuation of the existing 3 accounting principles for a relatively more conservative lifeworld—but the contribution to the molding and control of beliefs, values, and behaviors by way of the proposal of new accounting principles for a relatively more liberal lifeworld). And the reverse direction also holds true.

In relation to the stability-reaction principle, if there is stability (e.g., the relatively more stable condition of the 3 accounting principles as developed in the past few hundred years), there is reaction (e.g., the 3 accounting principles as developed in the past few hundred years then led to further changes later on, when nowadays there is the call for accounting reform to create new accounting principles to replace the old ones). And the reverse direction also holds true.

In relation to the functionality-nonfunctionality principle, if there is functionality (e.g., the presence of action for which something or someone is fitted, as shown in the establishment of the 3 accounting principles, which serves the function of providing an accounting groundwork for business transactions in the last few hundred years), there is nonfunctionality (e.g., the relative lack of action for which something or someone is fitted, as shown in the problems with the 3 accounting principles aforediscussed, such that the 3 principles are not as functional as the advocates would like us to believe). And the reverse direction also holds true.

In relation to the intentionality-nonintentionality principle, if there is intentionality (e.g., the planning or design of something for a certain outcome, as shown in the original intention of those who proposed the "Historical Cost model" in the older days so as to provide an accounting groundwork for the facilitation of business transactions), there is nonintentionality (e.g., the relative absence of planning or design of something for a certain outcome, such that the outcome can even be counter-intentional in being the opposite of what was originally planned, as shown in the creation of "the Constant Purchasing Power model" as "an alternative basic accounting model" to the "Historical Cost model" by the "International Accounting Standards Board" in 1989—contrary to the original intention of those who proposed the "Historical Cost model" in the older days so as to provide an accoutning groundwork for the facilitation of business transactions). And the reverse direction also holds true.

And in relation to the survivability-nonsurvivability principle, if there is survivability (e.g., the relative survival, nowadays, of the "Historical Cost model," as it is "still used in most accounting systems"), there is non-survivability (e.g., the relative non-survival, nowadays, of the "Historical Cost model" among those in the "International Accounting Standards Board" who proposed in 1989 "the Capital Maintenance in Units of Constant Purchasing Power model" as "an alternative basic accounting model" to the "Historical Cost model"). And the reverse direction also holds true.

Addition, Legal Traditions, and Accounting Standards

Another way to evaluate the possibility (or impossibility) and desirability (or undesirability) of addition from the perspective of culture concerns addition, legal traditions, and accounting standards.

An argument in the literature is that different countries have different accounting standards, so as to facilitate (or increase) business transactions which can fit in specific cultural (legal) traditions; for instance, "usually, countries practicing civil law system write standards into law and countries with English common law systems have private organizations to set the rules." (WK 2013jj)

Rationale for Uniform Accounting Standards

The term "accounting standard" refers to "accounting practices" which "require publicly traded companies to follow certain accounting rules when presenting financial statements so that the readers of the statements can easily compare different companies. Private companies are also often required by banks and shareholders, for example, to present information according to their specified rules." (WK 2013jj)

There is a good rationale for uniform accounting standards in a country, and nothing illustrates this better than "the lack of transparent accounting standards in some nations" which "has been cited as increasing the difficulty of doing business in them. In particular, the Asian financial meltdown in the late 1990s has been partially attributed due to the lack of detailed accounting standards. Giant firms in some Asian countries were able to take advantage of their poorly devised accounting standards to cover up immense debts and losses, which yielded a collective effect that eventually led the whole region into financial crisis." (WK 2013jj)

There are different "accounting standards" in the world, and some good examples include the following: (WK 2013jj)

- Ex: "generally accepted accounting principles (India)"
- Ex: "generally accepted accounting principles (United States)"
- Ex: "generally accepted accounting principles (United Kingdom)"
- Ex: "generally accepted accounting principles (France)"
- Ex: "grundsätze ordnungsmäßiger Buchführung (Germany)"
- Ex: "Chinese accounting standards (China)"

Surely, there can be others, so the examples above are illustrative, not exhaustive.

Case Study: United States Generally Accepted Accounting Principles

As a case study, consider the United States Generally Accepted Accounting Principles (or US GAAP), in what follows (as summarized in *Table 2.6*). (WK 2013ll)

(a) Brief History

Historically, "accounting standards have…been set by the American Institute of Certified Public Accountants (AICPA) subject to Securities and Exchange Commission regulations." (WK 2013ll)

In 1939, "the AICPA first created the Committee on Accounting Procedure…, and replaced that with the Accounting Principles Board in 1959. In 1973, the Accounting Principles Board was replaced by the Financial Accounting Standards Board (FASB) under the supervision of the Financial Accounting Foundation with the Financial Accounting Standards Advisory Council serving to advise and provide input on the accounting standards. Other organizations involved in determining United States accounting standards include the Governmental Accounting Standards Board (GASB), formed in 1984, and the Public Company Accounting Oversight Board (PCAOB)." (WK 2013ll)

By 2008, "the FASB issued the FASB Accounting Standards Codification, which reorganized the thousands of US GAAP pronouncements into roughly 90 accounting topics" (WK 2013ll)

(b) Basic Objectives

The basic objectives of US GAAP are to structure financial reporting, so that it should provide the following types of information: (WK 2013ll)

- "useful to present to potential investors and creditors and other users in making rational investment, credit, and other financial decisions"

- "helpful to present to potential investors and creditors and other users in assessing the amounts, timing, and uncertainty of prospective cash receipts
- helpful to understand "economic resources, the claims to those resources, and the changes in them"
- "helpful in making financial decisions"
- "helpful in making long-term decisions"
- "helpful in improving the performance of the business"
- "useful in maintaining records"

Of course, there can be other objectives, since the examples above are illustrative, not exhaustive.

(c) Basic Assumptions

To achieve its objectives, US GAAP has four basic assumptions, as shown below: (WK 2013ll)

- "Accounting Entity"—"assumes that the business is separate from its owners or other businesses. Revenue and expense should be kept separate from personal expenses."
- "Monetary Unit principle"—"assumes a stable currency is going to be the unit of record. The FASB accepts the nominal value of the US Dollar as the monetary unit of record unadjusted for inflation."
- "Going Concern"—"assumes that the business will be in operation indefinitely. This validates the methods of asset capitalization, depreciation, and amortization. Only when liquidation is certain this assumption is not applicable. The business will continue to exist in the unforeseeable future."
- "The Time-Period principle"—"implies that the economic activities of an enterprise can be divided into artificial time periods."

Surely, there can be other assumptions, since the examples above are illustrative, not exhaustive.

(d) Basic Principles

To achieve its objectives, US GAAP makes good use of four basic principles, as shown below: (WK 2013ll)

- "Historical Cost Principle"—"requires companies to account and report based on acquisition costs rather than fair market value for most assets and liabilities. This principle provides information that is reliable (removing opportunity to provide subjective and potentially biased market values), but not very relevant. Thus there is a trend to use fair values. Most debts and securities are now reported at market values."

- "Revenue Recognition Principle"—"requires companies to record when revenue is (1) realized or realizable and (2) earned, not when cash is received. This way of accounting is called accrual basis accounting."
- "Matching Principle"—requires that "expenses have to be matched with revenues as long as it is reasonable to do so. Expenses are recognized not when the work is performed, or when a product is produced, but when the work or the product actually makes its contribution to revenue. Only if no connection with revenue can be established, cost may be charged as expenses to the current period (e.g. office salaries and other administrative expenses). This principle allows greater evaluation of actual profitability and performance (shows how much was spent to earn revenue). Depreciation and Cost of Goods Sold are good examples of application of this principle."
- "Full Disclosure Principle"—requires that "amount and kinds of information disclosed should be decided based on trade-off analysis as a larger amount of information costs more to prepare and use. Information disclosed should be enough to make a judgment while keeping costs reasonable. Information is presented in the main body of financial statements, in the notes or as supplementary information."

Again, there can be other principles, since the examples above are illustrative, not exhaustive.

(e) Basic Constraints

And to achieve its objectives, US GAAP pays attention to four basic constraints, as shown below: (WK 2013ll)

- "Objectivity Principle"—means that "the company financial statements provided by the accountants should be based on objective evidence."
- "Consistency Principle"—means that "the company uses the same accounting principles and methods from period to period."
- "Materiality Principle"—means that "the significance of an item should be considered when it is reported. An item is considered significant when it would affect the decision of a reasonable individual."

- "Conservatism Principle"—means that "when choosing between two solutions, the one that will be least likely to overstate assets and income should be picked."

Once more, there can be other constraints, since the examples above are illustrative, not exhaustive.

Problems with the Study on Legal Traditions and Accounting Standards

However, to the critics, this study on addition, legal traditions, and accounting standards has some major problems. Consider, for illustration, a few examples below (as summarized in *Table 2.6*).

Firstly, one main criticism of the study on addition, legal traditions, and accounting standards is that accounting standards do not necessarily prevent accounting scandals.

For example, as the critics are quick to point out, "the accounting scandals of the early 21st century, which affected companies such as Worldcom and Enron, have shown the limitations of accounting standards in the United States." (WK 2013jj)

Secondly, another main criticism of the study on addition, legal traditions, and accounting standards is that there is a call for more global accounting standards.

For instance, "International Financial Reporting Standards (IFRS) are designed as a common global language for business affairs so that company accounts are understandable and comparable across international boundaries. They are a consequence of growing international shareholding and trade and are particularly important for companies that have dealings in several countries. They are progressively replacing the many different national accounting standards. The rules to be followed by accountants to maintain books of accounts which is comparable, understandable, reliable and relevant as per the users internal or external." (WK 2013kk) In the case of the U.S., "the SEC expressed their aim to fully adopt International Financial Reporting Standards in the U.S. by 2014." (WK 2013ll; A. Jeffers 2010)

Thirdly, still another main criticism of the study on addition, legal traditions, and accounting standards is that there is no general consensus on the exact number of constraints in US GAAP.

For instance, in 2010, "due to recent developments in the convergence of US GAAP and IFRS, SFAC No. 8 replaced SFAC No. 1 and 2 in September" and "Chapter 3 of SFAC No 8 includes only the following constraint," as shown below: (WK 2013ll)

- "Cost Constraint"—means that "the benefits of reporting financial information should justify and be greater than the costs imposed on supplying it. Conservatism is no longer a constraint, and materiality is a feature of relevance that is determined at the entity-specific level."

Of course, there may be more changes in the future on how to interpret constraints.

Fourthly, an additional main criticism of the study on addition, legal traditions, and accounting standards is that there is also no general consensus on the exact number of accounting principles which should be included.

For instance, as already discussed in the previous section (i.e., *Sec. 2.5.2* on "Addition, Values, and Accounting Principles"), the "Accounting for Value" website published an interesting article titled "Encouraging a Wider Discussion on the Nature of Accounting Principles," in which there are only three accounting principles to be addressed. (ACCO 2013) But in the article here, there are more than 10 principles listed. (WK 2013ll)

And fifthly, another additional main criticism of the study on addition, legal traditions, and accounting standards is that different works make different (even conflicting) claims about what really makes a principle to be an "assumption," a "requirement," or a "constraint" instead.

For instance, in the article aforecited, the "monetary unit principle" is classified as an "assumption, the "historical cost principle" is classified as a "requirement," and the "objectivity principle" is classified as a "constraint." (WK 2013ll) But this kind of classification seems quite arbitrary, when contrasted with two other works discussed earlier. The first contrasting work has to do with the "recent developments in the convergence of US GAAP and IFRS, SFAC No. 8 replaced SFAC No. 1 and 2 in September" and "Chapter 3 of SFAC No 8" deletes all the "constraints" aforecited (including the so-called "objectivity principle") and includes only the "cost constraint" (not even on the previous list of "constraints" in US GAAP), (WK 2013ll) And the second contrasting work has to do with my earlier analysis in *Sec. 2.3.2* on "Addition, Space-Time, and Account," where Sadao Takatera (2000) and Norio Sawabe claimed that the "historical cost principle" has two assumptions, namely, the "constant value of money" and the "constant time value of money" (but it is not listed as an assumption at all in the article in this section).

Beyond Legal Traditions and Accounting Standards

These criticisms of the study on addition, legal traditions, and accounting standards do not entail, however, a total rejection of its usefulness but serve to show us the two opposing sides of the debate, such that the possibility (or impossibility) and desirability (or undesirability) of addition (from the perspective of culture with addition, legal traditions, and accounting standards as a case study here) are not to the extent that the respective defenders would like us to believe.

In fact, the analysis of addition, legal traditions, and accounting standards can teach us something important about the ontological principles in existential dialectics, and good examples include the formalness-informalness principle, the absoluteness-relativeness principle, the partiality-totality principle, the predictability-unpredictability principle, the explicability-inexplicability principle, the fiction-reality principle, the cognitiveness-noncognitiveness principle, the finiteness-transfiniteness principle, the preciseness-vagueness principle, the simpleness-complicatedness principle, the openness-hiddenness principle, the denseness-emptiness principle, the rule-exception principle, the prototypicality-variation principle, the change-constancy principle, the order-chaos principle, the slowness-quickness principle, the expansion-contraction principle, the optimality-nonoptimality principle, the simultaneity-nonsimultaneity principle, the isolation-interaction principle, the theory-praxis principle, the convention-novelty principle, the evolution-transformation principle, the symmetry-asymmetry principle, the softness-hardness principle, the seriousness-playfulness principle, the activeness-inactivess principle, the selfness-otherness principle, the regression-progression principle, the same-difference principle, the stability-reaction principle, the functionality-nonfunctionality principle, the intentionality-nonintentionality principle, and the survivability-nonsurvivability principle.

For instance, in relation to the formalness-informalness principle, if there is formalness (e.g., the formal logical requirement of soundness, completeness, and consistency in a system of ideas, as in the argument on accounting standards), there is informalness (e.g., the non-formal existence of unsoundness, incompleteness, and inconsistency in the argument on accounting standards, as shown in the criticisms that accounting standards do not necessarily prevent accounting scandals, that there is a call for more global accounting standards, that there is no general consensus on the exact number of constraints in US GAAP, that there is also no general consensus on the exact number of accounting principles which should be included, and that different works make different (even conflicting) claims about what really makes a principle to be an "assumption," a "requirement," or a "constraint" instead). And the reverse direction also holds true.

In relation to the absoluteness-relativeness principle, if there is absoluteness (e.g., the absolute view on the "cost constraint" as the only constraint accepted by SFAC No. 8 in 2010, which replaced SFAC No. 1 and 2), there is relativeness (e.g., what is right for SFAC No. 8 in 2010 in regard to the "cost constraint" as the only constraint accepted is not necessarily so for SFAC No. 1 and 2, which did include 4 constraints aforecited, but not the "cost constraint"). And the reverse direction also holds true.

In relation to the partiality-totality principle, if there is partiality (e.g., the partial view on constraints by the advocates of SFAC No. 8 in 2010), there is totality (e.g., the more holistic view on constraints, such that the whole is not the sum of its parts, that is, it cannot be reduced to the sum of opposing views by the advocates of SFAC No. 8 in 2010, the advocates of SFAC No. 1 and 2 before 2010, etc., since there will emerge new views in the future not yet known today, just as there are already alternative views nowadays which do not side exclusively with any of them, as is the analysis here). And the reverse direction also holds true.

In relation to the predictability-unpredictability principle, if there is predictability (e.g., the predictable tendency that US GAAP will use SFAC No. 8 and includes only the "Cost Constraint" in the foreseeable future), there is unpredictability (e.g., the more difficult task to predict exactly the extent to which the "Cost Constraint" will remain to be the only constraint listed in international accountng standards in a particular future era). And the reverse direction also holds true.

In relation to the explicability-inexplicability principle, if there is explicability (e.g., the explanation, by those advocates of US GAAP as analyzed in this section, that, to achieve its objectives, US GAAP has four basic assumptions, including the "historical cost principle"), there is inexplicability (e.g., the lack of sufficient explanation, by those advocates of US GAAP as analyzed in this section, of why the four constraints are necessarily valid, since, to other scholars like Sadao Takatera and Norio Sawabe in *Sec. 2.3.2* on "Addition, Space-Time, and Account," the "historical cost principle" is embedded with two assumptions, namely, the "constant value of money" and the "constant time value of money," contrary to the view in the article here that it is not listed as an assumption at all). And the reverse direction also holds true.

In relation to the fiction-reality principle, if there is fiction (e.g., the fictional aspect of "constraints" in accounting standards, insofar as SFAC

No. 8 in 2010 replaced SFAC No. 1 and 2 and accepted the "cost constraint" as the only constraint, while rejecting the "objectivity principle," the "consistency principle," and the "conservatism principle" as constraints), there is reality (e.g., the realistic aspect of "constraints" in accounting standards, insofar as SFAC No. 8 in 2010 focused more on the factor of "cost" in reporting financial info). And the reverse direction also holds true.

In relation to the cognitiveness-noncognitiveness principle, if there is cognitiveness (e.g., a view of things on the basis of reason and evidences, as shown in the finding that "the SEC has a stated goal of moving from US GAAP to the International Financial Reporting Standards [IFRS]"), there is noncognitiveness (e.g., a view of things on the basis of non-cognitive factors like envy, jealousy, power, nationality, race, gender, age, class, greed, lust, status, faith, anger, sadness, joy, fear, wish, etc.—as shown in the bureaucratic "politics," such that, "although the SEC has a stated goal of moving from US GAAP to the International Financial Reporting Standards [IFRS],...progress has been slow and uncertain"). (WK 2013ll; K. Tysiac 2013) And the reverse direction also holds true.

In relation to the finiteness-transfiniteness principle, if there is finiteness (e.g., the finite number of the categories of concepts in US GAAP, like "assumptions," "principles," and "constraints"), there is transfiniteness (e.g., the transfinite number of all the individual instances of business transactions in human history that can be understood on the basis of such concepts like"assumptions," "principles," and "constraints"). And the reverse direction also holds true.

In relation to the preciseness-vagueness principle, if there is preciseness (e.g., the precise identification of 3 assumptions in US GAAP in the article aforecited), there is vagueness (e.g., the vagueness in the identification of the 3 assumptions in US GAAP in the article aforecited, since it is not clear why there must be only 3, not 4, 5, 6, and so on). And the reverse direction also holds true.

In relation to the simpleness-complicatedness principle, if there is simpleness (e.g., the relatively simple analysis of US GAAP by the advocates), there is complicatedness (e.g., the relatively more complicated analysis of US GAPP, by challenging the claims and assumptions, as shown in the problems pointed out by the critics as cited). And the reverse direction also holds true.

In relation to the openness-hiddenness principle, if there is openness (e.g., the open exploration of using concepts like "assumptions," "principles," and "constraints" for US GAAP), there is hiddenness (e.g., the hidden bias in using concepts like "assumptions," "principles," and "constraints" for US GAAP, since different research works make different [even conflicting] claims about what really makes a principle to be an "assumption," a "requirement," or a "constraint" in accounting standards). And the reverse direction also holds true.

In relation to the denseness-emptiness principle, if there is denseness (e.g., the relatively denser concentration of concerns with the "objectivity principle," the "consistency principle," and the "conservatism principle" as constraints among the advocates of SFAC No. 1 and 2), there is emptiness (e.g., the relatively less dense, or more empty, concentration of concerns with the "cost constraint" among the advocates of SFAC No. 1 and 2). And the reverse direction also holds true.

In relation to the rule-exception principle, if there are rules (e.g., the usual generalization, procedure, habit, or tendency that business enterprises in the U.S. are expected to follow US GAAP), there are exceptions (e.g., a case to which a rule does not apply, such as the exceptional case when, "under the AICPA's Code of Professional Ethics under Rule 203-Accounting Principles, a member must depart from GAAP if following it would lead to a material misstatement on the financial statements, or otherwise be misleading"). (WK 2013ll) And the reverse direction also holds true.

In relation to the prototypicality-variation principle, if there is prototypicality (e.g., the prototype, like the example about the original Committee on Accounting Procedure set by the AICPA in 1939), there is variation (e.g., different variants of the protoptye, as shown in many works since the original Committee on Accounting Procedure set by the AICPA in 1939—like the later ones such as "the Financial Accounting Standards Board" in 1973, "the Governmental Accounting Standards Board" in 1984, etc., though each with its own variation, contribution, and disagreement). And the reverse direction also holds true.

In relation to the change-constancy principle, if there is change (e.g., the ever new ways to study accounting, as shown in the work on US GAAP), there is constancy (e.g., the ever constant existence of different problems with the ever new ways to study accounting, as shown in the problems with the work on US GAAP pointed out by the critics). And the reverse direction also holds true.

In relation to the order-chaos principle, if there is order (e.g., the more or less orderly view on accounting standards, if looked from the sole vantage point of the well-established US GAAP), there is chaos (e.g., the more or less chaotic view on accounting standards, if looked from the disorderly vantage point of "the Asian financial meltdown in the late 1990s" which "has been partially attributed due to the lack of detailed accounting standards," as "giant firms…were able to take advantage of their poorly devised accounting standards to cover up immense debts and losses, which yielded a collective effect that eventually led the whole region into financial crisis"). (WK 2013jj). And the reverse direction also holds true.

In relation to the slowness-quickness principle, if there is slowness (e.g., the relatively slower speed for the U.S. to move "from US GAAP to the International Financial Reporting Standards," because "the provisions of which differ considerably from GAAP" so "progress has been slow and uncertain"), there is quickness (e.g., the relatively quicker speed for "the Securities and Exchange Commission" to simply "issue…a preliminary 'roadmap'" in 2008 to move "from US GAAP to the International Financial Reporting Standards"). And the reverse direction also holds true.

In relation to the expansion-contraction principle, if there is expansion (e.g., the relative spread of "the International Financial Reporting Standards" nowadays), there is contraction (e.g., the relative decline of the US GAAP nowadays, as it is in the process of moving to IFRS). And the reverse direction also holds true.

In relation to the optimality-nonoptimality principle, if there is optimality (e.g., the search for highly favorable options, like the quest for the specification of "constraints" in US GAAP to minimize costs and achieve better efficiency and effectiveness), there is nonoptimality (e.g., the existence of non-optimal alternatives to optimality, like the argument for more realistic studies by the critics, who pointed out different problems with US GAAP aforediscussed). And the reverse direction also holds true.

In relation to the simultaneity-nonsimultaneity principle, if there is simultaneity (e.g., the occurrence of different things more or less at the same time, such as the simultaneous co-existence of US GAAP and IFRS nowadays), there is nonsimultaneity (e.g., the occurrence of something after something else, such as the creation of US GAAP prior to the creation of IFRS). And the reverse direction also holds true.

In relation to the isolation-interaction principle, if there is isolation (e.g., the state of a situation that is separate from others, such as the separation of US GAAP from IFRS prior to 2008), there is interaction (e.g., the influence of entities on one another, such as the impact of IFRS on US GAAP after 2008, which aims to move to IFRS by 2014 or later). (WK 2013ll) And the reverse direction also holds true.

In relation to the theory-praxis principle, if there is theory (e.g., the theoretical construction of "a common global language for business affairs" like "IFRS" in the field of finance and accounting), there is praxis (e.g., the practical application of "a common global language for business affairs" like "IFRS" in the field of finance and accounting to countries around the world, since, "as of August 2008, more than 113 countries around the world" use it). (WK 2013kk) And the reverse direction also holds true.

In relation to the convention-novelty principle, if there is convention (e.g., the conventional wisdom about the need for US GAAP since 1939), there is novelty (e.g., the alternative novel challenge to the conventional wisdom about the need for US GAAP since 1939 —by the new idea in the late 2000s about the U.S. to abandon US GAAP and move to IFRS by 2014 or later). And the reverse direction also holds true.

In relation to the evolution-transformation principle, if there is evolution (e.g., the natural evolution of humans to exchange things with each other in the state of nature in a primitive way), there is transformation (e.g., the technical transformation of human ability to exchange things with each other by the invention of US GAAP for a "common…language for business affairs" in the U.S.). (WK 2013kk) And the reverse direction also holds true.

In relation to the symmetry-asymmetry principle, if there is symmetry (e.g., the co-existence of different organizations like "the American Institute of Certified Public Accountants [AICPA]" and its members like individual business enterprises), there is asymmetry (e.g., AICPA issues codes of professional ethics for individual business enterprises to follow—but individual business enterprises do not have the power to issue codes of professional ethics for AICPA to follow). And the reverse direction also holds true.

In relation to the softness-hardness principle, if there is softness (e.g., the defense of US GAAP by the advocates like "the Governmental Accounting Standards Board" and "the Public Company Accounting Oversight Board"), there is hardness (e.g., the critique of US GAAP, as shown in the problems pointed out by the critics aforecited, as there is a call for it to move to IFRS by 2014 or later). And the reverse direction also holds true.

In relation to the seriousness-playfulness principle, if there is seriousness (e.g., the serious business of those who tried to set up the necessary concepts for US GAAP in the older days), there is playfulness (e.g., the playful part of those who tried to set up the necessary concepts for US GAAP in the older days, when they play around with different "assumptions," "principles," and "constraints" over the decades, as discussed earlier). And the reverse direction also holds true.

In relation to the activeness-inactivess principle, if there is activeness (e.g., the relative activeness of those who try to implement IFRS around the world), there is inactiveness (e.g., the relative inactiveness of those who want to keep US GAAP as it was, since there is a call for it to move to IFRS by 2014 or later). And the reverse direction also holds true.

In relation to the selfness-otherness principle, if there is selfness (e.g., the tendency to be mostly centered on one's self or group, as shown in the bureaucratic inertia of US GAAP, since, "although the SEC has a stated goal of moving from US GAAP to the International Financial Reporting Standards,...progress has been slow and uncertain"), there is otherness (e.g., the tendency to be mostly centered on others, as shown in the academic analysis of both sides of the debate on accounting standards, as an advancement in the knowledge of accounting standards for humanity). And the reverse direction also holds true.

In relation to the regression-progression principle, if there is regression (e.g., the regression made by US GAAP, as shown in the problems pointed out by the critics), there is progression (e.g., the progress made by US GAAP, as shown in the contribution to a common langauge for business affairs in the U.S. aforecited). And the reverse direction also holds true.

In relation to the same-difference principle, if there is similarity in outcome (e.g., the contribution to the molding and control of beliefs, values, and behaviors, regardless of whether this be done by way of US GAAP, or by way of IFRS), there is difference in outcome (e.g., the contribution to the molding and control of beliefs, values, and behaviors by way of US GAAP for a relatively more regional lifeworld—but the contribution to the molding and control of beliefs, values, and behaviors by way of IFRS for a relatively more global lifeworld). And the reverse direction also holds true.

In relation to the stability-reaction principle, if there is stability (e.g., the relatively more stable condition of setting up US GAAP in the older days), there is reaction (e.g., US GAAP in the older days then led to further changes later on, when nowadays, there is a call for it to move to IFRS by 2014 or later). And the reverse direction also holds true.

In relation to the functionality-nonfunctionality principle, if there is functionality (e.g., the presence of action for which something or someone is fitted, as shown in the creation of US GAAP, which serves the function of providing a common language for business affairs in the U.S.), there is nonfunctionality (e.g., the relative lack of action for which something or someone is fitted, as shown in the problems with US GAAP aforediscussed, such that it is not as functional as the advocates would like us to believe). And the reverse direction also holds true.

In relation to the intentionality-nonintentionality principle, if there is intentionality (e.g., the planning or design of something for a certain outcome, as shown in the original intention of the "American Institute of Certified Public Accountants" to work on some accounting standards for the U.S., later known as US GAAP), there is nonintentionality (e.g., the relative absence of planning or design of something for a certain outcome, such that the outcome can even be counter-intentional in being the opposite of what was originally planned, as shown in the call nowadays to abandon US GAAP by adopting IFRS by 2014 or later—contrary to the original intention of the "American Institute of Certified Public Accountants" to work on US GAAP). And the reverse direction also holds true.

And in relation to the survivability-nonsurvivability principle, if there is survivability (e.g., the relative survival, nowadays, of IFRS), there is non-survivability (e.g., the relative non-survival, nowadays, of US GAAP, as it is in the process of moving to IFRS by 2014 or later). And the reverse direction also holds true.

The Misuse of Addition

This comprehensive analysis of the possibility (or impossibility) and desirability (or undesirability) of addition in the context of accounting, from the perspectives of the mind, nature, society, and culture, is important to show us the different ways in which addition is both possible and desirable, but not to the extent that the spokespersons from each side would like us to believe.

Addition thus has its misuse, just as it has its use too, as the other side of the same coin.

But this is only one part of a larger story, since there is the other side of the story, which concerns subtraction. Since this chapter already deals with addition, the next chapter will therefore address the issue of subtraction.

These dual analyses should not be underestimated, because those on the side of addition often downgrade subtraction merely as the other side of the story without really appreciating it from the vantage point of subtraction too—and vice versa.

With this in mind, let's now turn to Chapter Three for the study of subtraction and its dual attributes—for which we now turn to Chapter Three.

Table 2.1. Addition and its Two Faces

- **Addition and the Mind**
  - —Ex: addition, deception, and creative accounting
  - —Ex: addition, reputation, and intangible assets
- **Addition and Nature**
  - —Ex: addition, eco-management, and audit scheme
  - —Ex: addition, space-time, and accounting
- **Addition and Society**
  - —Ex: addition, financial performance, and auditing
  - —Ex: addition, government revenue, and the accounting of VAT
- **Addition and Culture**
  - —Ex: addition, values, and accounting principles
  - —Ex: addition, legal traditions, and accounting standards

Notes: The examples in the categories are solely illustrative (not exhaustive), and the comparison is relative (not absolute), nor are they necessarily mutually exclusive. And some can be easily re-classified elsewhere. As generalities, they allow exceptions.
Source: A summary of Ch.2 of *FPHACCO*

Table 2.2. Addition, Deception, and Creative Accounting

- **Characteristics of Creative Accounting**
  —Ex: "unsuitable revenue recognition"
  —Ex: "inappropriate accruals and estimates of liabilities"
  —Ex: "excessive provisions and generous reserve accounting"

- **Motivations of Creative Accounting**
  —Ex: "personal incentives"
  —Ex: "bonus-related pay"
  —Ex: "benefits from shares and share options"
  —Ex: "job security"
  —Ex: "personal satisfaction"

- **Cognitive Psychology to Detect Creative Accounting**
  —Ex: "decision making"
  —Ex: "numbers and behaviors"
  —Ex: "complexity of audits"
  —Ex: "personalities of folks"
  —Ex: "investor mentality"

- **Problems with the Argument on Deception and Creative Accounting**
  —Firstly, the problem with accounting scandals is not wholly the fault of creative accountants.
  —Secondly, the nature of business cycles in the free market makes it easier for creative accountants to cheat.
  —Thirdly, the identification of "fraud" from "error" is difficult in practice.
  —Fourthly, the analysis of "the mind's internal processes" in cognitive psychology to detect fraud is easier said than done.
  —Fifthly, extensive lobbying efforts in the business world can hinder accounting reform to prevent fraud.
  —And sixthly, the role of the government can be part of the problem.

Notes: The examples in the categories are solely illustrative (not exhaustive), and the comparison is relative (not absolute), nor are they necessarily mutually exclusive. And some can be easily re-classified elsewhere. As generalities, they allow exceptions.
Source: From *Sec. 2.2.1* of *FPHACCO*. See book for citations.

Table 2.3. Addition, Eco-Management, and Audit Scheme

- **Implementation of EMAS**
  —Ex: "environmental review"
  —Ex: "environmental policy"
  —Ex: "environmental programme"
  —Ex: "environmental management system"
  —Ex: "environmental audit"
  —Ex: "environmental statement"

- **Benefits of EMAS**
  —Ex: "environmental and financial performance through a systematic framework"
  —Ex: "risk and opportunity management"
  —Ex: "credibility, transparency and reputation"
  —Ex: "employee empowerment and motivation"

- **Problems with Eco-management and Audit Scheme**
  —Firstly, one main criticism of the argument on addition, eco-management, and audit scheme is that the required measurement to implement EMAS is not practical.
  –Secondly, another main criticism of the argument on addition, eco-management, and audit scheme is that the benefits of EMAs vary from country to country and from organization to organization.
  –Thirdly, still another main criticism of the argument on addition, eco-management, and audit scheme is that the implementation is not mandatory.
  —Fourthly, an additional main criticism of the argument on addition, eco-management, and audit scheme is that there are costs of EMAS.
  –And fifthly, another additional main criticism of the argument on addition, eco-management, and audit scheme is that there is no consensus on the audit tools to be used for EMAS.

Notes: The examples in the categories are solely illustrative (not exhaustive), and the comparison is relative (not absolute), nor are they necessarily mutually exclusive. And some can be easily re-classified elsewhere. As generalities, they allow exceptions.
Source: From *Sec. 2.3.1* of *FPHACCO*. See book for citations.

Table 2.4. Addition, Space-Time, and Accounting

- **Evolution of Accounting Conventions**
  —Ex: the pre-going concern period
  —Ex: the going concern period

- **Assumptions on Time in Accrual Accounting Conventions**
  —Ex: assumption 1: the constant value of money
  —Ex: assumption 2: the constant time value of money

- **Assumptions on Space in Accrual Accounting Conventions**
  —Ex: assumption 1: primary enclosing as a gatekeeper
  —Ex: assumption 2: secondary enclosing as demarcated space

- **Problems with the Study of Space-Time in Accounting**
  —Firstly, Takatera and Sawabe used a writing sytle which is too obstruse (and flowery) to be empirically useful for clear analysis.
  —Secondly, the claim by Takatera and Sawabe about the social construction of space-time in modern accrual accounting conventions is exaggeratedly misleading.
  –Thirdly, the claims in the article are often argumentatively relative to the theories that Takatera and Sawabe personally preferred.
  —Fourthly, Takatera and Sawabe confused "chaos theory" with "complexity theory," which then raises the thorny question about their scholarly qualification to use the two fields for the analysis of accounting.
  –And fifthly, Takatera and Sawabe are polemically driven by the discourse on "chaos" to analyze accounting in their idiosyncratic way (with the subsequent bias).

Notes: The examples in the categories are solely illustrative (not exhaustive), and the comparison is relative (not absolute), nor are they necessarily mutually exclusive. And some can be easily re-classified elsewhere. As generalities, they allow exceptions.
Source: From *Sec. 2.3.2* of *FPHACCO*. See book for citations.

Table 2.5. Additon, Values, and Accounting Principles

- **Three Accounting Principles**
  —Ex: the principle of accounting entity
  —Ex: the principle of the unit of measure
  —Ex: the principle of historical cost

- **Problems with the Study of Values and Accounting Principles**
  —Firstly, one main criticism of the argument on addition, values, and accounting principles is that there are opposing sides in the debate on "the principle of historical cost."
  —Secondly, another main criticism of the argument on addition, values, and accounting principles is that there are opposing sides in the debate on "the principle of the unit of measure."
  —Thirdly, still another main criticism of the argument on addition, values, and accounting principles is that there are opposing sides in the debate on "the principle of the accounting entity."
  —Fourthly, an additional main criticism of the argument on addition, values, and accounting principles is that any change to the principles is not costless, even with benefits.
  –And fifthly, another additional main criticism of the argument on addition, values, and accounting principles is that each principle, be it the old one (as defended by its advocates) or any new one (as suggested by the opponents of the old one), has both advantages and disadvantages.

Notes: The examples in the categories are solely illustrative (not exhaustive), and the comparison is relative (not absolute), nor are they necessarily mutually exclusive. And some can be easily re-classified elsewhere. As generalities, they allow exceptions.
Source: From *Sec. 2.5.1* of *FPHACCO.* See book for citations.

Table 2.6. Addition, Legal Traditions, and Accounting Standards

- **Different Accounting Standards**
  —Ex: "generally accepted accounting principles (United States)"
  —Ex: "Chinese accounting standards (China)"

- **Case Study: United States Generally Accepted Accounting Principles**
  —Basic assumptions"
    - Ex: "accounting entity"
    - Ex: "monetary unit principle"
    - Ex: "going concern"
    - Ex: "the time-period principle"

  —Basic Principles
    - Ex: "historical cost principle"
    - Ex: "revenue recognition principle"
    - Ex: "matching Principle"
    - Ex: "full disclosure principle"

  —Basic Constraints
    - Ex: "objectivity principle"
    - Ex: "consistency principle"
    - Ex: "materiality principle"
    - Ex: "conservatism principle"

- **Problems with the Study on Accounting Standards**
  —Firstly, accounting standards do not necessarily prevent accounting scandals.
  —Secondly, there is a call for more global accounting standards.
  —Thirdly, there is no general consensus on the exact number of constraints in US GAAP.
  —Fourthly, there is also no general consensus on the exact number of accounting principles which should be included.
  —And fifthly, different works make different (even conflicting) claims about what really makes a principle to be an "assumption," a "requirement," or a "constraint" instead.

Notes: The examples in the categories are solely illustrative (not exhaustive), and the comparison is relative (not absolute), nor are they necessarily mutually exclusive. And some can be easily re-classified elsewhere. As generalities, they allow exceptions.
Source: From *Sec. 2.5.2* of *FPHACCO*. See book for citations.

• PART THREE •

Subtraction

CHAPTER 3
SUBTRACTION AND ITS DUAL ATTRIBUTES

> Some of the biggest errors are flat-out...subtraction errors.
> —Sue Hales (FAM 2013)

The Application of Subtraction

The study of accounting involves both addition and subtraction—as already indicated in *Sec. 1.3*.

The relationship between addition and subtraction (in the context of accounting) is dialectic, since there is no analysis of addition without that of subtraction, and vice versa—to be eventually transcended altogether later.

Yet, a scholarly danger here is to reduce the analysis of one to that of the other; thus, this book is to devote a whole chapter on each, with the understanding, of course, that the two are related.

This dialectic relationship between addition and subtraction (in the context of accounting) can be revealed by showing how and why the analysis of addition (in Chapter Two) is inadequate without the inclusion of subtraction, just as the examination of subtraction (in Chapter Three) is insufficient without the consideration of addition.

The issues concerning subtraction in the study of accounting are to be addressed in this chapter (Chapter Three), whereas the issues concerning addition in the study of accounting were already analyzed in the previous chapter (Chapter Two).

With this dialectic treatment in mind—a good way to examine subtraction (in the context of accounting) is by way of the evaluation of the extent to which it is in fact both possible (or impossible) and desirable (or undesirable).

This can be done by way of a comprehensive analysis of subtraction (in the context of accounting) from the four perspectives of the mind, nature, society, and culture (in accordance to my sophisticated methodological holism, as explained in *Sec. 1.9*).

In other words, this chapter is organized in four main sections, namely, in relation to (3.2) subtraction and the mind, (3.3) subtraction and nature, (3.4) subtraction and society, and (3.5) subtraction and culture—to be addressed hereafter, in that order (and summarized in *Table 3.1*).

Subtraction and the Mind

Subtration, when examined from the perspective of the mind, can illuminate its possibility (or impossibility) and desirability (or undesirability), and this can be shown by way of two case studies, namely, (3.2.1) subtraction, learning, and control self-assessment, and (3.2.2) subtraction, cognitive biases, and mental accounting—to be addressed in what follows, respectively.

Subtraction, Learning, and Control Self-Assessment

A good way to evaluate the possibility (or impossibility) and desirability (or undesirability) of subtraction from the perspective of the mind is to explore subtraction, learning, and control self-assessment.

An argument in the literature is that individuals in an organization can "learn" by engaging in "control self-assessment" so as to minimize or reduce "the risk of error." (WK 2013mm)

Development of Control Self-Assessment

More precisely, the term "control self-assessment" refers to "a technique developed in 1987 that is used by a wide range of organisations including corporations, charities and government departments, to assess the effectiveness of their risk management and control processes. A 'control process' is a check or process performed to reduce or eliminate the risk of error....There are a number of ways a control self-assessment can be implemented but its key feature is that, in contrast to a traditional audit, the tests and checks are made by staff whose normal day-to-day responsibilities are within the business unit being assessed." (WK 2013mm; G. Joseph 2005)

Historically, in 1987, "control self-assessment was developed by Gulf Canada...when the company's General Auditor, Bruce McCuaig, was dissatisfied with the standard auditing techniques in use following the impact of the Watergate affair on the parent company, Gulf Oil

Corporation. The decision to fully implement control self-assessment at Gulf Canada was driven by a number of factors. These included the presence of a consent decree requiring the company to report on its internal controls and the difficulties it was facing in estimating its oil and gas reserves using more traditional audit measures." (WK 2013mm; IIA 1998)

In the following years, "Gulf Canada developed a framework to support the analysis and evaluation of control processes by operational staff. This included anonymous voting to ensure there was no impediment to staff expressing their views. The approach was first published in *Internal Auditor* in December 1990. Gulf Canada discontinued this facilitated meeting approach in 1997 although it continued with control self-assessment using different techniques." (WK 2013mm; IIA 1998)

Later, "following Gulf Canada's introduction of control self-assessment many private sector organizations implemented similar techniques. In the United States several states made reviews based on control self-assessment practices mandatory as did the Federal Deposit Insurance Corporation and the Canadian Deposit Insurance Corporation." (WK 2013mm; IIA 1998)

At the beginning, though, "external auditors ignored the benefits of control self-assessment even though it was effective at providing audit evidence around the 'soft' areas (such as staff morale) that are critical to the effectiveness of internal control systems." (WK 2013mm; G. Joseph 2001)

However, "after a number of financial scandals, notably the collapse of Robert Maxwell's publishing empire, the United Kingdom government commissioned Adrian Cadbury to chair an investigation into corporate governance. The committee published its report *The Financial Aspects of Corporate Governance* in 1992. In section 4, Reporting and Controls, Cadbury made a number of recommendations that led to the increased adoption of control self-assessment in the UK. In particular section 4.5 of the Code of Practice contained within the report required that the directors of a company should report on the effectiveness of the company's system of internal control in each annual report." (WK 2013mm)

By 2000, "the European Commission approved a white paper on reform that led to a major change in the way the Commission was managed. These changes included recommendations for each department to establish an effective internal control system. To support the implementation of the internal controls the Directorate General BUDGET's Central Financial Service developed a control self-assessment process. This first control self-assessment identified several areas for improvement in internal control across the Commission most notably the need to implement a more systematic approach to risk management. The outcome of this first self-assessment was the implementation of the requirement for every Directorate General to perform a control and risk self-assessment annually." (WK 2013mm; CEN 2002)

Then, "in 2007, the United States implemented the Sarbanes-Oxley Act. In order to comply with section 404 of the Act the company had to perform a top down risk assessment which necessitated the production of an 'internal control report' that affirmed 'the responsibility of management for establishing and maintaining an adequate internal control structure and procedures for financial reporting.' 15 U.S.C. § 7262(a). This report has to include an evaluation of the effectiveness of the internal controls and procedures that are related to financial reporting. To meet this requirement organizations increasingly began to perform a control self-assessment using a recognized standard methodology. The organization's external auditors, who are required to sign-off the internal control report, typically became more deeply involved in the control self-assessment process as it facilitated their later review of the internal control report." (WK 2013mm; T. Engel 2007)

And in 2011, "in the United Kingdom...the Financial Services Authority recognized in its recommendations for the improvement of operational risk management that the assessment of risks through a control self-assessment may be an important means of identifying risks. It also noted that for the assessment to be fully effective it had to be fully integrated into the financial organization's risk-management process." (WK 2013mm; FIN 2011)

Methodology of Control Self-Assessment

Individuals in an organization who want to learn control self-assessment are to be familiar with its "six basic methodologies," as shown below (and summarized in *Table 3.2*): (WK 2013mm; S. Bakshi 2004)

- "Internal Control Questionnaire (ICQ) self-audit"
- "customized questionnaires"
- "control guides"
- "interview techniques"
- "control model workshops"
- "interactive workshops"

Different organizations have different ways to use the methodologies. For example, "the National Institute of Standards and Technology control self-assessment methodology is based on customised questionnaires. It is an IT focused methodology suitable for assessing system based controls. It provides a cost-effective technique to determine the status of information security controls, identify any weaknesses and, where necessary, define an improvement plan. The methodology uses a questionnaire that contains specific control objectives and techniques against a system or group of systems can be tested and measured. The methodology was designed for United States federal agencies but can also be valuable for private sector organisations." (WK 2013mm; M. Swanson 2013)

In addition, "the COBIT methodology can be used for control self-assessment; like the NIST methodology it was designed for IT focused assessments. COBIT's Process Description component provides a reference model of an organisation's processes and their ownership. Its Control Objectives component provides a set of requirements considered necessary for effective control of each IT process with the organisation. Assessment and evaluation of these components using the Management Guidelines component provides an assessment mechanism that generates a maturity model indicating if the organisation is meeting its control objectives." (WK 2013mm; M. Swanson 2013)

And "the Institute of Internal Auditors based its control self-assessment methodology on the Total Quality Management approaches of the 1990s as well as the COSO's framework. The methodology became part of the International Standards for Professional Practice of Internal Auditing and was adopted by a large number of major organisations." (WK 2013mm; R. Moeller 2009)

Implementation of Control Self-Assessment

Control self-assessment is individual-based, so its learning process differs from that of "auditing."

For instance, "the first step in control self-assessment is to document the organization's control processes with the aim of identifying suitable ways of measuring or testing each control. The actual testing of the controls is performed by staff whose day-to-day role is within the area of the organization that is being examined as they have the greatest knowledge of how the processes operate. The two common techniques for performing the evaluations" are shown below: (WK 2013mm; J. McNally 2007; G. Joseph 2001)

- "Workshops"—"that may be but do not have to be independently facilitated, involving some or all staff from the business unit being tested"

- "Surveys or Questionnaires"—"completed independently by the staff."

Obviously, "both approaches are the opposite of formal audits where the auditors, not the business unit staff, will perform the assessment." (WK 2013mm)

In the end, "on completion of the assessment each control may be rated based on the responses received to determine the probability of its failure and the impact if a failure occurred. These ratings can be mapped to produce a heatmap showing potential areas of vulnerability." (WK 2013mm)

Benefits of Control Self-Assessment

The benefits of "control self-assessment" are multiple, as shown below (and summarized in *Table 3.2*): (WK 2013mm; J. McNally 2007; S. Pickett 2010)

- Ex: the creation of "a clear line of accountability for controls"
- Ex: the reduction of "the risk of fraud (by examining data that may flag unusual patterns of transactions)"
- Ex: the "result…in an organization with a lower risk profile"

Also, other "soft" benefits "have been claimed for organizations performing control self-assessment," and hereafter are some examples: (WK 2013mm; J. McNally 2007)

- Ex: "a better understanding of business operations (by both management and operational staff)"
- Ex: "stronger awareness of risk practices"
- Ex: "a reinforced corporate governance regime and internal audit efficiency improvements"

Of course, there can be other examples, so the ones above are illustrative, not exhaustive.

Problems with the Argument on Learning and Control Self-Assessment

However, to the critics, this argument on learning and control self-assessment has some major problems. Consider, for illustration, a few examples below (as summarized in *Table 3.2*).

Firstly, one main criticism of the argument on learning and control self-assessment is that there is no consensus on its methodologies.

For example, besides the 6 aforecited, "a number of other methodologies to standardise the control self-assessment have been published." (WK 2013mm; G. Alvarez 2004 & 2010)

Secondly, another main criticism of the argument on learning and control self-assessment is that there are alternative approaches to control self-assessment.

For instance, "auditing" is an alternative approach to control self-assessment (as already discussed in *Sec. 2.3.1* on "addition, eco-management, and audit scheme" and in *Sec. 2.4.1* on "addition, financial performance, and auditing").

Thirdly, still another main criticism of the argument on learning and control self-assessment is that it does not give sufficient attention to "downside risk." (WK 2013mm)

For instance, "some researchers have criticized control self-assessment as a flawed approach as the way risk is defined and measured is unsophisticated. In particular, control self-assessment may understate risk by not identifying extreme downside risk. An extreme downside risk is a highly improbable event that would have catastrophic consequences if it occurred. These risks should have a high overall risk score (generally calculated as a product of the probability of a risk occurring and the impact if it does occur on a scale of 1 to 5). Individuals performing the control self-assessment are consequently unable to significantly differentiate between risks leading to extreme low probability risks either being excluded from the analysis or grouped together with other more probable (but still unlikely) risks that have a less severe impact." (WK 2013mm; J. Lee 2005)

And fourthly, an additional main criticism of the argument on learning and control self-assessment is that it is too obsessed with the "costs" of risk (but not its "benefits").

For instance, "the continual focus on risk elimination that a control self-assessment can lead to has also been criticised. The process of continual evaluation of risks and making plans to mitigate and eliminate them may lead to an unbalanced corporate culture where risks are eliminated ignoring the risk-return ratio of different business choices." (WK 2013mm; J. Lee 2005)

Beyond Control Self-Assessment

These criticisms of the argument on subtraction, learning, and control self-assessment should not be misunderstood as a total rejection of its validity but serve to show us the two opposing sides of the debate, such that the possibility (or impossibility) and desirability (or undesirability) of subtraction (from the perspective of the mind with subtraction, learning, and control self-assessment as a case study here) are not to the extent that the respective defenders would like us to believe.

In fact, the analysis of subtraction, learning, and control self-assessment is insightful about the ontological principles in existential dialectics, and good examples include the formalness-informalness principle, the absoluteness-relativeness principle, the partiality-totality principle, the predictability-unpredictability principle, the explicability-inexplicability principle, the fiction-reality principle, the cognitiveness-noncognitiveness principle, the finiteness-transfiniteness principle, the preciseness-vagueness principle, the simpleness-complicatedness principle, the openness-hiddenness principle, the denseness-emptiness principle, the rule-exception principle, the prototypicality-variation principle, the change-constancy principle, the order-chaos principle, the slowness-quickness principle, the expansion-contraction principle, the optimality-nonoptimality principle, the simultaneity-nonsimultaneity principle, the isolation-interaction principle, the theory-praxis principle, the convention-novelty principle, the evolution-transformation principle, the symmetry-asymmetry principle, the softness-hardness principle, the seriousness-playfulness principle, the activeness-inactivess principle, the selfness-otherness principle, the regression-progression principle, the same-difference principle, the stability-reaction principle, the functionality-nonfunctionality principle, the intentionality-nonintentionality principle, and the survivability-nonsurvivability principle.

For instance, in relation to the formalness-informalness principle, if there is formalness (e.g., the formal logical requirement of soundness, completeness, and consistency in a system of ideas, as in the argument on control self-assessment), there is informalness (e.g., the non-formal existence of unsoundness, incompleteness, and inconsistency in the argument on control self-assessment, as shown in the criticisms that there is no consensus on its methodologies, that there are alternative approaches to control self-assessment, that it does not give sufficient attention to "downside risk," and that it is too obsessed with the "costs" of risk but not its "benefits"). And the reverse direction also holds true.

In relation to the absoluteness-relativeness principle, if there is absoluteness (e.g., the absolute view on auditing by the advocates), there is relativeness (e.g., what is right for the advocates in regard to auditing is not necessarily so for the critics, since there are alternative approaches like "control self-assessment"). And the reverse direction also holds true.

In relation to the partiality-totality principle, if there is partiality (e.g., the partial view on learning by the advocates of "control self-assessment"), there is totality (e.g., the more holistic view on learning, such that the whole is not the sum of its parts, that is, it cannot be reduced to the sum of opposing views by "control self-assessment," "external auditing," "internal auditing," etc., since there will emerge new views in the future not yet known today, just as there are already alternative views nowadays which do not side exclusively with any of them, as is the analysis here). And the reverse direction also holds true.

In relation to the predictability-unpredictability principle, if there is predictability (e.g., the predictable tendency that those for "control self-assessment" focus on "risk management"), there is unpredictability (e.g., the more difficult task to predict exactly the extent to which a particular "control self-assessment" will not end up understating or ignoring "extreme downside risk" on a particular future occasion). And the reverse direction also holds true.

In relation to the explicability-inexplicability principle, if there is explicability (e.g., the explanation, by those advocates of "control self-assessment," of its desirability on the basis of its reduction or elimination of "the risk of error"), there is inexplicability (e.g., the lack of sufficient explanation, by those advocates of "control self-assessment," of why it is necessarily desirable, since, to the critics, it does not give sufficient attention to "downside risk" and is too obsessed with the "costs" of risk but not its "benefits"). And the reverse direction also holds true.

In relation to the fiction-reality principle, if there is fiction (e.g., the fictional aspect of "control self-assessment," insofar as it does not give sufficient attention to "downside risk" and is too obsessed with the "costs" of risk but not its "benefits"), there is reality (e.g., the realistic aspect of "control self-assessment," insofar as it is "to reduce or eliminate the risk of error"). And the reverse direction also holds true.

In relation to the cognitiveness-noncognitiveness principle, if there is cognitiveness (e.g., a view of things on the basis of reason and evidences, as shown in the finding that "control self-assessment" can be used "to reduce or eliminate the risk of error"), there is noncognitiveness (e.g., a view of things on the basis of non-cognitive factors like envy, jealousy, power, nationality, race, gender, age, class, greed, lust, status, faith, anger, sadness, joy, fear, wish, etc.—as shown in the preference for "inertia," since, "initially external auditors ignored the benefits of control self-assessment even though it was effective at providing audit evidence around the 'soft' areas [such as staff morale] that are critical to the effectiveness of internal control systems"). (WK 2013mm) And the reverse direction also holds true.

In relation to the finiteness-transfiniteness principle, if there is finiteness (e.g., the finite number of the categories of approaches to risk management, like "control self-assessment," "external auditing," and "external auditing"), there is transfiniteness (e.g., the transfinite number of all the individual instances of "errors" in business that have ever existed in human history, regardless of how they can be detected by the available approaches to risk management at the time). And the reverse direction also holds true.

In relation to the preciseness-vagueness principle, if there is preciseness (e.g., the precise identification of 6 benefits of control self-assessment in the article aforecited), there is vagueness (e.g., the vagueness in the identification of the 6 benefits of control self-assessment in the article aforecited, since it is not clear why there must be only 6, not 7, 8, 9,, and so on). And the reverse direction also holds true.

In relation to the simpleness-complicatedness principle, if there is simpleness (e.g., the relatively simple analysis of "control self-assessment" by the advocates), there is complicatedness (e.g., the relatively more complicated analysis of "control self-assessment," by challenging the claims and assumptions, as shown in the problems pointed out by the critics as cited). And the reverse direction also holds true.

In relation to the openness-hiddenness principle, if there is openness (e.g., the open exploration of using "control self-assessment" for the reduction or elimination of "the risk of error"), there is hiddenness (e.g., the hidden bias in using "control self-assessment" for the reduction or elimination of "the risk of error," since, to the critics, it does not give sufficient attention to "downside risk" and is too obsessed with the "costs" of risk but not its "benefits"). And the reverse direction also holds true.

In relation to the denseness-emptiness principle, if there is denseness (e.g., the relatively denser concentration of concerns with the "costs" of risk by those for "control self-assessment"), there is emptiness (e.g., the relatively less dense, or more empty, concentration of concerns with the benefits" of risk by those for "control self-assessment"). And the reverse direction also holds true.

In relation to the rule-exception principle, if there are rules (e.g., the usual generalization, procedure, habit, or tendency that those in "control self-assessment" pay attention to "risk"), there are exceptions (e.g., a case to which a rule does not apply, such as the exceptional case that those in "control self-assessment" may "understate risk by not identifying extreme downside risk" at all). And the reverse direction also holds true.

In relation to the prototypicality-variation principle, if there is prototypicality (e.g., the prototype, like the example about the original "control self-assessment...developed by Gulf Canada in 1987"), there is variation (e.g., different variants of the protoptye, as shown in many studies since the original "control self-assessment...developed by Gulf Canada in 1987" —like the later ones such as the "control self-assessment practices" by "the Federal Deposit Insurance Corporation," "the Canadian Deposit Insurance Corporation," etc., though each with its own variation, contribution, and disagreement). And the reverse direction also holds true.

In relation to the change-constancy principle, if there is change (e.g., the ever new ways to study accounting, as shown in the work on "control self-assessment"), there is constancy (e.g., the ever constant existence of different problems with the ever new ways to study accounting, as shown in the problems with the work on "control self-assessment" pointed out by the critics). And the reverse direction also holds true.

In relation to the order-chaos principle, if there is order (e.g., the more or less orderly view on "risk," if looked from the sole vantage point of by the advocates for "control self-assessment," who may "understate risk by not identifying extreme downside risk"), there is chaos (e.g., the more or less chaotic view on "risk," if looked from the disorderly vantage point of "extreme downside risk," which is "a highly improbable event that would have catastrophic consequences if it occurred"). And the reverse direction also holds true.

In relation to the slowness-quickness principle, if there is slowness (e.g., the relatively slower speed for those in "control self-assessment" to ever detect "extreme downside risk"), there is quickness (e.g., the relatively quicker speed for those in "control self-assessment" to detect "normal" risk). And the reverse direction also holds true.

In relation to the expansion-contraction principle, if there is expansion (e.g., the relative spread of "control self-assessment" since 1987), there is contraction (e.g., the relative decline of those "external auditors" nowadays who still want to "ignore...the benefits of control self-assessment"). And the reverse direction also holds true.

In relation to the optimality-nonoptimality principle, if there is optimality (e.g., the search for highly favorable options, like the quest for the comprehensive way "to reduce or eliminate the risk of error," as in "control self-assessment"), there is nonoptimality (e.g., the existence of non-optimal alternatives to optimality, like the argument for more realistic studies by the critics, who pointed out different problems with "control self-assessment" aforecited). And the reverse direction also holds true.

In relation to the simultaneity-nonsimultaneity principle, if there is simultaneity (e.g., the occurrence of different things more or less at the same time, such as the simultaneous co-existence of different approaches nowadays, like "control self-assessment" and "auditing"), there is nonsimultaneity (e.g., the occurrence of something after something else, such as the creation of "control self-assessment" in 1987, long after the existence of "auditing"). And the reverse direction also holds true.

In relation to the isolation-interaction principle, if there is isolation (e.g., the state of a situation that is separate from others, such as the isolation of "auditing" from "control self-assessment" in the older days, when "external auditors ignored the benefits of control self-assessment"), there is interaction (e.g., the influence of entities on one another, such as the integration of "control self-assessment" into "auditing" nowadays). And the reverse direction also holds true.

In relation to the theory-praxis principle, if there is theory (e.g., the theoretical construction of "a framework to support the analysis and evaluation of control processes by operational staff" by Gulf Canada in the 1980s), there is praxis (e.g., the practical application of "a framework to support the analysis and evaluation of control processes by operational staff" to the public sphere like its application by "the Federal Deposit Insurance Corporation and the Canadian Deposit Insurance Corporation"). And the reverse direction also holds true.

In relation to the convention-novelty principle, if there is convention (e.g., the conventional wisdom about "external auditing" for risk management prior to the 1980s), there is novelty (e.g., the alternative novel challenge to the conventional wisdom about "external auditing" for risk management prior to the 1980s—by the new idea of using "control self-assessment" in 1987). And the reverse direction also holds true.

In relation to the evolution-transformation principle, if there is evolution (e.g., the natural evolution of humans to commit errors in the state of nature), there is transformation (e.g., the technical transformation of human ability to cope with errors by the invention of "control self-assessment"). And the reverse direction also holds true.

In relation to the symmetry-asymmetry principle, if there is symmetry (e.g., the co-existence of different entities like governments and individual businesses), there is asymmetry (e.g., "the United Kingdom government" in 1992 "required that the directors of a company should report on the effectiveness of the company's system of internal control in each annual report" —but individual companies do not have the authority to require the UK government how or what to do about its annual reports). And the reverse direction also holds true.

In relation to the softness-hardness principle, if there is softness (e.g., the defense of "control self-assessment" by Gulf Canada), there is hardness (e.g., the critique of control self-assessment," as shown in the problems pointed out by the critics aforecited). And the reverse direction also holds true.

In relation to the seriousness-playfulness principle, if there is seriousness (e.g., the serious business of those who try to understand the nature of "control self-assessment"), there is playfulness (e.g., the playful part of those who try to understand the nature of "control self-assessment," when they play around over the decades with different basic methodologies, like the 6 ones aforecited). And the reverse direction also holds true.

In relation to the activeness-inactivess principle, if there is activeness (e.g., the relative activeness of those auditors who make good use of "control self-assessment" nowadays), there is inactiveness (e.g., the relative inactiveness of those auditors who still "ignored the benefits of control self-assessment" nowadays). And the reverse direction also holds true.

In relation to the selfness-otherness principle, if there is selfness (e.g., the tendency to be mostly centered on one's self or group, as shown in the problem of greed in "financial scandals, notably the collapse of Robert Maxwell's publishing empire," which later "led to the increased adoption of control self-assessment in the UK"), there is otherness (e.g., the tendency to be mostly centered on others, as shown in the academic analysis of both sides of the debate on "control self-assessment," as an advancement in the knowledge of accounting for humanity). And the reverse direction also holds true.

In relation to the regression-progression principle, if there is regression (e.g., the regression made by "control self-assessment practices," as shown in the problems pointed out by the critics), there is progression (e.g., the progress made by "control self-assessment practices," as shown in the contribution to the reduction of the "risk of error" aforecited). And the reverse direction also holds true.

In relation to the same-difference principle, if there is similarity in outcome (e.g., the contribution to the molding and control of beliefs, values, and behaviors, regardless of whether this be done by way of person-to-person "interview techniques," or by way of IT-focused "customised questionnaires"), there is difference in outcome (e.g., the contribution to the molding and control of beliefs, values, and behaviors by way of person-to-person "interview techniques" for a relatively more people-interactive lifeworld—but the contribution to the molding and control of beliefs, values, and behaviors by way of IT-focused "customised questionnaires" for a relatively more impersonal lifeworld). And the reverse direction also holds true.

In relation to the stability-reaction principle, if there is stability (e.g., the relatively more stable condition of "external auditing" in the older days pior to the creation of "control self-assessment"), there is reaction (e.g., "external auditing" in the older days then led to further changes later on, when "control self-assessment" was created in 1987 as another approach). And the reverse direction also holds true.

In relation to the functionality-nonfunctionality principle, if there is functionality (e.g., the presence of action for which something or someone is fitted, as shown in the creation of "control self-assessment," which serves the function to reduce the "risk of error"), there is nonfunctionality (e.g., the relative lack of action for which something or someone is fitted, as shown in the problems with "control self-assessment," such that it is not as functional as the advocates would like us to believe). And the reverse direction also holds true.

In relation to the intentionality-nonintentionality principle, if there is intentionality (e.g., the planning or design of something for a certain outcome, as shown in the original intention of Gulf Canada to introduce control self-assessment to reduce "risk"), there is nonintentionality (e.g., the relative absence of planning or design of something for a certain outcome, such that the outcome can even be counter-intentional in being the opposite of what was originally planned, as shown in the problem with "control self-assessment," since it does not give sufficient attention to "downside risk," etc.—contrary to the original intention of Gulf Canada to introduce control self-assessment to reduce "risk"). And the reverse direction also holds true.

And in relation to the survivability-nonsurvivability principle, if there is survivability (e.g., the relative survival, nowadays, of "control self-assessment"), there is non-survivability (e.g., the relative non-survival, nowadays, of the habit of those auditors who "initially ignored the benefits of control self-assessment even though it was effective at providing audit evidence around the 'soft' areas [such as staff morale] that are critical to the effectiveness of internal control systems"). (WK 2013mm) And the reverse direction also holds true.

Subtraction, Cognitive Biases, and Mental Accounting

Another way to evaluate the possibility (or impossibility) and desirability (or undesirability) of subtraction from the perspective of the mind is to explore subtraction, cognitive biases, and mental accounting.

An argument in the literature is that people use "mental accounting" to "code, categorize and evaluate economic outcomes" so as to simplify (or subtract things from) their decision-making process.

Theory of Mental Accounting

Richard Thaler (1980) first proposed this "mental accounting" theory, which is "to describe the process whereby people code, categorize and evaluate economic outcomes….In mental accounting theory, *framing* means that the way a person subjectively frames a transaction in their mind will determine the utility they receive or expect. This concept is similarly used in prospect theory, and many mental accounting theorists adopt that theory as the value function in their analysis." (WK 2013o)

An alternative way "to understand mental accounting is that of modified utility function. There are two values attached to any transaction—acquisition value and transaction value. Acquisition value is the money that one is ready to part with for physically acquiring some good. Transaction value is the value one attaches to having a good deal. If the price that one is paying is equal to the mental reference price for the good, the transaction value is zero. If the price is lower than the reference price, the transaction utility is positive." (WK 2013a)

The cost of mental accounting can be understood as "a mental accounting cost or mental transaction cost," which is "a kind of transaction cost," that is, "the cost of making a useful decision, especially of a consumer making a useful decision to buy, and may set a lower bound on useful price granularity in a market." (WK 2013o)

In practice, "one detailed application of mental accounting, the behavioral life cycle hypothesis (Shefrin & Thaler, 1988), posits that people mentally frame assets as belonging to either current income, current wealth or future income and this has implications for their behavior as the accounts are largely non-fungible and marginal propensity to consume out of each account is different." (WK 2013o)

Types of Mental Accounting

There are different types of mental accounting. For illustration, consider a few of them below (as summarized in *Table 3.3*):

(a) Income Accounting

The first type of mental accounting has to do with "income accounting." (R. Thaler 1999: 196-197)

An interesting study is by P. Kooreman (1997), who "studies the spending behavior of families that receive child allowance payments from the Dutch government. He finds that spending on children's clothing is much more sensitive to changes in the designated child allowance than to other income sources....[T]he fact that the child allowance was labeled as such seemed to matter in the way people spent the money. Labeling effects are common." (R. Thaler 1999: 196-197)

(b) Tax Refunds

The second type of mental accounting has to do with "tax refunds." (J. Cross 2013)

An article by the Washington Post "found that where money comes from influences how we decide to spend it. A major example of this…pertains to money we did not expect to get. Generally speaking, human beings are more likely to be impulsive or reckless with unexpected money because, they reason, it was never factored in to their serious financial plans to begin with. Tax refunds are a case in point." (J. Cross 2013)

In other words, "most of us know a tax refund is coming, but rarely know the exact amount. Therefore, we tend to forget about it until it arrives and then regard it as a sort of windfall whose eventual disposal means nothing to the broader financial picture of our lives." (J. Cross 2013)

(c) Birthday Money

The third type of mental accounting has to do with "birthday money." (J. Cross 2013)

Consider this: "As kids…, we all remember getting cards stuffed with cash from our relatives on each birthday. Other than a few young geniuses with the foresight to save some of it, most of us eagerly rushed out and bought the most outlandish and impractical things we could. The reasoning? We weren't expecting the money. Sure, people can criticize your spending choices when it comes to recurring job income, but who could possibly second-guess what you did with a random, one-time birthday payoff?" (J. Cross 2013)

(d) Bonuses

The fourth type of mental accounting has to do with "bonuses." (J. Cross 2013)

In fact, "the very name —'bonus'—suggests a sum of money meant to be seen in a different light than mere ordinary income. Unsurprisingly, many employees are seen spending their bonus money on things they would never be able to justify (to their spouses or to themselves) spending 'regular' income on—boats, new cars or lavish vacations, for instance. Of course, nothing says that these types of purchases are bad or irrational. Far from it!" (J. Cross 2013)

(e) Money You Can Afford to Lose

The fifth type of mental accounting has to do with "Money You Can Afford to Lose." (J. Cross 2013)

Here is a pattern among "investors," who "are, as a group, highly prone to the 'money you can afford to lose' variant of mental accounting. Under this notion, investors view some arbitrary amount of their investment capital as 'play money' which they feel comfortable squandering on speculative and uncertain things. At first glance, this has the makings of sensible decision making. It seems prudent to clearly delineate between money that matters and money that doesn't." (J. Cross 2013)

(f) Safety Capital

The sixth type of mental accounting has to do with "safety capital." (J. Cross 2013)

The opposite of "money I can afford to lose" (in the previous example) is "safety capital," where, "true to the mental accounting fallacy, this is the money people regard as 'money they need.' Consequently, these funds are handled with the prudence and care befitting of money one expects to pay a mortgage, fund retirement accounts or other serious financial purposes." (J. Cross 2013)

In reality, of course, "there is no dividing line between money that matters (safety capital) and money you can afford to lose (risk capital). Any dividing line you honor is nothing more than a misleading mental illusion." (J. Cross 2013)

(g) Lottery Winnings

The seventh type of mental accounting has to do with "lottering winnings." (J. Cross 2013)

In fact, "countless lottery winners have managed to go bankrupt after spending their millions on dubious purchases that 'seemed' to be justified by the unexpected prize they had won. Had these fallen winners spent even slightly more in line with how they spent their job income prior to winning, they would likely still be rich today." (J. Cross 2013)

(h) Money Already Spent

The eighth type of mental accounting has to do with "money already spent." (J. Cross 2013)

This is related to "the sunk cost fallacy," as the Washington Post once published this story: "A man buys an expensive membership in a tennis club. Right after he puts down the money, which is nonrefundable, he hurts his ankle. He grits his teeth and continues to play through the pain — even though not playing would mean much less agony. Mental accounting is behind the problem. Playing is the only way to ensure that the tennis club membership remains in the man's mental category of money well spent. To not play would be to write off the membership cost as a loss, which is more painful to the man than the agony of hobbling through games on an injured ankle." (J. Cross 2013)

(i) Confusing Identical Purchases

And the ninth type of mental accounting has to do with "confusing identical purchases." (J. Cross 2013)

In a 1984 Princeton University study, "imagine you just arrived at a theater and as you reach into your pocket to pull out the $10 ticket you purchased in advance, you discover that it's missing. Would you fork over another $10 to see the movie? Compare that to a second scenario in which you did not buy the ticket in advance, but when you arrive at the theater, you discover you lost a $10 bill. Would you still buy a movie ticket?" (J. Cross 2013)

But the 1984 Princeton University study "discovered that only 46% of the study participants in scenario one said they would spend another $10 to buy another movie ticket. However, a whopping 88% of the subjects in scenario two said they would still spend $10 to buy a theater ticket." (J. Cross 2013)

Problems with the Study of Cognitive Biases and Mental Accounting

However, to the critics, the argument on subtraction, cognitive biases, and mental accounting has some major problems. Consider, for illustration, a few examples below (as summarized in *Table 3.3*).

Firstly, one main criticism of the argument on subtraction, cognitive biases, and mental accounting is that the issue of "rationality" in the debate on mental accounting can be relative.

For instance, in the fourth example about "bonuses," spending them (on "boats, new cars or lavish vacations" instead of something else more productive if the money comes from a different source) may appear questionable from the "rational" perspective of the principle of "fungibility," since "different mental accounts violates one of the basic rules of economics—that money is fungible, or interchangeable" (and should be treated the same in different mental accounts). (S. Vedantam 2007) But why should the economic principle of "fungibility" be so god-like that no violation is allowed or that it must be so desirable? As the poet Walt Whitman well put it, "Do I contradict myself? Very well then I contradict myself, (I am large, I contain multitudes.)" (QUO 2013)

Secondly, another main criticism of the argument on subtraction, cognitive biases, and mental accounting is that cognitive biases in mental accounting, if they exist in some cases, can be more lasting than previously thought.

For instance, to the critics, even when those with cognitive biases in "mental accounting" receive treatment from a CBM [Cognitive Bias Modification] program, "one concern is whether CBM modification procedures will achieve lasting benefits. This is not yet clear from research. Another concern is that participants with cognitive vulnerability are not able to maintain their new learning even within a CBM course, with their negative cognitive bias beginning to re-emerge over time." (WK 2013p)

Thirdly, still another main criticism of the argument on subtraction, cognitive biases, and mental accounting is that there can be missing factors not accounted for in those cases of mental accounting.

For instance, in the eighth example above (about "money already spent"), a possible alternative explanation is that the man simply loved playing tennis a lot, even though he also suffered some pain at the time because of the injury. And in the third example about "birthday money," a reader has this interesting reply: "However birthday money is another matter, if a friend or relative gives you money for your birthday then I think there is an expectation that this money should be spent on a present for yourself. Naturally you shouldn't get junk, maybe something you genuinely need is better suited." (J. Cross 2013)

Fourthly, an additional main criticism of the argument on subtraction, cognitive biases, and mental accounting is that it does not give sufficient attention to the "good" side of mental accounting (but its "bad" side).

For instance, it is not surprising that even the author Jay Cross (2013) conceded that "of course, nothing says that these types of purchases are bad or irrational. Far from it!" And R. Thaler (1999: 202-203) thus added: "A question that has not received much attention is whether mental accounting is good for us. What is the normative status of mental accounting?...Mental accounting procedures have evolved to economize on time and thinking costs and also to deal with self-control problems," although, "as is to be expected, the procedures do not work perfectly." Thus, mental accounting has its "good" side, just as it has its "bad" side, but one should not simply condemn it too negatively too quickly.

And fifthly, another additional main criticism of the argument on subtraction, cognitive biases, and mental accounting is that there are "no easy answers" on how to deal with mental accounting.

For instance, as R. Thaler (1999: 203) humbly put it, "it is not possible to say that the system [of mental accounting] is flawed without knowing how to fix it. Given that optimization is not feasible (too costly) repairing one problem may create another. For example, if we teach people to ignore sunk costs, do they stop abiding by the principle: 'waste not, want not'? If we stop being lured by good deals, do we stop paying attention to price altogether? There are no easy answers."

Beyond Mental Accounting

These criticisms of the argument on subtraction, cognitive biases, and mental accounting do not entail a total rejection of its usefulness but serve to show us the two opposing sides of the debate, such that the possibility (or impossibility) and desirability (or undesirability) of subtraction (from the perspective of the mind with subtraction, cognitive biases, and mental accounting as a case study here) are not to the extent that the respective defenders would like us to believe.

Furthermore, the analysis of subtraction, cognitive biases, and mental accounting can cast a bright light on the ontological principles in existential dialectics, and good examples include the formalness-informalness principle, the absoluteness-relativeness principle, the partiality-totality principle, the predictability-unpredictability principle, the explicability-inexplicability principle, the fiction-reality principle, the cognitiveness-noncognitiveness principle, the finiteness-transfiniteness principle, the preciseness-vagueness principle, the simpleness-complicatedness principle, the openness-hiddenness principle, the denseness-emptiness principle, the rule-exception principle, the prototypicality-variation principle, the change-constancy principle, the order-chaos principle, the slowness-quickness principle, the expansion-contraction principle, the optimality-nonoptimality principle, the simultaneity-nonsimultaneity principle, the isolation-interaction principle, the theory-praxis principle, the convention-novelty principle, the evolution-transformation principle, the symmetry-asymmetry principle, the softness-hardness principle, the seriousness-playfulness principle, the activeness-inactivess principle, the selfness-otherness principle, the regression-progression principle, the same-difference principle, the stability-reaction principle, the functionality-nonfunctionality principle, the intentionality-nonintentionality principle, and the survivability-nonsurvivability principle.

For instance, in relation to the formalness-informalness principle, if there is formalness (e.g., the formal logical requirement of soundness and completeness in a system of ideas, as in the argument on mental accounting), there is informalness (e.g., the non-formal existence of unsoundness and incompleteness in the argument on mental accounting, as shown in the criticisms that the issue of "rationality" in the debate on mental accounting can be relative, that cognitive biases in mental accounting [if they exist in some cases] can be more lasting than previously thought, that there can be missing factors not accounted for in those cases of mental accounting, that it does not give sufficient attention to the "good" side of mental accounting [but its "bad" side], and that there are "no easy answers" on how to deal with mental accounting). And the reverse direction also holds true.

In relation to the absoluteness-relativeness principle, if there is absoluteness (e.g., the absolute view on the principle of "fungibility" by the advocates), there is relativeness (e.g., what is right for the advocates in regard to the principle of "fungibility" is not necessarily so for those who are willing to violate it, since "nothing says that these types of purchases are bad or irrational. Far from it!"). And the reverse direction also holds true.

In relation to the partiality-totality principle, if there is partiality (e.g., the partial view on the principle of "fungibility" by the advocates), there is totality (e.g., the more holistic view on the principle of "fungibility," such that the whole is not the sum of its parts, that is, it cannot be reduced to the sum of opposing views by the advocates, the critics, etc., since there will emerge new views in the future not yet known today, just as there are already alternative views nowadays which do not side exclusively with any of them, as is the analysis here). And the reverse direction also holds true.

In relation to the predictability-unpredictability principle, if there is predictability (e.g., the predictable tendency that those with cognitive biases in "mental accounting" can receive treatment from a CBM [Cognitive Bias Modification] program), there is unpredictability (e.g., the more difficult task to predict exactly the extent to which a patient with treatment from a CBM program will fully recover on a particular future occasion, since, to the critics, "one concern is whether CBM modification procedures will achieve lasting benefits"). And the reverse direction also holds true.

In relation to the explicability-inexplicability principle, if there is explicability (e.g., the explanation, by those advocates of the principle of "fungibility," of its desirability on the basis of a more rational decision making, since people can learn to avoid buying "things they don't need because the deal is too good to pass up" or quitting "early on a good day," etc.), there is inexplicability (e.g., the lack of sufficient explanation, by those advocates of the principle of "fungibility," of why it is necessarily desirable, since, to the critics, "mental accounting procedures have evolved to economize on time and thinking costs and also to deal with self-control problems," among other benefits). (R. Thaler 1999: 203) And the reverse direction also holds true.

In relation to the fiction-reality principle, if there is fiction (e.g., the fictional aspect of mental accounting, insofar as "there is no dividing line between money that matters [safety capital] and money you can afford to lose [risk capital]," so "any dividing line you honor is nothing more than a misleading mental illusion"), there is reality (e.g., the realistic aspect of mental accounting, insofar as "mental accounting procedures...do not work perfectly"). And the reverse direction also holds true.

In relation to the cognitiveness-noncognitiveness principle, if there is cognitiveness (e.g., a view of things on the basis of reason and evidences, as shown in the finding that "the way a person subjectively frames a transaction in their mind will determine the utility they receive or expect"), there is noncognitiveness (e.g., a view of things on the basis of non-cognitive factors like envy, jealousy, power, nationality, race, gender, age, class, greed, lust, status, faith, anger, sadness, joy, fear, wish, etc.—as shown in the "recidivism" of some patients who "are not able to maintain their new learning even within a CBM course, with their negative cognitive bias beginning to re-emerge over time," even though they know that they should not do that). (WK 2013p) And the reverse direction also holds true.

In relation to the finiteness-transfiniteness principle, if there is finiteness (e.g., the finite number of the categories of mental accounting, like the 9 types aforedescribed), there is transfiniteness (e.g., the transfinite number of all the individual instances of mental accounting that have ever occurred in human history, regardless of which categories they belong to). And the reverse direction also holds true.

In relation to the preciseness-vagueness principle, if there is preciseness (e.g., the precise identification of 8 types of mental accounting by J. Cross in the article aforecited), there is vagueness (e.g., the vagueness in the identification of the 8 types of mental accounting by J.

Cross in the article aforecited, since it is not clear why there must be only 8, not 10, 11, 12, 13, and so on). And the reverse direction also holds true.

In relation to the simpleness-complicatedness principle, if there is simpleness (e.g., the relatively simple analysis of mental accounting by the advocates), there is complicatedness (e.g., the relatively more complicated analysis of mental accounting, by challenging the claims and assumptions, as shown in the problems pointed out by the critics as cited). And the reverse direction also holds true.

In relation to the openness-hiddenness principle, if there is openness (e.g., the open exploration of mental accounting to understand how people "subjectively frame…a transaction in their mind"), there is hiddenness (e.g., the hidden bias in the exploration of mental accounting to understand how people "subjectively frame…a transaction in their mind," since the critics point out that the issue of "rationality" in the debate on mental accounting can be relative, that it does not give sufficient attention to the "good" side of mental accounting [but its "bad" side], etc.). And the reverse direction also holds true.

In relation to the denseness-emptiness principle, if there is denseness (e.g., the relatively denser concentration of concerns with "cognitive biases" by those advocates of mental accounting), there is emptiness (e.g., the relatively less dense, or more empty, concentration of concerns with the "good" side of mental accounting by the advocates, since "a question that has not received much attention is whether mental accounting is good for us" in the literature). And the reverse direction also holds true.

In relation to the rule-exception principle, if there are rules (e.g., the usual generalization, procedure, habit, or tendency that "as kids…, we all remember getting cards stuffed with cash from our relatives on each birthday" and "most of us eagerly rushed out and bought the most outlandish and impractical things we could"), there are exceptions (e.g., a case to which a rule does not apply, such as the exceptional case about "a few young geniuses with the foresight to save some of it [birthday money]," while "most of us eagerly rushed out and bought the most outlandish and impractical things we could"). And the reverse direction also holds true.

In relation to the prototypicality-variation principle, if there is prototypicality (e.g., the prototype, like the example about the original work on mental accounting by Richard Thaler), there is variation (e.g., different variants of the protoptye, as shown in many studies since the original work on mental accounting by Richard Thaler—like the later ones by J. Cross, H. Shefrin, etc., though each with its own variation, contribution, and disagreement). And the reverse direction also holds true.

In relation to the change-constancy principle, if there is change (e.g., the ever new ways to study accounting, as shown in the research on mental accounting), there is constancy (e.g., the ever constant existence of different problems with the ever new ways to study accounting, as shown in the problems with the research on mental accounting pointed out by the critics). And the reverse direction also holds true.

In relation to the order-chaos principle, if there is order (e.g., the more or less orderly view on decision making about money, if looked from the sole vantage point of the principle of "fungibility" by the advocates), there is chaos (e.g., the more or less chaotic view on decision making about money, if looked from the multiple, conflicting vantage points of those who mess up their purchases because of different "mental accounts" based on illusions aforecited). And the reverse direction also holds true.

In relation to the slowness-quickness principle, if there is slowness (e.g., the relatively slower speed of those "young geniuses with the foresight to save some of it [birthday money]," to ever buy "the most outlandish and impractical things we could"), there is quickness (e.g., the relatively quicker speed of most "kids" to buy "the most outlandish and impractical things" if given "birthday money"). And the reverse direction also holds true.

In relation to the expansion-contraction principle, if there is expansion (e.g., the relative increased perception of money as "play money" when it is "money you can afford to lose"), there is contraction (e.g., the relative decreased perception of money as "play money" when it is "safety capital"). And the reverse direction also holds true.

In relation to the optimality-nonoptimality principle, if there is optimality (e.g., the search for highly favorable options, like the quest for "the comprehensive account frame" as a way to deal with mental accounting problems), there is nonoptimality (e.g., the existence of non-optimal alternatives to optimality, like the argument for more realistic studies by the critics, who pointed out different problems with any solutions to mental accounting, since "optimization is not feasible [too costly]," because "repairing one problem may create another"). (R. Thaler 1999: 186, 203) And the reverse direction also holds true.

In relation to the simultaneity-nonsimultaneity principle, if there is simultaneity (e.g., the occurrence of different things more or less at the same time, such as the simultaneous co-existence of different cognitive biases in people with "mental accounting"), there is nonsimultaneity (e.g., the occurrence of something after something else, such as the diagnosis of those with cognitive biases in "mental accounting" before it is recommended that they can receive treatment afterwards from a CBM

[Cognitive Bias Modification] program). And the reverse direction also holds true.

In relation to the isolation-interaction principle, if there is isolation (e.g., the state of a situation that is separate from others, such as the isolated mindset of someone with mental accounting, when he treats tax refund as "a sort of windfall whose eventual disposal means nothing to the broader financial picture" of his life), there is interaction (e.g., the influence of entities on one another, such as the situation of someone with mental accounting under "a CBM [Cognitive Bias Modification] program," when he learns how to integrate tax refund with "the broader financial picture" of his life). And the reverse direction also holds true.

In relation to the theory-praxis principle, if there is theory (e.g., the theoretical construction of "cognitive bias" in the field of psychology), there is praxis (e.g., the practical application of the idea of "cognitive bias" in the field of psychology to the field of accouunting, like the work on "mental accounts"). And the reverse direction also holds true.

In relation to the convention-novelty principle, if there is convention (e.g., the conventional wisdom about "mental accounting" as undesirable, as it often leads to cognitve biases), there is novelty (e.g., the alternative novel challenge to the conventional wisdom about "mental accounting" as undesirable—by the new idea of the critics that it does not give sufficient attention to the "good" side of mental accounting [but its "bad" side], etc.). And the reverse direction also holds true.

In relation to the evolution-transformation principle, if there is evolution (e.g., the natural evolution of humans to experience illusions in the state of nature), there is transformation (e.g., the technical transformation of human ability to cope with illusions by the invention of "CBM [Cognitive Bias Modification] program"). And the reverse direction also holds true.

In relation to the symmetry-asymmetry principle, if there is symmetry (e.g., the co-existence of both "good" and "bad" sides of mental accounting), there is asymmetry (e.g., in mainstream literature, "a question that has not received much attention is whether mental accounting is good for us"—but the literature has received much attention in regard to the "bad" side of mental accounting like cognitive biases which led to irresponsible purchases). And the reverse direction also holds true.

In relation to the softness-hardness principle, if there is softness (e.g., the defense of using CBM programs to treat those with cognitive biases in mental accounting by the advocates aforecited), there is hardness (e.g., the critique of using CBM programs to treat those with cognitive biases in mental accounting, as shown in the problems pointed out by the critics aforecited). And the reverse direction also holds true.

In relation to the seriousness-playfulness principle, if there is seriousness (e.g., the serious business of those who try to understand the nature of mental accounting), there is playfulness (e.g., the playful part of those who try to understand the nature of mental accounting, when they play around with its different possibilities over the decades, like the 9 types aforedescribed). And the reverse direction also holds true.

In relation to the activeness-inactivess principle, if there is activeness (e.g., the relative activeness of those who can make more decisions about money which are smart and responsible, without mental accounting), there is inactiveness (e.g., the relative inactiveness of those who suffer from "cognitive vulnerability" in mental accounting, so that they cannot engage in smart, responsible decision making about money). And the reverse direction also holds true.

In relation to the selfness-otherness principle, if there is selfness (e.g., the tendency to be mostly centered on one's self or group, as shown in those who received "birthday money" from relatives and "eagerly rushed out and bought the most outlandish and impractical things" for themselves), there is otherness (e.g., the tendency to be mostly centered on others, as shown in the academic analysis of both sides of the debate on mental accounting, as an advancement in the knowledge of accounting and psychology for humanity). And the reverse direction also holds true.

In relation to the regression-progression principle, if there is regression (e.g., the regression made by the CBM program, as shown in the problems pointed out by the critics), there is progression (e.g., the progress made by the CBM program, as shown in the contribution to the treatment of cognitive biases aforecited). And the reverse direction also holds true.

In relation to the same-difference principle, if there is similarity in outcome (e.g., the contribution to the molding and control of beliefs, values, and behaviors, regardless of whether this be done by way of the observance of the principle of "fungibility," or by way of the appeal to "mental accounts"), there is difference in outcome (e.g., the contribution to the molding and control of beliefs, values, and behaviors by way of the observance of the principle of "fungibility" for a relatively more realistic lifeworld—but the contribution to the molding and control of beliefs, values, and behaviors by way of the appeal to "mental accounts" for a relatively more illusory lifeworld). And the reverse direction also holds true.

In relation to the stability-reaction principle, if there is stability (e.g., the relatively more stable condition of the use of mental accounting among some people), there is reaction (e.g., the use of mental accounting among some people try then led to further changes later on, when nowadays, some propose the treatment of cognitive biases in mental accounting with a CBM program). And the reverse direction also holds true.

In relation to the functionality-nonfunctionality principle, if there is functionality (e.g., the presence of action for which something or someone is fitted, as shown in the proposal of the mental accounting theory, which serves the function of explainng how "a person subjectively frames a transaction in their mind" and "will determine the utility they receive or expect"), there is nonfunctionality (e.g., the relative lack of action for which something or someone is fitted, as shown in the problems with the mental accounting theory aforecited, such that it is not as functional as the advocates would like us to believe). And the reverse direction also holds true.

In relation to the intentionality-nonintentionality principle, if there is intentionality (e.g., the planning or design of something for a certain outcome, as shown in the original intention of those who proposed the CBM program so as to cure those with cognitive biases), there is nonintentionality (e.g., the relative absence of planning or design of something for a certain outcome, such that the outcome can even be counter-intentional in being the opposite of what was originally planned, as shown in the problem with the CBM program, since it was found that "participants with cognitive vulnerability are not able to maintain their new learning even within a CBM course, with their negative cognitive bias beginning to re-emerge over time"—contrary to the original intention of those who proposed the CBM program so as to cure those with cognitive biases). And the reverse direction also holds true.

And in relation to the survivability-nonsurvivability principle, if there is survivability (e.g., the relative survival, nowadays, of the mental accounting theory), there is non-survivability (e.g., the relative non-survival, nowadays, of the naïve belief that a CBM program can cure those with cognitive biases, since it is now recognized that "participants with cognitive vulnerability are not able to maintain their new learning even within a CBM course, with their negative cognitive bias beginning to re-emerge over time"). And the reverse direction also holds true.

Subtraction and Nature

Subtraction, when examined from the perspective of nature, can show us in a controversial way its possibility (or impossibility) and desirability (or undesirability), and this can be shown by way of two case studies, namely, (3.3.1) subtraction, sustainability, and carbon accounting, and (3.3.2) subtraction, the environment, and profit-loss accounting—to be addressed in what follows, respectively.

Subtraction, Sustainability, and Carbon Accounting

A good way to evaluate the possibility (or impossibility) and desirability (or undesirability) of subtraction from the perspective of nature is to explore subtraction, sustainability, and carbon accounting.

An argument in the literature is that "carbon accounting" can be used to "'measure' amounts of carbon dioxide equivalents emitted by an entity," so as to contribute to the reduction of carbon emissions.

Carbon Accounting, and Sustainability

The term "carbon accounting" refers to "processes undertaken to 'measure' amounts of carbon dioxide equivalents emitted by an entity. It is used *inter alia* by nation states, corporations and individuals. Correspondingly, examples for products based upon forms of carbon accounting can be found in national inventories, corporate environmental reports or carbon footprint calculators." (WK 2013s)

(a) Its Relation to Carbon Footprint

Carbon accounting is based on the larger concept of "carbon footprint," which means "the total sets of greenhouse gas emissions caused by an organization, event, product or person"—and "the conceptual name of the carbon footprint originates from ecological footprint discussion, which was developed by Rees and Wackernagel in the 1990s which estimates the number of 'earths' that would theoretically be required if everyone on the planet consumed resources at the same level as the person calculating their ecological footprint." (WK 2013nn)

In general, "greenhouse gases can be emitted through transport, land clearance, and the production and consumption of food, fuels, manufactured goods, materials, wood, roads, buildings, and services. For simplicity of reporting, it is often expressed in terms of the amount of carbon dioxide, or its equivalent of other GHGs, emitted. Most of the carbon footprint emissions for the average U.S. household come from 'indirect' sources, i.e. fuel burned to produce goods far away from the final consumer. These are distinguished from emissions which come from burning fuel directly in one's car or stove, commonly referred to as 'direct' sources of the consumer's carbon footprint." (WK 2013nn)

There are three ways to calculate "carbon footprint," as shown below: (WK 2013nn)

- "Of Products"—"several organizations have calculated carbon footprints of products. The US Environmental Protection Agency has addressed paper, plastic (candy wrappers), glass, cans, computers, carpet and tires....Evaluating the package of some products is key to figuring out the carbon footprint. The key way to determine a carbon footprint is to look at the materials used to make the item. For example, a juice carton is made of an aseptic carton, a beer can is made of aluminum, and some water bottles either made of glass or plastic. The larger the size, the larger the footprint will be."
- "Of Energy"—some "peer-reviewed studies of full life cycle emissions" showed "the carbon footprint of various forms of energy generation: Nuclear, Hydro, Coal, Gas, Solar Cell, Peat and Wind generation technology....These...studies thus concluded that hydroelectric, wind, and nuclear power produced the least CO_2 per kilowatt-hour of any other electricity sources. These figures do not allow for emissions due to accidents or terrorism. Wind power and solar power, emit no carbon from the operation, but do leave a footprint during construction phase and maintenance during operation. Hydropower from reservoirs also has large footprints from initial removal of vegetation and ongoing methane (stream detritus decays anaerobically to methane in bottom of reservoir, rather than aerobically to CO_2 if it had stayed in an unrestricted stream)."
- "Of Heat" and Other Sources—a good example is "the carbon footprint per kilowatt-hour of electricity generated, which is about half the world's man-made CO_2 output. The CO_2 footprint for heat is equally significant and research shows that using waste heat from power generation in combined heat and power district heating, chp/dh has the lowest carbon footprint. much lower than micro-power or heat pumps."

Of course, there can be other examples, so the ones above are illustrative, not exhaustive.

(b) Its Relation to Sustainability Measurement

As a kind of "sustainability measurement"—"carbon accounting is hoped to provide a factual ground for carbon-related decision-making," so as to further reduce carbon emissions in practice. (WK 2013s)

For instance, "carbon accounting can be used as part of sustainability accounting by for-profit organizations. A corporate or organizational 'carbon' or greenhouse gas (GHG) emissions assessment promises to quantify the greenhouse gases produced directly and indirectly from a business or organization's activities within a set of boundaries." (WK 2013s)

Nowadays, "corporate carbon accounting" can be done by using "mandatory GHG reporting in directors' reports, investment due diligence, shareholder and stakeholder communication, staff engagement, green messaging, and tender requirements for business and government contracts. Accounting for greenhouse gas emissions is increasingly framed as a standard requirement for business. As of June 2011, 60% of UK FTSE 100 companies had published environmental targets, with 53% of these 240+ targets relating to carbon, greenhouse gas emissions or energy reductions (representing 59% of the FTSE 100). In June 2012, the UK coalition government announced the introduction of mandatory carbon reporting, requiring around 1,100 of the UK's largest listed companies to report their greenhouse gas emissions every year. Deputy Prime Minister Nick Clegg confirmed that emission reporting rules would come into effect from April 2013 in his piece for The Guardian." (WK 2013s)

Sustainability Methodology

Sustainability measurement so understood depends on a general methodology which includes "three key dimension of sustainability," as shown below (and summarized in *Table 3.4*): (WK 2013oo)

- "Environmental Factors"
  –Ex: "energy"
  –Ex: "water"
  –Ex: "greenhouse gases"
  –Ex: "emissions"
  –Ex: "hazardous and non hazardous waste"
  –Ex: "recycling"
  –Ex: "packaging"
- "Social Factors"
  –Ex: "community investment"
  –Ex: "working conditions"

–Ex: "human rights and fair trade"
–Ex: "public policy"
–Ex: "diversity"
–Ex: "safety"
–Ex: "anticorruption"

- "Economic Factors"
  –Ex: "accountability/transparency"
  –Ex: "corporate governance"
  –Ex: "stakeholder value"
  –Ex: "economic performance"
  –Ex: "financial performance"

In this way, "carbon accounting" relies on a general "sustainability accounting" which "connects the companies' strategies from a sustainable framework by disclosing information on the three dimensional levels (environment, economical and social)." (WK 2013oo)

Sustainability Measurement

With these three dimensions of sustainability in mind—"in the last couple of decades there has arisen a crowded toolbox of quantitative methods used to assess sustainability," as shown below (and summarized in *Table 3.4*): (WK 2013pp)

- "Benchmarks"—"a benchmark is a point of reference for a measurement. Once a benchmark is established it is possible to assess trends and measure progress. Baseline global data on a range of sustainability parameters is available at list of global sustainability statistics."
- "Indices"—"a sustainability index is an aggregate sustainability indicator that combines multiple sources of data," like "Air Quality Index," "Environmental Performance Index," "Environmental Sustainability Index," etc.
- "Metrics"— "many environmental problems ultimately relate to the human effect on those global biogeochemical cycles that are critical to life. Over the last decade monitoring these cycles has become a more urgent target for research," and good examples include "water cycle," "carbon cycle," "phosphorus cycle," "nitrogen cycle," etc.
- "Auditing"—"sustainability auditing and reporting are used to evaluate the sustainability performance of a company, organization, or other entity using various performance indicators. Popular

auditing procedures available at the global level include...ISO 14000, ISO 14031," etc.

Again, there can be other examples, so the ones above are illustrative, not exhaustive.

Problems with the Study on Carbon Accounting

However, to the critics, this argument on subtraction, sustainability, and carbon accounting has some major problems. Consider, for illustration, a few examples below (as summarized in *Table 3.4*).

Firstly, one main criticism of the argument on subtraction, sustainability, and carbon accounting is that the practice of carbon accounting is not objective at all.

For instance, to the critics, "social scientific studies of accounting challenge this hope, pointing to the socially constructed character of carbon conversion factors or of the accountants' work practice which cannot implement abstract accounting schemes into reality." (WK 2013s; L. Lohmann 2009; I. Lippert 2012; D. MacKenzie 2009)

Secondly, another main criticism of the argument on subtraction, sustainability, and carbon accounting is that different carbon accountants can produce different results (in addition to the problem of "abstraction" in accounting schemes as mentioned above).

For instance, to the critics, "while natural sciences claim to know and measure carbon, for organisations it is usually easier to employ forms of carbon accounting to represent carbon. The trustworthiness of accounts of carbon emissions can easily be contested....Science and Technology Studies scholar Donna Haraway's pluralised concept of knowledge, i.e. knowledges, can well be used to understand better the status of knowledge produced by carbon accounting: carbon accounting produced a version of understanding of carbon emissions. Other carbon accountants would produce other results." (WK 2013s)

Thirdly, still another main criticism of the argument on subtraction, sustainability, and carbon accounting is that "the total carbon footprint cannot be calculated."

For instance, to the critics, "the total carbon footprint cannot be calculated because the large amount of data required and the fact that carbon dioxide can be produced by natural occurrences. It is for this reason that Wright, Kemp, and Williams, writing in the journal *Carbon Management*, have suggested" a more realistic assessment. (WK 2013nn)

Fourthly, an additional main criticism of the argument on subtraction, sustainability, and carbon accounting is that the different "dimensions" of sustainability accounting involves difficult trade-offs in implementation.

For instance, to the critics, "in practice, however, it is difficult to put together policies that promote simultaneously environmental, economic and social goals," as "multiple factors" are involved here, like "sustainability issues that materially affect a company's creation of value, risk and liabilities," "the need for business to appropriately respond to sustainable growth," and so on. (WK 2013oo)

Fifthly, another additional main criticism of the argument on subtraction, sustainability, and carbon accounting is that there are competing approaches to sustainability accounting.

For instance, to the critics, there are competing approaches to sustainbility accounting, and good examples include one approach by "measuring environmental-economic-social interrelationships," the other approach which is more "wealth-based," etc. (WK 2013oo)

And sixthly, still another additional main criticism of the argument on subtraction, sustainability, and carbon accounting is that organizations may not have the incentives to fully implement sustainability accounting.

For instance, as the critics are quick to point out, "it is unrealistic to expect business to voluntarily commit the resources required for full sustainable accounting implementation....The future direction of sustainability accounting and sustain economic development should continue to display the essential quality of diversity." (WK 2013oo)

Beyond Carbon Accounting

These criticisms of the argument on subtraction, sustainability, and carbon accounting do not mean a total rejection of its usefulness but serve to show us the two opposing sides of the debate, such that the possibility (or impossibility) and desirability (or undesirability) of subtraction (from the perspective of nature with subtraction, sustainability, and carbon accounting as a case study here) are not to the extent that the respective defenders would like us to believe.

Besides, the analysis of subtraction, sustainability, and carbon accounting can teach us something useful about the ontological principles in existential dialectics, and good examples include the formalness-informalness principle, the absoluteness-relativeness principle, the partiality-totality principle, the predictability-unpredictability principle, the explicability-inexplicability principle, the fiction-reality principle, the cognitiveness-noncognitiveness principle, the finiteness-transfiniteness principle, the preciseness-vagueness principle, the simpleness-complicatedness principle, the openness-hiddenness principle, the denseness-emptiness principle, the rule-exception principle, the prototypicality-variation principle, the change-constancy principle, the order-chaos principle, the slowness-quickness principle, the expansion-contraction principle, the optimality-nonoptimality principle, the simultaneity-nonsimultaneity principle, the isolation-interaction principle, the theory-praxis principle, the convention-novelty principle, the evolution-transformation principle, the symmetry-asymmetry principle, the softness-hardness principle, the seriousness-playfulness principle, the activeness-inactivess principle, the selfness-otherness principle, the regression-progression principle, the same-difference principle, the stability-reaction principle, the functionality-nonfunctionality principle, the intentionality-nonintentionality principle, and the survivability-nonsurvivability principle.

For instance, in relation to the formalness-informalness principle, if there is formalness (e.g., the formal logical requirement of soundness, completeness, and consistency in a system of ideas, as in the argument on carbon accounting), there is informalness (e.g., the non-formal existence of unsoundness, incompleteness, and inconsistency in the argument on carbon accounting, as shown in the criticisms that the practice of carbon accounting is not objective at all, that different carbon accountants can produce different results, that "the total carbon footprint cannot be calculated," that the different "dimensions" of sustainability accounting involves difficult trade-offs in implementation, that there are competing approaches to sustainability accounting, and that organizations may not have the incentives to fully implement sustainability accounting). And the reverse direction also holds true.

In relation to the absoluteness-relativeness principle, if there is absoluteness (e.g., the absolute view on sustainability measurement by the advocates in regard to the approach by "measuring environmental-economic-social interrelationships"), there is relativeness (e.g., what is right for the advocates in regard to the approach by "measuring environmental-economic-social interrelationships" is not necessarily so for the advocates of the other competing approach which is more "growth-based"). And the reverse direction also holds true.

In relation to the partiality-totality principle, if there is partiality (e.g., the partial view on sustainability measurement by the advocates in regard to the approach by "measuring environmental-economic-social interrelationships"), there is totality (e.g., the more holistic view on sustainability measurement, such that the whole is not the sum of its parts, that is, it cannot be reduced to the sum of opposing views by the advocates of the approach by "measuring environmental-economic-social interrelationships," the other advocates of the opposing approach which is more "growth-based," etc., since there will emerge new views in the future not yet known today, just as there are already alternative views nowadays which do not side exclusively with any of them, as is the analysis here). And the reverse direction also holds true.

In relation to the predictability-unpredictability principle, if there is predictability (e.g., the predictable tendency that different carbon accountants can produce different results), there is unpredictability (e.g., the more difficult task to predict exactly which particular version of carbon accounting result will become popular in a particular future era). And the reverse direction also holds true.

In relation to the explicability-inexplicability principle, if there is explicability (e.g., the explanation, by those advocates, of the desirability of sustainabiltiy measurement in terms of three main dimensions aforedisdescribed), there is inexplicability (e.g., the lack of sufficient explanation, by those advocates, of why sustainabiltiy measurement in terms of three main dimensions aforedisdescribed is necessarily desirable, since, to the critics, "in practice, however, it is difficult to put together policies that promote simultaneously environmental, economic and social goals," as "multiple factors" are involved here). And the reverse direction also holds true.

In relation to the fiction-reality principle, if there is fiction (e.g., the fictional aspect of sustainability measurement, insofar as, "in practice, however, it is difficult to put together policies that promote simultaneously environmental, economic and social goals," as "multiple factors" are involved here), there is reality (e.g., the realistic aspect of sustainabiltiy measurement, insofar as there are three main dimensions aforedescribed). And the reverse direction also holds true.

In relation to the cognitiveness-noncognitiveness principle, if there is cognitiveness (e.g., a view of things on the basis of reason and evidences, as shown in the finding that "carbon accounting is hoped to provide a factual ground for carbon-related decision-making"), there is noncognitiveness (e.g., a view of things on the basis of non-cognitive factors like envy, jealousy, power, nationality, race, gender, age, class, greed, lust, status, faith, anger, sadness, joy, fear, wish, etc.—as shown in the politics of "hope" in carbon accounting, in spite of the challenge by "social scientific studies of accounting" against "this hope, pointing to the socially constructed character of carbon conversion factors or of the accountants' work practice which cannot implement abstract accounting schemes into reality"). And the reverse direction also holds true.

In relation to the finiteness-transfiniteness principle, if there is finiteness (e.g., the finite number of "environmental factors" in sustainability measurement, like "energy," "water," "greenhouse gases," "emissions," "hazardous and non hazardous waste," "recycling," "packaging," etc.), there is transfiniteness (e.g., the transfinite number of all the individual instances of "energy," "water," "greenhouse gases," "emissions," "hazardous and non hazardous waste," "recycling," "packaging," etc. that have ever occurred in history). And the reverse direction also holds true.

In relation to the preciseness-vagueness principle, if there is preciseness (e.g., the precise identification of 3 dimensions of sustainability in the article aforecited), there is vagueness (e.g., the vagueness in the identification of the 3 dimensions of sustainability in the article aforecited, since it is not clear why there must be only 3, not 4, 5, 6, 7, and so on). And the reverse direction also holds true.

In relation to the simpleness-complicatedness principle, if there is simpleness (e.g., the relatively simple analysis of carbon accounting by the advocates), there is complicatedness (e.g., the relatively more complicated analysis of carbon accounting, by challenging the claims and assumptions, as shown in the problems pointed out by the critics as cited). And the reverse direction also holds true.

In relation to the openness-hiddenness principle, if there is openness (e.g., the open exploration of sustainability measurement for carbon accounting), there is hiddenness (e.g., the hidden bias in the exploration of sustainability measurement for carbon accounting, since, to the critics, "social scientific studies of accounting challenge this hope, pointing to the socially constructed character of carbon conversion factors or of the accountants' work practice which cannot implement abstract accounting schemes into reality"). And the reverse direction also holds true.

In relation to the denseness-emptiness principle, if there is denseness (e.g., the relatively denser concentration of concerns with "environmental-economic-social interrelationships" by the advocates), there is emptiness (e.g., the relatively less dense, or more empty, concentration of concerns with "growth-based" economic models by the advocates of the approach by "measuring environmental-economic-social interrelationships"). And the reverse direction also holds true.

In relation to the rule-exception principle, if there are rules (e.g., the usual generalization, procedure, habit, or tendency that "the Kyoto Protocol defines legally binding targets and timetables for cutting the GHG emissions of…countries that ratified the Kyoto Protocol"), there are exceptions (e.g., a case to which a rule does not apply, such as the exceptional situation when "nations which have failed to deliver their Kyoto emissions reductions obligations can enter Emissions Trading to purchase CERS and ERUs to cover their treaty shortfalls"). (WK 2013nn) And the reverse direction also holds true.

In relation to the prototypicality-variation principle, if there is prototypicality (e.g., the prototype, like the example about the original use of "Hubbert peak theory" for sustainabilty measurement "in relation to peak oil production"), there is variation (e.g., different variants of the protoptye, as shown in many studies since the original use of "Hubbert peak theory" for sustainabilty measurement "in relation to peak oil production"—like the later uses of "Hubbert peak theory" for sustainabilty measurement in relation to "other natural resources" like "natural gas," "coal," etc., though each with its own variation, contribution, and disagreement). (WK 2013pp) And the reverse direction also holds true.

In relation to the change-constancy principle, if there is change (e.g., the ever new ways to study accounting, as shown in the work on carbon accounting), there is constancy (e.g., the ever constant existence of different problems with the ever new ways to study accounting, as shown in the problems with the work on carbon accounting pointed out by the critics). And the reverse direction also holds true.

In relation to the order-chaos principle, if there is order (e.g., the more or less orderly view on the study of carbon accounting, if looked from the sole vantage point of the approach which is only "growth-based"), there is chaos (e.g., the more or less chaotic view on the study of carbon accounting, if looked from the multiple, conflicting vantage points of both the advocates and the critics, especially in regard to the problems aforecited, such that they do not add up to much of anything coherent). And the reverse direction also holds true.

In relation to the slowness-quickness principle, if there is slowness (e.g., the relatively slower speed of carbon dioxide emissions after the implementation of different projects to curb them under the Kyoto Protocol), there is quickness (e.g., the relatively quicker speed of carbon dioxide emissions before the implementation of different projects to curb them under the Kyoto Protocol). And the reverse direction also holds true.

In relation to the expansion-contraction principle, if there is expansion (e.g., the relative spread of "sustainable" economic models nowadays), there is contraction (e.g., the relative decline of "growth-based" economic models nowadays). And the reverse direction also holds true.

In relation to the optimality-nonoptimality principle, if there is optimality (e.g., the search for highly favorable options, like the quest for sustainability measurement like the use of "Hubbert peak theory…in relation to peak oil production"), there is nonoptimality (e.g., the existence of non-optimal alternatives to optimality, like the argument for more realistic studies by the critics, who pointed out different problems with sustainability measurement aforecited). And the reverse direction also holds true.

In relation to the simultaneity-nonsimultaneity principle, if there is simultaneity (e.g., the occurrence of different things more or less at the same time, such as the simultaneous co-existence of three different "dimensions" of sustainability aforecited), there is nonsimultaneity (e.g., the occurrence of something after something else, such as the existence of the approach which is more "growth-based," before the existence of the new approach by "measuring environmental-economic-social interrelationships"). And the reverse direction also holds true.

In relation to the isolation-interaction principle, if there is isolation (e.g., the state of a situation that is separate from others, such as the separation of "environmental" and "social" factors from the study of "economic factors" in the older days, before the construction of carbon accounting and sustainability accounting), there is interaction (e.g., the influence of entities on one another, such as the inclusion of "environmental" and "social" factors into the study of "economic factors" in the new approach by "measuring environmental-economic-social interrelationships"). And the reverse direction also holds true.

In relation to the theory-praxis principle, if there is theory (e.g., the theoretical construction of the "pluralised concept of knowledge" in the field of philosophy), there is praxis (e.g., the practical application of the "pluralised concept of knowledge" in the field of philosophy to the field of accounting, like the view that "knowledges…can well be used to understand better the status of knowledge produced by carbon accounting: carbon accounting produced a version of understanding of carbon emissions. Other carbon accountants would produce other results"). And the reverse direction also holds true.

In relation to the convention-novelty principle, if there is convention (e.g., the conventional wisdom about the approach to accounting which is more "growth-based"), there is novelty (e.g., the alternative novel challenge to the conventional wisdom about the approach to accounting which is more "growth-based"—by the new approach on the basis of "measuring environmental-economic-social interrelationships"). And the reverse direction also holds true.

In relation to the evolution-transformation principle, if there is evolution (e.g., the natural evolution of humans to encounter the "natural occurrences" of "carbon dioxide"), there is transformation (e.g., the technical transformation of human ability to deal with "carbon dioxide" by the invention of "carbon accounting"). And the reverse direction also holds true.

In relation to the symmetry-asymmetry principle, if there is symmetry (e.g., the co-existence of different entities, like companies and governments), there is asymmetry (e.g., "in June 2012, the UK coalition government announced the introduction of mandatory carbon reporting, requiring around 1,100 of the UK's largest listed companies to report their greenhouse gas emissions every year"—but these companies do not have the authority to require the British government to report its greenhouse gas emissions to them). (WK 2013s) And the reverse direction also holds true.

In relation to the softness-hardness principle, if there is softness (e.g., the defense of carbon accounting by the advocates aforecited), there is hardness (e.g., the critique of carbon accounting, as shown in the problems pointed out by the critics aforecited). And the reverse direction also holds true.

In relation to the seriousness-playfulness principle, if there is seriousness (e.g., the serious business of those who research on sustainability measurement), there is playfulness (e.g., the playful part of those who research on sustainability measurement, when they play around over the decades with different quantitative methods like "benchmarks," "indices," "metrics," "auditing," etc.). And the reverse direction also holds true.

In relation to the activeness-inactivess principle, if there is activeness (e.g., the relative activeness of those countries in the EU which have participated in "Emissions Trading Schemes" under the Kyoto Protocol, so they are active in trading emissions), there is inactiveness (e.g., the relative inactiveness of those countries like China, India and the United States, which have not participated in "Emissions Trading Schemes" under the Kyoto Protocol, so they are not active in trading emissions). And the reverse direction also holds true.

In relation to the selfness-otherness principle, if there is selfness (e.g., the tendency to be mostly centered on one's self or group, as shown in the finding that "it is unrealistic to expect business to voluntarily commit the resources required for full sustainable accounting implementation"), there is otherness (e.g., the tendency to be mostly centered on others, as shown in the academic analysis of both sides of the debate on sustainability accounting, as an advancement in the knowledge of accounting for humanity). And the reverse direction also holds true.

In relation to the regression-progression principle, if there is regression (e.g., the regression made by carbon accounting, as shown in the problems pointed out by the critics), there is progression (e.g., the progress made by carbon accounting, as shown in the contribution to the reduction of carbon emissions aforecited). And the reverse direction also holds true.

In relation to the same-difference principle, if there is similarity in outcome (e.g., the contribution to the molding and control of beliefs, values, and behaviors, regardless of whether this be done by way of "Emissions Trading," or by way of "the Verified Emission Reductions"), there is difference in outcome (e.g., the contribution to the molding and control of beliefs, values, and behaviors by way of "Emissions Trading," for a relatively mandatory lifeworld—but the contribution to the molding and control of beliefs, values, and behaviors by way of "the Verified Emission Reductions" for a relatively more voluntary lifeworld). (WK 2013nn) And the reverse direction also holds true.

In relation to the stability-reaction principle, if there is stability (e.g., the relatively more stable condition of "growth-based" economic models in the older days), there is reaction (e.g., "growth-based" economic models in the older days then led to further changes later on, when nowadays, there is the alternative idea of "sustainable" economic models). And the reverse direction also holds true.

In relation to the functionality-nonfunctionality principle, if there is functionality (e.g., the presence of action for which something or someone is fitted, as shown in the creation of "carbon accounting," which serves the function of providing a "sustainable" economic model to reduce carbon emissions), there is nonfunctionality (e.g., the relative lack of action for which something or someone is fitted, as shown in the problems with "carbon accounting" aforecited, such that it is not as functional as the advocates would like us to believe). And the reverse direction also holds true.

In relation to the intentionality-nonintentionality principle, if there is intentionality (e.g., the planning or design of something for a certain outcome, as shown in the original intention of M. King Hubbert to propose "Hubbert peak theory" to analyze "oil production"), there is nonintentionality (e.g., the relative absence of planning or design of something for a certain outcome, such that the outcome can even be counter-intentional in being the opposite of what was originally planned, as shown in the application of Hubbert peak theory to non-oil "natural resources" like "natural gas," "coal," etc.—contrary to the original intention of M. King Hubbert to propose "Hubbert peak theory" to analyze "oil production"). (WK 2013pp) And the reverse direction also holds true.

And in relation to the survivability-nonsurvivability principle, if there is survivability (e.g., the relative survival, nowadays, of the approach to sustainbility accounting by "measuring environmental-economic-social interrelationships"), there is non-survivability (e.g., the relative non-survival, nowadays, of the "growth-based" economic approach). And the reverse direction also holds true.

Subtraction, the Environment, & Profit-Loss Accounting

Another good way to evaluate the possibility (or impossibility) and desirability (or undesirability) of subtraction from the perspective of nature is to explore subtraction, the environment, and profit-loss accounting.

An argument in the literature is that "an environmental profit and loss" (E P&L) accounting can help a company evaluate its environmental impacts, so as to reduce "the cost of direct and indirect negative impacts on the environment." (WK 2013qq)

Background of the E P &L Accounting

The term "environmental profit and loss" accounting refers to "a company's monetary valuation and analysis of its environmental impacts including its business operations and its supply chain from cradle-to-gate. An E P&L internalizes externalities and monetizes the cost of business to nature by accounting for the ecosystem services a business depends on to operate in addition to the cost of direct and indirect negative impacts on the environment. The primary purpose of an E P&L is to allow managers and stakeholders to see the magnitude of these impacts and where in the supply chain they occur." (WK 2103qq)

More precisely, "the E P&L analysis provides a metric to measure and monitor the footprint of the company's operations and suppliers all the way to the initial raw materials. It is a tool to build awareness of the importance of nature to the sustainability of businesses; enhance visibility across a company's supply chain and deepen understanding to focus sustainability efforts and implement better-informed operational decisions; improve specificity for risk management regarding environmental dependencies and impacts; and support a more holistic view of a company's performance, while bringing clarity and transparency to stakeholders at all levels and identifying new opportunities to enhance the sustainability of a company's products." (WK 2013qq)

Historically, it was "conceived by PUMA Chairman, Jochen Zeitz, and launched by Sportlifestyle company PUMA and its parent company's sustainability initiative (PPR HOME)," and "the first-ever E P&L was conducted on 2010 data and released in two phases. In May 2011, the valuation of PUMA's 2010 Greenhouse Gas Emissions (GHG) and water usage was announced, followed in November 2011, by PUMA's overall E P&L, which also included valuation results for other forms of air pollution, land conversion and waste. Simultaneously, the PPR Group announced in November 2011 that a Group E P&L would be implemented across its Luxury and Sport & Lifestyle brands by 2015." (WK 2013qq; R. Anderson 2011; BBC 2011; GUA 2011; L. Guevarra 2011; P. Guest 2011)

Methodology of the E P&L Accounting

The methodology of the E P&L accounting was "developed with the support of PricewaterhouseCoopers LLP and Trucost PLC." (WK 2013qq)

For instance, "the E P&L used existing input-output models and developed new valuation methodologies, building on a large volume of work in the fields of environmental and natural resource economics such as TEEB, the UN study on The Economics of Ecosystems and Biodiversity. The TEEB study is hosted by the United Nations Environment Programme (UNEP) with financial support from the European Commission, Germany, the United Kingdom, Netherlands, Norway, Sweden and Japan." (WK 2013qq; J. Ayling 2011)

Ecosystem Services in the E P&L Accounting

The E P&L accounting focuses especially on "ecosystem services" that "a business depends on to operate in addition to the cost of direct and indirect negative impacts on the environment." (WK 2013qq)

After all, "humankind benefits from a multitude of resources and processes that are supplied by ecosystems. Collectively, these benefits are known as ecosystem services and include products like clean drinking water and processes such as the decomposition of wastes. While scientists and environmentalists have discussed ecosystem services for decades, these services were popularized and their definitions formalized by the United Nations 2005 Millennium Ecosystem Assessment [MILL 2005), a four-year study involving more than 1,300 scientists worldwide." (WK 2013rr)

In general, "ecosystem services" are classified into four broad categories as shown below (and summarized in *Table 3.5*): (WK 2013rr;MILL 2005)

- "Supporting Services"—"that are necessary for the production of all other ecosystem services"
  –Ex: "nutrient dispersal and cycling"
  –Ex: "seed dispersal"
  –Ex: "primary production"
- "Provisioning Services"—that produce "products obtained from ecosystems"
  –Ex: "food (including seafood and game), crops, wild foods, and spices"
  –Ex: "water"
  –Ex: "minerals (including diatomite)"
  –Ex: "pharmaceuticals, biochemicals, and industrial products"
  –Ex: "energy (hydropower, biomass fuels)"
- "Regulating Services"—are "benefits obtained from the regulation of ecosystem processes"
  –Ex: "carbon sequestration and climate regulation"
  –Ex: "waste decomposition and detoxification"
  –Ex: "purification of water and air"
  –Ex: "crop pollination"
  –Ex: "pest and disease control"
- "Cultural Services"—are "nonmaterial benefits people obtain from ecosystems through spiritual enrichment, cognitive development, reflection, recreation, and aesthetic experiences"
  –Ex: "cultural, intellectual and spiritual inspiration"
  –Ex: "recreational experiences (including ecotourism)"
  –Ex: "scientific discovery"

Surely, the examples above are illustrative, not exhaustive.

Implementation of the E P&L Accounting

The E P&L accounting has been implemented in different countries.

For instance, "the UK government used the PUMA E P&L as a case study for sustainable business in the Department for Environment, Food and Rural Affair (DEFRA) Natural Environment White Paper in June 2011." (WK 2013qq; JOUR 2011)

Also, "in July 2011, Pavan Sukhdev who was the Study Leader of TEEB and the Special Advisor and Head of UNEP's Green Economy Initiative, referred to the PUMA E P&L in his TED presentation. Sustainability authority, John Elkington includes the PUMA E P&L in his 'The Future Quotient: 50 Stars in Seriously Long-Term Innovation.' In the October issue of The Harvard Business Review, the PUMA E P&L is included in 'The Sustainable Economy' by Yvon Chouinard, Jib Ellison, and Rick Ridgeway." (WK 2013qq; TED 2011; J. Elkington 2013)

And "in the Winter 2012 issue, the Stanford Social Innovation Review published 'Connecting Heart to Head' by Ram Nidumolu [2011], Kevin Kramer, & Jochen Zeitz. The PUMA E P&L is included as a business case study." (WK 2013qq)

Problems with the E P & L Accounting

However, to the critics, this study on the E P & L accounting has some major problems. Consider, for illustration, a few examples below (as summarized in *Table 3.5*).

Firstly, one main criticism of the study on the E P & L accounting is that there are competing definitons of "ecosystem services." (E. Ojea 2010)

For example, an alternative definition which differs from the one above is "the most recent revision by TEEB to synthesize work in this field and prevent double counting in ecosystem services audits," which "has revised the MA definition to remove 'Supporting Services' and replace it on the one hand with 'Habitat Services' and on the other hand with 'Ecosystem Functions' that 'are defined as a subset of the interactions between ecosystem structure and processes that underpin the capacity of an ecosystem to provide goods and services.'" (WK 2013rr; TEED 2010)

Secondly, another main criticism of the study on the E P & L accounting is that the understanding of ecosystem services is culturally relative.

For instance, to the critics, the understanding of ecosystem services can be culturally relative in three ways, namely, (a) "pivotal cultural values attaching to the natural/cultivated environment rely on an area's unique character that cannot be addressed by methods that use universal scientific parameters to determine ecological structures and functions," (b) "if a natural/cultivated environment has symbolic meanings and cultural values, the object of these values are not ecosystems but shaped lifeworldly phenomena like mountains, lakes, forests, and, mainly, symbolic landscapes," and (c) "those cultural values do result not from properties produced by ecosystems but are the product of a specific way of seeing within the given cultural framework of symbolic experience." (WK 2013rr; T. Kirchhoff 2012; D. Cosgrove 1985)

Thirdly, still another main criticism of the study on the E P & L accounting is that there are also competing ways to operationalize the concept of "ecosystem services."

For instance, there are disagreements "as to how the concept of cultural ecosystem services can be operationalized," and a good example is to review different "approaches in landscape aesthetics, cultural heritage, outdoor recreation, and spiritual significance to define and assess cultural values of our environment" by T. Daniel (2012) and others. (WK 2013rr)

Fourthly, an additional main criticism of the study on the E P & L accounting is that ecosystem services can be more complex than previously thought.

For instance, the critics are quick to point out that "the complexity of Earth's ecosystems poses a challenge for scientists as they try to understand how relationships are interwoven among organisms, processes and their surroundings....From defining the problems to finding solutions that can be applied in practical and sustainable ways, there is much to overcome. Considering options must balance present and future human needs, and decision-makers must frequently work from valid but incomplete information. Existing legal policies are often considered insufficient since they typically pertain to human health-based standards that are mismatched with necessary means to protect ecosystem health and services." (WK 2013rr)

Fifthly, another additional main criticism of the study on the E P & L accounting is that there are alternative approaches to explain ecosystem services.

For instance, there are alternative approaches known as "the redundancy hypothesis" (which "assumes that more than one species performs a given role within an ecosystem"),"the portfolio effect" (which "compares biodiversity to stock holdings, where diversification minimizes the volatility of the investment, or in this case, the risk in stability of ecosystem services"), etc. (WK 2013rr; B. Walker 1992; D. Tilman 1998)

Sixthly, still another additional main criticism of the study on the E P & L accounting is that the empirical validity of each study is limited.

For instance, to the critics, "an especially challenging aspect of this process is that interpreting ecological information collected from one spatial-temporal scale does not necessarily mean it can be applied at another....Science alone cannot usually drive ecosystem services decisions as the scientific data and analytical models are uncertain." (WK 2013rr; R. DeFries 2004)

And seventhly, one last main criticism of the study on the E P & L accounting is that there are disagreements about the true value of ecosystem services.

For instance, "a peer-reviewed study published in 1997 estimated the value of the world's ecosystem services and natural capital to be between US$16–54 trillion per year, with an average of US$33 trillion per year. However, Salles (2011) indicates 'The total value of biodiversity is infinite....'" (WK 2013rr; R. Costanza 1997)

Beyond the E P & L Accounting

These criticisms of the argument on subtraction, the environment, and profit-loss accounting do not suggest a total rejection of its usefulness but serve to show us the two opposing sides of the debate, such that the possibility (or impossibility) and desirability (or undesirability) of subtraction (from the perspective of nature with subtraction, the environment, and profit-loss accounting as a case study here) are not to the extent that the respective defenders would like us to believe.

Furthermore, the analysis of subtraction, the environment, and profit-loss accounting can reveal the inner working of the ontological principles in existential dialectics, and good examples include the formalness-informalness principle, the absoluteness-relativeness principle, the partiality-totality principle, the predictability-unpredictability principle, the explicability-inexplicability principle, the fiction-reality principle, the cognitiveness-noncognitiveness principle, the finiteness-transfiniteness principle, the preciseness-vagueness principle, the simpleness-complicatedness principle, the openness-hiddenness principle, the denseness-emptiness principle, the rule-exception principle, the prototypicality-variation principle, the change-constancy principle, the order-chaos principle, the slowness-quickness principle, the expansion-contraction principle, the optimality-nonoptimality principle, the simultaneity-nonsimultaneity principle, the isolation-interaction principle, the theory-praxis principle, the convention-novelty principle, the evolution-transformation principle, the symmetry-asymmetry principle, the softness-hardness principle, the seriousness-playfulness principle, the activeness-inactivess principle, the selfness-otherness principle, the regression-progression principle, the same-difference principle, the stability-reaction principle, the functionality-nonfunctionality principle, the intentionality-nonintentionality principle, and the survivability-nonsurvivability principle.

For instance, in relation to the formalness-informalness principle, if there is formalness (e.g., the formal logical requirement of soundness, completeness, and consistency in a system of ideas, as in the argument on the E P & L accounting), there is informalness (e.g., the non-formal existence of unsoundness, incompleteness, and inconsistency in the argument on the E P & L accounting, as shown in the criticisms that there are competing definitons of "ecosystem services," that the understanding of ecosystem services is culturally relative, that there are also competing ways to operationalize the concept of "ecosystem services," that ecosystem services can be more complex than previously thought, that there are alternative approaches to explain ecosystem services, that the empirical validity of each study is limited, and that there are disagreements about the true value of ecosystem services). And the reverse direction also holds true.

In relation to the absoluteness-relativeness principle, if there is absoluteness (e.g., the absolute view on the "ecological information collected from one spatial-temporal scale" by its advocates), there is relativeness (e.g., what is right for the advocates in regard to the "ecological information collected from one spatial-temporal scale" is not necessarily so for the critics, since "interpreting ecological information collected from one spatial-temporal scale does not necessarily mean it can be applied at another"). And the reverse direction also holds true.

In relation to the partiality-totality principle, if there is partiality (e.g., the partial view on ecosystem services by the advocates of "the redundancy hypothesis"), there is totality (e.g., the more holistic view on ecosystem services, such that the whole is not the sum of its parts, that is, it cannot be reduced to the sum of opposing views by the advocates of "the redundancy hypothesis," "the portfolio effect," etc., since there will emerge new views in the future not yet known today, just as there are already alternative views nowadays which do not side exclusively with any of them, as is the analysis here). And the reverse direction also holds true.

In relation to the predictability-unpredictability principle, if there is predictability (e.g., the predictable tendency that "ecosystem services decisions require making complex choices at the intersection of ecology, technology, society and the economy"), there is unpredictability (e.g., the more difficult task to predict exactly the extent to which a particular "ecosystem services decision" will be correct on a particular future occasion, since "complicating the process, these decisions are often spatial, always multi-objective, and based on uncertain data, models, and estimates"). (WK 2013rr) And the reverse direction also holds true.

In relation to the explicability-inexplicability principle, if there is explicability (e.g., the explanation, by those advocates of "the redundancy hypothesis," of ecosystem services on the basis that "more than one species performs a given role within an ecosystem"), there is inexplicability (e.g., the lack of sufficient explanation, by those advocates of "the redundancy hypothesis," of why it is necessarily plausible, since, to the critics, there are alternative approaches like "the portfolio effect," which "compares biodiversity to stock holdings, where diversification minimizes the volatility of the investment, or in this case, the risk in stability of ecosystem services"). And the reverse direction also holds true.

In relation to the fiction-reality principle, if there is fiction (e.g., the fictional aspect of the understanding of ecosystem services, insofar as it can be culturally relative in that "pivotal cultural values attaching to the natural/cultivated environment rely on an area's unique character that cannot be addressed by methods that use universal scientific parameters to determine ecological structures and functions"), there is reality (e.g., the realistic aspect of the understanding of ecosystem services, insofar as there are "nonmaterial benefits people obtain from ecosystems through spiritual enrichment, cognitive development, reflection, recreation, and aesthetic experiences"). And the reverse direction also holds true.

In relation to the cognitiveness-noncognitiveness principle, if there is cognitiveness (e.g., a view of things on the basis of reason and evidences, as shown in the finding that there are different studies on ecosystem services aforecited), there is noncognitiveness (e.g., a view of things on the basis of non-cognitive factors like envy, jealousy, power, nationality, race, gender, age, class, greed, lust, status, faith, anger, sadness, joy, fear, wish, etc.—as shown in the politics of cultural relativism, since "pivotal cultural values attaching to the natural/cultivated environment rely on an area's unique character that cannot be addressed by methods that use universal scientific parameters to determine ecological structures and functions"). And the reverse direction also holds true.

In relation to the finiteness-transfiniteness principle, if there is finiteness (e.g., the finite number of the categories of ecosystem services, like the 4 examples aforedescribed), there is transfiniteness (e.g., the transfinite number of all the individual instances of ecosystem services that have ever existed in history, such that, in one study by Salles [2011], "the total value of biodiversity is infinite"). And the reverse direction also holds true.

In relation to the preciseness-vagueness principle, if there is preciseness (e.g., the precise identification of 4 categories of ecosystem services in the article aforecited), there is vagueness (e.g., the vagueness in the identification of the 4 categories of ecosystem services in the article aforecited, since it is not clear why there must be only 4, not 5, 6, 7, 8, and so on). And the reverse direction also holds true.

In relation to the simpleness-complicatedness principle, if there is simpleness (e.g., the relatively simple analysis of the E P & L accounting by the advocates), there is complicatedness (e.g., the relatively more complicated analysis of the E P & L accounting, by challenging the claims and assumptions, as shown in the problems pointed out by the critics as cited). And the reverse direction also holds true.

In relation to the openness-hiddenness principle, if there is openness (e.g., the open exploration of 4 categories of ecosystem services in the article aforecited), there is hiddenness (e.g., the hidden bias in the exploration of the 4 categories of ecosystem services in the article aforecited, since "the most recent revision by TEEB...has revised the…definition" by removing "Supporting Services" and replacing it "on the one hand with 'Habitat Services' and on the other hand with 'Ecosystem Functions'"). And the reverse direction also holds true.

In relation to the denseness-emptiness principle, if there is denseness (e.g., the relatively denser concentration of concerns with the view that "more than one species performs a given role within an ecosystem" by the advocates of "the redundancy hypothesis"), there is emptiness (e.g., the relatively less dense, or more empty, concentration of concerns with "the portfolio effect" by the advocates of "the redundancy hypothesis"). And the reverse direction also holds true.

In relation to the rule-exception principle, if there are rules (e.g., the usual generalization, procedure, habit, or tendency that "environmental awareness is rapidly improving in our contemporary world" nowadays), there are exceptions (e.g., a case to which a rule does not apply, such as the exceptional case that there are some people who still "suffer from the so-called 'tragedy of the commons'"). (WK 2013rr) And the reverse direction also holds true.

In relation to the prototypicality-variation principle, if there is prototypicality (e.g., the prototype, like the example about the original idea by Marsh in 1864, "when he challenged the idea that Earth's natural resources are unbounded by pointing out changes in soil fertility in the Mediterranean"), there is variation (e.g., different variants of the protoptye, as shown in many studies since the original idea by Marsh in 1864, "when he challenged the idea that Earth's natural resources are unbounded by pointing out changes in soil fertility in the Mediterranean"—like the later ones by authors like Osborn [1948], Vogt [1948], Leopold [1949], etc., though each with its own variation, contribution, and disagreement). (WK 2013rr) And the reverse direction also holds true.

In relation to the change-constancy principle, if there is change (e.g., the ever new ways to study accounting, as shown in the work on the E P & L accounting), there is constancy (e.g., the ever constant existence of different problems with the ever new ways to study accounting, as shown in the problems with the work on the E P & L accounting pointed out by the critics). And the reverse direction also holds true.

In relation to the order-chaos principle, if there is order (e.g., the more or less orderly view on the E P & L accounting, if looked from the sole vantage point of the advocates), there is chaos (e.g., the more or less chaotic view on the E P & L accounting, if looked from the multiple, conflicting vantage points of both the advocates and the critics, especially in regard to the problems aforecited, such that they do not add up to much of anything coherent, and "further complicating the process, these decisions are often spatial, always multi-objective, and based on uncertain data, models, and estimates"). (WK 2013rr) And the reverse direction also holds true.

In relation to the slowness-quickness principle, if there is slowness (e.g., the relatively slower speed of "wild bees" to "provide partial or complete pollination services" in the presence of "intensified agricultural practices" which "can quickly erode pollination services through the loss of species," in California's agricultural region as studied by Kremen [2005]), there is quickness (e.g., the relatively quicker speed of "wild bees" to "provide partial or complete pollination services" in the absence of any "intensified agricultural practices," in California's agricultural region as studied by Kremen [2005]). (WK 2013rr) And the reverse direction also holds true.

In relation to the expansion-contraction principle, if there is expansion (e.g., the relative spread of "environmental awareness" nowadays), there is contraction (e.g., the relative decline of the practice of the "tragedy of the commons" nowadays). (WK 2013rr; G. Hardin 1968) And the reverse direction also holds true.

In relation to the optimality-nonoptimality principle, if there is optimality (e.g., the search for highly favorable options, like the quest for a comprehensive "environmental profit and loss" [E P&L] accounting which can provide "a company's monetary valuation and analysis of its environmental impacts including its business operations and its supply chain from cradle-to-gate," so as to give "a more holistic view of a company's performance"), there is nonoptimality (e.g., the existence of non-optimal alternatives to optimality, like the argument for more realistic studies by the critics, who pointed out different problems with the E P & L accounting aforecited). And the reverse direction also holds true.

In relation to the simultaneity-nonsimultaneity principle, if there is simultaneity (e.g., the occurrence of different things more or less at the same time, such as the more or less simultaneous announcements in

November 2011, by PUMA in regard to its "overall E P&L" on "air pollution, land conversion and waste" and by the PPR Group in regard to "a Group E P&L" which "would be implemented across its Luxury and Sport & Lifestyle brands by 2015"), there is nonsimultaneity (e.g., the occurrence of something after something else, such as "the first-ever E P&L" which "was conducted on 2010 data and released in two phases," well before the decision by the PPR Group in regard to "a Group E P&L" which "would be implemented across its Luxury and Sport & Lifestyle brands by 2015"). (WK 2013qq) And the reverse direction also holds true.

In relation to the isolation-interaction principle, if there is isolation (e.g., the state of a situation that is separate from others, such as the "tragedy of the commons" in the older days, without "environmental awareness"), there is interaction (e.g., the influence of entities on one another, such as the "environmental awareness" nowadays, which includes or considers "the ecosystem services a business depends on to operate in addition to the cost of direct and indirect negative impacts on the environment"). And the reverse direction also holds true.

In relation to the theory-praxis principle, if there is theory (e.g., the theoretical construction of "ecosystem" in the field of ecology), there is praxis (e.g., the practical application of "ecosystem" in the field of ecology to the field of accounting, like the E P & L accounting). And the reverse direction also holds true.

In relation to the convention-novelty principle, if there is convention (e.g., the conventional wisdom about the "tragedy of the commons" in the older days), there is novelty (e.g., the alternative novel challenge to the conventional wisdom about the "tragedy of the commons" in the older days—by the new idea of the "environmental awareness" nowadays, which includes "the ecosystem services a business depends on to operate in addition to the cost of direct and indirect negative impacts on the environment"). And the reverse direction also holds true.

In relation to the evolution-transformation principle, if there is evolution (e.g., the natural evolution of humans to encounter "ecosystem services" in the state of nature in a primitive way), there is transformation (e.g., the technical transformation of human ability to deal with "ecosystem services" by the invention of the E P & L accounting). And the reverse direction also holds true.

In relation to the symmetry-asymmetry principle, if there is symmetry (e.g., the co-existence of different entities like animals and humans), there is asymmetry (e.g.,humans have damaged the environment in a much more severe way than animals, since "it has been reported that approximately 40-50% of Earth's ice-free land surface has been heavily transformed or degraded by anthropogenic activities, 66% of marine fisheries are either overexploited or at their limit, atmospheric CO_2 has increased more than 30% since the advent of industrialization, and nearly 25% of Earth's bird species have gone extinct in the last two thousand years"). (WK 2013rr) And the reverse direction also holds true.

In relation to the softness-hardness principle, if there is softness (e.g., the defense of the E P & L accounting by the advocates aforecited), there is hardness (e.g., the critique of the E P & L accounting, as shown in the problems pointed out by the critics aforecited). And the reverse direction also holds true.

In relation to the seriousness-playfulness principle, if there is seriousness (e.g., the serious business of those who try to understand the nature of ecosystem services), there is playfulness (e.g., the playful part of those who try to understand the nature of ecosystem services, when they play around with different definitions over the decades, as shown in the comflicting definitions aforedescribed). And the reverse direction also holds true.

In relation to the activeness-inactivess principle, if there is activeness (e.g., the relative activeness of those who act in accordance to the environmental awareness nowadays, because this is increasingly encouraged), there is inactiveness (e.g., the relative inactiveness of those who act in accordance to the "tragedy of the commons" nowadays, because this is increasingly condemned). And the reverse direction also holds true.

In relation to the selfness-otherness principle, if there is selfness (e.g., the tendency to be mostly centered on one's self or group, as shown by those who act in accordance to the "tragedy of the commons"), there is otherness (e.g., the tendency to be mostly centered on others, as shown in the preaching of "the environmental awareness" nowadays, which "is rapidly improving in our contemporary world" in relation to nature). (WK 2013rr) And the reverse direction also holds true.

In relation to the regression-progression principle, if there is regression (e.g., the regression made by the E P & L accounting, as shown in the problems pointed out by the critics), there is progression (e.g., the progress made by the E P & L accounting, as shown in the contribution to the understanding of "the ecosystem services a business depends on to operate in addition to the cost of direct and indirect negative impacts on the environment" aforecited). And the reverse direction also holds true.

In relation to the same-difference principle, if there is similarity in outcome (e.g., the contribution to the molding and control of beliefs, values, and behaviors, regardless of whether this be done by way of the "tragedy of the commons," or by way of the "environmental awareness"), there is difference in outcome (e.g., the contribution to the molding and control of beliefs, values, and behaviors by way of the "tragedy of the commons" for a relatively more individualist lifeworld—but the contribution to the molding and control of beliefs, values, and behaviors by way of the "environmental awareness" for a relatively more collectivist lifeworld). And the reverse direction also holds true.

In relation to the stability-reaction principle, if there is stability (e.g., the relatively more stable condition of the practice of the "tragedy of the commons" in the older days), there is reaction (e.g., the practice of the "tragedy of the commons" in the older days then led to further changes later on, when nowadays, there is the "environmental awareness"). And the reverse direction also holds true.

In relation to the functionality-nonfunctionality principle, if there is functionality (e.g., the presence of action for which something or someone is fitted, as shown in the creation of the E P & L accounting, which serves the function of providing an evaluation of a company's environmental impacts, so as to reduce "the cost of direct and indirect negative impacts on the environment"), there is nonfunctionality (e.g., the relative lack of action for which something or someone is fitted, as shown in the problems with the E P & L accounting, such that it is not as functional as the advocates would like us to believe). And the reverse direction also holds true.

In relation to the intentionality-nonintentionality principle, if there is intentionality (e.g., the planning or design of something for a certain outcome, as shown in the original intention of Marsh in 1864 to challenge "the idea that Earth's natural resources are unbounded" for better environmental awareness), there is nonintentionality (e.g., the relative absence of planning or design of something for a certain outcome, such that the outcome can even be counter-intentional in being the opposite of what was originally planned, as shown in the criticism that the understanding of ecosystem services is culturally relative in a way that "cannot be addressed by methods that use universal scientific parameters to determine ecological structures and functions"—contrary to the original intention of Marsh in 1864 to challenge "the idea that Earth's natural resources are unbounded" for better environmental awareness). And the reverse direction also holds true.

And in relation to the survivability-nonsurvivability principle, if there is survivability (e.g., the relative survival, nowadays, of the "environmental awareness"), there is non-survivability (e.g., the relative non-survival, nowadays, of the practice of the "tragedy of the commons"). And the reverse direction also holds true.

Subtraction and Society

Subtraction, when examined from the perspective of society, can reveal to us some powerful features in relation to its possibility (or impossibility) and desirability (or undesirability), and this can be shown by way of two case studies, namely, (3.4.1) subtraction, computer programming, and bookkeeping, and (3.4.2) subtraction, corporate misdeeds, and internal control—to be addressed hereafter.

Subtraction, Computer Programming, and Bookkeeping

A good way to evaluate the possibility (or impossibility) and desirability (or undesirability) of subtraction from the perspective of society is to explore subtraction, computer programming, and bookkeeping.

An argument in the literature is that computer programming can be used to make bookkeeping easier, especially with the reduction of the need for "boilerplate." (WK 2013u)

Overview of Computing for Bookkeeping

The term "boilerplate" (or "boilerplate code") in computer programming refers to "the sections of code that have to be included in many places with little or no alteration. It is more often used when referring to languages that are considered verbose, i.e. the programmer must write a lot of code to do minimal jobs." (WK 2013u)

The need for boilerplate can be reduced in several ways, as shown below (and summarized in *Table 3.6*): (WK 2013u)

- "Metaprogramming"—"which has the computer automatically write the needed boilerplate text"
- "Convention over Configuration"—"which provides good default values, reducing the need to specify program details in every project"
- "Model-Driven Engineering"—"which uses models and model-to-code generators, eliminating the need for boilerplate manual code"

For example, hereafter is "a C++ programming language boilerplate, #include guard": (WK 2013u)

```
#ifndef _MYINTERFACE_H
#define _MYINTERFACE_H
...
#endif
```

In this example, the boilerplate "sets up, and checks, a global flag to tell the compiler whether the file myinterface.h has already been included. As many interdepending files may be involved in the compilation of a module, this avoids processing the same header multiple times (which would lead to errors due to multiply defined elements)." (WK 2013u)

The Case of Object-Oriented Programming

There are different computer programs to make bookkeeping easier, and a good elaborated example is the use of "object-oriented programming" (or "OOP" in short)," which is "a programming paradigm that represents concepts as 'objects' that have data fields (attributes that describe the object) and associated procedures known as methods. Objects, which are usually instances of classes, are used to interact with one another to design applications and computer programs. C++, Objective-C, Smalltalk, Java, C#, Perl, Python, Ruby and PHP are examples of object-oriented programming languages." (WK 2013u & 2013v; E. Kindler 2011; J. Lewis 2008)

In OOP, "an object has state (data) and behavior (code). Objects correspond to things found in the real world. So for example, a graphics program will have objects such as *circle*, *square*, *menu*. An online shopping system will have objects such as *shopping cart*, *customer*, *product*. The shopping system will support behaviors such as *place order*, *make payment*, and *offer discount*. The objects are designed as class hierarchies. So for example with the shopping system there might be high level classes such as *electronics product*, *kitchen product*, and *book*. There may be further refinements for example under electronic products: *CD Player*, *DVD player*, etc. These classes and subclasses correspond to sets and subsets in mathematical logic." (WK 2013v)

In general, "the goals of object-oriented programming" are shown below (and summarized in *Table 3.6*): (WK 2013v)

- "Increased understanding"
- "Ease of maintenance"
- "Ease of evolution"

Historically, "terminology invoking 'objects' and 'oriented' in the modern sense of object-oriented programming made its first appearance at MIT in the late 1950s and early 1960s. In the environment of the artificial intelligence group, as early as 1960, 'object' could refer to identified items (LISP atoms) with properties (attributes); Alan Kay was later to cite a detailed understanding of LISP internals as a strong influence on his thinking in 1966. Another early MIT example was Sketchpad created by Ivan Sutherland in 1960–61; in the glossary of the 1963 technical report based on his dissertation about Sketchpad, Sutherland defined notions of 'object' and 'instance' (with the class concept covered by 'master' or 'definition'), albeit specialized to graphical interaction." (WK 2013v; J. McCarthy 1960 & 1962; I. Sutherland 1963)

However, "more recently, a number of languages have emerged that are primarily object-oriented, but that are also compatible with procedural methodology. Two such languages are Python and Ruby....Just as procedural programming led to refinements of techniques such as structured programming, modern object-oriented software design methods include refinements such as the use of design patterns, design by contract, and modeling languages (such as UML)." (WK 2013v)

Features of Object-Oriented Programming

To identify the fundamental features of OOP is not easy; for example, "Benjamin C. Pierce [2002] and some other researchers view any attempt to distill OOP to a minimal set of features as futile. He nonetheless identifies fundamental features that support the OOP programming style in most object-oriented languages," as shown below (and summarized in *Table 3.6*): (WK 2013v)

- "Dynamic Dispatch"—refers to the situation "when a method is invoked on an object, the object itself determines what code gets executed by looking up the method at run time in a table associated with the object. This feature distinguishes an object from an abstract data type (or module), which has a fixed (static) implementation of the operations for all instances. It is a programming methodology that gives modular component development while at the same time being very efficient."
- "Encapsulation"—is also known as "multi-methods, in which case the state is kept separate."
- "Subtype Polymorphism"
- "Object Inheritance (or Delegation)"

- "Open Recursion"—refers to "a special variable (syntactically it may be a keyword), usually called *this or self*, that allows a method body to invoke another method body of the same object. This variable is *late-bound*; it allows a method defined in one class to invoke another method that is defined later, in some subclass thereof."

In addition, there are other features in OOP, as shown below: (WK 2013v)

- "Classes of objects"
- "Instances of classes"
- "Methods which act on the attached objects"
- "Message passing"
- "Abstraction"

Surely, there can be other features, so the ones above are illustrative, not exhaustive.

Design Patterns of Object-Oriented Programming

Yet, there are "challenges of object-oriented design" which "are addressed by several methodologies. Most common is known as the design patterns....More broadly, the term 'design patterns' can be used to refer to any general, repeatable solution to a commonly occurring problem in software design. Some of these commonly occurring problems have implications and solutions particular to object-oriented development," and some good examples are shown below.

(a) Inheritance and Behavioral Subtyping

The first challenge of "object-oriented design" has to do with "inheritance and behavioral subtyping."

In other words, "it is intuitive to assume that inheritance creates a semantic 'is a' relationship, and thus to infer that objects instantiated from subclasses can always be *safely* used instead of those instantiated from the superclass. This intuition is unfortunately false in most OOP languages, in particular in all those that allow mutable objects. Subtype polymorphism as enforced by the type checker in OOP languages (with mutable objects) cannot guarantee behavioral subtyping in any context. Behavioral subtyping is undecidable in general, so it cannot be implemented by a program (compiler). Class or object hierarchies must be carefully designed, considering possible incorrect uses that cannot be detected syntactically. This issue is known as the Liskov substitution principle." (WK 2013v)

(b) Gang of Four Design Patterns

The second challenge of "object-oriented design" has to do with "gang of Four design patterns."

This refers to the 1995 work titled *Design Patterns: Elements of Reusable Object-Oriented Software* by Erich Gamma, Richard Helm, Ralph Johnson, and John Vlissides, so it is "often referred to humorously as the 'Gang of Four.' Along with exploring the capabilities and pitfalls of object-oriented programming, it describes 23 common programming problems and patterns for solving them," and some of them are shown below: (WK 2013v)

- "Creational Patterns"—"Factory method pattern, Abstract factory pattern, Singleton pattern, Builder pattern, Prototype pattern"
- "Structural Patterns"—"Adapter pattern, Bridge pattern, Composite pattern, Decorator pattern, Facade pattern, Flyweight pattern, Proxy pattern"
- "Behavioral Patterns"—"Chain-of-responsibility pattern, Command pattern, Interpreter pattern, Iterator pattern, Mediator pattern, Memento pattern, Observer pattern, State pattern, Strategy pattern, Template method pattern, Visitor pattern"

Again, there can be other examples, so the ones above are illustrative, not exhaustive.

(c) Gang Object-orientation and databases

The third challenge of "object-oriented design" has to do with "object-orientation and databases."

This refers to the problem when "both object-oriented programming and relational database management systems (RDBMSs) are extremely common in software today. Since relational databases don't store objects directly (though some RDBMSs have object-oriented features to approximate this), there is a general need to bridge the two worlds. The problem of bridging object-oriented programming accesses and data patterns with relational databases is known as object-relational impedance mismatch....One of the most common approaches is object-relational mapping, as found in libraries like Java Data Objects and Ruby on Rails' ActiveRecord." (WK 2013v)

(d) Real-World Modeling and Relationships

The fourth challenge of "object-oriented design" has to do with "real-world modeling and relationships."

In other words, "OOP can be used to associate real-world objects and processes with digital counterparts," and "Niklaus Wirth...said of OOP in his paper, 'Good Ideas through the Looking Glass': 'This paradigm closely reflects the structure of systems 'in the real world', and it is therefore well suited to model complex systems with complex behaviours.'" (WK 2013v)

(e) OOP and Control Flow

The fifth challenge of "object-oriented design" has to do with "OOP and control flow."

For example, "OOP was developed to increase the reusability and maintainability of source code. Transparent representation of the control flow had no priority and was meant to be handled by a compiler. With the increasing relevance of parallel hardware and multithreaded coding, developing transparent control flow becomes more important...." (WK 2013v; S. Ambler 1998)

(f) Responsibility- vs. Data-Driven Design

And the sixth challenge of "object-oriented design" has to do with "responsibility- vs. data-driven design."

In other words, in OOP, "responsibility-driven design defines classes in terms of a contract, that is, a class should be defined around a responsibility and the information that it shares. This is contrasted by Wirfs-Brock and Wilkerson with data-driven design, where classes are defined around the data-structures that must be held." (WK 2013v)

Problems with Computer Programming and Bookkeeping

However, to the critics, this argument on subtraction, computer programming, and bookkeeping has some major problems. Consider, for illustration, a few examples below (as summarized in *Table 3.6*).

Firstly, one main criticism of this argument on subtraction, computer programming, and bookkeeping is that computer programming often does not model the "real world."

For instance, to the critics, "not everyone agrees that OOP facilitates direct real-world mapping...; Bertrand Meyer argues in *Object-Oriented Software Construction* that a program is not a model of the world but a model of some part of the world; 'Reality is a cousin twice removed.' At the same time, some principal limitations of OOP had been noted. For example, the Circle-ellipse problem is difficult to handle using OOP's concept of inheritance." (WK 2013v; M. Trofimov 1993; B. Meyer 1997) And "Rich Hickey, creator of Clojure, described object systems as overly simplistic models of the real world. He emphasized the inability of OOP to model time properly, which is getting increasingly problematic as software systems become more concurrent." (WK 2013v; R. Hickey 2009)

Secondly, another main criticism of this argument on subtraction, computer programming, and bookkeeping is that OOP is too obsessed with "objects" (or "nouns") but not other things (like "verbs" or "actions," etc.).

For instance, the critic Alexander Stepanov (2010) "compares object orientation unfavourably to multimethods," because "OOP…attempts to decompose the world in terms of interfaces that vary on a single type [for objects]…. It claims that everything is an object. Even if it is true it is not very interesting—saying that everything is an object is saying nothing at all…." (WK 2013v) And "Steve Yegge [2006], making a roundabout comparison with Functional programming, writes, 'Object Oriented Programming puts the Nouns first and foremost. Why would you go to such lengths to put one part of speech on a pedestal? Why should one kind of concept take precedence over another? It's not as if OOP has suddenly made verbs less important in the way we actually think. It's a strangely skewed perspective.'" (WK 2013v)

Thirdly, still another main criticism of the study on subtraction, computer programming, and bookkeeping is that there are disagreements about the fundamental features of OOP.

For instance, unlike the 5 ones by Benjamin C. Pierce and others as cited above, "in his 2003 book, *Concepts in programming languages*, John C. Mitchell identifies 4 main features: dynamic dispatch, abstraction, subtype polymorphism, and inheritance. Michael Lee Scott in *Programming Language Pragmatics* considers only encapsulation, inheritance and dynamic dispatch." (WK 2013v)

Fourthly, an additional main criticism of the study on subtraction, computer programming, and bookkeeping is that there are trade-offs in computer programming for bookkeeping.

For instance, in relation to "object-orientation and databases,…there are a number of approaches to cope with this problem, but no general solution without downsides." (WK 2013v; T. Neward 2006)

Fifthly, another additional main criticism of the study on subtraction, computer programming, and bookkeeping is that OOP is not effective to deal with the issue of transparency.

For instance, in regard to "OOP and control flow," what is problematic here is that "transparent representation of the control flow had no priority and was meant to be handled by a compiler," but "with the increasing relevance of parallel hardware and multithreaded coding, developing transparent control flow becomes more important, something hard to achieve with OOP." (WK 2013v; A. Shelly 2008 & 2009a; J. James 2007; R. Harper 2011)

And sixthly, still another additional main criticism of the study on subtraction, computer programming, and bookkeeping is that OOP has not yet been proved to be better than other approaches in practice.

For instance, "a study by Potok et al.[1999] has shown no significant difference in productivity between OOP and procedural approaches." (WK 2013v)

Beyond Computing Programming and Bookkeeping

These criticisms of subtraction, computer programming, and bookkeeping do not suggest a total rejection of its usefulness but serve to show us the two opposing sides of the debate, such that the possibility (or impossibility) and desirability (or undesirability) of subtraction (from the perspective of society with subtraction, computer programming, and bookkeeping as a case study here) are not to the extent that the respective defenders would like us to believe.

In fact, the analysis of subtraction, computer programming, and bookkeeping is to reveal to us the ontological principles in existential dialectics, and good examples include the formalness-informalness principle, the absoluteness-relativeness principle, the partiality-totality principle, the predictability-unpredictability principle, the explicability-inexplicability principle, the fiction-reality principle, the cognitiveness-noncognitiveness principle, the finiteness-transfiniteness principle, the preciseness-vagueness principle, the simpleness-complicatedness principle, the openness-hiddenness principle, the denseness-emptiness principle, the rule-exception principle, the prototypicality-variation principle, the change-constancy principle, the order-chaos principle, the slowness-quickness principle, the expansion-contraction principle, the optimality-nonoptimality principle, the simultaneity-nonsimultaneity principle, the isolation-interaction principle, the theory-praxis principle, the convention-novelty principle, the evolution-transformation principle, the symmetry-asymmetry principle, the softness-hardness principle, the seriousness-playfulness principle, the activeness-inactivess principle, the selfness-otherness principle, the regression-progression principle, the same-difference principle, the stability-reaction principle, the functionality-nonfunctionality principle, the intentionality-nonintentionality principle, and the survivability-nonsurvivability principle.

For instance, in relation to the formalness-informalness principle, if there is formalness (e.g., the formal logical requirement of soundness, completeness, and consistency in a system of ideas, as in the argument on computer programming and bookkeeping), there is informalness (e.g., the non-formal existence of unsoundness, incompleteness, and inconsistency in the argument on computer programming and bookkeeping, as shown in the criticisms that computer programming often does not model the "real world," that OOP is too obsessed with "objects" [or "nouns"] but not other things [like "verbs" or "actions," etc.], that there are disagreements about the fundamental features of OOP, that there are trade-offs in computer programming for bookkeeping, that OOP is not effective to deal with the issue of transparency, and that OOP has not yet been proved to be better than other approaches in practice). And the reverse direction also holds true.

In relation to the absoluteness-relativeness principle, if there is absoluteness (e.g., the absolute view on the focus on "nouns" and "objects" by the advocates of OOP), there is relativeness (e.g., what is right for the advocates of OOP in regard to the focus on "nouns" and "objects" is not necessarily so for the critic like Alexander Stepanov who argued instead that "OOP…claims that everything is an object. Even if it is true it is not very interesting—saying that everything is an object is saying nothing at all"). And the reverse direction also holds true.

In relation to the partiality-totality principle, if there is partiality (e.g., the partial view on "responsibility-driven design" by the advocates), there is totality (e.g., the more holistic view on "design patterns," such that the whole is not the sum of its parts, that is, it cannot be reduced to the sum of opposing views by the advocates of "responsibility-driven design," the other advocates of "data-driven design," etc., since there will emerge new views in the future not yet known today, just as there are already alternative views nowadays which do not side exclusively with any of them, as is the analysis here). And the reverse direction also holds true.

In relation to the predictability-unpredictability principle, if there is predictability (e.g., the predictable tendency that OOP will remain relevant in the foreseeable future), there is unpredictability (e.g., the more difficult task to predict exactly the extent to which OOP will remain influentail in a particular future era, siince there are already alternative approaches nowadays, like "procedural programming," etc.). And the reverse direction also holds true.

In relation to the explicability-inexplicability principle, if there is explicability (e.g., the explanation, by those advocates of OOP, of its desirability on the basis of its integration of the two concepts of "data" and "code" into one, such that "an object has state [data] and behavior [code]"), there is inexplicability (e.g., the lack of sufficient explanation, by those advocates of OOP, of why its focus on "objects" and "nouns" is necessarily desirable, since, to the critics like Alexander Stepanov, "OOP…claims that everything is an object. Even if it is true it is not very interesting—saying that everything is an object is saying nothing at all"). And the reverse direction also holds true.

In relation to the fiction-reality principle, if there is fiction (e.g., the fictional aspect of OOP, insofar as, to the critics, "not everyone agrees that OOP facilitates direct real-world mapping…; Bertrand Meyer argues…that a program is not a model of the world" and "Rich Hickey…emphasized the inability of OOP to model time properly, which is getting increasingly problematic as software systems become more concurrent"), there is reality (e.g., the realistic aspect of OOP, insofar as it has mechanisms to reduce the need for boilerplate codes like "metaprogramming," "convention over configuration," and "model-driven engineering"). And the reverse direction also holds true.

In relation to the cognitiveness-noncognitiveness principle, if there is cognitiveness (e.g., a view of things on the basis of reason and evidences, as shown in the finding that "the goals of object-oriented programming" are for "increased understanding," "ease of maintenance," and "ease of evolution"), there is noncognitiveness (e.g., a view of things on the basis of non-cognitive factors like envy, jealousy, power, nationality, race, gender, age, class, greed, lust, status, faith, anger, sadness, joy, fear, wish, etc.—as shown in the politics of defending OOP, in spite of the damaging criticisms that computer programming often does not model the "real world," that OOP is too obsessed with "objects" [or "nouns"] but not other things [like "verbs" or "actions," etc.], that OOP is not effective to deal with the issue of transparency, and that OOP has not yet been proved to be better than other approaches in practice). And the reverse direction also holds true.

In relation to the finiteness-transfiniteness principle, if there is finiteness (e.g., the finite number of the concepts of objects, like the 2 concerning "state [data] and behavior [code]" in OOP), there is transfiniteness (e.g., the transfinite number of all the individual instances of "state [data] and behavior [code]" of objects that have ever existed in history, regardless of whether or not they are captured by any OOP). And the reverse direction also holds true.

In relation to the preciseness-vagueness principle, if there is preciseness (e.g., the precise identification of 5 features of OOP by Benjamin C. Pierce and others as cited above), there is vagueness (e.g., the vagueness in the identification of the 5 features of OOP by Benjamin C. Pierce and others as cited above, since it is not clear why there must be only 5, not 4, 3, 6, 7, 8, and so on, since others like John C. Mitchell identified only 4, and Michael Lee Scott considered only 3, as cited earlier). And the reverse direction also holds true.

In relation to the simpleness-complicatedness principle, if there is simpleness (e.g., the relatively simple analysis of OOP by the advocates), there is complicatedness (e.g., the relatively more complicated analysis of OOP, by challenging the claims and assumptions, as shown in the problems pointed out by the critics as cited). And the reverse direction also holds true.

In relation to the openness-hiddenness principle, if there is openness (e.g., the open exploration of 5 features of OOP by Benjamin C. Pierce and others as cited above), there is hiddenness (e.g., the hidden bias in exploration of the 5 features of OOP by Benjamin C. Pierce and others as cited above, since others like John C. Mitchell identified only 4, and Michael Lee Scott considered only 3, as cited earlier). And the reverse direction also holds true.

In relation to the denseness-emptiness principle, if there is denseness (e.g., the relatively denser concentration of concerns with "objects" and "nouns" by those advocates of OOP), there is emptiness (e.g., the relatively less dense, or more empty, concentration of concerns with "verbs" and "actions" by those advocates of OOP). And the reverse direction also holds true.

In relation to the rule-exception principle, if there are rules (e.g., the usual generalization, procedure, habit, or tendency that, as shown in a survey by Deborah J. Armstrong, there is "a number of fundamental concepts…found in the large majority of definitions of OOP"), there are exceptions (e.g., a case to which a rule does not apply, such as the exceptional cases where "not all of these concepts [of OOP] appear in all object-oriented programming languages," as shown in the example that "object-oriented programming that uses classes is sometimes called class-based programming, while prototype-based programming does not typically use classes"). (WK 2013v) And the reverse direction also holds true.

In relation to the prototypicality-variation principle, if there is prototypicality (e.g., the prototype, like the example about the original works on OOP "invoking 'objects' and 'oriented' in the modern sense of object-oriented programming…at MIT in the late 1950s and early 1960s"), there is variation (e.g., different variants of the protoptye, as shown in many studies since the original works on OOP "invoking 'objects' and 'oriented' in the modern sense of object-oriented programming…at MIT in the late 1950s and early 1960s"—like the later ones which "have emerged that are primarily object-oriented, but that are also compatible with procedural methodology," like "Python and Ruby", etc., though each with its own variation, contribution, and disagreement). And the reverse direction also holds true.

In relation to the change-constancy principle, if there is change (e.g., the ever new ways to study bookkeeping, as shown in the work on boilerplate codes like OOP), there is constancy (e.g., the ever constant existence of different problems with the ever new ways to study bookkeeping, as shown in the problems with the work on boilerplate codes like OOP pointed out by the critics). And the reverse direction also holds true.

In relation to the order-chaos principle, if there is order (e.g., the more or less orderly view on OOP, if looked from the sole vantage point of "objects" and "nouns" by the advocates), there is chaos (e.g., the more or less chaotic view on OOP, if looked from the multiple, conflicting vantage points of both the advocates and the critics, especially in regard to the problems aforecited, such that they do not add up to much of anything coherent). And the reverse direction also holds true.

In relation to the slowness-quickness principle, if there is slowness (e.g., the relatively slower speed of a programmer to finish writing codes when he has to rely on the need for boilerplate manual codes without any "metaprogramming," "convention over configuration," and "model-driven engineering"), there is quickness (e.g., the relatively quicker speed of a programmer to finish writing codes when the need for boilerplate manual code can be reduced by way of "metaprogramming," "convention over configuration," and "model-driven engineering"). And the reverse direction also holds true.

In relation to the expansion-contraction principle, if there is expansion (e.g., the relative spread, nowadays, of high-level mechanisms like "metaprogramming," "convention over configuration," and "model-driven engineering" to reduce the need for boilerplate manual codes nowadays), there is contraction (e.g., the relative decline, nowadays, of those types of computer proamming where "the programmer must write a lot of code to do minimal jobs," like the dependence on boilerplate manual codes). And the reverse direction also holds true.

In relation to the optimality-nonoptimality principle, if there is optimality (e.g., the search for highly favorable options, like the quest for the best "design patterns" which "can be used" as "any general, repeatable solution to a commonly occurring problem in software design"), there is nonoptimality (e.g., the existence of non-optimal alternatives to optimality, like the argument for more realistic studies by the critics, who pointed out different problems with OOP and its "design patterns" aforedisucussed). And the reverse direction also holds true.

In relation to the simultaneity-nonsimultaneity principle, if there is simultaneity (e.g., the occurrence of different things more or less at the same time, such as the simultaneous co-existence of "state [data] and behavior [code]" when they are integrated by "an object-oriented system" into an "object"), there is nonsimultaneity (e.g., the occurrence of something after something else, such as the development of "object-oriented programming…at MIT in the late 1950s and early 1960s," before it was recently "refined," like "the use of design patterns, design by contract, and modeling languages"). (WK 2013v) And the reverse direction also holds true.

In relation to the isolation-interaction principle, if there is isolation (e.g., the state of a situation that is separate from others, such as the separate construction of computer programs as "code" and "data" in the older days, prior to the emergence of OOP), there is interaction (e.g., the influence of entities on one another, such as the integration in "an object-oriented system" of "the two ['code' and 'data'] using the concept of an 'object,'" so that "an object has state [data] and behavior [code]"). And the reverse direction also holds true.

In relation to the theory-praxis principle, if there is theory (e.g., the theoretical construction of "inheritance" in social sciences), there is praxis (e.g., the practical application of "inheritance" in social sciences to the field of computing, like "object inheritance" in OOP). (WK 2013v) And the reverse direction also holds true.

In relation to the convention-novelty principle, if there is convention (e.g., the conventional wisdom about the separate construction of computer programs as "code" and "data" in the older days), there is novelty (e.g., the alternative novel challenge to the conventional wisdom about the separate construction of computer programs as "code" and "data" in the older days—by the new idea of integrating in "an object-oriented system" "the two ['code' and 'data'] using the concept of an 'object,'" so that "an object has state [data] and behavior [code]"). And the reverse direction also holds true.

In relation to the evolution-transformation principle, if there is evolution (e.g., the natural evolution of humans to encounter "objects" in the state of nature), there is transformation (e.g., the technical transformation of human ability to deal with "objects" by the invention of OOP). And the reverse direction also holds true.

In relation to the symmetry-asymmetry principle, if there is symmetry (e.g., the co-existence of different features of OOP, like "Inheritance,

Object, Class, Encapsulation, Method, Message Passing, Polymorphism, Abstraction," etc.), there is asymmetry (e.g., according to Deborah Armstrong, in descending order of popularity, the features of OOP are: "Inheritance, Object, Class, Encapsulation, Method, Message Passing, Polymorphism, Abstraction"). (WK 2013v) And the reverse direction also holds true.

In relation to the softness-hardness principle, if there is softness (e.g., the defense of OOP by the advocates aforecited), there is hardness (e.g., the critique of OOP, as shown in the problems pointed out by the critics aforecited). And the reverse direction also holds true.

In relation to the seriousness-playfulness principle, if there is seriousness (e.g., the serious business of those who try to understand the features of OOP), there is playfulness (e.g., the playful part of those who try to understand the features of OOP, when they play around with different possibilities over the decades, like the 5 features by Benjamin C. Pierce, the 4 features by John C. Mitchell, and the 3 features by Michael Lee Scott). And the reverse direction also holds true.

In relation to the activeness-inactivess principle, if there is activeness (e.g., the relative activeness of those programmers who have to rely on the need for boilerplate manual codes without any "metaprogramming," "convention over configuration," and "model-driven engineering"), there is inactiveness (e.g., the relative inactiveness of those programmers who reduce the need for boilerplate manual codes by way of "metaprogramming," "convention over configuration," and "model-driven engineering"). And the reverse direction also holds true.

In relation to the selfness-otherness principle, if there is selfness (e.g., the tendency to be mostly centered on one's self or group, as shown in the criticism by Paul Graham [2013], who "suggested that OOP's popularity within large companies is due to 'large [and frequently changing] groups of mediocre programmers"), there is otherness (e.g., the tendency to be mostly centered on others, as shown in the academic analysis of both sides of the debate on OOP, as an advancement in the knowledge of computing and bookkeeping for humanity). And the reverse direction also holds true.

In relation to the regression-progression principle, if there is regression (e.g., the regression made by OOP, as shown in the problems pointed out by the critics), there is progression (e.g., the progress made by OOP, as shown in the contribution to the understanding of boilerplate codes aforecited). And the reverse direction also holds true.

In relation to the same-difference principle, if there is similarity in outcome (e.g., the contribution to the molding and control of beliefs, values, and behaviors, regardless of whether this be done by way of

"OOP," or by way of "procedural programming"), there is difference in outcome (e.g., the contribution to the molding and control of beliefs, values, and behaviors by way of "OOP" for a relatively more object-oriented worldview—but the contribution to the molding and control of beliefs, values, and behaviors by way of "procedural programming" for a relatively more procedural worldview). And the reverse direction also holds true.

In relation to the stability-reaction principle, if there is stability (e.g., the relatively more stable condition of "object-oriented software design methods" in the 1950s and 1960s), there is reaction (e.g., "object-oriented software design methods" in the 1950s and 1960s then led to further changes later on, when nowadays, they are "refined," with "the use of design patterns, design by contract, and modeling languages [such as UML]"). (WK 2013v) And the reverse direction also holds true.

In relation to the functionality-nonfunctionality principle, if there is functionality (e.g., the presence of action for which something or someone is fitted, as shown in the development of OOP, which serves the function of dealing with the problem of bolierplate codes in computing), there is nonfunctionality (e.g., the relative lack of action for which something or someone is fitted, as shown in the problems with OOP aforecited, such that it is not as functional as the advocates would like us to believe). And the reverse direction also holds true.

In relation to the intentionality-nonintentionality principle, if there is intentionality (e.g., the planning or design of something for a certain outcome, as shown in the original intention of designing OOP to offer an alternative approach to procedural programming to deal with boilerplate codes), there is nonintentionality (e.g., the relative absence of planning or design of something for a certain outcome, such that the outcome can even be counter-intentional in being the opposite of what was originally planned, as shown in the "study by Potok et al. [1999]" which "has shown no significant difference in productivity between OOP and procedural approaches"—contrary to the original intention of designing OOP to offer an alternative approach to procedural programming to deal with boilerplate codes). And the reverse direction also holds true.

And in relation to the survivability-nonsurvivability principle, if there is survivability (e.g., the relative survival, nowadays, of the need for high-level mechanism like "metaprogramming," "convention over configuration," and "model-driven engineering"), there is non-survivability (e.g., the relative non-survival, nowadays, of the need for boilerplate manual codes, without any use of high-level mechanism like "metaprogramming," "convention over configuration," and "model-driven engineering," such that "the programmer must write a lot of code to do minimal jobs"). (WK 2013u) And the reverse direction also holds true.

Subtraction, Corporate Misdeeds, and Internal Control

Another good way to evaluate the possibility (or impossibility) and desirability (or undesirability) of subtraction from the perspective of society is to explore subtraction, corporate misdeeds, and internal control.

An argument in the literature is that corporate misdeeds (say, with the use of "understated earnings") can be minimzed or reduced if appropriate internal controls are carried out reguarly to detect any wrongdoing early enough.

Levels of Corporate Misdeeds

In the corporate world, there are "business scandals which arise with the disclosure of financial misdeeds by trusted executives of corporations or governments. Such misdeeds typically involve complex methods for misusing or misdirecting funds, overstating revenues, understating expenses, overstating the value of corporate assets or underreporting the existence of liabilities, sometimes with the cooperation of officials in other corporations or affiliates." (WK 2013ss)

These corporate misdeeds "can amount to fraud, and investigations are typically launched by government oversight agencies, such as the Securities and Exchange Commission (SEC) in the United States." (WK 2013ss)

Of course, there are different levels of corporate misdeeds; for illustration, consider the following three levels of corporate misdeeds, namely, (a) top executives, (b) managers, and (c) ordinary employees.

(a) Top Executives

The first level of corporate misdeeds has to do with "top executives."

After all, "it is fairly easy for a top executive to reduce the price of his/her company's stock—due to information asymmetry. The executive can accelerate accounting of expected expenses, delay accounting of expected revenue, engage in off balance sheet transactions to make the company's profitability appear temporarily poorer, or simply promote and report severely conservative (e.g. pessimistic) estimates of future earnings. Such seemingly adverse earnings news will be likely to (at least temporarily) reduce share price. (This is again due to information asymmetries since it is more common for top executives to do everything they can to window dress their company's earnings forecasts). There are typically very few legal risks to being 'too conservative' in one's accounting and earnings estimates." (WK 2013ss)

This corporate trick of "a reduced share price makes a company an easier takeover target. When the company gets bought out (or taken private)—at a dramatically lower price—the takeover artist gains a windfall from the former top executive's actions to surreptitiously reduce share price. This can represent tens of billions of dollars (questionably) transferred from previous shareholders to the takeover artist. The former top executive is then rewarded with a golden handshake for presiding over the firesale that can sometimes be in the hundreds of millions of dollars for one or two years of work. (This is nevertheless an excellent bargain for the takeover artist, who will tend to benefit from developing a reputation of being very generous to parting top executives.)" (WK 2013ss)

The same situation occurs, "when a publicly held asset or non-profit organization undergoes privatization. Top executives often reap tremendous monetary benefits when a government-owned or non-profit entity is sold to private hands. Just as in the example above, they can facilitate this process by making the entity appear to be in financial crisis —this reduces the sale price (to the profit of the purchaser), and makes non-profits and governments more likely to sell. It can also contribute to a public perception that private entities are more efficiently run, thereby reinforcing the political will to sell off public assets. Again, due to asymmetric information, policy makers and the general public see a government-owned firm that was a financial 'disaster'—miraculously turned around by the private sector (and typically resold) within a few years." (WK 2013ss)

(b) Managers

The second level of corporate misdeeds has to do with "managers."

Quite often, "managers…are pressured or willingly alter financial statements for the personal benefit of the individuals over the company. Managerial opportunism plays a large role in these scandals. For example, managers who would be compensated more for short-term results would report inaccurate information, since short-term benefits outweigh the long-term ones such as pension obligations." (WK 2013ss)

(c) Ordinary Employees

And the third level of corporate misdeeds has to do with "employees."

In fact, both "managers and employees are pressured or willingly alter financial statements for the personal benefit of the individuals over the company," since "most employees see the scandals as affecting the economy, and some say investor demands are weakening the corporate focus on meeting customer needs. The feelings with respect to the economy are very strong: More than nine out of ten employees (91%) agree that the scandals are a serious or very serious problem for the nation's economy, and nearly two out of three (62%) rank the scandals a 'very serious' problem for the economy. Feelings about customer focus are almost as strong: Nearly three-fourths (72%) agree that recent pressure to satisfy shareholder expectations has led publicly held companies to focus more on meeting the demands of investors than customers. A significant minority (41%) say they have observed this trend in their own company." (WK 2013ss)

Cases of Corporate Misdeeds

In the past few decades alone, there are numerous cases of corporate misdeeds, and hereafter are a few examples for illustration: (WK 2013ss; D. Reece 2004; PBS 1994; CNN 2001; P. Patsuris 2002)

- "ZZZZ Best" (in 1986)—"United States Ponzi scheme run by Barry Minkow"
- "Barlow Clowes" (in 1988)—"United Kingdom Gilts management service. £110 million missing."
- "Phar-Mor" (in 1992)—"Coopers & Lybrand United States mail fraud, wire fraud, bank fraud, and transportation of funds obtained by theft or fraud"
- "Waste Management, Inc." (in 1999)—"Arthur Andersen United States Financial misstatements"
- "Lernout & Hauspie" (in 2000)—"KPMG Belgium Fictitious transactions in Korea and improper accounting methodologies elsewhere"
- "Xerox" (in 2000)—"KPMG United States Falsifying financial results"
- "AOL" (in 2002)—"Ernst & Young United States Inflated sales"
- "Bristol-Myers Squibb" (in 2002)—"PricewaterhouseCoopers United States Inflated revenues"
- "Mirant" (in 2002)—"KPMG United States Overstated assets and liabilities"
- "Nicor" (in 2002)—"Arthur Andersen United States Overstated assets, understated liabilities"
- "Peregrine Systems" (in 2002)—"KPMG United States Overstated sales"

In fact, the case of "Freddie Mac" in 2002 is most relevant here, because it "revealed that it had understated earnings by almost $5 billion, one of the largest corporate restatements in U.S. history. As a result, in November, it was fined $125 million—an amount called 'peanuts' by Forbes." (WK 2013tt; Forbes 2003)

In addition, "on April 18, 2006, Freddie Mac was fined $3.8 million, by far the largest amount ever assessed by the Federal Election Commission, as a result of illegal campaign contributions. Freddie Mac was accused of illegally using corporate resources between 2000 and 2003 for 85 fundraisers that collected about $1.7 million for federal candidates. Much of the illegal fund raising benefited members of the House Financial Services Committee, a panel whose decisions can affect Freddie Mac. Notably, Freddie Mac held more than 40 fundraisers for House Financial Services Chairman Michael Oxley, R-Ohio." (WK 2013tt; Forbes 2003)

Internal Control to Minimize Corporate Misdeeds

Of course, corporate misdeeds can be minimized or reduced if appropriate "internal controls" are carried out regularly to detect any wrongdoing early enough.

In accounting, the term "internal control" refers to "a process affected by an organization's structure, work and authority flows, people and management information systems, designed to help the organization accomplish specific goals or objectives. It is a means by which an organization's resources are directed, monitored, and measured. It plays an important role in detecting and preventing fraud and protecting the organization's resources, both physical (e.g., machinery and property) and intangible (e.g., reputation or intellectual property such as trademarks)." (WK 2013uu; COSO 2013)

In general, internal control can take three main forms, as described below.

(a) Objective or Assertions Categorization

The first type of internal control has to do with "objective or assertions categorization."

In this internal control, "controls may be defined against the particular financial statement assertion to which they relate. There are five such assertions": (WK 2013uu)

- "Existence / Occurrence / Validity"—"only valid or authorized transactions are processed."
- "Completeness"—"all transactions are processed that should be."

- "Rights and Obligations"—"assets are the rights of the organization and the liabilities are its obligations as of a given date."
- "Valuation"—"transactions are valued accurately using the proper methodology, such as a specified means of computation or formula."
- "Presentation and Disclosure"—"accounts and disclosures are properly described in the financial statements of the organization."

A good example of "a validity control objective" may look like this: "Payments are made only for authorized products and services received"; and "a typical control procedure would be: 'The payable system compares the purchase order, receiving record, and vendor invoice prior to authorizing payment.' Management is responsible for implementing appropriate controls that apply to all transactions in their areas of responsibility." (WK 2013uu)

(b) Activity Categorization

The second type of internal control has to do with "activity categorization."

In this internal control, "control activities may…be explained by the type or nature of activity. These include (but are not limited to)" the following: (WK 2013uu)

- "Segregation of Duties"—"separating authorization, custody, and record keeping roles to prevent fraud or error by one person"
- "Authorization of Transactions"—"review of particular transactions by an appropriate person"
- "Retention of Records"—"maintaining documentation to substantiate transactions"
- "Supervision or Monitoring of Operations"—"observation or review of ongoing operational activity"
- "Physical Safeguards"—"usage of cameras, locks, physical barriers, etc. to protect property, such as merchandise inventory"
- "Top-Level Reviews"—"analysis of actual results versus organizational goals or plans, periodic and regular operational reviews, metrics, and other key performance indicators (KPIs)"
- "IT General Controls"—"controls related to: (a) security, to ensure access to systems and data is restricted to authorized personnel, such as usage of passwords and review of access logs; and (b) change management, to ensure program code is properly controlled, such as separation of production and test environments, system and user testing of changes prior to acceptance, and controls over migration of code into production"

- "IT Application Controls"—"controls over information processing enforced by IT applications, such as edit checks to validate data entry, accounting for transactions in numerical sequences, and comparing file totals with control accounts"

Surely, there can be other activities, so the one aboves are illustrative, not exhaustive.

(c) Control Precision

And the third type of internal control has to do with "control precision."

In this internal control, the task here concerns "the alignment or correlation between a particular control procedure and a given control objective or risk. A control with direct impact on the achievement of an objective (or mitigation of a risk) is said to be more precise than one with indirect impact on the objective or risk. Precision is distinct from sufficiency; that is, multiple controls with varying degrees of precision may be involved in achieving a control objective or mitigating a risk." (WK 2013uu)

The factor of "precision" is thus important here in "performing a SOX 404 top-down risk assessment. After identifying specific financial reporting material misstatement risks, management and the external auditors are required to identify and test controls that mitigate the risks. This involves making judgments regarding both precision and sufficiency of controls required to mitigate the risks." (WK 2013uu)

Problems with the Study on Corporate Misdeeds and Internal Control

However, to the critics, this study on subtraction, corporate misdeeds, and internal control has some major problems. Consider, for illustration, a few examples below.

Firstly, one main criticism of the study on subtraction, corporate misdeeds, and internal control is that internal control does not guarantee the realization of objectives.

For example, to the critics, "internal control can provide reasonable, not absolute, assurance that the objectives of an organization will be met. The concept of reasonable assurance implies a high degree of assurance" but is "constrained by the costs and benefits of establishing incremental control procedures." (WK 2013uu)

Secondly, another main criticism of the study on subtraction, corporate misdeeds, and internal control is that internal control has no control over external factors which can affect the performance of an organization.

For instance, to the critics, there exist other "factors outside the enterprise, such as competition or technological innovation. These factors are outside the scope of internal control; therefore, effective internal control provides only timely information or feedback on progress towards the achievement of operational and strategic objectives" but is not immune from outside distortion. (WK 2013uu)

Thirdly, still another main criticism of the study on subtraction, corporate misdeeds, and internal control is that there are competing definitions or views of "internal control."

For instance, to the critics, "there are many definitions of internal control, as it affects the various constituencies (stakeholders) of an organization in various ways and at different levels of aggregation." (WK 2013uu)

Fourthly, an additional main criticism of the study on subtraction, corporate misdeeds, and internal control is that there are limitations in the verifiability of data in internal control (or in accounting in general).

For instance, the critics are quick to point out some general limitations of "verifiability" of data in accounting in 3 major ways, namely, (a) "verifiability doesn't have to do with determining the truthfulness of the data a company provides, but rather with making sure its results logically flow from the data. If a company tells you that total revenue is $200,000, total costs are $125,000 and profit is $75,000, the math is verifiable—but if the revenue and cost figures are incorrect, the profit figure is not accurate," (b) "verifiability also doesn't pass judgment on whether the assumptions made are correct or even appropriate, just whether the result matches the assumptions," and (c) "verifiability is silent on the interpretation of accounting results. A company may tell shareholders that it's strong because revenue rose 25 percent last year. The claim about revenue increasing can be tested for verifiability; what the increase says about the company's prospects cannot." (C. Merritt 2013)

Fifthly, another additional main criticism of the study on subtraction, corporate misdeeds, and internal control is that the problem with accounting scandals is not wholly the fault of corporate criminals.

For instance, "part of the difficulty lies in the accepted recognition that there is no such thing as a single 'right' earnings figure," so this tempts "creative accountants" to offer alternative interpretations of earnings figures in accountng scandals. (WK 2013q)

And sixthly, still another additional main criticism of the study on subtraction, corporate misdeeds, and internal control is that the nature of business cycles in the free market makes it easier for accounting scandals to occur.

For instance, to the critics, "conformance to acceptable accounting practices is a matter of personal integrity. Aggressive earnings management becomes more probable when a company is affected by a downturn in business," which makes it easier for accounting scandals to occur (WK 2013q)

Beyond Internal Control

These criticisms of subtraction, corporate misdeeds, and internal control should not be misconceived as a total rejection of its usefulness but serve to show us the two opposing sides of the debate, such that the possibility (or impossibility) and desirability (or undesirability) of subtraction (from the perspective of society with subtraction, corporate misdeeds, and internal control as a case study here) are not to the extent that the respective defenders would like us to believe.

Furthermore, the analysis of subtraction, corporate misdeeds, and internal control can cast a bright light on the ontological principles in existential dialectics, and good examples include the formalness-informalness principle, the absoluteness-relativeness principle, the partiality-totality principle, the predictability-unpredictability principle, the explicability-inexplicability principle, the fiction-reality principle, the cognitiveness-noncognitiveness principle, the finiteness-transfiniteness principle, the preciseness-vagueness principle, the simpleness-complicatedness principle, the openness-hiddenness principle, the denseness-emptiness principle, the rule-exception principle, the prototypicality-variation principle, the change-constancy principle, the order-chaos principle, the slowness-quickness principle, the expansion-contraction principle, the optimality-nonoptimality principle, the simultaneity-nonsimultaneity principle, the isolation-interaction principle, the theory-praxis principle, the convention-novelty principle, the evolution-transformation principle, the symmetry-asymmetry principle, the softness-hardness principle, the seriousness-playfulness principle, the activeness-inactivess principle, the selfness-otherness principle, the regression-progression principle, the same-difference principle, the stability-reaction principle, the functionality-nonfunctionality principle, the intentionality-nonintentionality principle, and the survivability-nonsurvivability principle.

For instance, in relation to the formalness-informalness principle, if there is formalness (e.g., the formal logical requirement of soundness, completeness, and consistency in a system of ideas, as in the argument on corproate misdeeds and internal control), there is informalness (e.g., the non-formal existence of unsoundness, incompleteness, and inconsistency in the argument on corproate misdeeds and internal control, as shown in the criticisms that internal control does not guarantee the realization of objectives, that internal control has no control over external factors which can affect the performance of an organization, that there are competing definitions or views of "internal control," that there are limitations in the verifiability of data in internal control, that the problem with accounting scandals is not wholly the fault of corporate criminals, and that the nature of business cycles in the free market makes it easier for accounting scandals to occur). And the reverse direction also holds true.

In relation to the absoluteness-relativeness principle, if there is absoluteness (e.g., the absolute view on corporate misdeeds as "fraudulent" and thus "bad' by government oversight agencies like the Securities and Exchange Commission in the United States), there is relativeness (e.g., what is right for the Securities and Exchange Commission in the United States in regard to corporate misdeeds as "fraudulent" and thus "bad" is not necessarily so for those corporate criminals who are determined to cheat). And the reverse direction also holds true.

In relation to the partiality-totality principle, if there is partiality (e.g., the partial view on corporate misdeeds by the Securities and Exchange Commission in the United States), there is totality (e.g., the more holistic view on corporate misdeeds, such that the whole is not the sum of its parts, that is, it cannot be reduced to the sum of opposing views by the Securities and Exchange Commission in the United States, the corporate criminals, etc., since there will emerge new views in the future not yet known today, just as there are already alternative views nowadays which do not side exclusively with any of them, as is the analysis here). And the reverse direction also holds true.

In relation to the predictability-unpredictability principle, if there is predictability (e.g., the predictable tendency that "internal control can provide reasonable…assurance that the objectives of an organization will be met"), there is unpredictability (e.g., the more difficult task to predict exactly the extent to which a particular internal control will be effective in preventing corporate misdeeds on a particular future occasion, since internal control does not provide "absolute" assurance that "the objectives of an organization will be met" and also "has no control over external factors which can affect the performance of an organization"). And the reverse direction also holds true.

In relation to the explicability-inexplicability principle, if there is explicability (e.g., the explanation, by the Securities and Exchange Commission in the United States, of the need to prosecute those corporate criminals on the basis of "fraud" which is "bad" and not legal), there is inexplicability (e.g., the lack of sufficient explanation, by the Securities and Exchange Commission in the United States, of why "corporate misdeeds" are necessarily undesirable, since those corporate criminals think otherwise and are determined to cheat). And the reverse direction also holds true.

In relation to the fiction-reality principle, if there is fiction (e.g., the fictional aspect of "earnings report," insofar as the case of "Freddie Mac" in 2002 "revealed that it had understated earnings by almost $5 billion, one of the largest corporate restatements in U.S. history"), there is reality (e.g., the realistic aspect of "earnings report," insofar as there are companies which are truthful in their reports but others which are not). And the reverse direction also holds true.

In relation to the cognitiveness-noncognitiveness principle, if there is cognitiveness (e.g., a view of things on the basis of reason and evidences, as shown in the finding that the Securities and Exchange Commission in the United States prosecutes those corporate criminals for their "fraudulent" activities), there is noncognitiveness (e.g., a view of things on the basis of non-cognitive factors like envy, jealousy, power, nationality, race, gender, age, class, greed, lust, status, faith, anger, sadness, joy, fear, wish, etc.—as shown in the "temptation" of corporate criminals to cheat, in spite of the understanding that this is "bad" and of the danger that they may be caught and prosecuted by the Securities and Exchange Commission in the United States). And the reverse direction also holds true.

In relation to the finiteness-transfiniteness principle, if there is finiteness (e.g., the finite number of the "methods" of accounting scandals, like "misusing or misdirecting funds," "overstating revenues," "understating expenses," "overstating the value of corporate assets," "underreporting the existence of liabilities," etc.), there is transfiniteness (e.g., the transfinite "amounts of money" that have lost in history, due to accounting scandals, regardless of which methods the cases belong to). (WK 2013ss) And the reverse direction also holds true.

In relation to the preciseness-vagueness principle, if there is preciseness (e.g., the precise identification of 3 types of internal control in the article aforecited), there is vagueness (e.g., the vagueness in the identification of the 3 types of internal control in the article aforecited, since it is not clear why there must be only 3, not 2, 4, 5, 6, 7, and so on, since there are competing definitions of internal control). And the reverse direction also holds true.

In relation to the simpleness-complicatedness principle, if there is simpleness (e.g., the relatively simple analysis of internal control by the advocates), there is complicatedness (e.g., the relatively more complicated analysis of internal control, by challenging the claims and assumptions, as shown in the problems pointed out by the critics as cited). And the reverse direction also holds true.

In relation to the openness-hiddenness principle, if there is openness (e.g., the open exploration of "corporate misdeeds" as "fraudulent" by the Securities and Exchange Commission in the United States), there is hiddenness (e.g., the hidden bias in the exploration of "corporate misdeeds" as "fraudulent" by the Securities and Exchange Commission in the United States, since the critics pointed out that the problem with accounting scandals is not wholly the fault of corporate criminals and that the nature of business cycles in the free market makes it easier for accounting scandals to occur). And the reverse direction also holds true.

In relation to the denseness-emptiness principle, if there is denseness (e.g., the relatively denser concentration of concerns with "internal control" by the advocates), there is emptiness (e.g., the relatively less dense, or more empty, concentration of concerns with "external auditing" by the advocates of "internal control"). And the reverse direction also holds true.

In relation to the rule-exception principle, if there are rules (e.g., the usual generalization, procedure, habit, or tendency that "internal control can provide reasonable…assurance that the objectives of an organization will be met"), there are exceptions (e.g., a case to which a rule does not apply, such as the exceptional cases when internal control does not provide "absolute" assurance that "the objectives of an organization will be met" and also "has no control over external factors which can affect the performance of an organization"). And the reverse direction also holds true.

In relation to the prototypicality-variation principle, if there is prototypicality (e.g., the prototype, like the example about the original "Foreign Corrupt Practices Act [FCPA] of 1977" for internal control in the U.S.), there is variation (e.g., different variants of the protoptye, as shown in many works since the original "Foreign Corrupt Practices Act [FCPA] of 1977" for internal control in the U.S.—like the later ones such as "the Sarbanes–Oxley Act of 2002," the guideliens by "the AICPA, IIA, and ACFE" in 2008, etc., though each with its own variation, contribution, and disagreement). (WK 2013uu) And the reverse direction also holds true.

In relation to the change-constancy principle, if there is change (e.g., the ever new ways to study accounting, as shown in the work on internal control), there is constancy (e.g., the ever constant existence of different problems with the ever new ways to study accounting, as shown in the problems with the work on internal control pointed out by the critics). And the reverse direction also holds true.

In relation to the order-chaos principle, if there is order (e.g., the more or less orderly view on internal control, if looked from the sole vantage point by the advocates that "internal control can provide reasonable…assurance that the objectives of an organization will be met"), there is chaos (e.g., the more or less chaotic view on internal control, if looked from the multiple, conflicting vantage points of both the advocates and the critics, especially in regard to the problems aforecited, such that they do not add up to much of anything coherent, especially since internal control does not provide "absolute" assurance that "the objectives of an organization will be met" and also "has no control over external factors which can affect the performance of an organization"). And the reverse direction also holds true.

In relation to the slowness-quickness principle, if there is slowness (e.g., the relatively slower speed of accounting scandals to occur, in the presence of "internal control" which "plays an important role in the prevention and detection of fraud"), there is quickness (e.g., the relatively quicker speed of accounting scandals to occur, in the absence of any "internal control" or any government oversight agencies like the Securities and Exchange Commission in the United States). (WK 2013uu; Z. Rezaee 2002) And the reverse direction also holds true.

In relation to the expansion-contraction principle, if there is expansion (e.g., "inflated" revenues in the accounting scandal involving Bristol-Myers Squibb in 2002), there is contraction (e.g., "understated" earnings in the accounting scandal involving Freddie Mac in 2002). And the reverse direction also holds true.

In relation to the optimality-nonoptimality principle, if there is optimality (e.g., the search for highly favorable options, like the quest for the comprehensive three internal controls to "play..an important role in the prevention and detection of fraud"), there is nonoptimality (e.g., the existence of non-optimal alternatives to optimality, like the argument for more realistic proposals by the critics, who pointed out different problems with internal control aforedescribed). And the reverse direction also holds true.

In relation to the simultaneity-nonsimultaneity principle, if there is simultaneity (e.g., the occurrence of different things more or less at the same time, such as the simultaneous co-existence of different statutes nowadays, like "the Foreign Corrupt Practices Act [FCPA] of 1977" and "the Sarbanes–Oxley Act of 2002"), there is nonsimultaneity (e.g., the occurrence of something after something else, such as the passing of the "Foreign Corrupt Practices Act [FCPA] of 1977" before the passing of "the Sarbanes–Oxley Act of 2002"). And the reverse direction also holds true.

In relation to the isolation-interaction principle, if there is isolation (e.g., the state of a situation that is separate from others, such as the separation of internal control in companies from any legal requirement in 18th century America), there is interaction (e.g., the influence of entities on one another, such as the legal requirement, under the Sarbanes-Oxley Act of 2002, of companies "to perform a fraud risk assessment and assess related controls" in the U.S.). (WK 2013uu) And the reverse direction also holds true.

In relation to the theory-praxis principle, if there is theory (e.g., the theoretical construction of "internal control procedures" in the field of accounting), there is praxis (e.g., the practical application of "internal control procedures" in the field of accounting to the public sphere, like the adoption of "internal control procedures" under "the Foreign Corrupt Practices Act (FCPA) of 1977 and the Sarbanes–Oxley Act of 2002" in the U.S.). (WK 2013uu) And the reverse direction also holds true.

In relation to the convention-novelty principle, if there is convention (e.g., the conventional wisdom about internal control to play "an important role in the prevention and detection of fraud"), there is novelty (e.g., the alternative novel challenge to the conventional wisdom about internal control to play "an important role in the prevention and detection of fraud"—by the new idea that internal control does not provide "absolute" assurance that "the objectives of an organization will be met" and also "has no control over external factors which can affect the performance of an organization"). And the reverse direction also holds true.

In relation to the evolution-transformation principle, if there is evolution (e.g., the natural evolution of humans to make mistakes, either intentionally or unintentionally, in the state of nature), there is transformation (e.g., the technical transformation of human ability to deal with making mistakes, either intentionally or unintentionally, by the invention of internal control). And the reverse direction also holds true.

In relation to the symmetry-asymmetry principle, if there is symmetry (e.g., the co-existence of different entities like government oversight agencies and individual companies), there is asymmetry (e.g., government oversight agencies like the Securities and Exchange Commission in the

United States have the authority to prosecute individual companies for corporate misdeeds—but individual companies do not have the authority to prosecute government oversight agencies for their own misdeeds). And the reverse direction also holds true.

In relation to the softness-hardness principle, if there is softness (e.g., the voluntary compliance of individual companies with the Securities and Exchange Commission in the United States in regard to internal control), there is hardness (e.g., the prosecution of individual companies by the Securities and Exchange Commission in the United States for corporate misdeeds). And the reverse direction also holds true.

In relation to the seriousness-playfulness principle, if there is seriousness (e.g., the serious business of those who try to understand the nature of internal control), there is playfulness (e.g., the playful part of those who try to understand the nature of internal control, when they play around with different definitions over the decades, like the 3 types aforediscussed). And the reverse direction also holds true.

In relation to the activeness-inactivess principle, if there is activeness (e.g., the relative activeness of internal controllers when they investigate an organization for "misusing or misdirecting funds," "overstating revenues," "understating expenses," "overstating the value of corporate assets," "underreporting the existence of liabilities," etc.), there is inactiveness (e.g., the relative inactiveness of internal controllers, when they have no control over external factors which can affect the performance of an organization). And the reverse direction also holds true.

In relation to the selfness-otherness principle, if there is selfness (e.g., the tendency to be mostly centered on one's self or group, as shown in the corporate misdeeds by those in the accounting scandal involving Freddie Mac aforecited), there is otherness (e.g., the tendency to be mostly centered on others, as shown in the academic analysis of both sides of the debate on internal control, as an advancement in the knowledge of accounting for humanity). And the reverse direction also holds true.

In relation to the regression-progression principle, if there is regression (e.g., the regression made by internal control, as shown in the problems pointed out by the critics), there is progression (e.g., the progress made by internal control, as shown in the contribution to "the prevention and detection of fraud"). And the reverse direction also holds true.

In relation to the same-difference principle, if there is similarity in outcome (e.g., the contribution to the molding and control of beliefs, values, and behaviors, regardless of whether this be done by way of corporate misdeeds committed by creative accountants, or by way of prosecution against individual companies by government oversight agencies), there is difference in outcome (e.g., the contribution to the molding and control of beliefs, values, and behaviors by way of corporate misdeeds committed by creative accountants for a relatively more selfish lifeworld—but the contribution to the molding and control of beliefs, values, and behaviors by way of prosecution by government oversight agencies against individual companies for a relatively more legalistic lifeworld). And the reverse direction also holds true.

In relation to the stability-reaction principle, if there is stability (e.g., the relatively more stable condition of "the Foreign Corrupt Practices Act [FCPA] of 1977 in the older days), there is reaction (e.g., "the Foreign Corrupt Practices Act [FCPA] of 1977 in the older days then led to further changes later on, when it was later refined by "the Sarbanes–Oxley Act of 2002"). And the reverse direction also holds true.

In relation to the functionality-nonfunctionality principle, if there is functionality (e.g., the presence of action for which something or someone is fitted, as shown in the creation of "internal control," which serves the function to "prevent…and detect…fraud"), there is nonfunctionality (e.g., the relative lack of action for which something or someone is fitted, as shown in the problems with "internal control" aforediscussed, such that it is not as functional as the advocates would like us to believe). And the reverse direction also holds true.

In relation to the intentionality-nonintentionality principle, if there is intentionality (e.g., the planning or design of something for a certain outcome, as shown in the original intention of the creation of internal control for "the prevention and detection of fraud"), there is nonintentionality (e.g., the relative absence of planning or design of something for a certain outcome, such that the outcome can even be counter-intentional in being the opposite of what was originally planned, as shown in the criticisms nowadays that internal control does not guarantee the realization of objectives, that internal control has no control over external factors which can affect the performance of an organization, that there are competing views of "internal control," that there are limitations in the verifiability of data in internal control, that the problem with accounting scandals is not wholly the fault of corporate criminals, and that the nature of business cycles in the free market makes it easier for accounting scandals to occur—contrary to the original intention of the creation of internal control for "the prevention and detection of fraud"). And the reverse direction also holds true.

And in relation to the survivability-nonsurvivability principle, if there is survivability (e.g., the relative survival, nowadays, of "the Sarbanes–Oxley Act of 2002" in the U.S.), there is non-survivability (e.g., the relative non-survival, nowadays, of the "dual administration" as it existed in "Hellenistic Egypt," "with one set of bureaucrats charged with collecting taxes and another with supervising them"). (WK 2013uu) And the reverse direction also holds true.

Subtraction and Culture

Subtraction, when examined from the perspective of culture, can cast an illuminating light on its possibility (or impossibility) and desirability (or undesirability), and this can be shown by way of two case studies, namely, (3.5.1) subtraction, accounting scandals, and ethics, and (3.5.2) subtraction, accounting standards, and history—to be addressed hereafter, in that order.

Subtraction, Accounting Scandals, and Ethics

An excellent case study here to understand subtraction from the perspective of culture has to do with subtraction, accounting scandals, and ethics.

An argument in the literature is that there is accounting ethics which sets up standards for accountants to follow, so as to minimize or reduce accounting scandals.

Importance of Accounting Ethics

The term "accounting ethics" refers to "a field of applied ethics, the study of moral values and judgments as they apply to accountancy. It is an example of professional ethics. Accounting ethics were first introduced by Luca Pacioli, and later expanded by government groups, professional organizations, and independent companies. To combat…and prevent fraudulent accounting, various accounting organizations and governments have developed regulations and remedies for improved ethics among the accounting profession." (WK 2013w)

The important of accounting ethics is not difficult to understand, since "the nature of the work carried out by accountants and auditors requires a high level of ethics. Shareholders, potential shareholders, and other users of the financial statements rely heavily on the yearly financial statements of a company as they can use this information to make an informed decision about investment. They rely on the opinion of the accountants who prepared the statements, as well as the auditors that verified it, to present a true and fair view of the company. Knowledge of ethics can help accountants and auditors to overcome ethical dilemmas, allowing for the right choice that, although it may not benefit the company, will benefit the public who relies on the accountant/auditor's reporting." (WK 2013w; D. Alexander 2004; D. Dietz 2002; R. Duska 2003)

History of Accounting Ethics

In 1494, "Luca Pacioli...wrote on accounting ethics in his first book *Summa de arithmetica, geometria, proportioni, et proportionalita*.... Ethical standards have since then been developed through government groups, professional organizations, and independent companies. These various groups have led accountants to follow several codes of ethics to perform their duties in a professional work environment. Accountants must follow the code of ethics set out by the professional body of which they are a member. United States accounting societies such as the Association of Government Accountants, Institute of Internal Auditors, and the National Association of Accountants all have codes of ethics, and many accountants are members of one or more of these societies." (WK 2013w; M. Smith 2008; S. Loeb 1988)

Then, "in 1887, the American Association of Public Accountants (AAPA) was created; it was the first step in developing professionalism in the United States accounting industry. By 1905, the AAPA's first ethical codes were formulated to educate its members. During its twentieth anniversary meeting in October 1907, ethics was a major topic of the conference among its members. As a result of discussions, a list of professional ethics was incorporated into the organization's bylaws. However, because membership to the organization was voluntary, the association could not require individuals to conform to the suggested behaviors. Other accounting organizations, such as the Illinois Institute of Accountants, also pursued discussion on the importance of ethics for the field. The AAPA was renamed several times throughout its history, before becoming the American Institute of Certified Public Accountants (AICPA) as its named today. The AICPA developed five divisions of ethical principles that its members should follow: 'independence, integrity, and objectivity'; 'competence and technical standards'; 'responsibilities to clients'; 'responsibilities to colleagues'; as well as 'other responsibilities and practices.' Each of these divisions provided guidelines on how a Certified Public Accountant (CPA) should act as a professional. Failure to comply with the guidelines could have caused an accountant to be barred from practicing. When developing the ethical principles, the AICPA also considered how the profession would be viewed by those outside of the accounting industry." (WK 2013w; A. Preston 2006; CHI 1907)

Ethical Violations in Accountng Scandals

However, "from the 1980s to the present, there have been multiple accounting scandals that were widely reported on by the media and resulted in fraud charges, bankruptcy protection requests, and the closure of companies and accounting firms. The scandals were the result of creative accounting, misleading financial analysis, as well as bribery. Various companies had issues with fraudulent accounting practices, including Nugan Hand Bank, Phar-Mor, WorldCom, and AIG." (WK 2013w)

In fact, "one of the most widely reported violation of accounting ethics involved Enron, a multinational company, that for several years had not shown a true or fair view of their financial statements. Their auditor Arthur Andersen, an accounting firm considered one of the 'Big Five,' signed off on the validity of the accounts despite the inaccuracies in the financial statements. When the unethical activities were reported, not only did Enron dissolve but Arthur Andersen also went out of business. Enron's shareholders lost $25 billion as a result of the company's bankruptcy. Although only a fraction of Arthur Anderson's employees were involved with the scandal, the closure of the firm resulted in the loss of 85,000 jobs." (WK 2013w; I. Dey 2009; NEW 2009; M. Rosenwald 2007; D. Alexander 2002)

(a) Reasons of Ethical Failures

So a good question to ask is: Why are there ethical failures in accounting?

Well, "fraudulent accounting can arise from a variety of issues. These problems usually come to light eventually and could ruin not only the company but also the auditors for not discovering or revealing the misstatements. Several studies have proposed that a firm's corporate culture as well as the values it stresses may negatively alter an accountant's behavior. This environment could contribute to the degradation of ethical values that were learned from universities." (WK 2013w; P. Clikeman 2003; S. Appelbaum 2005)

More precisely, "a 2007 article in *Managerial Auditing Journal* determined the top nine factors that contributed to ethical failures for accountants based on a survey of 66 members of the International Federation of Accountants," and these "factors include (in order of most significant) the following (as summarized in *Table 3.7*): (WK 2013w; B. Jackling 2007)

- Ex: "self-interest"

- Ex: "failure to maintain objectivity and independence"
- Ex: "inappropriate professional judgment"
- Ex: "lack of ethical sensitivity"
- Ex: "improper leadership and ill-culture"
- Ex: "failure to withstand advocacy threats"
- Ex: "lack of competence"
- Ex: "lack of organizational and peer support:
- Ex: "lack of professional body support"

The most important one is the first one, namely, "self-interest," which is "the motivation by an accountant to act in his/her best interest or when facing a conflict of interest. For example, if an auditor has an issue with an account he/she is auditing, but is receiving financial incentives to ignore these issues, the auditor may act unethically." (WK 2013w; B. Jackling 2007)

(b) The Case of Parmalat

The accounting scandal involving Parmalat in 2003 is a good case in point.

At the beginning, "in 1997, Parmalat jumped into the world of financial markets in a big way, financing several international acquisitions, especially in the Western Hemisphere, with debt. But by 2001, many of the new divisions were producing losses, and the company financing shifted largely to the use of derivatives, apparently at least in part with the intention of hiding the extent of its losses and debt. In February 2003, chief financial officer Fausto Tonna unexpectedly announced a new €500 million bond issue. This came as a surprise both to the markets and to the CEO, Calisto Tanzi. Tanzi forced Tonna to resign and replaced him with Alberto Ferraris. According to an interview he later gave *Time* magazine, Ferraris was surprised to discover that, though now CFO, he still did not have access to some of the corporate books, which were being handled by chief accounting officer Luciano Del Soldato. He began making some inquiries and began to suspect that the company's total debt was more than double that on the balance sheet." (WK 2013ww)

Then, in September 2003, "the plan for a €300M fundraising effort was dropped...and the company's shares depreciated significantly as a result of the publicised concerns raised over transactions with mutual fund Epicurum, another Cayman-based company linked to Parmalat by November. Ferraris resigned less than a week after the public fall-out and was replaced by Del Soldato. Del Soldato resigned the next month, unable to get cash from Epicurum fund, needed to pay debts and make bond payments totalling at least €150M. Enrico Bondi was called in to help the company as it went into administration, while Tanzi himself resigned as chairman and CEO. Bonlat was a subsidiary of Parmalat set up in the Cayman Islands. Bonlat's bank, Bank of America, then released a document showing €3.95 billion in Parmalat's bank account as a forgery. Prime Minister Silvio Berlusconi initiated a fraud investigation and appointed Bondi to administer the company's rescue. Hundreds of thousands of investors lost their money and would never recover it. The company officially went bankrupt, though the Italian government used the legal mean 'commissariamento' to save the trademark." (WK 2013ww)

Then, in December 2003, "Calisto Tanzi, once a symbol of unlimited success, was detained hours after the firm was declared officially insolvent…and admitted that there was a hole of €8bn in Parmalat's accounts, but denied any cover-up. The arrest of five further executives of the company followed. The auditors of the administration eventually determined that the debts amounted to €14.3bn, which was almost eight times the sum originally stated. Several of the company's subsidiaries subsequently went insolvent, including its Brazilian and American operations and its football club in Parma, before Parmalat sued several multi-national banks for huge sums of money." (WK 2013ww)

In the end, "Tanzi was eventually charged with financial fraud and money laundering. Italians were shocked that such a vast and established empire could crumble so quickly. Among the questionable accounting practices used by Parmalat: it sold itself credit-linked notes, in effect placing a bet on its own credit worthiness in order to conjure up an asset out of thin air. After his arrest, Tanzi reportedly admitted during questioning at Milan's San Vittore prison, that he diverted funds from Parmalat into Parmatour and elsewhere. The family football and tourism enterprises were financial disasters; as well as Tanzi's attempt to rival Berlusconi by buying Odeon TV, only to sell it at a loss of about €45 million. Tanzi was sentenced to 10 years in prison for fraud relating to the collapse of the dairy group. The other seven defendants, including executives and bankers, were acquitted. Another eight defendants settled out of court in September 2008." (WK 2013ww; BBC 2008)

Ethical Responses to Accounting Scandals

Because of these accounting scandals within the profession of acounting, "critics of accountants have stated that when asked by a client 'what does two plus two equal?' the accountant would be likely to respond 'what would you like it to be?.' This thought process along with other criticisms of the profession's issues with conflict of interest, have led to various increased standards of professionalism while stressing ethics in the work environment." (WK 2013w; L. Berton 1984a)

Since then, there have been different reforms, and two main categories are worthy to be mentioned below (as summarized in *Table 3.7*): (WK 2012w; S. Dellaportas 2006; B. Mader 2002; B. Jackling 2007; S. Labaton 2002)

- Education—"by educating accountants on ethics before entering the workforce, such as through higher education or initial training at companies, it is believed it will help to improve the credibility of the accounting profession. Companies and accounting organizations have expanded their assistance with educators by providing education materials to assist professors in educating students."
- Regulations—"new regulations in response to the scandals include the Corporate Law Economic Reform Program Act 2004 in Australia as well as the Sarbanes-Oxley Act of 2002, developed by the United States. Sarbanes-Oxley limits the level of work which can be carried out by accounting firms. In addition, the Act put a limit on the fee which a firm can receive from one client as a percentage of their total fees. This ensures that companies are not wholly reliant on one firm for its income, in the hope that they do not need to act unethically to keep a steady income. The act also protects whistleblowers and requires senior management in public companies to sign off on the accuracy of its company's accounting records. In 2002, the five members of the Public Oversight Board (POB), which oversaw ethics within the accounting profession, resigned after critics deemed the board ineffective and the SEC proposed developing a new panel, the Public Company Accounting Oversight Board (PCAOB)."

Surely, there can be other examples, so the ones above are illustrative, not exhaustive.

Problems with the Study of Accounting Scandals and Ethics

However, to the critics, the study of subtraction, accounting scandals, and ethics has some major problems. Consider, for illustration, a few examples below (as summarized in *Table 3.7*).

Firstly, one main criticism of the study of subtraction, accounting scandals, and ethics is that there have been business scandals in spite of the existence of accounting ethics.

For instance, to the critics, "due to range of accounting services and recent corporate collapses, attention has been drawn to ethical standards accepted within the accounting profession. These collapses have resulted in a widespread disregard for the reputation of the accounting profession." (WK 2013w; B. Jackling 2007; S. Dellaportas 2006)

Secondly, another main criticism of the study of subtraction, accounting scandals, and ethics is that different countries have different legal standards and regulations in regard to the enforcement of accounting practices.

For instance, "most countries have differing focuses on enforcing accounting laws. In Germany, accounting legislation is governed by 'tax law'; in Sweden, by 'accounting law'; and in the United Kingdom, by the 'company law.' In addition, countries have their own organizations which regulate accounting. For example, Sweden has the Bokföringsnämden (BFN - Accounting Standards Board), Spain the Instituto de Comtabilidad y Auditoria de Cuentas (ICAC), and the United States the Financial Accounting Standards Board (FASB)." (WK 2013w; C. Gowthorpe 1998)

Thirdly, still another main criticism of the study of subtraction, accounting scandals, and ethics is that there are competing approaches to ethical principles in accounting ethics.

For instance, there are two main camps in the debate on accounting ethics, with those who favor "principles-based" standards, whereas others favor "rules-based" standards. (WK 2013w)

Fourthly, an additional main criticism of the study of subtraction, accounting scandals, and ethics is that morality can be "relative," not "universal." (WK 2013vv)

For instance, to the critics, "separatism is based on a relativist conception of morality that there can be different, equally valid moral codes that apply to different sections of society and differences in codes between societies....If moral universalism is ascribed to, then this would be inconsistent with the view that professions can have a different moral code, as the universalist holds that there is only one valid moral code for all." (WK 2013vv; A. Gewirth 1986)

And fifthly, another additional main criticism of the study of subtraction, accounting scandals, and ethics is that there are the larger institutional factors which contribute to the occurrence of accounting scandals (not just due to the fault of creative accountants).

For instance, to the critics, "until 1977, ethics rules prevented accounting and auditing firms from advertising to clients. When the rules were lifted, spending by the largest CPA firms on advertisements rose from US$4 million in the 1980s to more than $100 million in the 2000s. Critics claimed that, by allowing the firms to advertise, the business side overstepped the professional side of the profession, which led to a conflict of interest. This focus allowed for occurrences of fraud, and caused the firms, according to Arthur Bowman, 'to offer services that made them more consultants and business advisers than auditors.' As accounting firms became less interested in the lower-paying audits due to more focus on higher earning services such as consulting, problems arose. This disregard for the lack of time spent on audits resulted in a lack of attention to catching creative and fraudulent accounting." (WK 2013w; L. Berton 1984; A. Crenshaw 1998; L. Somerville 2003)

Beyond Accounting Scandals and Ethics

These criticisms of subtraction, accounting scandals, and ethics, however, do not mean a total rejection of its usefulness but serve to show us the two opposing sides of the debate, such that the possibility (or impossibility) and desirability (or undesirability) of subtraction (from the perspective of culture with subtraction, accounting scandals, and ethics as a case study here) are not to the extent that the respective defenders would like us to believe.

Moreover, the analysis of subtraction, accounting scandals, and ethics reveals the refreshing nature of the ontological principles in existential dialectics, and good examples include the formalness-informalness principle, the absoluteness-relativeness principle, the partiality-totality principle, the predictability-unpredictability principle, the explicability-inexplicability principle, the fiction-reality principle, the cognitiveness-noncognitiveness principle, the finiteness-transfiniteness principle, the preciseness-vagueness principle, the simpleness-complicatedness principle, the openness-hiddenness principle, the denseness-emptiness principle, the rule-exception principle, the prototypicality-variation principle, the change-constancy principle, the order-chaos principle, the slowness-quickness principle, the expansion-contraction principle, the optimality-nonoptimality principle, the simultaneity-nonsimultaneity principle, the isolation-interaction principle, the theory-praxis principle, the convention-novelty principle, the evolution-transformation principle, the symmetry-asymmetry principle, the softness-hardness principle, the seriousness-playfulness principle, the activeness-inactivess principle, the selfness-otherness principle, the regression-progression principle, the same-difference principle, the stability-reaction principle, the functionality-nonfunctionality principle, the intentionality-nonintentionality principle, and the survivability-nonsurvivability principle.

For instance, in relation to the formalness-informalness principle, if there is formalness (e.g., the formal logical requirement of soundness, completeness, and consistency in a system of ideas, as in the argument on accounting ethics), there is informalness (e.g., the non-formal existence of unsoundness, incompleteness, and inconsistency in the argument on accounting ethics, as shown in the criticisms that there have been business scandals in spite of the existence of accounting ethics, that different countries have different legal standards and regulations in regard to the enforcement of accounting practices, that there are competing approaches to ethical principles in accounting ethics, that morality can be "relative" [not "universal"], and that there are the larger institutional factors which contribute to the occurrence of accounting scandals [not just due to the fault of creative accountants]). And the reverse direction also holds true.

In relation to the absoluteness-relativeness principle, if there is absoluteness (e.g., the absolute view on "principle-based" accounting standards by the International Financial Reporting Standards [IFRS]), there is relativeness (e.g., what is right for the International Financial Reporting Standards [IFRS] in regard to "principle-based" accounting standards is not necessarily so for the critics like those who prefer "the United States Generally Accepted Accounting Principles [GAAP]," which are "largely rule-based"). (WK 2013w; K. Scannell 2008; IAS 2009) And the reverse direction also holds true.

In relation to the partiality-totality principle, if there is partiality (e.g., the partial view on accounting ethics by the advocates of moral universalism), there is totality (e.g., the more holistic view on accounting ethics, such that the whole is not the sum of its parts, that is, it cannot be reduced to the sum of opposing views by the advocates of moral relativism, the advocates of moral universalism, etc., since there will emerge new views in the future not yet known today, just as there are already alternative views nowadays which do not side exclusively with any of them, as is the analysis here). And the reverse direction also holds true.

In relation to the predictability-unpredictability principle, if there is predictability (e.g., the predictable tendency that there will be more education and regulations about accounting ethics in the foreseeable future), there is unpredictability (e.g., the more difficult task to predict exactly the extent to which a particular education and regulation about accounting ethics will be effective on a particular future occasion, since there have been business scandals in spite of the existence of different accounting reforms). And the reverse direction also holds true.

In relation to the explicability-inexplicability principle, if there is explicability (e.g., the explanation, by those advocates of moral universalism in accounting ethics, of its validity on the basis that "there is only one valid moral code for all" in accounting ethics, like the one against corruption and fraud), there is inexplicability (e.g., the lack of sufficient explanation, by those advocates of moral universalism in accounting ethics, of why it is necessarily valid, since, to the critics, "separatism is based on a relativist conception of morality that there can be different, equally valid moral codes that apply to different sections of society and differences in codes between societies"). And the reverse direction also holds true.

In relation to the fiction-reality principle, if there is fiction (e.g., the fictional aspect of accounting ethics, insofar as those advocates of moral universalism in accounting ethics claim "there is only one valid moral code for all," since, to the critics, "separatism is based on a relativist conception of morality that there can be different, equally valid moral codes that apply to different sections of society and differences in codes between societies"), there is reality (e.g., the realistic aspect of accounting ethics, insofar as different countries have different legal standards and regulations in regard to the enforcement of accounting practices). And the reverse direction also holds true.

In relation to the cognitiveness-noncognitiveness principle, if there is cognitiveness (e.g., a view of things on the basis of reason and evidences, as shown in the finding that "new reforms, regulations, and calls for increased higher education have been introduced to combat the dangers of unethical behavior" in accounting), there is noncognitiveness (e.g., a view of things on the basis of non-cognitive factors like envy, jealousy, power, nationality, race, gender, age, class, greed, lust, status, faith, anger, sadness, joy, fear, wish, etc.—as shown in the "temptation" of creative accountants to cheat due to "self-interest" or "greed," in spite of all the "reforms, regulations, and calls for increased higher education…to combat the dangers of unethical behavior" in accounting). (WK 2013w; H. Johnsson 2005) And the reverse direction also holds true.

In relation to the finiteness-transfiniteness principle, if there is finiteness (e.g., the finite number of the reasons of ethical failures, like the 9 reasons aforecited), there is transfiniteness (e.g., the transfinite number of all the amounts of money that have lost in history, regardless of which reasons are responsible for the losses). And the reverse direction also holds true.

In relation to the preciseness-vagueness principle, if there is preciseness (e.g., the precise identification of 9 reasons of ethical failures in accounting in the article aforecited), there is vagueness (e.g., the vagueness in the identification of the 9 reasons of ethical failures in accounting in the article aforecited, since it is not clear why there must be only 9, not 10, 11, 12, 13, and so on). And the reverse direction also holds true.

In relation to the simpleness-complicatedness principle, if there is simpleness (e.g., the relatively simple analysis of accounting ethics by the advocates), there is complicatedness (e.g., the relatively more complicated analysis of accounting ethics, by challenging the claims and assumptions, as shown in the problems pointed out by the critics as cited). And the reverse direction also holds true.

In relation to the openness-hiddenness principle, if there is openness (e.g., the open exploration of "principle-based" accounting standards to be used for the accounting profession by the International Financial Reporting Standards [IFRS]), there is hiddenness (e.g., the hidden bias in the exploration of "principle-based" accounting standards to be used for the accounting profession by the International Financial Reporting Standards [IFRS]," since the critics like those who favor "the United States Generally Accepted Accounting Principles [GAAP]" prefer those which are "largely rule-based" instead). And the reverse direction also holds true.

In relation to the denseness-emptiness principle, if there is denseness (e.g., the relatively denser concentration of concerns with "principle-based" accounting standards by those for the International Financial Reporting Standards [IFRS]), there is emptiness (e.g., the relatively less dense, or more empty, concentration of concerns with "rule-based" accounting standards by those for the International Financial Reporting Standards [IFRS]). And the reverse direction also holds true.

In relation to the rule-exception principle, if there are rules (e.g., the usual generalization, procedure, habit, or tendency that "the International Financial Reporting Standards [IFRS]...are used by over 115 countries" around the world and are "principle-based"), there are exceptions (e.g., a case to which a rule does not apply, such as the exceptional cases that some countries like the U.S. do not use "the International Financial Reporting Standards [IFRS]" and prefer those which are "largely rule-based" instead, at least for now). And the reverse direction also holds true.

In relation to the prototypicality-variation principle, if there is prototypicality (e.g., the prototype, like the example about the original work on accounting ethics by Luca Pacioli in 1494), there is variation (e.g., different variants of the protoptye, as shown in many works since the original work on accounting ethics by Luca Pacioli in 1494—like the later ones "expanded by government groups, professional organizations, and independent companies" like "the International Financial Reporting Standards [IFRS]," "the United States Generally Accepted Accounting

Principles [GAAP]," etc., though each with its own variation, contribution, and disagreement). And the reverse direction also holds true.

In relation to the change-constancy principle, if there is change (e.g., the ever new ways to study accounting, as shown in the work on accounting ethics), there is constancy (e.g., the ever constant existence of different problems with the ever new ways to study accounting, as shown in the problems with the work on accounting ethics pointed out by the critics). And the reverse direction also holds true.

In relation to the order-chaos principle, if there is order (e.g., the more or less orderly view on acounting standards, if looked from the sole vantage point of "the International Financial Reporting Standards [IFRS]" which "are used by over 115 countries" around the world and are "principle-based"), there is chaos (e.g., the more or less chaotic view on acounting standards, if looked from the multiple, conflicting vantage points of both the advocates and the critics, especially in regard to different problems aforecited, such that they do not add up to much of anything coherent). And the reverse direction also holds true.

In relation to the slowness-quickness principle, if there is slowness (e.g., the relatively slower speed of corruption and fraud to occur, in the presence of accounting reforms like "education" and "regulations" as described earlier), there is quickness (e.g., the relatively quicker speed of corruption and fraud to occur, in the absence of any accounting reforms like "education" and "regulations" as described earlier, together with the institutional indulgence of "largest CPA firms" to spend "on advertisements" which "rose from US$4 million in the 1980s to more than $100 million in the 2000s"). (WK 2013w) And the reverse direction also holds true.

In relation to the expansion-contraction principle, if there is expansion (e.g., the relative expansion of "education" and "regulations" against corruption and fraud nowadays), there is contraction (e.g., the relative decline, nowadays, of the institutional indulgence of "largest CPA firms" to spend "on advertisements" which "rose from US$4 million in the 1980s to more than $100 million in the 2000s"). And the reverse direction also holds true.

In relation to the optimality-nonoptimality principle, if there is optimality (e.g., the search for highly favorable options, like the quest for the most comprehensive accounting reform against corruption and fraud "in July 2010 when President Obama signed 'The Dodd-Frank Wall Street Reform and Consumer Protection Act'" which "covers a broad range of changes"), there is nonoptimality (e.g., the existence of non-optimal alternatives to optimality, like the argument for more realistic works by the critics, who pointed out different problems with accounting reforms aforecited). (WK 2013w) And the reverse direction also holds true.

In relation to the simultaneity-nonsimultaneity principle, if there is simultaneity (e.g., the occurrence of different things more or less at the same time, such as the simultaneous co-existence of different accounting standards, like the U.S. GAAP and the London-based IFRS), there is nonsimultaneity (e.g., the occurrence of something after something else, such as the decision in 2008 by the Securities and Exchange Commission in the U.S. to "issue…a preliminary 'roadmap' that may lead the U.S. to abandon GAAP in the future…and to join…the London-based IFRS," after "more than 100 countries around the world" had already joined). (WK 2013ll) And the reverse direction also holds true.

In relation to the isolation-interaction principle, if there is isolation (e.g., the state of a situation that is separate from others, such as the isolation of US GAAP from the London-based International Financial Reporting Standards in the older days), there is interaction (e.g., the influence of entities on one another, such as the steady integration of US GAAP into the London-based International Financial Reporting Standards nowadays, after the decision in 2008 by the Securities and Exchange Commission in the U.S. to "issue…a preliminary 'roadmap' that may lead the U.S." to eventually join the international standards). And the reverse direction also holds true.

In relation to the theory-praxis principle, if there is theory (e.g., the theoretical construction of "ethical principles" in the field of ethics), there is praxis (e.g., the practical application of "ethical principles" in the field of ethics to the field of accounting, like "accounting ethics"). And the reverse direction also holds true.

In relation to the convention-novelty principle, if there is convention (e.g., the conventional wisdom about US GAAP as the accounting standards to be used in the U.S., which are "ruled-based"), there is novelty (e.g., the alternative novel challenge to the conventional wisdom about US GAAP as the accounting standards to be used in the U.S., which are "ruled-based"—by the new idea of the Securities and Exchange Commission in the U.S. to "issue…a preliminary 'roadmap'" in 2008 "that may lead the U.S." to eventually join the international standards IFRS which are "principle-based"). And the reverse direction also holds true.

In relation to the evolution-transformation principle, if there is evolution (e.g., the natural evolution of humans to behave badly sometimes in the state of nature), there is transformation (e.g., the technical transformation of human ability to deal with bad behaviors by the invention of "accounting ethics"). And the reverse direction also holds true.

In relation to the symmetry-asymmetry principle, if there is symmetry (e.g., the co-existence of different entities like the Italian court system and individual companies like Parmalat), there is asymmetry (e.g., . the Italian court system sentenced the CEO of Parmalat, Calisto Tanzi, to "10 years in prison for fraud relating to the collapse of the dairy group"—but Tanzi cannot sentence the Italian court system to "10 years in prison"). And the reverse direction also holds true.

In relation to the softness-hardness principle, if there is softness (e.g., the defense of Parmalat by the advocates like Tanzi), there is hardness (e.g., the critique of Parmalat, as shown in the conviction of Tanzi by the Italian court system aforecited). And the reverse direction also holds true.

In relation to the seriousness-playfulness principle, if there is seriousness (e.g., the serious business of those who try to understand the reasons of ethical failures in accounting), there is playfulness (e.g., the playful part of those who try to understand the reasons of ethical failures in accounting, when they play around with different possibiliites over the decades, like the 9 possible reasons aforecited). And the reverse direction also holds true.

In relation to the activeness-inactivess principle, if there is activeness (e.g., the relative activeness of those like Tanzi in Parmalat who actively engaged in fraudulent activities when he was the CEO), there is inactiveness (e.g., the relative inactiveness of those like Tanzi in Parmalat who later spent "10 years in prison for fraud," so he could not engage in corporate misdeeds anymore). And the reverse direction also holds true.

In relation to the selfness-otherness principle, if there is selfness (e.g., the tendency to be mostly centered on one's self or group, as shown in the corruption and fraud by Tanzi in Parmalat), there is otherness (e.g., the tendency to be mostly centered on others, as shown in the academic analysis of both sides of the debate on accounting ethics, as an advancement in the knowledge of professional ethics for humanity). And the reverse direction also holds true.

In relation to the regression-progression principle, if there is regression (e.g., the regression made by accounting ethics, as shown in the problems pointed out by the critics), there is progression (e.g., the progress made by accounting ethics, as shown in the contribution to the prevention of corruption and fraud aforecited). And the reverse direction also holds true.

In relation to the same-difference principle, if there is similarity in outcome (e.g., the contribution to the molding and control of beliefs, values, and behaviors, regardless of whether this be done by way of corporate misdeeds by Tanzi, or by way of the law enforcement by the Italian court system against Tanzi), there is difference in outcome (e.g., the contribution to the molding and control of beliefs, values, and behaviors by way of corporate misdeeds by Tanzi for a relatively more amoral lifeworld—but the contribution to the molding and control of beliefs, values, and behaviors by way of the law enforcemeny by the Italian court system against Tanzi for a relatively more moral lifeworld). And the reverse direction also holds true.

In relation to the stability-reaction principle, if there is stability (e.g., the relatively more stable condition of accounting scandals in the last few decades of the 20th century), there is reaction (e.g., accounting scandals in the last few decades of the 20th century then led to further changes later on, when nowadays, there are different accounting reforms like "education" and "regulations" aforecited). And the reverse direction also holds true.

In relation to the functionality-nonfunctionality principle, if there is functionality (e.g., the presence of action for which something or someone is fitted, as shown in the construction of accounting ethics, which serves the function of preventing corruption and fraud in the profession of accounting), there is nonfunctionality (e.g., the relative lack of action for which something or someone is fitted, as shown in the criticism that there have been business scandals in spite of the existence of accounting ethics, such that it is not as functional as the advocates would like us to believe). And the reverse direction also holds true.

In relation to the intentionality-nonintentionality principle, if there is intentionality (e.g., the planning or design of something for a certain outcome, as shown in the original intention of those to construct professional ethics to combat corruption and fraud), there is nonintentionality (e.g., the relative absence of planning or design of something for a certain outcome, such that the outcome can even be counter-intentional in being the opposite of what was originally planned, as shown in the different criticisms by the critics that there have been business scandals in spite of the existence of accounting ethics, that there are competing approaches to ethical principles in accounting ethics, that morality can be "relative" [not "universal"], that there are the larger institutional factors which contribute to the occurrence of accounting scandals [not just due to the fault of creative accountants], that different countries have different legal standards and regulations in regard to the enforcement of accounting practices, etc.—contrary to the original intention of those to construct professional ethics to combat corruption and fraud). And the reverse direction also holds true.

And in relation to the survivability-nonsurvivability principle, if there is survivability (e.g., the relative survival, nowadays, of the international standards like IFRS), there is non-survivability (e.g., the relative non-survival, nowadays, of US GAAP, since the Securities and Exchange Commission in the U.S. "issued a preliminary 'roadmap'" in 2008 "that may lead the U.S." to abandon US GAAP and to join IFRS instead). And the reverse direction also holds true.

Subtraction, Accounting Standards, and History

Another way to evaluate the possibility (or impossibility) and desirability (or undesirability) of subtraction from the perspective of culture concerns subtraction, accounting standards, and history.

An argument in the literature is that the history of accounting standards has evolved to the point where there is a need for the convergence of different accounting standards in our time, so as to reduce costs (among other benefits).

Convergence of Accounting Standards

The term "convergence of accounting standards" refers to "the goal of establishing a single set of accounting standards that will be used internationally, and in particular the effort to reduce the differences between the US Generally Accepted Accounting Principles (US GAAP), and the International Financial Reporting Standards (IFRS). Convergence in some form has been taking place for several decades, and efforts today include projects that aim to reduce the differences between accounting standards." (WK 2013xx; FASB 2012 & 2012a; AICPA 2008)

In particular, "the Financial Accounting Standards Board (FASB) and the International Accounting Standards Board (IASB)...are currently working towards completing the convergence programme laid out by a 2006 memorandum of understanding, which was updated in 2008. Efforts include projects that aim to improve the respective accounting standards, and those that aim to reduce the differences between them." (WK 2013xx; AICPA 2008; IASB 2012)

For example, "in a joint report published in 2012, the IASB and FASB stated that most of the short-term projects outlined in the memorandum of understanding had been completed, and that greater priority was now being placed on long-term projects. Short term projects involve the amendment of one of the boards' standards to better align them with the other board's, jointly issuing new standards. Some short-term projects and corresponding action taken are listed below": (WK 2013xx; IASB 2012)

- "Segment Reporting (Completed)"—"a new standard, IFRS 8 Segment Reporting, was issued in 2006."
- "Fair Value Option (Completed)"—"US GAAP was amended to include the fair value option in 2007."
- "Joint Ventures (Completed)"—IFRS 11 Joint Arrangements was issued in 2011."
- "Income Tax (Priority Lowered)"—"A joint exposure draft was published in 2009."

Later, "an update to the memorandum of understanding in 2008 introduced long-term convergence projects, including the following": (WK 2013xx; IASB 2012)

- "Derecognition (Completed)"—"both boards issued amendments"
- "Fair Value Measurement (Completed)"—"FASB Statement No. 257 and IFRS 13 were issued in 2011."
- "Financial Instruments with the Characteristics of Equity (Priority Lowered)"—"a joint discussion paper was released."
- "Revenue Recognition (in Progress)"—"the boards issued joint proposals in 2010."

In addition, "convergence is also taking place in other countries, with 'all major economies' planning to either adopt the IFRS or converge towards it, 'in the near future.'" (WK 2013xx)

Motivations for the Convergence of Accounting Standards

There are different motivations for the convergence of accounting standards. For example, "a 2008 report by PricewaterhouseCoopers (PwC) stated that convergence of accounting standards would contribute to the flow of international investment and benefit 'all capital markets stakeholders' because it": (WK 2013xx; PRICE 2007)

- "renders international investments more comparable to investors"
- "reduces the cost of complying with accounting requirements for global businesses"
- "potentially establishes a more transparent accounting system with greater accountability"
- "reduces 'operational challenges' for accounting firms"

- "gives standard-setters the opportunity to 'improve the reporting model.'"

Also, "a survey conducted by the International Federation of Accountants found that 89% of accounting profession leaders who responded expressed that convergence was either very important or important for economic growth for their respective countries." (WK 2013xx; AICPA 2008)

History of the Evolution of Accounting Standards

Historically, the evolution of accounting standards to the current effort of convergence can be summarized in terms of 3 main periods, as explained below.

(a) 1950s to 1960s

The first period of the history of the evolution of accounting standards covers the years from the 1950s to the 1960s.

At the beginning, "the idea of convergence has roots in the 1950s, and was a response to greater economic integration and international capital flows after the second world war. Before the 1990s, convergence took the form of harmonization, the reduction of differences between the various accounting standards used internationally." (WK 2013xx; FASB 2012a)

Later, in 1962, "at the 8th International Congress of Accountants hosted by the American Institute of Certified Public Accountants..., many participants expressed the need for the development of accounting standards on an international basis; in the same year the AICPA reactivated the Committee on International Relations, that aimed to improve cooperation among accountants globally." (WK 2013xx; FASB 2012a)

(b) 1970s to 1990s

The second period of the history of the evolution of accounting standards covers the years from the 1970s to the 1990s.

In 1973, "the International Accounting Standards Committee (IASC), the predecessor to the International Accounting Standards Board (IASB) was established...with the goal of developing accounting standards and promoting them internationally; by 1987 the IASC had issued 25 standards, and by the late 1980s there was 'worldwide interest' in the need for convergence." (WK 2013xx; FASB 2012a)

In 1991, "the FASB formally set out the goal of developing an internationally-used set of accounting standards, and in subsequent years the FASB and IASC undertook various projects to lay the groundwork for convergence. In 1996, the National Securities Markets Improvement Act became law; the act expressed support for convergence efforts and required the SEC to report to congress on progress towards convergence." (WK 2013xx; FASB 2012a)

(c) 2000s to the Present

And the third period of the history of the evolution of accounting standards covers the years from the 2000s to the present.

In 2001, "the IASC was reconstituted into the IASB…, and the FASB and IASB began working towards convergence in 2002, expressing their commitment to convergence in the Norwalk agreement and pledging to make their respective standards 'compatible as soon as is practicable' and to maintain compatibility by coordinating future programs. Their commitment to working towards convergence was reaffirmed at meetings in 2005." (WK 2013xx; FASB 2012a; AICPA 2008)

In 2006, "the IASB and FASB signed a memorandum of understanding…which laid guidelines on their convergence projects and set short-term goals such as to issue converged standards on business combinations by 2008. Work towards the goals were reviewed in 2008, and a progress report published that also set out subsequent steps for each convergence topic. The FASB and IASB met again in 2009 and agreed to 'intensify their efforts' in working towards the goals of the memorandum of understanding, while laying down future plans and targets." (WK 2013xx; FASB 2008 & 2008a)

And more recently, "convergence of accounting standards is also taking place in other jurisdictions; the IFRS Foundation has stated that 'all major economies' have plans to either converge towards or adopt the IFRS 'in the near future,' and that the Group of 20 Leaders (G20) have encouraged accounting bodies globally to increase efforts to achieve convergence." (WK 2013xx; IFRS 2012)

Problems with the Study of Accounting Standards and History

However, to the critics, this study of subtraction, accounting standards, and history has some major problems. Consider, for illustration, a few examples below (as summarized in *Table 3.8*).

Firstly, one main criticism of the study of subtraction, accounting standards, and history is that the convergence of different accounting standards is costly.

For example, to the critics, "criticisms of convergence include its cost," and "the cost and pace of adoption have been cited as the most common criticism of the SEC's 2008 convergence roadmap." (WK 2013xx; AICPA 2009)

Secondly, another main criticism of the study of subtraction, accounting standards, and history is that the claim of "comparability" is questionable.

For instance, to the critics, "criticisms of convergence include…the idea that the link between convergence and comparability may not be strong….In particular, Shyam Sunder [2011] of the Yale School of Management has called the link between convergence and comparability 'overblown.'" (WK 2013xx)

Thirdly, still another main criticism of the study of subtraction, accounting standards, and history is that the convergence process is too slow.

For instance, to the critics, "criticisms of convergence include its…pace," (WK 2013xx; AICPA 2009) A good reminder here is that "these efforts [of convergence] have stalled recently, with IASB chairman Hans Hoogervorst suggesting that convergence is no longer a top priority for the IFRS standard-setter." (WK 2013xx; CFO 2012)

Fourthly, an additional main criticism of the study of subtraction, accounting standards, and history is that different countries comply with the convergence requirements in different ways.

For example, "Canada required all listed entities to use the IFRS from January 1, 2012" but "Japan permitted the use of IFRS for certain multinational companies from 2010…."' (WK 2013xx)

And fifthly, another additional main criticism of the study of subtraction, accounting standards, and history is that it is not as popular as conventionally assumed.

For instance, to the critics, in 2013, "a new accounting proposal on leasing equipment looks like it will bring about change in global accounting standards….The proposal was issued jointly by the FASB and IASB. As quoted by Hans Hoogervorst, chairman of the IASB, 'This is going to be one of the least popular standards.' Under this proposal, companies will have to report having more assets and liabilities on their balance sheet than previously required. This could have extreme effects on the construction industry." (WK 2013xx; FORC 2013)

Beyond Accounting Standards and History

These criticisms of subtraction, accounting standards, and history, however, should not be misunderstood as a total rejection of its usefulness but serve to show us the two opposing sides of the debate, such that the possibility (or impossibility) and desirability (or undesirability) of subtraction (from the perspective of culture with subtraction, accounting standards, and history as a case study here) are not to the extent that the respective defenders would like us to believe.

In addition, the analysis of subtraction, accounting standards, and history can teach us a valuable lesson on the ontological principles in existential dialectics, and good examples include the formalness-informalness principle, the absoluteness-relativeness principle, the partiality-totality principle, the predictability-unpredictability principle, the explicability-inexplicability principle, the fiction-reality principle, the cognitiveness-noncognitiveness principle, the finiteness-transfiniteness principle, the preciseness-vagueness principle, the simpleness-complicatedness principle, the openness-hiddenness principle, the denseness-emptiness principle, the rule-exception principle, the prototypicality-variation principle, the change-constancy principle, the order-chaos principle, the slowness-quickness principle, the expansion-contraction principle, the optimality-nonoptimality principle, the simultaneity-nonsimultaneity principle, the isolation-interaction principle, the theory-praxis principle, the convention-novelty principle, the evolution-transformation principle, the symmetry-asymmetry principle, the softness-hardness principle, the seriousness-playfulness principle, the activeness-inactivess principle, the selfness-otherness principle, the regression-progression principle, the same-difference principle, the stability-reaction principle, the functionality-nonfunctionality principle, the intentionality-nonintentionality principle, and the survivability-nonsurvivability principle.

For instance, in relation to the formalness-informalness principle, if there is formalness (e.g., the formal logical requirement of soundness and completeness in a system of ideas, as in the argument on the convergence of accounting standards), there is informalness (e.g., the non-formal existence of unsoundness and incompleteness in the argument on the convergence of accounting standards, as shown in the criticisms that the convergence of different accounting standards is costly, that the claim of "comparability" is questionable, that the convergence process is too slow, that different countries comply with the convergence requirements in different ways, and that it is not as popular as conventionally assumed). And the reverse direction also holds true.

In relation to the absoluteness-relativeness principle, if there is absoluteness (e.g., the absolute view on the convergence of accounting standards by the FASB and IASB), there is relativeness (e.g., what is right for the FASB and IASB in regard to the convergence of accounting standards is not necessarily so for the critic like Shyam Sunder, who "has called the link between convergence and comparability 'overblown'"). And the reverse direction also holds true.

In relation to the partiality-totality principle, if there is partiality (e.g., the partial view on the convergence of accounting standards by the advocates like the FASB and IASB), there is totality (e.g., the more holistic view on the convergence of accounting standards, such that the whole is not the sum of its parts, that is, it cannot be reduced to the sum of opposing views by the advocates like the FASB and IASB, the critics like Shyam Sunder, etc., since there will emerge new views in the future not yet known today, just as there are already alternative views nowadays which do not side exclusively with any of them, as is the analysis here). And the reverse direction also holds true.

In relation to the predictability-unpredictability principle, if there is predictability (e.g., the predictable tendency that the FASB and IASB will continue working on the convergence of accounting standards in the foreseeable future), there is unpredictability (e.g., the more difficult task to predict exactly the extent to which the convergence of accounting standards will occur in a particular future era). And the reverse direction also holds true.

In relation to the explicability-inexplicability principle, if there is explicability (e.g., the explanation, by those advocates of the convergence of accounting standards, of its desirability on the basis of "comparability"), there is inexplicability (e.g., the lack of sufficient explanation, by those advocates of the convergence of accounting standards, of why "comparability" is necessarily the case, since the critics like Shyam Sunder dismissed "the link between convergence and comparability" as "overblown"). And the reverse direction also holds true.

In relation to the fiction-reality principle, if there is fiction (e.g., the fictional aspect of the convergence of accounting standards, insofar as the critics like Shyam Sunder dismissed "the link between convergence and comparability" as "overblown"), there is reality (e.g., the realistic aspect of the convergence of accounting standards, insofar as the aim is to strive for certain degree of "comparability"). And the reverse direction also holds true.

In relation to the cognitiveness-noncognitiveness principle, if there is cognitiveness (e.g., a view of things on the basis of reason and evidences, as shown in the finding that the FASB and IASB have worked on the convergence of accounting standards so as to "render international investments more comparable to investors"), there is noncognitiveness (e.g., a view of things on the basis of non-cognitive factors like envy, jealousy, power, nationality, race, gender, age, class, greed, lust, status, faith, anger, sadness, joy, fear, wish, etc.—as shown in the "politics" of "convergence," such that the project continues, even though the critics like Shyam Sunder dismissed "the link between convergence and comparability" as "overblown," together with other criticisms that the convergence of different accounting standards is costly, that the convergence process is too slow, and that it is not as popular as conventionally assumed). And the reverse direction also holds true.

In relation to the finiteness-transfiniteness principle, if there is finiteness (e.g., the finite number of different accounting standards in the world, like the Chinese, the American, the Indian, the French, etc.), there is transfiniteness (e.g., the transfinite number of all the individual instances of accounting standards that have ever been applied in history, regardless of which accounting standards they actually belong to). And the reverse direction also holds true.

In relation to the preciseness-vagueness principle, if there is preciseness (e.g., the precise identification of 5 motivations for the convergence of accounting standards in the article aforecited), there is vagueness (e.g., the vagueness in the identification of the 5 motivations for the convergence of accounting standards in the article aforecited, since it is not clear why there must be only 5, not 6, 7, 8, and so on). And the reverse direction also holds true.

In relation to the simpleness-complicatedness principle, if there is simpleness (e.g., the relatively simple analysis of the convergence of accounting standards by the advocates), there is complicatedness (e.g., the relatively more complicated analysis of the convergence of accounting standards, by challenging the claims and assumptions, as shown in the problems pointed out by the critics as cited). And the reverse direction also holds true.

In relation to the openness-hiddenness principle, if there is openness (e.g., the open exploration of the convergence of accounting standards by the FASB and IASB), there is hiddenness (e.g., the hidden bias in the exploration of the convergence of accounting standards by the FASB and IASB, since the critics pointed out that the claim of "comparability" is questionable, that the convergence process is costly, that it is not as popular as conventionally assumed, etc.). And the reverse direction also holds true.

In relation to the denseness-emptiness principle, if there is denseness (e.g., the relatively denser concentration of concerns with the convergence of accounting standards by the FASB and IASB), there is emptiness (e.g., the relatively less dense, or more empty, concentration of concerns with the convergence of accounting standards by the critics like Shyam Sunder). And the reverse direction also holds true.

In relation to the rule-exception principle, if there are rules (e.g., the usual generalization, procedure, habit, or tendency that "convergence in some form has been taking place for several decades"), there are exceptions (e.g., a case to which a rule does not apply, such as the exceptional case that in 2006 "senior partners at PricewaterhouseCoopers …called for convergence to be 'shelved indefinitely' in a draft paper" and thus did not support it). (WK 2013xx) And the reverse direction also holds true.

In relation to the prototypicality-variation principle, if there is prototypicality (e.g., the prototype, like the example about the original work on "convergence" which "took the form of harmonization, the reduction of differences between the various accounting standards used internationally," before the 1990s), there is variation (e.g., different variants of the protoptye, as shown in many studies since the original work on "convergence" which "took the form of harmonization, the reduction of differences between the various accounting standards used internationally," before the 1990s —like the later ones in the 2000s by the FASB, the IASB, etc., though each with its own variation, contribution, and disagreement). And the reverse direction also holds true.

In relation to the change-constancy principle, if there is change (e.g., the ever new ways to study accounting, as shown in the work on the convergence of accounting standards), there is constancy (e.g., the ever constant existence of different problems with the ever new ways to study accounting, as shown in the problems with the work on the convergence of accounting standards pointed out by the critics). And the reverse direction also holds true.

In relation to the order-chaos principle, if there is order (e.g., the more or less orderly view on the convergence of accounting standards, if looked from the sole vantage point of "comparability" by the advocates), there is chaos (e.g., the more or less chaotic view on the convergence of accounting standards, if looked from the multiple, conflicting vantage points of both the advocates and the critics, especially since the critics like Shyam Sunder dismissed "the link between convergence and comparability" as "overblown," together with other problems, such that all of these views do not add up to much of anything coherent). And the reverse direction also holds true.

In relation to the slowness-quickness principle, if there is slowness (e.g., the relatively slower speed of communication among "investors, companies, auditors, and other participants" if handicapped by "incomparability" among different accounting standards), there is quickness (e.g., the relatively quicker speed of communication among "investors, companies, auditors, and other participants" if helped by "comparability" in the convergence of accounting standards). And the reverse direction also holds true.

In relation to the expansion-contraction principle, if there is expansion (e.g., the relative spread of the convergence of accounting standards nowadays), there is contraction (e.g., the relative decline, nowadays, of individual country accounting standards with no interaction with each other). And the reverse direction also holds true.

In relation to the optimality-nonoptimality principle, if there is optimality (e.g., the search for highly favorable options, like the quest for the most extensive "comparability" in the convergence of accounting standards by the FASB and IASB), there is nonoptimality (e.g., the existence of non-optimal alternatives to optimality, like the argument for more realistic studies by the critics, who pointed out different problems with the convergence of accounting standards). And the reverse direction also holds true.

In relation to the simultaneity-nonsimultaneity principle, if there is simultaneity (e.g., the occurrence of different things more or less at the same time, such as the simultaneous co-existence of different accounting standards nowadays, like the US GAAP and the IFRS), there is nonsimultaneity (e.g., the occurrence of something after something else, such as the effort for the convergence of accounting standards by the FASB and IASB, after the existence of the US GAAP and the IFRS). And the reverse direction also holds true.

In relation to the isolation-interaction principle, if there is isolation (e.g., the state of a situation that is separate from others, such as the isolated existence of the US GAAP in the older days without any interaction with the IFRS), there is interaction (e.g., the influence of entities on one another, such as the effort to integrate the US GAAP into the IFRS, as part of the project for the convergence of accounting standards by the FASB and IASB). And the reverse direction also holds true.

In relation to the theory-praxis principle, if there is theory (e.g., the theoretical construction of "fair value" in the field of economics), there is praxis (e.g., the practical application of "fair value" in the field of economics to the realm of the convergence of accounting standards by the FASB and IASB, as shown in their "fair value option" project in 2007). And the reverse direction also holds true.

In relation to the convention-novelty principle, if there is convention (e.g., the conventional wisdom about the need for the convergence of accounting standards by the FASB and IASB), there is novelty (e.g., the alternative novel challenge to the conventional wisdom about the need for the convergence of accounting standards by the FASB and IASB—by the new idea of Shyam Sunder, who "has called the link between convergence and comparability 'overblown'" in his questioning of the project for the convergence of accounting standards). And the reverse direction also holds true.

In relation to the evolution-transformation principle, if there is evolution (e.g., the natural evolution of humans to count and report things in the state of nature in a primitive way), there is transformation (e.g., the technical transformation of human ability to deal with counting and reporting by the invention of "the convergence of accounting standards" for "comparability" nowadays). And the reverse direction also holds true.

In relation to the symmetry-asymmetry principle, if there is symmetry (e.g., the co-existence of different entities like the standard-setters FASB and IASB, and individual companies), there is asymmetry (e.g., the standard-setters like the FASB and the IASB can set up standards for the convergence of accounting standards for individual companies—but individual companies cannot set up standards for the FASB and the IASB to follow). And the reverse direction also holds true.

In relation to the softness-hardness principle, if there is softness (e.g., the defense of the convergence of accounting standards by the FASB and the IASB), there is hardness (e.g., the critique of the convergence of accounting standards, as shown in the problems pointed out by the critics aforecited). And the reverse direction also holds true.

In relation to the seriousness-playfulness principle, if there is seriousness (e.g., the serious business of those who work on the convergence of accounting standards), there is playfulness (e.g., the playful part of those who work on the convergence of accounting standards, when they play around with different convergence projects over the decades, like the 8 examples aforecited). And the reverse direction also holds true.

In relation to the activeness-inactivess principle, if there is activeness (e.g., the relative activeness of those who work on the convergence of accounting standards, since it is demanded around the world), there is inactiveness (e.g., the relative inactiveness of those who still work in the old-fashioned way of individual country accounting standards which have no interaction with the IFRS, as they become more and more out-of-date).

In relation to the selfness-otherness principle, if there is selfness (e.g., the tendency to be mostly centered on one's self or group, as shown by those in the U.S. who still prefer the US GAAP over the IFRS), there is otherness (e.g., the tendency to be mostly centered on others, as shown in the academic analysis of both sides of the debate on the convergence of accounting standards, as an advancement in the knowledge of the evolution of accounting standards for humanity). And the reverse direction also holds true.

In relation to the regression-progression principle, if there is regression (e.g., the regression made by the convergence of accounting standards, as shown in the problems pointed out by the critics), there is progression (e.g., the progress made by the convergence of accounting standards, as shown in the contribution to "comparability" aforecited). And the reverse direction also holds true.

In relation to the same-difference principle, if there is similarity in outcome (e.g., the contribution to the molding and control of beliefs, values, and behaviors, regardless of whether this be done by way of individual country accounting standards, or by way of the convergence of accounting standards), there is difference in outcome (e.g., the contribution to the molding and control of beliefs, values, and behaviors by way of individual country accounting standards for a relatively more nationalistic lifeworld—but the contribution to the molding and control of beliefs, values, and behaviors by way of the convergence of accounting standards for a relatively more globalized lifeworld). And the reverse direction also holds true.

In relation to the stability-reaction principle, if there is stability (e.g., the relatively more stable condition of the US GAAP in the older days), there is reaction (e.g., the US GAAP in the older days then led to further changes later on, when nowadays, there is a call for the reduction of "the differences between the US GAAP and the IFRS for the ultimate convergence of accounting standards). And the reverse direction also holds true.

In relation to the functionality-nonfunctionality principle, if there is functionality (e.g., the presence of action for which something or someone is fitted, as shown in the convergence of accounting standards, which serves the function of providing "comparability" and other benefits aforecited), there is nonfunctionality (e.g., the relative lack of action for which something or someone is fitted, as shown in the problems with the convergence of accounting standards, such that it is not as functional as the advocates would like us to believe). And the reverse direction also holds true.

In relation to the intentionality-nonintentionality principle, if there is intentionality (e.g., the planning or design of something for a certain outcome, as shown in the original intention of the FASB and IASB to focus on the convergence of accounting standards), there is nonintentionality (e.g., the relative absence of planning or design of something for a certain outcome, such that the outcome can even be counter-intentional in being the opposite of what was originally planned, as shown in the recent announcement of the IASB chairman Hans Hoogervorst, when he suggested in 2012 that "convergence is no longer a top priority for the IFRS standard-setter"—contrary to the original intention of the FASB and IASB to focus on the convergence of accounting standards). And the reverse direction also holds true.

And in relation to the survivability-nonsurvivability principle, if there is survivability (e.g., the relative survival, nowadays, of the efforts for the convergence of accounting standards), there is non-survivability (e.g., the relative non-survival, nowadays, of individiual country accounting standards like the US GAAP, since it is to be eventually abandoned by joining the IFRS, under the convergence of accounting standards). And the reverse direction also holds true.

The Misapplication of Subtraction

This comprehensive analysis of the possibility (or impossibility) and desirability (or undesirability) of subtraction in the context of accounting, from the perspectives of the mind, nature, society, and culture, is important to show us the different ways in which subtraction is both possible and desirable, but not to the extent that the spokespersons from each side would like us to believe.

Subtraction thus has its misapplication, just as it has its application too, as the other side of the same coin.

But this is only one part of a larger story, since there is the other side of the story, which concerns addition, which was already addressed in the previous chapter (Chapter Two).

Now that we have reached the end of the long intellectual journey of exploring addition and subtraction in accounting, what then can be concluded about the future accounting?

With this question in mind, let's now turn to Chapter Four for the study of the future of accounting.

Table 3.1. Subtraction and its Dual Attributes

- **Subtraction and the Mind**
  —Ex: subtraction, learning, and control self-assessment
  —Ex: subtraction, cognitive biases, and mental accounting

- **Subtraction and Nature**
  —Ex: subtraction, sustainability, and carbon accounting
  —Ex: subtraction, the environment, and profit-loss accounting

- **Subtraction and Society**
  —Ex: subtraction, computer programming, and bookkeeping
  —Ex: subtraction, corporate misdeeds, and internal control

- **Subtraction and Culture**
  —Ex: subtraction, accounting scandals, and ethics
  —Ex: subtraction, accounting standards, and history

Notes: The examples in the categories are solely illustrative (not exhaustive), and the comparison is relative (not absolute), nor are they necessarily mutually exclusive. And some can be easily re-classified elsewhere. As generalities, they allow exceptions.
Source: A summary of Ch.3 of *FPHACCO*

Table 3.2. Subtraction, Learning, and Control Self-Assessment

- **Methodology of Control Self-Assessment**
  —Ex: "internal control questionnaire (ICQ) self-audit"
  —Ex: "customized questionnaires"
  —Ex: "control guides"
  —Ex: "interview techniques"
  —Ex: "control model workshops"
  —Ex: "interactive workshops

- **Benefits of Control Self-Assessment**
  —Ex: the creation of "a clear line of accountability for controls"
  —Ex: the reduction of "the risk of fraud (by examining data that may flag unusual patterns of transactions)"
  —Ex: the "result…in an organization with a lower risk profile"

- **Problems with the Argument for Control Self-Assessment**
  —Firstly, one main criticism of the argument on learning and control self-assessment is that there is no consensus on its methodologies.
  —Secondly, another main criticism of the argument on learning and control self-assessment is that there are alternative approaches to control self-assessment.
  —Thirdly, still another main criticism of the argument on learning and control self-assessment is that it does not give sufficient attention to "downside risk." (WK 2013mm)
  —And fourthly, an additional main criticism of the argument on learning and control self-assessment is that it is too obsessed with the "costs" of risk (but not its "benefits").

Notes: The examples in the categories are solely illustrative (not exhaustive), and the comparison is relative (not absolute), nor are they necessarily mutually exclusive. And some can be easily re-classified elsewhere. As generalities, they allow exceptions.
Source: From *Sec. 3.2.1* of *FPHACCO*. See book for citations.

Table 3.3. Subtraction, Cognitive Biases, and Mental Accounting

- **Types of Mental Accounting**
  —Ex: "income accounting"
  —Ex: "tax refunds"
  —Ex: "birthday money"
  —Ex: "bonuses"
  —Ex: "money you can afford to lose"
  —Ex: "safety capital"
  —Ex: "lottery winnings"
  —Ex: "money already spent"
  —Ex: "confusing identical purchases"

- **Problems with the Study of Cognitive Biases and Mental Accounting**
  —Firstly, one main criticism of the argument on subtraction, cognitive biases, and mental accounting is that the issue of "rationality" in the debate on mental accounting can be relative.
  —Secondly, another main criticism of the argument on subtraction, cognitive biases, and mental accounting is that cognitive biases in mental accounting, if they exist in some cases, can be more lasting than previously thought.
  —Thirdly, still another main criticism of the argument on subtraction, cognitive biases, and mental accounting is that there can be missing factors not accounted for in those cases of mental accounting.
  —Fourthly, an additional main criticism of the argument on subtraction, cognitive biases, and mental accounting is that it does not give sufficient attention to the "good" side of mental accounting (but its "bad" side).
  —And fifthly, another additional main criticism of the argument on subtraction, cognitive biases, and mental accounting is that there are "no easy answers" on how to deal with mental accounting.

Notes: The examples in the categories are solely illustrative (not exhaustive), and the comparison is relative (not absolute), nor are they necessarily mutually exclusive. And some can be easily re-classified elsewhere. As generalities, they allow exceptions.
Source: From *Sec. 3.2.2* of *FPHACCO.* See book for citations.

Table 3.4. Subtraction, Sustainability, and Carbon Accounting

- **Carbon Accounting, and Sustainability**
  —Ex: its relation to carbon footprint
  —Ex: its relation to sustainability measurement

- **Sustainability Methodology**
  —Ex: "environmental factors"
  —Ex: "social factors"
  —Ex: "economic factors"

- **Sustainability Measurement**
  —Ex: "benchmarks"
  —Ex: "indices"
  —Ex: "metrics"
  —Ex: "auditing"

- **Problems with the Study on Carbon Accounting**
  —Firstly, the practice of carbon accounting is not objective at all.
  —Secondly, different carbon accountants can produce different results (in addition to the problem of "abstraction" in accounting schemes as mentioned above).
  —Thirdly, "the total carbon footprint cannot be calculated."
  —Fourthly, the different "dimensions" of sustainability accounting involves difficult trade-offs in implementation.
  —Fifthly, there are competing approaches to sustainability accounting.
  —And sixthly, organizations may not have the incentives to fully implement sustainability accounting.

Notes: The examples in the categories are solely illustrative (not exhaustive), and the comparison is relative (not absolute), nor are they necessarily mutually exclusive. And some can be easily re-classified elsewhere. As generalities, they allow exceptions.
Source: From *Sec. 3.3.1* of *FPHACCO*. See book for citations.

Table 3.5. Subtraction, the Environment, & Profit-Loss Accounting

- **Ecosystem Services in the E P&L Accounting**
  —Ex: "supporting services"
  —Ex: "provisioning services"
  —Ex: "regulating services"
  —Ex: "cultural services"

- **Problems with the E P & L Accounting**
  —Firstly, one main criticism of the study on the E P & L accounting is that there are competing definitons of "ecosystem services." (E. Ojea 2010)
  —Secondly, another main criticism of the study on the E P & L accounting is that the understanding of ecosystem services is culturally relative.
  —Thirdly, still another main criticism of the study on the E P & L accounting is that there are also competing ways to operationalize the concept of "ecosystem services."
  —Fourthly, an additional main criticism of the study on the E P & L accounting is that ecosystem services can be more complex than previously thought.
  —Fifthly, another additional main criticism of the study on the E P & L accounting is that there are alternative approaches to explain ecosystem services.
  —Sixthly, still another additional main criticism of the study on the E P & L accounting is that the empirical validity of each study is limited.
  —And seventhly, one last main criticism of the study on the E P & L accounting is that there are disagreements about the true value of ecosystem services.

Notes: The examples in the categories are solely illustrative (not exhaustive), and the comparison is relative (not absolute), nor are they necessarily mutually exclusive. And some can be easily re-classified elsewhere. As generalities, they allow exceptions.
Source: From *Sec. 3.3.2* of *FPHACCO*. See book for citations.

Table 3.6. Subtraction, Computer Programming, and Bookkeeping

- **Ways to Reduce the need for Boilerplate**
  —Ex: "metaprogramming"
  —Ex: "convention over configuration"
  —Ex: "model-driven engineering"

- **Goals of Object-Oriented Programming**
  —Ex: "increased understanding"
  —Ex: "ease of maintenance"
  —Ex: "ease of evolution"

- **Features of Object-Oriented Programming**
  —Ex: "dynamic dispatch"
  —Ex: "encapsulation"
  —Ex: "subtype polymorphism"
  —Ex: "object inheritance (or delegation)"

- **Problems with Computer Programming and Bookkeeping**
  —Firstly, computer programming often does not model the "real world."
  —Secondly, OOP is too obsessed with "objects" (or "nouns") but not other things (like "verbs" or "actions," etc.).
  —Thirdly, there are disagreements about the fundamental features of OOP.
  —Fourthly, there are trade-offs in computer programming for bookkeeping.
  —Fifthly, OOP is not effective to deal with the issue of transparency.
  —And sixthly, OOP has not yet been proved to be better than other approaches in practice.

Notes: The examples in the categories are solely illustrative (not exhaustive), and the comparison is relative (not absolute), nor are they necessarily mutually exclusive. And some can be easily re-classified elsewhere. As generalities, they allow exceptions.
Source: From *Sec. 3.4.1* of *FPHACCO*. See book for citations.

Table 3.7. Subtraction, Accounting Scandals, and Ethics

- **Reasons of Ethical Failures**
  - —Ex: "self-interest"
  - —Ex: "failure to maintain objectivity and independence"
  - —Ex: "inappropriate professional judgment"
  - —Ex: "lack of ethical sensitivity"
  - —Ex: "improper leadership and ill-culture"
  - —Ex: "failure to withstand advocacy threats"
  - —Ex: "lack of competence"
  - —Ex: "lack of organizational and peer support"

- **Ethical Responses to Accounting Scandals**
  - —Ex: education
  - —Ex: regulations

- **Problems with the Study of Accounting Scandals and Ethics**
  - —Firstly, there have been business scandals in spite of the existence of accounting ethics.
  - —Secondly, different countries have different legal standards and regulations in regard to the enforcement of accounting practices.
  - —Thirdly, there are competing approaches to ethical principles in accounting ethics.
  - —Fourthly, morality can be "relative," not "universal." (WK 2013vv)
  - —And fifthly, there are the larger institutional factors which contribute to the occurrence of accounting scandals (not just due to the fault of creative accountants).

Notes: The examples in the categories are solely illustrative (not exhaustive), and the comparison is relative (not absolute), nor are they necessarily mutually exclusive. And some can be easily re-classified elsewhere. As generalities, they allow exceptions.
Source: From *Sec. 3.5.1* of *FPHACCO*. See book for citations.

Table 3.8. Subtraction, Accounting Standards, and History

- **Motivations for the Convergence of Accounting Standards**
  —Ex: "renders international investments more comparable to investors"
  —Ex: "reduces the cost of complying with accounting requirements for global businesses"
  —Ex: "potentially establishes a more transparent accounting system with greater accountability"
  —Ex: "reduces 'operational challenges' for accounting firms"
  —Ex: "gives standard-setters the opportunity to 'improve the reporting model.'"

- **Problems with the Study of Accounting Standards and History**
  —Firstly, one main criticism of the study of subtraction, accounting standards, and history is that the convergence of different accounting standards is costly.
  —Secondly, another main criticism of the study of subtraction, accounting standards, and history is that the claim of "comparability" is questionable.
  —Thirdly, still another main criticism of the study of subtraction, accounting standards, and history is that the convergence process is too slow.
  —Fourthly, an additional main criticism of the study of subtraction, accounting standards, and history is that different countries comply with the convergence requirements in different ways.
  —And fifthly, another additional main criticism of the study of subtraction, accounting standards, and history is that it is not as popular as conventionally assumed.

Notes: The examples in the categories are solely illustrative (not exhaustive), and the comparison is relative (not absolute), nor are they necessarily mutually exclusive. And some can be easily re-classified elsewhere. As generalities, they allow exceptions.
Source: From *Sec. 3.5.2* of *FPHACCO*. See book for citations.

• Part Four •

Conclusion

CHAPTER 4
CONCLUSION—THE FUTURE OF ACCOUNTING

> We have the most crude accounting tools. It's tragic because our accounts and our national arithmetic doesn't tell us the things that we need to know.
> __Susan George (BRAIN 2014)

Beyond Addition and Subtraction

The analyses of addition (in Chapter Two) and subtraction (in Chapter Three) show the extent of their possibility (or impossibility) and desirability (or undesirability) in the study of accounting (and related fields).

These two analyses are valuable for the evaluation of the opposing views (as already introduced in *Sec. 1.5* and discussed in the rest of the book), such that accounting (in relation to addition and subtraction) are neither possible (or impossible) nor desirable (or undesirable) to the extent that the respective ideologues (on different sides) would like us to believe.

Of course, this reexamination of different opposing views on accounting does not mean that the study of addition and subtraction is useless, or that those fields (related to accounting)—like bookkeeping, auditing, forensics, finance, info management, philosophy of accounting, accounting ethics, lean accounting, mental accounting, environmental audit, creative accounting, carbon accounting, social accounting, and so on—are unimportant. (WK 2013) In fact, neither of these extreme views is plausible.

Rather, this book offers an alternative (better) way to understand the future of accounting in regard to the dialectic relationship between addition and subtraction—while learning from different approaches in the literature but without favoring any one of them (nor integrating them, since they are not necessarily compatible with each other).

More specifically, this book offers a new theory (that is, *the double-sided theory of accounting*) to go beyond the existing approaches in a novel way.

My *double-sided theory of accounting* has four major distinct features.

Firstly, my theory makes good use of all theoretical approaches to accounting in relation to addition and subtraction, be they about the mental argument, the natural argument, the societal argument, or the cultural argument—especially from the most comprehensive combined perspectives of the mind, nature, society, and culture (as already dealt with in Chapter Two and Chapter Three).

Secondly, just like many other theories of mine in my previous books, my theory here does not heavily favor any specific theory over others in the literature, nor trying to integrate them (as they are not necessarily compatible with each other).

Thirdly, my theory re-interprets accounting in the distinctive dialectic context of no analysis of addition without that of subtraction, and vice versa—to be eventually transcended altogether later.

And fourthly, it contains thirty-six major theses, namely, (a) the first thesis: the formalness-informalness principle, (b) the second thesis: the absoluteness-relativeness principle, (c) the third thesis: the partiality-totality principle, (d) the fourth thesis: the predictability-unpredictability principle, (e) the fifth thesis: the explicability-inexplicability principle, (f) the sixth thesis: the fiction-reality principle, (g) the seventh thesis: the cognitiveness-noncognitiveness principle, (h) the eighth thesis: the finiteness-transfiniteness principle, (i) the ninth thesis: the preciseness-vagueness principle, (j) the tenth thesis: the simpleness-complicatedness principle, (k) the eleventh thesis: the openness-hiddenness principle, (l) the twelfth thesis: the denseness-emptiness principle, (m) the thirteenth thesis: the rule-exception principle, (n) the fourteenth thesis: the prototypicality-variation principle, (o) the fifteenth thesis: the change-constancy principle, (p) the sixteenth thesis: the order-chaos principle, (q) the seventeenth thesis: the slowness-quickness principle, (r) the eighteenth thesis: the expansion-contraction principle, (s) the nineteenth thesis: the optimality-nonoptimality principle, (t) the twentieth thesis: the simultaneity-nonsimultaneity principle, (u) the twenty-first thesis: the isolation-interaction principle, (v) the twenty-second thesis: the theory-praxis principle, (w) the twenty-third thesis: the convention-novelty principle, (x) the twenty-fourth thesis: the evolution-transformation principle, (y) the twenty-fifth thesis: the symmetry-asymmetry principle, (z) the twenty-sixth thesis: the softness-hardness principle, (aa) the twenty-seventh thesis:

the seriousness-playfulness principle, (bb) the twenty-eighth thesis: the activeness-inactiveness principle, (cc) the twenty-ninth thesis: the selfness-otherness principle, (dd) the thirtieth thesis: the regression-progression principle, (ee) the thirty-first thesis: the sameness-difference principle, (ff) the thirty-second thesis: the stability-reaction principle, (gg) the thirty-third thesis: the functionality-nonfunctionality principle, (hh) the thirty-fourth thesis: the intentionality-nonintentionality principle, (ii) the thirty-fifth thesis: the survivability-non-survivability principle, and (jj) the thirty-sixth thesis: the post-human double-sidedness—to be elaborated in the rest of the book.

Because of the last thesis on "the post-human double-sidedness," the word "double-sidedness" in the title of my theory is suggestive, because it refers to the dual analyses of both addition and subtraction in the history of accounting—which then require a multi-faceted inquiry of the dialectic relationship between addition and subtraction, as well as an understanding of the different views regarding the relationships in accounting (together with other debates as already discussed in the book)—and in the end it calls for a need to transcend both addition and subtraction by future advanced humans and later post-humans in the context of the great transformations of accounting in the future worlds (as will be discussed in this chapter).

With this clarification in mind—the seminal project here will fundamentally change the way that we think about addition and subtraction (especially, though not solely, in the context of accounting), together with other debates as already discussed in the book, from the combined perspectives of the mind, nature, society, and culture, with enormous implications for the human future and what I originally called its "post-human" fate.

In Relation to Method

The first seven theses are in relation to method in existential dialectics, as shown below.

1st Thesis: The Formalness-Informalness Principle

The first thesis refers to the formalness-informalness principle (on the formal requirements of logical systems) in the category of "method" in existential dialectics, which was first proposed in *FPHML* (and later, *FPHH*, *FPHGAM*, *FPHSEM*, and other books of mine), in that the formal requirements of a logical system (e.g., consistency, soundness, and completeness) have both usefulness and non-usefulness, to the extent that, if there are formal systems requiring them, there are alternative ones which do not.

The formalness-informalness principle reveals something interesting in existential dialectics, in that it goes beyond both classical and non-classical logics (while learning something from both).

Thus, it does *not* exclude classical logics but simply goes beyond both classical and non-classical logics, while learning something from each. There is no formalness without informalness—and vice versa.

For illustration, existential dialectics can make use of both classical logics under certain conditions (e.g., especially, though not exclusively, when they are clear-cut, etc.) and non-classical logics under alternative conditions (especially, though not exclusively, when they are "unknown," "irrelevant," "ambiguous," "possible," with "different degrees of truth," empirically inconsistent in a desirable way, etc.). (WK 2008u)

For instance, on the one hand, there is formalness (e.g., the formal logical requirement of soundness and completeness in a system of ideas, as in the argument on deception and creative accounting, in *Sec. 2.2.1*; the formal logical requirement of soundness, completeness, and consistency in a system of ideas, as in the argument on reputation capital and management, in *Sec. 2.2.2*; the formal logical requirement of soundness, completeness, and consistency in a system of ideas, as in the argument on EMAS, in *Sec. 2.3.1*; the formal logical requirement of soundness and completeness in a system of ideas, as in the argument on space-time in accounting, in *Sec. 2.3.2*; the formal logical requirement of soundness, completeness, and consistency in a system of ideas, as in the argument on auditing, in *Sec. 2.4.1*; the formal logical requirement of soundness and completeness in a system of ideas, as in the argument on VAT, in *Sec. 2.4.2*; the formal logical requirement of soundness and completeness in a system of ideas, as in the argument on accounting principles, in *Sec. 2.5.1*; the formal logical requirement of soundness, completeness, and consistency in a system of ideas, as in the argument on accounting standards, in *Sec. 2.5.2*; the formal logical requirement of soundness, completeness, and consistency in a system of ideas, as in the argument on control self-assessment, in *Sec. 3.2.1*; the formal logical requirement of soundness and completeness in a system of ideas, as in the argument on mental accounting, in *Sec. 3.2.2*; the formal logical requirement of soundness, completeness, and consistency in a system of ideas, as in the argument on carbon accounting, in *Sec. 3.3.1*; the formal logical requirement of soundness, completeness, and consistency in a system of ideas, as in the argument on the E P & L accounting, in *Sec. 3.3.2*; the formal logical requirement of soundness, completeness, and consistency in a system of ideas, as in the argument on computer programming and bookkeeping, in *Sec. 3.4.1*; the formal logical requirement of soundness, completeness, and consistency in a system of ideas, as in the argument on corproate misdeeds and internal control, in *Sec. 3.4.2*; the formal logical requirement of soundness, completeness, and consistency in a system of ideas, as in the argument on accounting ethics, in *Sec. 3.5.1*; and the formal logical requirement of soundness and completeness in a system of ideas, as in the argument on the convergence of accounting standards in *Sec. 3.5.2*).

On the other hand, there is also informalness (e.g., the non-formal existence of unsoundness and incompleteness in the argument on deception and creative accounting, as shown in the criticisms that the problem with accounting scandals is not necessarily the fault of creative accountants, that the nature of business cycles in the free market makes it easier for creative accountants to cheat, that the identification of "fraud" from "error" is difficult in practice, that the analysis of "the mind's internal processes" in cognitive psychology to detect fraud is easier said than done, that extensive lobbying efforts in the business world can hinder accounting reform to prevent fraud, and that the role of the government can be part of the problem, in *Sec. 2.2.1*; the non-formal existence of unsoundness, incompleteness, and inconsistency in the argument on reputation capital and management, as shown in the criticisms that the calculation of "reputation" is not objective, that there are competing methods to account for goodwill over the decades, that there are conflicting models of reputation, that there are different approaches to reputation management, that the obsession with reputation can be a psychological problem instead, and that reputation building is easier said than done, in *Sec. 2.2.2*; the non-formal existence of unsoundness, incompleteness, and inconsistency in the argument on EMAS, as shown in the criticisms that the required measurement to implement EMAS is not practical, that the benefits of EMAs vary from country to country and from organization to organization, that the implementation is not mandatory, that there are costs of EMAS, and that there is no consensus on the audit tools to be used for EMAS, in *Sec. 2.3.1*; the non-formal existence of unsoundness and incompleteness in the argument on space-time in accounting, as shown in the criticisms that Takatera and Sawabe used a writing sytle which is too obstruse and flowery to be empirically useful for clear analysis, that the claim by Takatera and Sawabe about the social construction of space-time in modern accrual accounting conventions is exaggeratedly misleading, that the claims in the article are often argumentatively relative to the theories that Takatera and Sawabe personally preferred, that Takatera and Sawabe confused "chaos theory" with "complexity theory" and this thus raises the thorny question about their scholarly qualification to use the two fields for the analysis of accounting, and that Takatera and Sawabe are polemically driven by the discourse on "chaos" to analyze accounting in their idiosyncratic way with the subsequent bias, in *Sec. 2.3.2*; the non-formal existence of unsoundness, incompleteness, and inconsistency in the argument on auditing, as shown in the criticisms that an audit cannot guarantee a financial statement to be "free from material error," that even the conventional standard of "reasonable assurance" is no longer sufficient for some people, that auditing does not necessarily prevent fraudulent activities, that the history of auditing is not necessarily consistent in its implementation of certain standards and goals, and that the evaluation of the success or failure of audit functions itself can be difficult to measure, in *Sec. 2.4.1*; the non-formal existence of unsoundness and incompleteness in the argument on VAT, as shown in the criticisms that there is a trade-off between the national government and sub-national governments in regard to the revenue collected through VAT, that the introduction of VAT can lead to a new kind of fraud, that it is not necessarily "efficient" for the economy, that the implementation of VAT varies from country to country and is subject to different loopholes, that there is no general consensus on the right rate of VAT, that there is also no consensus on the need for VAT, that it can be a "regressive tax," and that it can be costly and thus lead to "higher prices," in *Sec. 2.4.2*; the non-formal existence of unsoundness and incompleteness in the argument on accounting principles, as shown in the criticisms that there are opposing sides in the debate on "the principle of historical cost," that there are opposing sides in the debate on "the principle of the unit of measure," that there are opposing sides in the debate on "the principle of the accounting entity," that any change to the principles is not costless [even with benefits], and that each principle [be it the old one as defended by its advocates or any new one as suggested by the opponents of the old one] has both advantages and disadvantages, in *Sec. 2.5.1*; the non-formal existence of unsoundness, incompleteness, and inconsistency in the argument on accounting standards, as shown in the criticisms that accounting standards do not necessarily prevent accounting scandals, that there is a call for more global accounting standards, that there is no general consensus on the exact number of constraints in US GAAP, that there is also no general consensus on the exact number of accounting principles which should be included, and that different works make different (even conflicting) claims about what really makes a principle to be an "assumption," a "requirement," or a "constraint" instead, in *Sec. 2.5.2*; the non-formal existence of unsoundness, incompleteness, and inconsistency in the argument on control self-assessment, as shown in the criticisms that there is no consensus on its methodologies, that there are alternative approaches to control self-assessment, that it does not give sufficient attention to "downside risk," and that it is too obsessed with the "costs" of risk but not its "benefits," in *Sec. 3.2.1*; the non-formal existence of unsoundness and incompleteness in the argument on mental accounting, as shown in the criticisms that the issue of "rationality" in the debate on mental accounting can be relative, that cognitive biases in mental accounting [if they exist in some cases] can be more lasting than previously thought, that there can be missing factors not accounted for in those cases of mental accounting, that it does not give sufficient attention to the "good" side of mental accounting [but its "bad" side], and that there are "no easy answers" on how to deal with mental accounting, in *Sec. 3.2.2*; the non-formal existence of unsoundness, incompleteness, and inconsistency in the argument on carbon accounting, as shown in the criticisms that the practice of carbon accounting is not objective at all, that different carbon accountants can produce different results, that "the total carbon footprint cannot be calculated," that the different "dimensions" of sustainability accounting involves difficult trade-offs in implementation, that there are competing approaches to sustainability accounting, and that organizations may not have the incentives to fully implement sustainability accounting, in *Sec. 3.3.1*; the non-formal existence of unsoundness, incompleteness, and inconsistency in the argument on the E P & L accounting, as shown in the criticisms that there are competing definitons of "ecosystem services," that the understanding of ecosystem services is culturally relative, that there are also competing ways to operationalize the concept of "ecosystem services," that ecosystem services can be more complex than previously thought, that there are alternative approaches to explain ecosystem services, that the empirical validity of each study is limited, and that there are disagreements about the true value of ecosystem services, in *Sec. 3.3.2*; the non-formal existence of unsoundness, incompleteness, and inconsistency in the argument on computer programming and bookkeeping, as shown in the criticisms that computer programming often does not model the "real world," that OOP is too obsessed with "objects" [or "nouns"] but not other things [like "verbs" or "actions," etc.], that there are disagreements about the fundamental features of OOP, that there are trade-offs in computer programming for bookkeeping, that OOP is not effective to deal with the issue of transparency, and that OOP has not yet been proved to be better than other approaches in practice, in *Sec. 3.4.1*; the non-formal existence of unsoundness, incompleteness, and inconsistency in the argument on corproate misdeeds and internal control, as shown in the criticisms that internal control does not guarantee the realization of objectives, that internal control has no control over external factors which can affect the performance of an organization, that there are competing definitions or views of "internal control," that there are limitations in the verifiability of data in internal control, that the problem with accounting scandals is not wholly the fault of corporate criminals, and that the nature of business cycles in the free market makes it easier for accounting scandals to occur, in *Sec. 3.4.2*; the non-formal existence of unsoundness, incompleteness, and inconsistency in the argument on accounting ethics, as shown in the criticisms that there have been business scandals in spite of the existence of accounting ethics, that different countries have different legal standards and regulations in regard to the enforcement of accounting practices, that there are competing approaches to ethical principles in accounting ethics, that morality can be "relative" [not "universal"], and that there are the larger institutional factors which contribute to the occurrence of accounting scandals [not just due to the fault of creative accountants], in *Sec. 3.5.1*; and the non-formal existence of unsoundness and incompleteness in the argument on the convergence of accounting standards, as shown in the criticisms that the convergence of different accounting standards is costly, that the claim of "comparability" is questionable, that the convergence process is too slow, that different countries comply with the convergence requirements in different ways, and that it is not as popular as conventionally assumed, in *Sec. 3.5.2*).

And the reverse direction also holds true.

However, it should be stressed (as this is something that I regularly did in my previous books, *almost verbatim*) that there are different shades of gray (or different degrees of truth) in the two opposites, and the classification is not necessarily mutually exclusive either. In fact, even when some combinations of the two occur, they only end up sharing the same dialectic relationship, but in a different degree.

Also, there are some other relationships (as a kind of family resemblance) which have something in common with the principle, but they are not exactly the same but only more or less comparable, which varies from case to case.

Good examples of family resemblance in relation to the principle are logicalness-nonlogicalness, rationality-nonrationality, etc.

2nd Thesis: The Absoluteness-Relativeness Principle

The second thesis refers to the absoluteness-relativeness principle (on the multiplicity of things) in the category of "method" in existential dialectics, which was first proposed in *FPHK*, in that there is the multiplicity of things in reality, be they about entities, qualities (or properties), and relationships. If there is something absolute, there is likewise something relative. And there is no absoluteness without relativeness—and vice versa.

Both absoluteness and relativeness here are also relevant to different modalities often cited in the literature on ontology, such as possibility (e.g., something "can" happen) and its opposite (e.g., impossibility) in alethic modalities, probability (e.g., something "will" happen) and its opposite (e.g., improbability) in temporal modalities, and necessity (e.g., something "should" happen) and its opposite (e.g., contingency) in alethic modalities.

For instance, on the one hand, there is absoluteness (e.g., the absolute view on "external influences" by the advocates of "behavioral psychology," in *Sec. 2.2.1*; the absolute view on "reputation capital and management" by the advocates in business, in *Sec. 2.2.2*; the absolute view on the implementation of EMAS by those organizations which "voluntarily decide to implement the scheme," in *Sec. 2.3.1*; the absolute view on "the status of an organization as an independent economic entity" by the advocates of the "accrual accounting system," in *Sec. 2.3.2*; the absolute view on "reasonable assurance" by the advocates, in *Sec. 2.4.1*; the absolute view on VAT by the advocates like some national governments for its implementation, in *Sec. 2.4.2*; the absolute view on the fundamental change of the 3 accounting principles by the advocates like Jeremy Nicholls of the SROI Network, in *Sec. 2.5.1*; the absolute view on the "cost constraint" as the only constraint accepted by SFAC No. 8 in 2010, which replaced SFAC No. 1 and 2, in *Sec. 2.5.2*; the absolute view on auditing by the advocates, in *Sec. 3.2.1*; the absolute view on the principle of "fungibility" by the advocates, in *Sec. 3.2.2*; the absolute view on sustainability measurement by the advocates in regard to the approach by "measuring environmental-economic-social interrelationships," in *Sec. 3.3.1*; the absolute view on the "ecological information collected from one spatial-temporal scale" by its advocates, in *Sec. 3.3.2*; the absolute view on the focus on "nouns" and "objects" by the advocates of OOP, in *Sec. 3.4.1*; the absolute view on corporate misdeeds as "fraudulent" and thus "bad' by government oversight agencies like the Securities and Exchange Commission in the United States, in *Sec. 3.4.2*; the absolute view on "principle-based" accounting standards by the International Financial Reporting Standards [IFRS], in *Sec. 3.5.1*; andthe absolute view on the convergence of accounting standards by the FASB and IASB, in *Sec. 3.5.2*).

On the other hand, there is relativeness (e.g., what is right for the advocates of "behavioral psychology" in regard to "external influences" is not necessarily so for "cognitive psychologists," since the latter focuses on "internal processes" instead, in *Sec. 2.2.1*; what is right for the advocates in regard to "reputation capital and management" in business is not necessarily so for the critics who view the obsession with reputation as a psychological problem instead, that is, as "a human fault, exaggerated in importance due to the fragile nature of the human ego," in *Sec. 2.2.2*; what is right for those organizations which "voluntarily decide to implement the scheme" is not necessarily so for other rganizations which do not voluntarily decide to implement the scheme," because "a common misconception is that ISO 14001 certification automatically implies legal compliance," in *Sec. 2.3.1*; what is right for the advocates in regard to "the status of an organization as an independent economic entity" is not necessarily so for the critics in "transaction cost economics," since "the status of an organization as an independent economic entity is an artificially created image that is contestable if we take different views from those of transaction cost economics," in *Sec. 2.3.2*; what is right for the advocates in regard to "reasonable assurance" is not necessarily so for the critics, since "recently, the argument that auditing should go beyond just true and fair is gaining momentum" and "the US Public Company Accounting Oversight Board has come out with a concept release on the same," in *Sec. 2.4.1*; what is right for the advocates in regard to the implementation of VAT is not necessarily so for the critics who "consider it to be a regressive tax, meaning that the poor pay more, as a percentage of their income, than the rich," in *Sec. 2.4.2*; what is right for the advocates like Jeremy Nicholls of the SROI Network in regard to the fundamental change of the 3 accounting principles is not necessarily so for the defenders of the principles, in *Sec. 2.5.1*; what is right for SFAC No. 8 in 2010 in regard to the "cost constraint" as the only constraint accepted is not necessarily so for SFAC No. 1 and 2, which did include 4 constraints aforecited, but not the "cost constraint," in *Sec. 2.5.2*; what is right for the advocates in regard to auditing is not necessarily so for the critics, since there are alternative approaches like "control self-assessment," in *Sec. 3.2.1*; what is right for the advocates in regard to the principle of "fungibility" is not necessarily so for those who are willing to violate it, since "nothing says that these types of purchases are bad or irrational. Far from it!," in *Sec. 3.2.2*; what is right for the advocates in regard to the approach by "measuring environmental- economic- social interrelationships" is not necessarily so for the advocates of the other competing approach which is more "growth-based," in *Sec. 3.3.1*; what is right for the advocates in regard to the "ecological information collected from one spatial-temporal scale" is not necessarily so for the critics, since "interpreting ecological information collected from one spatial-temporal scale does not necessarily mean it can be applied at another," in *Sec. 3.3.2*; what is right for the advocates of OOP in regard to the focus on "nouns" and "objects" is not necessarily so for the critic like Alexander Stepanov who argued instead that "OOP…claims that everything is an object. Even if it is true it is not very interesting—saying that everything is an object is saying nothing at all," in *Sec. 3.4.1*; what is right for the Securities and Exchange Commission in the United States in regard to corporate misdeeds as "fraudulent" and thus "bad" is not necessarily so for those corporate criminals who are determined to cheat, in *Sec. 3.4.2*; what is right for the International Financial Reporting Standards [IFRS] in regard to "principle-based" accounting standards is not necessarily so for the critics like those who prefer "the United States Generally Accepted Accounting Principles [GAAP]," which are "largely rule-based," in *Sec. 3.5.1*; and what is right for the FASB and IASB in regard to the convergence of accounting standards is not necessarily so for the critic like Shyam Sunder, who "has called the link between convergence and comparability 'overblown,'" in *Sec. 3.5.2*).

And the reverse direction also holds true.

However, it should be stressed (as this is something that I regularly did in my previous books, *almost verbatim*) that there are different shades of gray (or different degrees of truth) in the two opposites, and the classification is not necessarily mutually exclusive either. In fact, even when some combinations of the two occur, they only end up sharing the same dialectic relationship, but in a different degree.

Also, there are some other relationships (as a kind of family resemblance) which have something in common with the principle, but they are not exactly the same but only more or less comparable, which varies from case to case.

Good examples of family resemblance in relation to the principle are internalness-externalness, immanence-transcendence, independence-dependence, internality-externality, autonomy-dependency, universalness-relativeness, and so forth.

3rd Thesis: The Partiality-Totality Principle

The third thesis refers to the partiality-totality principle (on the relationships between whole and parts) in the category of "method" in existential dialectics, which was first proposed in *FC* and later *FPHGEOL*, based on earlier works like *FHC* and *FPHC*), in that any analysis of a phenomenon in the world requires both the examination of the individual parts and the whole of them (with the whole as being not the sum of its parts), without privileging the whole as more important than the parts, or vice-versa (that is, the parts as more important than the whole).

In fact, this principle targets two kinds of misleading methodologies as used over the ages, that is, again in the absence of better words, what I called (in *FHC*, *FPHC* and *FC*) "reductionism" and its opposite, "reverse-reductionism," which can take four major forms in relation to concept, theory, methodology, and ontology (as summarized in *Sec. 1.8*).

Thus, reductionism and reverse-reductionism can be (i) conceptual, (ii) theoretical, (iii) methodological, and (iv) ontological. And the whole is not the sum of its parts, and there is no partiality without totality—and vice versa.

Works on emergence (as in *FC*), gestalt psychology (as in *FPHGEOL*), and the hermeneutic circle (as in *FPHDA*), just to name a few of them, have contributed to the understanding of whole and parts.

For instance, on the one hand, there is partiality (e.g., the partial view on "external influences" by the advocates of "behavioral psychology," in *Sec. 2.2.1*; the partial view on "reputation" in accordance to a specific "model," in *Sec. 2.2.2*; the partial view on audit tools by the advocates of "large, highly detailed and prescriptive protocols," in *Sec. 2.3.1*; the partial view on the evolutionary process of modern accounting conventions in the last 5 centuries by Takatera and Sawabe in terms of two periods, namely, "the pre-going concern" and "the going concern," in *Sec. 2.3.2*; the partial view on "reasonable assurance" by the advocates, in *Sec. 2.4.1*; the partial view on VAT by the advocates like some national governments for its implementation, in *Sec. 2.4.2*; the partial view on the fundamental change of the 3 accounting principles by the advocates like Jeremy Nicholls, in *Sec. 2.5.1*; the partial view on constraints by the advocates of SFAC No. 8 in 2010, in *Sec. 2.5.2*; the partial view on learning by the advocates of "control self-assessment," in *Sec. 3.2.1*; the partial view on the principle of "fungibility" by the advocates, in *Sec. 3.2.2*; the partial view on sustainability measurement by the advocates in regard to the approach by "measuring environmental-economic-social interrelationships," in *Sec. 3.3.1*; the partial view on ecosystem services by the advocates of "the redundancy hypothesis," in *Sec. 3.3.2*; the partial view on "responsibility-driven design" by the advocates, in *Sec. 3.4.1*; the partial view on corporate misdeeds by the Securities and Exchange Commission in the United States, in *Sec. 3.4.2*; the partial view on accounting ethics by the advocates of moral universalism, in *Sec. 3.5.1*; and the partial view on the convergence of accounting standards by the advocates like the FASB and IASB, in *Sec. 3.5.2*).

On the other hand, there is also totality (e.g., the more holistic view on psychology in regard to deception, such that the whole is not the sum of its parts, that is, it cannot be reduced to the sum of opposing views by "behavioral psychologists," "cognitive psychologists," etc., since there will emerge new views in the future not yet known today, just as there are already alternative views nowadays which do not side exclusively with any of them, as is the analysis here, in *Sec. 2.2.1*; the more holistic view on "reputation," such that the whole is not the sum of its parts, that is, it cannot be reduced to the sum of "often conflicting models of reputation," according to Kevin Money and Carola Hillenbrand, since there will emerge new views in the future not yet known today, just as there are already alternative views nowadays which do not side exclusively with any of them, as is the analysis here, in *Sec. 2.2.2*; the more holistic view on audit tools, such that the whole is not the sum of its parts, that is, it cannot be reduced to the sum of opposing views by the advocates of "large, highly detailed and prescriptive protocols," the other advocates of "more flexible protocols," etc., since there will emerge new views in the future not yet known today, just as there are already alternative views nowadays which do not side exclusively with any of them, as is the analysis here, in *Sec. 2.3.1*; the more holistic view on the evolutionary process of modern accounting conventions in the last 5 centuries, such that the whole is not the sum of its parts, that is, it cannot be reduced to the sum of opposing views by Takatera and Sawabe, and their critics, since there will emerge new views in the future not yet known today, just as there are already alternative views nowadays which do not side exclusively with any of them, as is the analysis here, in *Sec. 2.3.2*; the more holistic view on "reasonable assurance," such that the whole is not the sum of its parts, that is, it cannot be reduced to the sum of opposing views by its advocates, its critics like the US Public Company Accounting Oversight Board, etc., since there will emerge new views in the future not yet known today, just as there are already alternative views nowadays which do not side exclusively with any of them, as is the analysis here, in *Sec. 2.4.1*; the more holistic view on VAT, such that the whole is not the sum of its parts, that is, it cannot be reduced to the sum of those opposing views by the advocates, the critics, etc. as discussed earlier, since there will emerge new views in the future not yet known today, just as there are already alternative views nowadays which do not side exclusively with any of them, as is the analysis here, in *Sec. 2.4.2*; the more holistic view on the fundamental change of the 3 accounting principles, such that the whole is not the sum of its parts, that is, it cannot be reduced to the sum of opposing views by the defenders, the advocates, etc., as discussed earlier, since there will emerge new views in the future not yet known today, just as there are already alternative views nowadays which do not side exclusively with any of them, as is the analysis here, in *Sec. 2.5.1*; the more holistic view on constraints, such that the whole is not the sum of its parts, that is, it cannot be reduced to the sum of opposing views by the advocates of SFAC No. 8 in 2010, the advocates of SFAC No. 1 and 2 before 2010, etc., since there will emerge new views in the future not yet known today, just as there are already alternative views nowadays which do not side exclusively with any of them, as is the analysis here, in *Sec. 2.5.2*; the more holistic view on learning, such that the whole is not the sum of its parts, that is, it cannot be reduced to the sum of opposing views by "control self-assessment," "external auditing," "internal auditing," etc., since there will emerge new views in the future not yet known today, just as there are already alternative views nowadays which do not side exclusively with any of them, as is the analysis here, in *Sec. 3.2.1*; the more holistic view on the principle of "fungibility," such that the whole is not the sum of its parts, that is, it cannot be reduced to the sum of opposing views by the advocates, the critics, etc., since there will emerge new views in the future not yet known today, just as there are already alternative views nowadays which do not side exclusively with any of them, as is the analysis here, in *Sec. 3.2.2*; the more holistic view on sustainability measurement, such that the whole is not the sum of its parts, that is, it cannot be reduced to the sum of opposing views by the advocates of the approach by "measuring environmental-economic-social interrelationships," the other advocates of the opposing approach which is more "growth-based," etc., since there will emerge new views in the future not yet known today, just as there are already alternative views nowadays which do not side exclusively with any of them, as is the analysis here, in *Sec. 3.3.1*; the more holistic view on ecosystem services, such that the whole is not the sum of its parts, that is, it cannot be reduced to the sum of opposing views by the advocates of "the redundancy hypothesis," "the portfolio effect," etc., since there will emerge new views in the future not yet known today, just as there are already alternative views nowadays which do not side exclusively with any of them, as is the analysis here, in

Sec. 3.3.2; the more holistic view on "design patterns," such that the whole is not the sum of its parts, that is, it cannot be reduced to the sum of opposing views by the advocates of "responsibility-driven design," the other advocates of "data-driven design," etc., since there will emerge new views in the future not yet known today, just as there are already alternative views nowadays which do not side exclusively with any of them, as is the analysis here, in *Sec. 3.4.1*; the more holistic view on corporate misdeeds, such that the whole is not the sum of its parts, that is, it cannot be reduced to the sum of opposing views by the Securities and Exchange Commission in the United States, the corporate criminals, etc., since there will emerge new views in the future not yet known today, just as there are already alternative views nowadays which do not side exclusively with any of them, as is the analysis here, in *Sec. 3.4.2*; the more holistic view on accounting ethics, such that the whole is not the sum of its parts, that is, it cannot be reduced to the sum of opposing views by the advocates of moral relativism, the advocates of moral universalism, etc., since there will emerge new views in the future not yet known today, just as there are already alternative views nowadays which do not side exclusively with any of them, as is the analysis here, in *Sec. 3.5.1*; and the more holistic view on the convergence of accounting standards, such that the whole is not the sum of its parts, that is, it cannot be reduced to the sum of opposing views by the advocates like the FASB and IASB, the critics like Shyam Sunder, etc., since there will emerge new views in the future not yet known today, just as there are already alternative views nowadays which do not side exclusively with any of them, as is the analysis here, in *Sec. 3.5.2*).

And the reverse direction also holds true.

However, it should be stressed (as this is something that I regularly did in my previous books, *almost verbatim*) that there are different shades of gray (or different degrees of truth) in the two opposites, and the classification is not necessarily mutually exclusive either. In fact, even when some combinations of the two occur, they only end up sharing the same dialectic relationship, but in a different degree.

Also, there are some other relationships (as a kind of family resemblance) which have something in common with the principle, but they are not exactly the same but only more or less comparable, which varies from case to case.

Good examples of family resemblance in relation to the principle are individualisticness-holisticness, partness-wholeness, analysis-synthesis, and the like.

4th Thesis: The Predictability-Unpredictability Principle

The fourth thesis is called the predictability-unpredictability principle (on the occurrence of events) in the category of "method" in existential dialectics (which was first proposed in *FC* and, later, other books of mine), in that both predictability and unpredictability have a major role to play in the occurrence of things, so that neither determinism nor indeterminism wins the centuries-old fight. There is no predictability without unpredictability—and vice versa.

There are events which are predictable, just as there are those which are not. Or what is regarded as unpredictable at one point in time may turn out to be predictable later, and, conversely, what is deemed as predictable may turn out to not be so predictable. Even in predictability, outcomes are subject to uncertainty, the degree of which varies from case to case.

Works on ecological uncertainty (as in *BNR*), complexity theory (as in *FC*), quantum mechanics (as in *FPHST* and *FPHG*), subjective (or Bayesian) probability (as in *FPHPROB*), "special causes" (as in *FPHGAM*), measurement uncertainty (as in *FPHPROB*), the butterfly effect (as in *FPHAERO*), and unintended consequences (as in *FPHAERO*), just to name a few of them, have contributed to the understanding of uncertainity.

The predictability-unpredictability principle is not to be confused with the absoluteness-relativeness principle; although both can address the issue of probability, the former is concerned with the extent of certainty in prediction, whereas the latter focuses on the relativity of viewpoints.

For instance, on the one hand, there is predictability (e.g., the predictable tendency of "high-risk takers" to "be overly confident,...lack self-control and often have a propensity for 'memory loss' following financial failures," according to the study on cognitive psychology and creative accounting aforecited, in *Sec. 2.2.1*; the predictable tendency of firms nowadays to pay attention to "reputation capital and management," in *Sec. 2.2.2*; the predictable tendency that "the benefits of EMAs vary from country to country and from organization to organization," in *Sec. 2.3.1*; the predictable tendency of the supporters of Takatera and Sawabe to argue about the evolutionary process of modern accounting conventions in the last 5 centuries in terms of two periods, namely, "the pre-going concern" and "the going concern," in *Sec. 2.3.2*; the predictable tendency that, in the U.S., "an audit must adhere to generally accepted standards established by governing bodies" like "the US Public Company Accounting Oversight Board," etc., in *Sec. 2.4.1*; the predictable tendency that Hong Kong or Libya is VAT-free in the foreseeable future, in *Sec.*

2.4.2; the predictable tendency that the advocates like Jeremy Nicholls favor the fundamental change of the 3 accounting principles in the foreseeable future, in *Sec. 2.5.1*; the predictable tendency that US GAAP will use SFAC No. 8 and includes only the "Cost Constraint" in the foreseeable future, in *Sec. 2.5.2*; the predictable tendency that those for "control self-assessment" focus on "risk management," in *Sec. 3.2.1*; the predictable tendency that those with cognitive biases in "mental accounting" can receive treatment from a CBM [Cognitive Bias Modification] program, in *Sec. 3.2.2*; the predictable tendency that different carbon accountants can produce different results, in *Sec. 3.3.1*; the predictable tendency that "ecosystem services decisions require making complex choices at the intersection of ecology, technology, society and the economy," in *Sec. 3.3.2*; the predictable tendency that OOP will remain relevant in the foreseeable future, in *Sec. 3.4.1*; the predictable tendency that "internal control can provide reasonable…assurance that the objectives of an organization will be met," in *Sec. 3.4.2*; the predictable tendency that there will be more education and regulations about accounting ethics in the foreseeable future, in *Sec. 3.5.1*; and the predictable tendency that the FASB and IASB will continue working on the convergence of accounting standards in the foreseeable future, in *Sec. 3.5.2*).

On the other hand, there is unpredictability (e.g., the more difficult task to predict exactly the extent to which a particular "high-risk taker" may end up actually committing fraud on a particular future occasion, in *Sec. 2.2.1*; the more difficult task to predict exactly the extent to which a particular firm will be sucessful in building "reputation" on a particular future occasion, since reputation building is easier said than done, as not everyone can make it and "reputation recovery is the long and arduous path to rebuilding equity in a company's good name," in *Sec. 2.2.2*; the more difficult task to predict exactly the extent to which, or the specific way in which, a particular country or a particular organization will be benefited from EMAS on a particular future occasion, in *Sec. 2.3.1*; the more difficult task to predict exactly the extent to which the argument by Takatera and Sawabe about the evolutionary process of modern accounting conventions in the last 5 centuries in terms of two periods will be well accepted in a particular future era, in *Sec. 2.3.2*; the more difficult task to predict exactly the extent to which an audit of a particular firm can prevent fraudulent activities on a particular future occasion, in *Sec. 2.4.1*; the more difficult task to predict exactly the extent to which Hong Kong or Libya will be VAT-free in a particular distant future era, in *Sec. 2.4.2*; the more difficult task to predict exactly in which particular ways any change of the accounting principles will be in a particular future era, in *Sec. 2.5.1*; the more difficult task to predict exactly the extent to which the "Cost Constraint" will remain to be the only constraint listed in international accountng standards in a particular future era, in *Sec. 2.5.2*; the more difficult task to predict exactly the extent to which a particular "control self-assessment" will not end up understating or ignoring "extreme downside risk" on a particular future occasion, in *Sec. 3.2.1*; the more difficult task to predict exactly the extent to which a patient with treatment from a CBM program will fully recover on a particular future occasion, since, to the critics, "one concern is whether CBM modification procedures will achieve lasting benefits," in *Sec. 3.2.2*; the more difficult task to predict exactly which particular version of carbon accounting result will become popular in a particular future era, in *Sec. 3.3.1*; the more difficult task to predict exactly the extent to which a particular "ecosystem services decision" will be correct on a particular future occasion, since "complicating the process, these decisions are often spatial, always multi-objective, and based on uncertain data, models, and estimates," in *Sec. 3.3.2*; the more difficult task to predict exactly the extent to which OOP will remain influentail in a particular future era, siince there are already alternative approaches nowadays, like "procedural programming," etc., in *Sec. 3.4.1*; the more difficult task to predict exactly the extent to which a particular internal control will be effective in preventing corporate misdeeds on a particular future occasion, since internal control does not provide "absolute" assurance that "the objectives of an organization will be met" and also "has no control over external factors which can affect the performance of an organization," in *Sec. 3.4.2*; the more difficult task to predict exactly the extent to which a particular education and regulation about accounting ethics will be effective on a particular future occasion, since there have been business scandals in spite of the existence of different accounting reforms, in *Sec. 3.5.1*; and the more difficult task to predict exactly the extent to which the convergence of accounting standards will occur in a particular future era, in *Sec. 3.5.2*).

And the reverse direction also holds true.

Again, it should be stressed, however, that there are different shades of gray (or different degrees of truth) in the two opposites, and the classification is not necessarily mutually exclusive either. In fact, even when some combinations of the two occur, they only end up sharing the same dialectic relationship, but in a different degree.

Also, there are some other relationships (as a kind of family resemblance) which have something in common with the principle, but they are not exactly the same but only more or less comparable, which varies from case to case.

Good examples of family resemblance in relation to the principle are sureness-arbitrariness, deterministicness-randomness, likeliness-riskiness, sureness-doubtfulness, deterministicness-indeterministicness, certainty-uncertainty, reliability-unreliability, and so forth.

5th Thesis: The Explicability-Inexplicability Principle

The fifth thesis is called the explicability-inexplicability principle (on the underlying mechanisms of things) in the category of "method" in existential dialectics (which was proposed in *FPHU* and other books of mine), in that both explicability and inexplicability are part of the understanding of things. There is no explicability without inexplicability—and vice versa.

This principle tells us the dual nature of the research dilemma, in that, if reality can be explained in some ways, it also has its other ways which are not quite explainable, at a given point in time.

For instance, on the one hand, there is explicability (e.g., the explanation, by "behavioral psychologists," of deception, on the basis of "external influences," in *Sec. 2.2.1*; the explanation, by those advocates of "reputation capital and management," of its desirability on the basis of "the benefit" to all "stakeholders," in *Sec. 2.2.2*; the explanation, by those advocates of "large, highly detailed and prescriptive protocols," of their desirability on the basis of "easier review by other parties," in *Sec. 2.3.1*; the explanation, by those advocates of the "accrual accounting system," of "the status of an organization as an independent economic entity" on the basis of its assumptions on space aforecited, in *Sec. 2.3.2*; the explanation, by those advocates of the Sarbanes-Oxley Act of 2002, of its desirability on the basis that "the profession's exposure and value was enhanced, as many internal auditors possessed the skills required to help companies meet the requirements of the law," in *Sec. 2.4.1*; the explanation, by those advocates of VAT, of its desirability on the basis of the increase of "government revenue," in *Sec. 2.4.2*; the explanation, by those advocates of the principle of the historical cost, of its desirability on the basis of its advantages, in *Sec. 2.5.1*; the explanation, by those advocates of US GAAP as analyzed in this section, that, to achieve its objectives, US GAAP has four basic assumptions, including the "historical cost principle," in *Sec. 2.5.2*; the explanation, by those advocates of "control self-assessment," of its desirability on the basis of its reduction or elimination of "the risk of error," in *Sec. 3.2.1*; the explanation, by those advocates of the principle of "fungibility," of its desirability on the basis of a more rational decision making, since people can learn to avoid buying "things they don't need because the deal is too good to pass up" or quitting "early on a good day," etc., in *Sec. 3.2.2*; the explanation, by those advocates, of the desirability of sustainabiltiy measurement in terms of three main dimensions aforedisdescribed, in *Sec. 3.3.1*; the explanation, by those advocates of "the redundancy hypothesis," of ecosystem services on the basis that "more than one species performs a given role within an ecosystem," in *Sec. 3.3.2*; the explanation, by those advocates of OOP, of its desirability on the basis of its integration of the two concepts of "data" and "code" into one, such that "an object has state [data] and behavior [code]," in *Sec. 3.4.1*; the explanation, by the Securities and Exchange Commission in the United States, of the need to prosecute those corporate criminals on the basis of "fraud" which is "bad" and not legal, in *Sec. 3.4.2*; the explanation, by those advocates of moral universalism in accounting ethics, of its validity on the basis that "there is only one valid moral code for all" in accounting ethics, like the one against corruption and fraud, in *Sec. 3.5.1*; and the explanation, by those advocates of the convergence of accounting standards, of its desirability on the basis of "comparability," in *Sec. 3.5.2*).

On the other hand, there is inexplicability (e.g., the lack of sufficient explanation, by "behavioral psychologists," of why "external influences" are necessarily useful to explain deception, since, to "cognitive psychologists," "internal processes" in the mind are more relevant, in *Sec. 2.2.1*; the lack of sufficient explanation, by those advocates of "reputation capital and management," of why it is necessarily desirable, since, to the critics, the obsession with reputation is a psychological problem instead, "a human fault, exaggerated in importance due to the fragile nature of the human ego," in *Sec. 2.2.2*; the lack of sufficient explanation, by those advocates of of "large, highly detailed and prescriptive protocols," of why they are necessarily desirable, since, to the critics, "there is a long standing debate among environmental audit professionals on the value of large, highly detailed and prescriptive protocols…versus more flexible protocols," in *Sec. 2.3.1*; the lack of sufficient explanation, by those advocates of the "accrual accounting system," of why "the status of an organization as an independent economic entity" on the basis of its assumptions on space is necessarily plausible, since, to the critics, "the status of an organization as an independent economic entity is an artificially created image that is contestable if we take different views from those of transaction cost economics," in *Sec. 2.3.2*; the lack of sufficient explanation, by those advocates of the Sarbanes-Oxley Act of

2002 of why it is necessarily desirable, since, to the critics, "the focus by internal audit departments of publicly traded companies on SOX related financial policy and procedures derailed progress made by the profession in the late 20th century toward Larry Sawyer's vision for internal audit" in a broader sense, in *Sec. 2.4.1*; the lack of sufficient explanation, by those advocates of VAT, of why it is necessarily desirable, since, to the critics, "because the price for someone rises, the quantity of goods traded decreases," so "correspondingly, some people are worse off by more than the government is made better off by tax income," in *Sec. 2.4.2*; the lack of sufficient explanation, by those advocates of the principle of the historical cost, of why it is necessarily desirable, since, to the critics, there are disadvantages, such that it should be replaced by the "Constant Purchasing Power model" instead, in *Sec. 2.5.1*; the lack of sufficient explanation, by those advocates of US GAAP as analyzed in this section, of why the four constraints are necessarily valid, since, to other scholars like Sadao Takatera and Norio Sawabe in one of the previous sections on "Addition, Space-Time, and Account," the "historical cost principle" is embedded with two assumptions, namely, the "constant value of money" and the "constant time value of money," contrary to the view in the article here that it is not listed as an assumption at all, in *Sec. 2.5.2*; the lack of sufficient explanation, by those advocates of "control self-assessment," of why it is necessarily desirable, since, to the critics, it does not give sufficient attention to "downside risk" and is too obsessed with the "costs" of risk but not its "benefits," in *Sec. 3.2.1*; the lack of sufficient explanation, by those advocates of the principle of "fungibility," of why it is necessarily desirable, since, to the critics, "mental accounting procedures have evolved to economize on time and thinking costs and also to deal with self-control problems," among other benefits, in *Sec. 3.2.2*; the lack of sufficient explanation, by those advocates, of why sustainabiltiy measurement in terms of three main dimensions aforedisdescribed is necessarily desirable, since, to the critics, "in practice, however, it is difficult to put together policies that promote simultaneously environmental, economic and social goals," as "multiple factors" are involved here, in *Sec. 3.3.1*; the lack of sufficient explanation, by those advocates of "the redundancy hypothesis," of why it is necessarily plausible, since, to the critics, there are alternative approaches like "the portfolio effect," which "compares biodiversity to stock holdings, where diversification minimizes the volatility of the investment, or in this case, the risk in stability of ecosystem services," in *Sec. 3.3.2*; the lack of sufficient explanation, by those advocates of OOP, of why its focus on "objects" and "nouns" is necessarily desirable, since, to the critics like

Alexander Stepanov, "OOP…claims that everything is an object. Even if it is true it is not very interesting—saying that everything is an object is saying nothing at all," in *Sec. 3.4.1*; the lack of sufficient explanation, by the Securities and Exchange Commission in the United States, of why "corporate misdeeds" are necessarily undesirable, since those corporate criminals think otherwise and are determined to cheat, in *Sec. 3.4.2*; the lack of sufficient explanation, by those advocates of moral universalism in accounting ethics, of why it is necessarily valid, since, to the critics, "separatism is based on a relativist conception of morality that there can be different, equally valid moral codes that apply to different sections of society and differences in codes between societies," in *Sec. 3.5.1*; and the lack of sufficient explanation, by those advocates of the convergence of accounting standards, of why "comparability" is necessarily the case, since the critics like Shyam Sunder dismissed "the link between convergence and comparability" as "overblown," in *Sec. 3.5.2*).

And the reverse direction also holds true.

Once more, it should be stressed, however, that there are different shades of gray (or different degrees of truth) in the two opposites, and the classification is not necessarily mutually exclusive either. In fact, even when some combinations of the two occur, they only end up sharing the same dialectic relationship, but in a different degree.

Also, there are some other relationships (as a kind of family resemblance) which have something in common with the principle, but they are not exactly the same but only more or less comparable, which varies from case to case.

Good examples of family resemblance in relation to the principle are underlyingness-regularness, causation-regularity, causation-correlation, explanation-description, explanation-narration, explanation-interpretation, relation-nonrelation, and so forth.

6th Thesis: The Fiction-Reality Principle

The sixth thesis is called the fiction-reality principle (on the different faces of reality) in the category of "method" in existential dialectics (as first proposed in *FPHLIT*), in that both fiction and reality co-exist. There is no reality without fiction—and vice versa.

This principle tells us the different categories of reality, in that, if there is reality, there is also some fiction in it—just as, if there is fiction, there is also some reality in it.

For instance, on the one hand, there is fiction (e.g., the fictional aspect of information, insofar as decision makers "cook the books" for private gains, such as the information in question is fraudulent, in *Sec. 2.2.1*; the fictional aspect of "reputation," insofar as the calculation of "reputation" is not objective, since "goodwill" in business is psycholgically tied to the "opinion," "gossip," etc. about an entity like a company, in *Sec. 2.2.2*; the fictional aspect of ISO 14001 certification, insofar as there is "a common misconception… that ISO 14001 certification automatically implies legal compliance," in *Sec. 2.3.1*; the fictional aspect of "money" in accounting, insofar as it is treated from the perspective of the "constant value of money," which is a "fiction," according to Takatera and Sawabe, in *Sec. 2.3.2*; the fictional aspect of "reasonable assurance," insofar as the standard is no longer deemed sufficient for some people, contrary to a conventional claim, since "recently, the argument that auditing should go beyond just true and fair is gaining momentum" and "the US Public Company Accounting Oversight Board has come out with a concept release on the same," in *Sec. 2.4.1*; the fictional aspect of VAT in regard to "government revenue," insofar as, "in many developing countries such as India,…there is strong opposition to this [VAT] by many sub-national governments as it leads to an overall reduction in the revenue they collect as well as of some autonomy," in *Sec. 2.4.2*; the fictional aspect of the principle of the historical cost, insofar as "historical cost accounts do not measure the loss of value of monetary assets as a result of inflation," in *Sec. 2.5.1*; the fictional aspect of "constraints" in accounting standards, insofar as SFAC No. 8 in 2010 replaced SFAC No. 1 and 2 and accepted the "cost constraint" as the only constraint, while rejecting the "objectivity principle," the "consistency principle," and the "conservatism principle" as constraints, in *Sec. 2.5.2*; the fictional aspect of "control self-assessment," insofar as it does not give sufficient attention to "downside risk" and is too obsessed with the "costs" of risk but not its "benefits," in *Sec. 3.2.1*; the fictional aspect of mental accounting, insofar as "there is no dividing line between money that matters [safety capital] and money you can afford to lose [risk capital]," so "any dividing line you honor is nothing more than a misleading mental illusion," in *Sec. 3.2.2*; the fictional aspect of sustainabiltiy measurement, insofar as, "in practice, however, it is difficult to put together policies that promote simultaneously environmental, economic and social goals," as "multiple factors" are involved here, in *Sec. 3.3.1*; the fictional aspect of the understanding of ecosystem services, insofar as it can be culturally relative in that "pivotal cultural values attaching to the natural/cultivated environment rely on an area's unique character that cannot be addressed by methods that use universal scientific parameters to determine ecological structures and functions," in *Sec. 3.3.2*; the fictional aspect of OOP, insofar as, to the critics, "not everyone agrees that OOP facilitates direct real-world mapping...; Bertrand Meyer argues...that a program is not a model of the world" and "Rich Hickey...emphasized the inability of OOP to model time properly, which is getting increasingly problematic as software systems become more concurrent," in *Sec. 3.4.1*; the fictional aspect of "earnings report," insofar as the case of "Freddie Mac" in 2002 "revealed that it had understated earnings by almost $5 billion, one of the largest corporate restatements in U.S. history," in *Sec. 3.4.2*; the fictional aspect of accounting ethics, insofar as those advocates of moral universalism in accounting ethics claim "there is only one valid moral code for all," since, to the critics, "separatism is based on a relativist conception of morality that there can be different, equally valid moral codes that apply to different sections of society and differences in codes between societies," in *Sec. 3.5.1*; and the fictional aspect of the convergence of accounting standards, insofar as the critics like Shyam Sunder dismissed "the link between convergence and comparability" as "overblown," in *Sec. 3.5.2*).

On the other hand, there is reality (e.g., the realistic aspect of information, insofar as decision makers open their books for auditing with no errors found, in *Sec. 2.2.1*; the realistic aspect of "reputation," insofar as "nationalities differ with regard to how valued specific aspects of the company's brand identity are in the respective national culture," in *Sec. 2.2.2*; the realistic aspect of ISO 14001 certification, insofar as "ISO 14001 demands the organization to evaluate its compliance with legal requirements," in *Sec. 2.3.1*; the realistic aspect of "money" in accounting,, insofar as it can be understood from the perspective of a more historical accounting, e.g., "inflating or deflating by general price index," in *Sec. 2.3.2*; the realistic aspect of "reasonable assurance," insofar as, "due to constraints, an audit seeks to provide only reasonable assurance that the statements are free from material error" and "hence, statistical sampling is often adopted in audits," in *Sec. 2.4.1*; the realistic aspect of VAT, insofar as it can increase "national" government revenue, in *Sec. 2.4.2*; the realistic aspect of the principle of the historical cost, insofar as "historical cost accounts do not record gains until they are realized," in *Sec. 2.5.1*; the realistic aspect of "constraints" in accounting standards, insofar as SFAC No. 8 in 2010 focused more on the factor of "cost" in reporting financial info, in *Sec. 2.5.2*; the realistic aspect of "control self-assessment," insofar as it is "to reduce or eliminate the risk of error," in *Sec. 3.2.1*; the realistic aspect of mental accounting, insofar as "mental accounting procedures...do not work perfectly," in *Sec. 3.2.2*; the realistic aspect of sustainability measurement, insofar as there are three main dimensions aforedescribed, in *Sec. 3.3.1*; the realistic aspect of the understanding of ecosystem services, insofar as there are "nonmaterial benefits people obtain from ecosystems through spiritual enrichment, cognitive development, reflection, recreation, and aesthetic experiences," in *Sec. 3.3.2*; the realistic aspect of OOP, insofar as it has mechanisms to reduce the need for boilerplate codes like "metaprogramming," "convention over configuration," and "model-driven engineering," in *Sec. 3.4.1*; the realistic aspect of "earnings report," insofar as there are companies which are truthful in their reports but others which are not, in *Sec. 3.4.2*; the realistic aspect of accounting ethics, insofar as different countries have different legal standards and regulations in regard to the enforcement of accounting practices, in *Sec. 3.5.1*; and the realistic aspect of the convergence of accounting standards, insofar as the aim is to strive for certain degree of "comparability," in *Sec. 3.5.2*).

And the reverse direction also holds true.

Once more, it should be stressed, however, that there are different shades of gray (or different degrees of truth) in the two opposites, and the classification is not necessarily mutually exclusive either. In fact, even when some combinations of the two occur, they only end up sharing the same dialectic relationship, but in a different degree.

Also, there are some other relationships (as a kind of family resemblance) which have something in common with the principle, but they are not exactly the same but only more or less comparable, which varies from case to case.

Good examples of family resemblance in relation to the principle are fiction-nonfiction, fiction-fact, imagination-fact, imagination-nonfiction, subjectivity-objectivity, inaccuracy-accuracy, arbitrariness-reality, error-reality, misunderstanding-correctness, and so forth.

7th Thesis: The Cognitiveness-Noncognitiveness Principle

The seventh thesis is called the cognitiveness-noncognitiveness principle (on the understanding of meaning) in the category of "method" in existential dialectics (as first proposed in *BNR*), in that, if a view of things is based on reason and evidences, it is also affected by non-cognitive factors (such that it is neither true nor false in the latter case)—and vice versa. In other words, there is no cognitiveness without noncognitiveness—and vice versa.

Noncognitive factors can come in all shapes and sizes, and good examples include envy, jealousy, power, nationality, race, gender, age, class, greed, lust, status, faith, anger, sadness, joy, fear, wish, etc.

Communication, even in its ideal type, is seldom based on reason and evidences only, as falsely assumed during the Age of the Enlightenment and its continued supporters nowadays. What is understood about something is often twisted on the basis of something else too. In these cases, what is at stake is often neither true nor false (which is something that classical logic has sidelined and ignored).

For instance, on the one hand, there is cognitiveness (e.g., a view of things on the basis of reason and evidences, as shown in the finding that governments can be asked to carry out "accounting reforms," in *Sec. 2.2.1*; a view of things on the basis of reason and evidences, as shown in the finding that "some individuals and organizations hire reputation management companies to attempt to hide truthful but unflattering information about themselves," in *Sec. 2.2.2*; a view of things on the basis of reason and evidences, as shown in the finding that there are different "audit tools" available for EMAS implementation, in *Sec. 2.3.1*; a view of things on the basis of reason and evidences, as shown in the finding about the "accrual accounting system" in modern times, in *Sec. 2.3.2*; a view of things on the basis of reason and evidences, as shown in the finding that "an audit seeks to provide only reasonable assurance that the statements are free from material error," in *Sec. 2.4.1*; a view of things on the basis of reason and evidences, as shown in the finding that EU countries support VAT, in *Sec. 2.4.2*; a view of things on the basis of reason and evidences, as shown in the finding that there is a debate on the desirability of the 3 accounting principles, in *Sec. 2.5.1*; a view of things on the basis of reason and evidences, as shown in the finding that "the SEC has a stated goal of moving from US GAAP to the International Financial Reporting Standards [IFRS]," in *Sec. 2.5.2*; a view of things on the basis of reason and evidences, as shown in the finding that "control self-assessment" can be used "to reduce or eliminate the risk of error," in *Sec. 3.2.1*; a view of things on the basis of reason and evidences, as shown in the finding that "the way a person subjectively frames a transaction in their mind will determine the utility they receive or expect," in *Sec. 3.2.2*; a view of things on the basis of reason and evidences, as shown in the finding that "carbon accounting is hoped to provide a factual ground for carbon-related decision-making," in *Sec. 3.3.1*; a view of things on the basis of reason and evidences, as shown in the finding that there are different studies on ecosystem services aforecited, in *Sec. 3.3.2*; a view of things on the basis of reason and evidences, as shown in the finding that "the goals of object-oriented programming" are for "increased understanding," "ease of maintenance," and "ease of evolution," in *Sec. 3.4.1*; a view of things on the basis of reason and evidences, as shown in the finding that the

Securities and Exchange Commission in the United States prosecutes those corporate criminals for their "fraudulent" activities, in *Sec. 3.4.2*; a view of things on the basis of reason and evidences, as shown in the finding that "new reforms, regulations, and calls for increased higher education have been introduced to combat the dangers of unethical behavior" in accounting, in *Sec. 3.5.1*; and a view of things on the basis of reason and evidences, as shown in the finding that the FASB and IASB have worked on the convergence of accounting standards so as to "render international investments more comparable to investors," in *Sec. 3.5.2*).

On the other hand, there is noncognitiveness (e.g., a view of things on the basis of non-cognitive factors like envy, jealousy, power, nationality, race, gender, age, class, greed, lust, status, faith, anger, sadness, joy, fear, wish, etc.—as shown in the politics of corruption, when "one barrier to accounting reform are governments themselves. They have the authority to determine what are accepted accounting principles, while using questionable accounting practices themselves," in *Sec. 2.2.1*; a view of things on the basis of non-cognitive factors like envy, jealousy, power, nationality, race, gender, age, class, greed, lust, status, faith, anger, sadness, joy, fear, wish, etc.—as shown in the psychological pathology of the obsession with reputation, as "a human fault, exaggerated in importance due to the fragile nature of the human ego," regardless of whether the person in question really loses his reputation in reality or not, in *Sec. 2.2.2*; a view of things on the basis of non-cognitive factors like envy, jealousy, power, nationality, race, gender, age, class, greed, lust, status, faith, anger, sadness, joy, fear, wish, etc.—as shown in the politics of "preference," when some auditors prefer "large, highly detailed and prescriptive protocols," in spite of other alternative protocols available too, like the "more flexible protocols that rely on the expertise and knowledge of experienced auditors and source documents [regulations, permits, etc.] directly," in *Sec. 2.3.1*; a view of things on the basis of non-cognitive factors like envy, jealousy, power, nationality, race, gender, age, class, greed, lust, status, faith, anger, sadness, joy, fear, wish, etc.—as shown in the power of "institutional norms," in that "periodic income is calculated as a basis for income distribution because periodic income is accepted as periodic income by one person because he or she knows that periodic income is accepted as periodic income by others," according to Takatera and Sawabe, in *Sec. 2.3.2*; a view of things on the basis of non-cognitive factors like envy, jealousy, power, nationality, race, gender, age, class, greed, lust, status, faith, anger, sadness, joy, fear, wish, etc.—as shown in the "political correctness" of the requirement of "reasonable assurance," in spite of the fact that the standard is no longer deemed sufficient for some people, since "recently, the argument that auditing should go beyond just true and fair is gaining momentum," in *Sec. 2.4.1*; a view of things on the basis of non-cognitive factors like envy, jealousy, power, nationality, race, gender, age, class, greed, lust, status, faith, anger, sadness, joy, fear, wish, etc.—as shown in the "politics" of supporting VAT in EU countries, in spite of the various criticisms against "loophole," "fraud, "regressive tax," "costliness," "inefficiency," etc. as discussed earlier, in *Sec. 2.4.2*; a view of things on the basis of non-cognitive factors like envy, jealousy, power, nationality, race, gender, age, class, greed, lust, status, faith, anger, sadness, joy, fear, wish, etc.—as shown in the subjective factor of "boredom," such that some people find the discussion of the 3 accounting principles "very, very boring," in spite of the fact that "the principles upon which accounting and reporting are based have a massive influence on the world we live in," in *Sec. 2.5.1*; a view of things on the basis of non-cognitive factors like envy, jealousy, power, nationality, race, gender, age, class, greed, lust, status, faith, anger, sadness, joy, fear, wish, etc.—as shown in the bureaucratic "politics," such that, "although the SEC has a stated goal of moving from US GAAP to the International Financial Reporting Standards [IFRS],...progress has been slow and uncertain," in *Sec. 2.5.2*; a view of things on the basis of non-cognitive factors like envy, jealousy, power, nationality, race, gender, age, class, greed, lust, status, faith, anger, sadness, joy, fear, wish, etc.—as shown in the preference for "inertia," since, "initially external auditors ignored the benefits of control self-assessment even though it was effective at providing audit evidence around the 'soft' areas [such as staff morale] that are critical to the effectiveness of internal control systems," in *Sec. 3.2.1*; a view of things on the basis of non-cognitive factors like envy, jealousy, power, nationality, race, gender, age, class, greed, lust, status, faith, anger, sadness, joy, fear, wish, etc.—as shown in the "recidivism" of some patients who "are not able to maintain their new learning even within a CBM course, with their negative cognitive bias beginning to re-emerge over time," even though they know that they should not do that, in *Sec. 3.2.2*; a view of things on the basis of non-cognitive factors like envy, jealousy, power, nationality, race, gender, age, class, greed, lust, status, faith, anger, sadness, joy, fear, wish, etc.—as shown in the politics of "hope" in carbon accounting, in spite of the challenge by "social scientific studies of accounting" against "this hope, pointing to the socially constructed character of carbon conversion factors or of the accountants' work practice which cannot implement abstract accounting schemes into reality," in *Sec. 3.3.1*; a view of things on the basis of non-cognitive factors like envy, jealousy, power, nationality, race, gender, age, class, greed, lust, status, faith, anger, sadness, joy, fear, wish, etc.—as shown in the politics of cultural relativism, since "pivotal cultural values attaching to the natural/cultivated environment rely on an area's unique character that cannot be addressed by methods that use universal scientific parameters to determine ecological structures and functions," in *Sec. 3.3.2*; a view of things on the basis of non-cognitive factors like envy, jealousy, power, nationality, race, gender, age, class, greed, lust, status, faith, anger, sadness, joy, fear, wish, etc.—as shown in the politics of defending OOP, in spite of the damaging criticisms that computer programming often does not model the "real world," that OOP is too obsessed with "objects" [or "nouns"] but not other things [like "verbs" or "actions," etc.], that OOP is not effective to deal with the issue of transparency, and that OOP has not yet been proved to be better than other approaches in practice, in *Sec. 3.4.1*; a view of things on the basis of non-cognitive factors like envy, jealousy, power, nationality, race, gender, age, class, greed, lust, status, faith, anger, sadness, joy, fear, wish, etc.—as shown in the "temptation" of corporate criminals to cheat, in spite of the understanding that this is "bad" and of the danger that they may be caught and prosecuted by the Securities and Exchange Commission in the United States, in *Sec. 3.4.2*; a view of things on the basis of non-cognitive factors like envy, jealousy, power, nationality, race, gender, age, class, greed, lust, status, faith, anger, sadness, joy, fear, wish, etc.—as shown in the "temptation" of creative accountants to cheat due to "self-interest" or "greed," in spite of all the "reforms, regulations, and calls for increased higher education…to combat the dangers of unethical behavior" in accounting, in *Sec. 3.5.1*; and a view of things on the basis of non-cognitive factors like envy, jealousy, power, nationality, race, gender, age, class, greed, lust, status, faith, anger, sadness, joy, fear, wish, etc.—as shown in the "politics" of "convergence," such that the project continues, even though the critics like Shyam Sunder dismissed "the link between convergence and comparability" as "overblown," together with other criticisms that the convergence of different accounting standards is costly, that the convergence process is too slow, and that it is not as popular as conventionally assumed, in *Sec. 3.5.2*).

And the reverse direction also holds true.

Once more, it should be stressed, however, that there are different shades of gray (or different degrees of truth) in the two opposites, and the classification is not necessarily mutually exclusive either. In fact, even when some combinations of the two occur, they only end up sharing the same dialectic relationship, but in a different degree.

Also, there are some other relationships (as a kind of family resemblance) which have something in common with the principle, but they are not exactly the same but only more or less comparable, which varies from case to case.

Good examples of family resemblance in relation to the principle are proposition-attitude, cognition-emotion, cognition-prescription, theory-extratheory, justification-nonjustification, recognition-nonrecogntion, reason-perception, and so forth.

In Relation to Structure

The next six theses are in relation to structure in existential dialectics, as shown below.

8th Thesis: The Finiteness-Transfiniteness Principle

The eighth thesis is called the finiteness-transfiniteness principle (on the nature of numbers) in the category of "structure" in existential dialectics (as first proposed in *FPHG*), in that, if there are finite things, there are likewise transfinite ones—and vice versa.

To avoid confusion, my usage of the word "transfinite" here differs radically from the one used by Cantor (and other mathematicians) for "relative" infinity—and is more limited, in light of the problems confronting any attempt to understand the idea of infinity, be it by intuition, imagination, and conception (as detailedly analyzed in *Sec. 2.2.3* of *FPHG*).

Instead, by "transfinity," I allow numbers which can be many times larger—or smaller, for that matter—than the finite things that we encounter in daily life, but they do not have to be related to the idea of infinity at all (which may not exist).

Of course, there may be some *borderline* cases, in which it is not clear whether the number in question is transfinite (in my usage) or simply a mathematical convenience. A good example of a borderline case is the Planck unit of length for "the smallest space possibly measured in nature," which is "less than billionths of trillionths of trillionths of an inch" (or something like 1.6×10^{-35} meters). (P. Baofu 2006a; N. McAleer 1987: 219; WKV 2008; D. Corbett 2008a)

That said—my usage of "transfinity" can also resolve (or better, dissolve) an age-old problem in philosophy known as "Zeno's paradoxes" (as already explained in *Sec. 2.2.3* of *FPHG*).

In any event, in relation to the finiteness-transfiniteness principle, on the other hand, there is finiteness (e.g., the finite number of the categories of people who commit fraud in business, like "creative accountants," "earnings managers," etc., in *Sec. 2.2.1*; the finite number of fields in which the concern with reputation is important, like "business, education, online communities," etc., in *Sec. 2.2.2*; the finite number of the categories of the environmental impact by organizations to be identified for environmental audit, like "activities," "products," "services," etc., in *Sec. 2.3.1*; the finite number of the categories of entities in an accounting system, like "receipts and expenditures," in *Sec. 2.3.2*; the finite number of the types of auditors, like the 5 examples aforedescribed, in *Sec. 2.4.1*; the finite number of the categories of taxation, like "sales tax," "VAT," etc., in *Sec. 2.4.2*; finite number of the principles in accounting, like the 3 main ones aforediscussed, in *Sec. 2.5.1*; the finite number of the categories of concepts in US GAAP, like "assumptions," "principles," and "constraints," in *Sec. 2.5.2*; the finite number of the categories of approaches to risk management, like "control self-assessment," "external auditing," and "external auditing," in *Sec. 3.2.1*; the finite number of the categories of mental accounting, like the 9 types aforedescribed, in *Sec. 3.2.2*; the finite number of "environmental factors" in sustainability measurement, like "energy," "water," "greenhouse gases," "emissions," "hazardous and non hazardous waste," "recycling," "packaging," etc., in *Sec. 3.3.1*; the finite number of the categories of ecosystem services, like the 4 examples aforedescribed, in *Sec. 3.3.2*; the finite number of the concepts of objects, like the 2 concerning "state [data] and behavior [code]" in OOP, in *Sec. 3.4.1*; the finite number of the "methods" of accounting scandals, like "misusing or misdirecting funds," "overstating revenues," "understating expenses," "overstating the value of corporate assets," "underreporting the existence of liabilities," etc., in *Sec. 3.4.2*; the finite number of the reasons of ethical failures, like the 9 reasons aforecited, in *Sec. 3.5.1*; and the finite number of different accounting standards in the world, like the Chinese, the American, the Indian, the French, etc., in *Sec. 3.5.2*).

On the other hand, there is transfinitty (e.g., the transfinite amount of all the money that has ever been stolen or taken away due to all the "fraudulent" activities in human history, in *Sec. 2.2.1*; the transfinite number of all the individual instances that have ever occurred in human hisory, where the concern with reputation becomes important "in business, education, online communities, and many other fields," in *Sec. 2.2.2*; the transfinite number of all the individual "activities," "products," "services," etc. that have ever produced in human history and have enviromental impact, in *Sec. 2.3.1*; the transfinite number of all the individual instances of "receipts and expenditures" that have ever existed in history, regardless of whether they followed the accounting conventions in the older days or nowadays, in *Sec. 2.3.2*; the transfinite number of all the individual items that have ever been examiend by all auditors in human history, regardless of which types they belong to, in *Sec. 2.4.1*; the transfinite number of all the individuals who have ever taxed in history, regardless of which categories of taxation the taxes belong to, in *Sec. 2.4.2*; the transfinite number of all the individual instances of business transactions that have ever been affected by the 3 accounting principles in human history, in *Sec. 2.5.1*; the transfinite number of all the individual instances of business transactions in human history that can be understood on the basis of such concepts like"assumptions," "principles," and "constraints," in *Sec. 2.5.2*; the transfinite number of all the individual instances of "errors" in business that have ever existed in human history, regardless of how they can be detected by the available approaches to risk management at the time, in *Sec. 3.2.1*; the transfinite number of all the individual instances of mental accounting that have ever occurred in human history, regardless of which categories they belong to, in *Sec. 3.2.2*; the transfinite number of all the individual instances of "energy," "water," "greenhouse gases," "emissions," "hazardous and non hazardous waste," "recycling," "packaging," etc. that have ever occurred in history, in *Sec. 3.3.1*; the transfinite number of all the individual instances of ecosystem services that have ever existed in history, such that, in one study by Salles [2011], "the total value of biodiversity is infinite," in *Sec. 3.3.2*; the transfinite number of all the individual instances of "state [data] and behavior [code]" of objects that have ever existed in history, regardless of whether or not they are captured by any OOP, in *Sec. 3.4.1*; the transfinite "amounts of money" that have lost in history, due to accounting scandals, regardless of which methods the cases belong to, in *Sec. 3.4.2*; the transfinite number of all the amounts of money that have lost in history, regardless of which reasons are responsible for the losses, in *Sec. 3.5.1*; and the transfinite number of all the individual instances of accounting standards that have ever been applied in history, regardless of which accounting standards they actually belong to, in *Sec. 3.5.2*).

And the reverse direction also holds true.

Once more, it should be stressed, however, that there are different shades of gray (or different degrees of truth) in the two opposites, and the classification is not necessarily mutually exclusive either. In fact, even when some combinations of the two occur, they only end up sharing the same dialectic relationship, but in a different degree.

Also, there are some other relationships (as a kind of family resemblance) which have something in common with the principle, but they are not exactly the same but only more or less comparable, which varies from case to case.

Good examples of family resemblance in relation to the principle are boundedness-quasiunboundedness, smallness-largeness, microness-macroness, boundedness-unboundedness, and the like.

9th Thesis: The Preciseness-Vagueness Principle

The ninth thesis is called the preciseness-vagueness principle (on the refinement of things) in the category of "structure" in existential dialectics (which was first out worked out in *FIA* and later, *FPHMORP*), in that both preciseness and vagueness are important, not that one is better than the other, but that both are used, in different degrees of preference, in accordance to the contextual application from the perspectives of nature, the mind, culture, and society. There is no preciseness without vagueness—and vice versa.

For illustration, in relation to taxonomy, preciseness has its taxonomic clarity, just as vagueness has its classificatory flexibility. And in relation to network, vagueness has its explorative liberty, just as preciseness has its conceptual definitiveness.

For instance, on the one hand, there is preciseness (e.g., the precise identification of 5 areas where cognitive psychology is applied to the field of creative accounting in the article aforecited, in *Sec. 2.2.1*; the precise identification of 3 levels of analyzing "the difference between image and reputation," in the article aforecited, in *Sec. 2.2.2*; the precise identification of 4 benefits of EMAS in the article aforecited, in *Sec. 2.3.1*; the precise identification of 2 assumptions on time and space in the article aforecited, in *Sec. 2.3.2*; the precise identification of 5 types of auditors in the article aforecited, in *Sec. 2.4.1*; the precise identification of 2 types of the method of collection and 2 types of the timing of collection in the article aforecited, in *Sec. 2.4.2*; the precise identification of 3 accounting principles in the article aforecited, in *Sec. 2.5.1*; the precise identification of 3 assumptions in US GAAP in the article aforecited, in *Sec. 2.5.2*; the precise identification of 6 benefits of control self-assessment in the article aforecited, in *Sec. 3.2.1*; the precise identification of 8 types of mental accounting by J. Cross in the article aforecited, in *Sec. 3.2.2*; the precise identification of 3 dimensions of sustainability in the article aforecited, in *Sec. 3.3.1*; the precise identification of 4 categories of ecosystem services in the article aforecited, in *Sec. 3.3.2*; the precise identification of 5 features of OOP by Benjamin C. Pierce and others as cited above, in *Sec. 3.4.1*; the precise identification of 3 types of internal control in the article aforecited, in *Sec. 3.4.2*; the precise identification of 9 reasons of ethical failures in accounting in the article aforecited, in *Sec. 3.5.1*; and the precise identification of 5 motivations for the convergence of accounting standards in the article aforecited, in *Sec. 3.5.2*).

On the other hand, there is vagueness (e.g., the vagueness in the identification of the 5 areas where cognitive psychology is applied to the field of creative accounting in the article aforecited, since it is not clear why there must be only 5, not 6, 7, 8, 9, and so on, in *Sec. 2.2.1*; the vagueness in the identification of the 3 levels of analyzing "the difference between image and reputation" in the article aforecited, since it is not clear why there must be only 3, not 4, 5, 6, 7, and so on, in *Sec. 2.2.2*; the vagueness in the identification of the 4 benefits of EMAS in the article aforecited, since it is not clear why there must be only 4, not 5, 6, 7, 8, and so on, in *Sec. 2.3.1*; the vagueness in the identification of the 2 assumptions on time and space in the article aforecited, since it is not clear why there must be only 2, not 1, 3, 4, 5, 6, and so on, in *Sec. 2.3.2*; the vagueness in the identification of the 5 types of auditors in the article aforecited, since it is not clear why there must be only 5, not 6, 7, 8, 9, and so on, in *Sec. 2.4.1*; the vagueness in the identification of the 2 types of the method of collection and 2 types of the timing of collection in the article aforecited, since it is not clear why there must be only 2, not 3, 4, 5, 6, and so on, in *Sec. 2.4.2*; the vagueness in the identification of the 3 accounting principles in the article aforecited, since it is not clear why there must be only 3, not 4, 5, 6, 7, and so on, in *Sec. 2.5.1*; the vagueness in the identification of the 3 assumptions in US GAAP in the article aforecited, since it is not clear why there must be only 3, not 4, 5, 6, and so on, in *Sec. 2.5.2*; the vagueness in the identification of the 6 benefits of control self-assessment in the article aforecited, since it is not clear why there must be only 6, not 7, 8, 9,, and so on, in *Sec. 3.2.1*; the vagueness in the identification of the 8 types of mental accounting by J. Cross in the article aforecited, since it is not clear why there must be only 8, not 10, 11, 12, 13, and so on, in *Sec. 3.2.2*; the vagueness in the identification of the 3 dimensions of sustainability in the article aforecited, since it is not clear why there must be only 3, not 4, 5, 6, 7, and so on, in *Sec. 3.3.1*; the vagueness in the identification of the 4 categories of ecosystem services in the article aforecited, since it is not clear why there must be only 4, not 5, 6, 7, 8, and so on, in *Sec. 3.3.2*; the vagueness in the identification of the 5 features of OOP by Benjamin C. Pierce and others as cited above, since it is not clear why there must be only 5, not 4, 3, 6, 7, 8, and so on, since others like John C. Mitchell identified only 4, and Michael Lee Scott considered only 3, as cited earlier, in *Sec. 3.4.1*; the vagueness in the identification of the 3 types of internal control in the article aforecited, since it is not clear why there must be only 3, not 2, 4, 5, 6, 7, and so on, since there are competing definitions of internal control, in *Sec. 3.4.2*; the vagueness in the identification of the 9 reasons of ethical failures in accounting in the article aforecited, since it is not clear why there must be only 9, not 10, 11, 12, 13, and so on, in *Sec. 3.5.1*; and the vagueness in the identification of the 5 motivations for the convergence of accounting standards in the article aforecited, since it is not clear why there must be only 5, not 6, 7, 8, and so on, in *Sec. 3.5.2*).

And the reverse direction also holds true.

Once more, it should be stressed, however, that there are different shades of gray (or different degrees of truth) in the two opposites, and the classification is not necessarily mutually exclusive either. In fact, even when some combinations of the two occur, they only end up sharing the same dialectic relationship, but in a different degree.

Also, there are some other relationships (as a kind of family resemblance) which have something in common with the principle, but they are not exactly the same but only more or less comparable, which varies from case to case.

Good examples of family resemblance in relation to the principle are clarity-ambiguity, directness-indirectness, quantitativeness-qualitativeness, specificity-obscurity, describability-nondescribability, concretness-abstractness, translatability-untranslatability, specificity-generality, clearcutness-transitionalness, sharpness-roughness, discreteness-mixedness, discreteness-vagueness, detailedness-broadness, and the like.

10th Thesis: The Simpleness-Complicatedness Principle

The tenth thesis refers to the simpleness-complicatedness principle (on the interconnection among things) in the category of "structure" in existential dialectics (which was first out worked out in *FIA*, *FPHPROB*, nd other books), in that both simpleness and complicatedness are vital, without favoring one over the other, and each is utilized, depending on the basis of the perspectives of nature, the mind, culture, and society. There is no simpleness without complicatedness—and vice versa.

As an illustration, in relation to taxonomy, simpleness has its heuristic usefulness, just as complicatedness has its realistic representation. In relation to network, simpleness has its economical attractiveness, just as complicatedness has its practical reliability. And in relation to logic, there is the simple zeroth-order logic, just as there is the complicated higher-order logic.

For instance, on the one hand, there is simpleness (e.g., the relatively simple analysis of the "fault" of creative accountants in the article aforecited, in *Sec. 2.2.1*; the relatively simple analysis of goodwill in accounting by the advocates, especially in relation to its psychological connectin to reputation, in *Sec. 2.2.2*; the relatively simple analysis of EMAS by the advocates, in *Sec. 2.3.1*; the relatively simple analysis of space-time in accounting by Takatera and Sawabe, in *Sec. 2.3.2*; the relatively simple analysis of auditing by the advocates, in *Sec. 2.4.1*; the relatively simple analysis of VAT by the advocates, in *Sec. 2.4.2*; the relatively simple analysis of the 3 accounting principles by the advocates, in *Sec. 2.5.1*; the relatively simple analysis of US GAAP by the advocates, in *Sec. 2.5.2*; the relatively simple analysis of "control self-assessment" by the advocates, in *Sec. 3.2.1*; the relatively simple analysis of mental accounting by the advocates, in *Sec. 3.2.2*; the relatively simple analysis of carbon accounting by the advocates, in *Sec. 3.3.1*; the relatively simple analysis of the E P & L accounting by the advocates, in *Sec. 3.3.2*; the relatively simple analysis of OOP by the advocates, in *Sec. 3.4.1*; the relatively simple analysis of internal control by the advocates, in *Sec. 3.4.2*; the relatively simple analysis of accounting ethics by the advocates, in *Sec. 3.5.1*; and the relatively simple analysis of the convergence of accounting standards by the advocates, in *Sec. 3.5.2*).

On the other hand, there is complicatedness (e.g., the relatively more complicated analysis of the "fault" of creative accountants, by challenging the claims and assumptions, as shown in the problems pointed out by the critics as cited, in *Sec. 2.2.1*; the relatively more complicated analysis of "goodwill," especially in relation to its psychological connectin to reputation, by challenging the claims and assumptions, as shown in the problems pointed out by the critics as cited, in *Sec. 2.2.2*; the relatively more complicated analysis of EMAS, by challenging the claims and assumptions, as shown in the problems pointed out by the critics as cited, in *Sec. 2.3.1*; the relatively more complicated analysis of space-time in accounting, by challenging the claims and assumptions, as shown in the problems pointed out by the critics as cited, in *Sec. 2.3.2*; the relatively more complicated analysis of auditing, by challenging the claims and assumptions, as shown in the problems pointed out by the critics as cited, in *Sec. 2.4.1*; the relatively more complicated analysis of VAT, by challenging the claims and assumptions, as shown in the problems pointed out by the critics as cited, in *Sec. 2.4.2*; the relatively more complicated analysis of the 3 accounting principles, by challenging the claims and assumptions, as shown in the problems pointed out by the critics as cited, in *Sec. 2.5.1*; the relatively more complicated analysis of US GAPP, by challenging the claims and assumptions, as shown in the problems pointed out by the critics as cited, in *Sec. 2.5.2*; the relatively more complicated analysis of "control self-assessment," by challenging the claims and assumptions, as shown in the problems pointed out by the critics as cited, in *Sec. 3.2.1*; the relatively more complicated analysis of mental accounting, by challenging the claims and assumptions, as shown in the problems pointed out by the critics as cited, in *Sec. 3.2.2*; the relatively more complicated analysis of carbon accounting, by challenging the claims and assumptions, as shown in the problems pointed out by the critics as cited, in *Sec. 3.3.1*; the relatively more complicated analysis of the E P & L accounting, by challenging the claims and assumptions, as shown in the problems pointed out by the critics as cited, in *Sec. 3.3.2*; the relatively more complicated analysis of OOP, by challenging the claims and assumptions, as shown in the problems pointed out by the critics as cited, in *Sec. 3.4.1*; the relatively more complicated analysis of internal control, by challenging the claims and assumptions, as shown in the problems pointed out by the critics as cited, in *Sec. 3.4.2*; the relatively more complicated analysis of accounting ethics, by challenging the claims and assumptions, as shown in the problems pointed out by the critics as cited, in *Sec. 3.5.1*; and the relatively more complicated analysis of the convergence of accounting standards, by challenging the claims and assumptions, as shown in the problems pointed out by the critics as cited, in *Sec. 3.5.2*).

And the reverse direction also holds true.

Again, it should be stressed, however, that there are different shades of gray (or different degrees of truth) in the two opposites, and the classification is not necessarily mutually exclusive either. In fact, even when some combinations of the two occur, they only end up sharing the same dialectic relationship, but in a different degree.

Also, there are some other relationships (as a kind of family resemblance) which have something in common with the principle, but they are not exactly the same but only more or less comparable, which varies from case to case.

Good examples of family resemblance in relation to the principle are inflexibility-flexibility, standardization-specialization, imperfectness-perfectness, superficiality-depth, shallowness-deepness, economicalness-elaboratedness, plainness-circumspection, onesidedness-multisidedness, succinctness-duplicatedness, simpleness-sophisticatedness, crudeness-subtleness, straightness-bendedness, and the like.

11th Thesis: The Openness-Hiddenness Principle

The eleventh thesis refers to the openness-hiddenenss principle (on the detection of things) in the category of "structure" in existential dialectics (which was already worked out in my previous books, especially in *FPHU*), in that reality has its hidden face, just as it is open to outside view in some other ways. There is no openness without hiddenness—and vice versa.

For instance, on the one hand, there is openness (e.g., the open availability of "financial reporting" and "structuring transactions" to be used by creative accountants, in *Sec. 2.2.1*; the open exploration of the "benefit" of reputation capital and management for a company, in *Sec. 2.2.2*; the open exploration of using "large, highly detailed and prescriptive protocols" as an audit tool, in *Sec. 2.3.1*; the open exploration of using "transaction cost economics" to question the "accrual accounting system" in regard to "the status of an organization as an independent economic entity," in *Sec. 2.3.2*; the open "implementation in the United States of the Sarbanes-Oxley Act of 2002" so as to enhance "the profession's exposure and value" in regard to auditing, in *Sec. 2.4.1*; the open exploration of using VAT to increase government revenue, in *Sec. 2.4.2*; the open exploration of using the 3 principles for accounting in the last few hundred years, in *Sec. 2.5.1*; the open exploration of using concepts like "assumptions," "principles," and "constraints" for US GAAP, in *Sec. 2.5.2*; the open exploration of using "control self-assessment" for the reduction or elimination of "the risk of error," in *Sec. 3.2.1*; the open exploration of mental accounting to understand how people "subjectively frame…a transaction in their mind," in *Sec. 3.2.2*; the open exploration of sustainability measurement for carbon accounting, in *Sec. 3.3.1*; the open exploration of 4 categories of ecosystem services in the article aforecited, in *Sec. 3.3.2*; the open exploration of 5 features of OOP by Benjamin C. Pierce and others as cited above, in *Sec. 3.4.1*; the open exploration of "corporate misdeeds" as "fraudulent" by the Securities and Exchange Commission in the United States, in *Sec. 3.4.2*; the open exploration of "principle-based" accounting standards to be used for the accounting profession by the International Financial Reporting Standards [IFRS], in *Sec. 3.5.1*; and the open exploration of the convergence of accounting standards by the FASB and IASB, in *Sec. 3.5.2*).

On the other hand, there is hiddenness (e.g., the hidden bias in the use of "financial reporting" and "structuring transactions" by creative accountants, because "managers" can then misuse them "to alter financial reports to either mislead some stakeholders about the underlying economic performance of a company or influence contractual outcomes that depend on reported accounting numbers," in *Sec. 2.2.1*; the hidden bias in the exploration of the "benefit" of reputation capital and management for a company, since it can become a psychological pathology instead, "exaggerated in importance due to the fragile nature of the human ego," in *Sec. 2.2.2*; the hidden bias in using "large, highly detailed and prescriptive protocols" as an audit tool, since, to the critics, "there is a long standing debate among environmental audit professionals on the value of large, highly detailed and prescriptive protocols...versus more flexible protocols," in *Sec. 2.3.1*; the hidden bias in the use of "transaction cost economics" to question the "accrual accounting system" in regard to "the status of an organization as an independent economic entity," since Takatera and Sawabe did not question the institutional approach itself, in *Sec. 2.3.2*; the hidden bias in the implementation of the Sarbanes-Oxley Act of 2002, since, to the critics, "the focus by internal audit departments of publicly traded companies on SOX related financial policy and procedures derailed progress made by the profession in the late 20th century toward Larry Sawyer's vision for internal audit," in *Sec. 2.4.1*; the hidden bias in using VAT to increase government revenue, because of different problems aforedescribed, like "regressive tax," "inefficiency," etc., in *Sec. 2.4.2*; the hidden bias in the use of the 3 principles for accounting in the last few hundred years, as shown in the different problems pointed out by the critics aforecited, in *Sec. 2.5.1*; the hidden bias in using concepts like "assumptions," "principles," and "constraints" for US GAAP, since different research works make different [even conflicting] claims about what really makes a principle to be an "assumption," a "requirement," or a "constraint" in accounting standards, in *Sec. 2.5.2*; the hidden bias in using "control self-assessment" for the reduction or elimination of "the risk of error," since, to the critics, it does not give sufficient attention to "downside risk" and is too obsessed with the "costs" of risk but not its "benefits," in *Sec. 3.2.1*; the hidden bias in the exploration of mental accounting to understand how people "subjectively frame...a transaction in their mind," since the critics point out that the issue of "rationality" in the debate on mental accounting can be relative, that it does not give sufficient attention to the "good" side of mental accounting [but its "bad" side], etc., in *Sec. 3.2.2*; the hidden bias in the exploration of sustainability measurement for carbon accounting, since, to the critics, "social scientific studies of accounting challenge this hope, pointing to the socially constructed character of carbon conversion factors or of the accountants' work practice which cannot implement abstract accounting schemes into reality," in *Sec. 3.3.1*; the hidden bias in the exploration of the 4 categories of ecosystem services in the article aforecited, since "the most recent revision by TEEB...has revised the...definition" by removing "Supporting Services" and replacing it "on the one hand with 'Habitat Services' and on the other hand with 'Ecosystem Functions,'" in *Sec. 3.3.2*; the hidden bias in exploration of the 5 features of OOP by Benjamin C. Pierce and others as cited above, since others like John C. Mitchell identified only 4, and Michael Lee Scott considered only 3, as cited earlier, in *Sec. 3.4.1*; the hidden bias in the exploration of "corporate misdeeds" as "fraudulent" by the Securities and Exchange Commission in the United States, since the critics pointed out that the problem with accounting scandals is not wholly the fault of corporate criminals and that the nature of business cycles in the free market makes it easier for accounting scandals to occur, in *Sec. 3.4.2*; the hidden bias in the exploration of "principle-based" accounting standards to be used for the accounting profession by the International Financial Reporting Standards [IFRS]," since the critics like those who favor "the United States Generally Accepted Accounting Principles [GAAP]" prefer those which are "largely rule-based" instead, in *Sec. 3.5.1*; and the hidden bias in the exploration of the convergence of accounting standards by the FASB and IASB, since the critics pointed out that the claim of "comparability" is questionable, that the convergence process is costly, that it is not as popular as conventionally assumed, etc. in *Sec. 3.5.2*).

And the reverse direction also holds true.

Again, it should be stressed, however, that there are different shades of gray (or different degrees of truth) in the two opposites, and the classification is not necessarily mutually exclusive either. In fact, even when some combinations of the two occur, they only end up sharing the same dialectic relationship, but in a different degree.

Also, there are some other relationships (as a kind of family resemblance) which have something in common with the principle, but they are not exactly the same but only more or less comparable, which varies from case to case.

Good examples of family resemblance in relation to the principle are overtness-covertness, participation-obstruction, publicness-privateness, openness-closedness, surface-depth, transparency-secrecy, openness-biasedness, , exposition-illusion, and so on.

12th Thesis: The Denseness-Emptiness Principle

The twelfth thesis concerns the denseness-emptiness principle (on the distribution of entities in space) in the category of "structure" in existential dialectics, which is first proposed in *FPHUP*, in that both density and void are needed, in relation to the mind, nature, culture, and society, albeit in different ways. There is no denseness without emptiness—and vice versa.

Lest any misunderstanding occurs, the term "void" is used here only as an approximation of emptiness (depending on the degree of the lack of density), since, in physics, it is well known that "empty" space is not really empty all the way, because it can be full of energy (e.g., random quantum fluctuations at the sub-atomic level, and, for that matter, dark energy in the universe) and matter (e.g., different versions of sub-atomic particles, and, for that matter, dark matter in the universe). (F. Wilczek 2008)

For instance, on the one hand, there is denseness (e.g., the relatively denser concentration of concerns with finding "errors" by auditors, in *Sec. 2.2.1*; the relatively denser concentration of concerns with "reputation capital and management" among those businesses with "public relations department" nowadays, in *Sec. 2.2.2*; the relatively denser concentration of concerns with "easier reviews" by those advocates of using "large, highly detailed and prescriptive protocols" as an audit tool, in *Sec. 2.3.1*; the relatively denser concentration of concerns with "accounting income" to be "determined completely by the market" in the "pre-going concern" period, in *Sec. 2.3.2*; the relatively denser concentration of concerns with "reasonable assurance" by its advocates, in *Sec. 2.4.1*; the relatively denser concentration of concerns with the rates of VAT to be implemented by those in EU countries, in *Sec. 2.4.2*; the relatively denser concentration of concerns with "historical cost" by those in the traditional Historical Cost Accounting model, in *Sec. 2.5.1*; the relatively denser concentration of concerns with the "objectivity principle," the "consistency principle," and the "conservatism principle" as constraints among the advocates of SFAC No. 1 and 2, in *Sec. 2.5.2*; the relatively denser concentration of concerns with the "costs" of risk by those for "control self-assessment," in *Sec. 3.2.1*; the relatively denser concentration of concerns with "cognitive biases" by those advocates of mental accounting, in *Sec. 3.2.2*; the relatively denser concentration of concerns with "environmental-economic-social interrelationships" by the advocates, in *Sec. 3.3.1*; the relatively denser concentration of concerns with the view that "more than one species performs a given role within an ecosystem" by the advocates of "the redundancy hypothesis," in *Sec. 3.3.2*; the relatively denser concentration of concerns with "objects" and "nouns" by those advocates of OOP, in *Sec. 3.4.1*; the relatively denser concentration of concerns with "internal control" by the advocates, in *Sec. 3.4.2*; the relatively denser concentration of concerns with "principle-based" accounting standards by those for the International Financial Reporting Standards [IFRS], in *Sec. 3.5.1*; and the relatively denser concentration of concerns with the convergence of accounting standards by the FASB and IASB, in *Sec. 3.5.2*).

On the other hand, there is emptiness (e.g., the relatively less dense, or more empty, concentration of concerns with "fraudulent" reports by creative accountants, in *Sec. 2.2.1*; the relatively less dense, or more empty, concentration of concerns with "reputation" as a psychological pathology, "exaggerated in importance due to the fragile nature of the human ego," among those businesses with "public relations department" nowadays, in *Sec. 2.2.2*; the relatively less dense, or more empty, concentration of concerns with "more flexible protocols" by those advocates of using "large, highly detailed and prescriptive protocols" as an audit tool, in *Sec. 2.3.1*; the relatively less dense, or more empty, concentration of concerns with "accounting income" to be "determined completely by the market" in the "going concern" period, in *Sec. 2.3.2*; the relatively less dense, or more empty, concentration of concerns with "reasonable assurance" by the critics like the US Public Company Accounting Oversight Board, who favor "the argument that auditing should go beyond just true and fair is gaining momentum," in *Sec. 2.4.1*; the relatively less dense, or more empty, concentration of concerns with the rates of VAT to be implemented by those in the U.S., in *Sec. 2.4.2*; the relatively less dense, or more empty, concentration of concerns with "the loss of value of monetary assets as a result of inflation" by those in the traditional Historical Cost Accounting model, in *Sec. 2.5.1*; the relatively less dense, or more empty, concentration of concerns with the "cost constraint" among the advocates of SFAC No. 1 and 2, in *Sec. 2.5.2*; the relatively less dense, or more empty, concentration of concerns with the benefits" of risk by those for "control self-assessment," in *Sec. 3.2.1*; the relatively less dense, or more empty, concentration of concerns with the "good" side of mental accounting by the advocates, since "a question that has not received much attention is whether mental accounting is good for us" in the literature, in *Sec. 3.2.2*; the relatively less dense, or more empty, concentration of concerns with "growth-based" economic models by the advocates of the approach by "measuring environmental-economic-social interrelationships," in *Sec. 3.3.1*; the relatively less dense, or more empty, concentration of concerns with "the portfolio effect" by the advocates of "the redundancy hypothesis," in *Sec. 3.3.2*; the relatively less dense, or more empty, concentration of concerns with "verbs" and "actions" by those advocates of OOP, in *Sec. 3.4.1*; the relatively less dense, or more empty, concentration of concerns with "external auditing" by the advocates of "internal control," in *Sec. 3.4.2*; the relatively less dense, or more empty, concentration of concerns with "rule-based" accounting standards by those for the International Financial Reporting Standards [IFRS], in *Sec. 3.5.1*; and the relatively less dense, or more empty, concentration of concerns with the convergence of accounting standards by the critics like Shyam Sunder, in *Sec. 3.5.2*).

And the reverse direction also holds true.

Again, it should be stressed, however, that there are different shades of gray (or different degrees of truth) in the two opposites, and the classification is not necessarily mutually exclusive either. In fact, even when some combinations of the two occur, they only end up sharing the same dialectic relationship, but in a different degree.

Also, there are some other relationships (as a kind of family resemblance) which have something in common with the principle, but they are not exactly the same but only more or less comparable, which varies from case to case.

Good examples of family resemblance in relation to the principle are fullness-voidness, nearness-farness, concentration-dispersion, denseness-nothingness, something-nothing, heaviness-lightness, composition-emptiness, solidness-void, presence-absence, focus-unfocus, emphasis-disemphasis, and the like.

13th Thesis: The Rule-Exception Principle

The thirteenth thesis concerns the rule-exception principle (on the extent of typicality) in the category of "structure" in existential dialectics, which is first proposed in *FPHTHAN*, in that both rules and their exceptions exist—in relation to the mind, nature, culture, and society, albeit in different ways. There is no rule without exception—and vice versa.

As the old saying goes, "There is an exception to the rule"; however, to avoid misuse, there are the following four important aspects of the relationship between rules and exceptions to be clarified:

- If there is an exception to the rule, there can be an exception to the.rule itself that there is an exception to the rule, as shown in some cases where no exception is allowed—but this regression can continue further (e.g., there is an exception to the rule that there can be an exception to the rule itself that there is an exception to the rule).

- To be an exception to the rule does not mean that one can say something nonsensual, either jokingly or seriously, and then expects others to allow an exception.
- To be an exception to the rule does not mean that one can appeal to an invalid generalization, either literally or figuratively, and then expects others to allow an exception.
- An exception to a rule means that the exception implies the existence of a rule, which does not apply in the special case under consideration—but the exception does not automatically justify that the rule is desirable, good, right, etc.

In any event, on the one hand, there are rules (e.g., the usual generalization, procedure, habit, or tendency that people normally "conform…to acceptable accounting practices" due to the pressure of "personal integrity," in *Sec. 2.2.1*; the usual generalization, procedure, habit, or tendency that assets can be amortized, in *Sec. 2.2.2*; the usual generalization, procedure, habit, or tendency that there are benefits and costs of EMAS, in *Sec. 2.3.1*; the usual generalization, procedure, habit, or tendency that accounting conventions can be understood in terms of two main periods, namely, "pre-going concern" and "going-concern," in *Sec. 2.3.2*; the usual generalization, procedure, habit, or tendency that "as a general rule, audits should always be an independent evaluation that will include some degree of quantitative and qualitative analysis," in *Sec. 2.4.1*; the usual generalization, procedure, habit, or tendency that "most countries today use the invoice method," in *Sec. 2.4.2*; the usual generalization, procedure, habit, or tendency that the issue of accounting princples is "very, very boring" to many ordinary people, in *Sec. 2.5.1*; the usual generalization, procedure, habit, or tendency that business enterprises in the U.S. are expected to follow US GAAP, in *Sec. 2.5.2*; the usual generalization, procedure, habit, or tendency that those in "control self-assessment" pay attention to "risk," in *Sec. 3.2.1*; the usual generalization, procedure, habit, or tendency that "as kids…, we all remember getting cards stuffed with cash from our relatives on each birthday" and "most of us eagerly rushed out and bought the most outlandish and impractical things we could," in *Sec. 3.2.2*; the usual generalization, procedure, habit, or tendency that "the Kyoto Protocol defines legally binding targets and timetables for cutting the GHG emissions of…countries that ratified the Kyoto Protocol," in *Sec. 3.3.1*; the usual generalization, procedure, habit, or tendency that "environmental awareness is rapidly improving in our contemporary world" nowadays, in *Sec. 3.3.2*; the usual generalization, procedure, habit, or tendency that, as shown in a survey by Deborah J. Armstrong, there is "a number of fundamental concepts...found in the large majority of definitions of OOP," in *Sec. 3.4.1*; the usual generalization, procedure, habit, or tendency that "internal control can provide reasonable...assurance that the objectives of an organization will be met," in *Sec. 3.4.2*; the usual generalization, procedure, habit, or tendency that "the International Financial Reporting Standards [IFRS]...are used by over 115 countries" around the world and are "principle-based," in *Sec. 3.5.1*; and the usual generalization, procedure, habit, or tendency that "convergence in some form has been taking place for several decades, "in *Sec. 3.5.2*).

On the other hand, there are exceptions (e.g., a case to which a rule does not apply, such as the exceptional case when "aggressive earnings management becomes more probable when a company is affected by a downturn in business," contrary to the rule that people normally "conform...to acceptable accounting practices" due to the pressure of "personal integrity," in *Sec. 2.2.1*; a case to which a rule does not apply, such as the exceptional case that, "according to International Financial Reporting Standards, goodwill is never amortized" since 2001, in *Sec. 2.2.2*; a case to which a rule does not apply, such as the exceptional cass that "some countries offer substantial reductions of waste fees, lower permitting costs, faster licensing procedures, etc.," in *Sec. 2.3.1*; a case to which a rule does not apply, such as the exceptional case that the two periods of accounting conventins do not apply to historical periods before the previous 5 centuries, since Takatera and Sawabe did not address them, in *Sec. 2.3.2*; a case to which a rule does not apply, such as the exceptional case about "an assessment," which implies "a less independent and more consultative approach," since "the outcome of the assessment should relate to the norms that were set for the task, product or event," in *Sec. 2.4.1*; a case to which a rule does not apply, such as the exceptional case of "Japan, which uses the accounts method" instead, in *Sec. 2.4.2*; a case to which a rule does not apply, such as the exceptional case that the issue of accounting princples "has been left to the experts, our modern day temple priests—the accountancy profession," in *Sec. 2.5.1*; a case to which a rule does not apply, such as the exceptional case when, "under the AICPA's Code of Professional Ethics under Rule 203-Accounting Principles, a member must depart from GAAP if following it would lead to a material misstatement on the financial statements, or otherwise be misleading," in *Sec. 2.5.2*; a case to which a rule does not apply, such as the exceptional case that those in "control self-assessment" may "understate risk by not identifying extreme downside risk" at all, in *Sec. 3.2.1*; a case to which a rule does not apply, such as the exceptional case about "a few young geniuses with the foresight to save some of it [birthday money]," while "most of us eagerly rushed out and bought the most outlandish and impractical things we could," in *Sec. 3.2.2*; a case to which a rule does not apply, such as the exceptional situation when "nations which have failed to deliver their Kyoto emissions reductions obligations can enter Emissions Trading to purchase CERS and ERUs to cover their treaty shortfalls," in *Sec. 3.3.1*; a case to which a rule does not apply, such as the exceptional case that there are some people who still "suffer from the so-called 'tragedy of the commons,'" in *Sec. 3.3.2*; a case to which a rule does not apply, such as the exceptional cases where "not all of these concepts [of OOP] appear in all object-oriented programming languages," as shown in the example that "object-oriented programming that uses classes is sometimes called class-based programming, while prototype-based programming does not typically use classes," in *Sec. 3.4.1*; a case to which a rule does not apply, such as the exceptional cases when internal control does not provide "absolute" assurance that "the objectives of an organization will be met" and also "has no control over external factors which can affect the performance of an organization," in *Sec. 3.4.2*; a case to which a rule does not apply, such as the exceptional cases that some countries like the U.S. do not use "the International Financial Reporting Standards [IFRS]" and prefer those which are "largely rule-based" instead, at least for now, in *Sec. 3.5.1*; and a case to which a rule does not apply, such as the exceptional case that in 2006 "senior partners at PricewaterhouseCoopers...called for convergence to be 'shelved indefinitely' in a draft paper" and thus did not support it, in *Sec. 3.5.2*).

And the reverse direction also holds true.

Again, it should be stressed, however, that there are different shades of gray (or different degrees of truth) in the two opposites, and the classification is not necessarily mutually exclusive either. In fact, even when some combinations of the two occur, they only end up sharing the same dialectic relationship, but in a different degree.

Also, there are some other relationships (as a kind of family resemblance) which have something in common with the principle, but they are not exactly the same but only more or less comparable, which varies from case to case.

Good examples of family resemblance in relation to the principle are generalization-exception, custom-deviance, procedure-exception, pattern-deviance, typicality-exception, and the like.

In Relation to Process

The next six theses are in relation to process in existential dialectics, as shown below.

14th Thesis: The Prototypicality-Variation Principle

The fourteenth thesis concerns the prototypicality-variation principle (on the nature of duplication) in the category of "process" in existential dialectics, which was first proposed in *FPHMORPH*, in that, if there are prototypes, there are also variants of the copies. There is no prototypicality without variation—and vice versa.

There are the following four important aspects of the relationship between prototypess and their copies, as shown in the fields of morphology (e.g., "proto-languages" in *FPHMORP*), information architecture (e.g., "taxonomies" in *FIA*), and evolutionry biology (e.g., "mutation" and "variation" in *BNN*), just to cite three well-documented sources:

- Prototypes and their copies are not static and may change or die out over time, but new ones may emerge.
- Some copies may have multiple prototypes, not just one.
- It can be very challenging to trace some copies back to their prototypes, under certain difficult historical conditions.
- Copies are not solely identical duplications of prototypes but have their own variations (mutations).

In any event, on the one hand, there are prototypes (e.g., the prototype, like the example about the original usage of the term "creative accounting" in 1968 in the film *The Producers* by Mel Brooks, in *Sec. 2.2.1*; the prototype, like the example about the original U.S. Generally Accepted Accounting Principles [FAS 141] on goodwill, in *Sec. 2.2.2*; the prototype, like the example about the original EMAS Regulation [EMAS I] in 1993, in *Sec. 2.3.1*; the prototype, like the example about the original works by D. R. Scott and Thorstein Veblen on institutional economics, in *Sec. 2.3.2*; the prototype, like the example about the original theory of internal auditing by Lawrence Sawyer, in *Sec. 2.4.1*; the prototype, like the example about the original concept of VAT by Wilhelm von Siemens in 1918, in *Sec. 2.4.2*; the prototype, like the example about the original "accounting structure" built over the last few hundred years, in *Sec. 2.5.1*;

the prototype, like the example about the original Committee on Accounting Procedure set by the AICPA in 1939, in *Sec. 2.5.2*; the prototype, like the example about the original "control self-assessment...developed by Gulf Canada in 1987," in *Sec. 3.2.1*; the prototype, like the example about the original work on mental accounting by Richard Thaler, in *Sec. 3.2.2*; the prototype, like the example about the original use of "Hubbert peak theory" for sustainabilty measurement "in relation to peak oil production," in *Sec. 3.3.1*; the prototype, like the example about the original idea by Marsh in 1864, "when he challenged the idea that Earth's natural resources are unbounded by pointing out changes in soil fertility in the Mediterranean," in *Sec. 3.3.2*; the prototype, like the example about the original works on OOP "invoking 'objects' and 'oriented' in the modern sense of object-oriented programming...at MIT in the late 1950s and early 1960s," in *Sec. 3.4.1*; the prototype, like the example about the original "Foreign Corrupt Practices Act [FCPA] of 1977" for internal control in the U.S., in *Sec. 3.4.2*; the prototype, like the example about the original work on accounting ethics by Luca Pacioli in 1494, in *Sec.3.5.1*; and the prototype, like the example about the original work on "convergence" which "took the form of harmonization, the reduction of differences between the various accounting standards used internationally," before the 1990s, in *Sec.3.5.2*).

On the other hand, there are variants (e.g., different variants of the protoptye, as shown in many studies since the original usage of the term "creative accounting" in 1968 in the film *The Producers* by Mel Brooks—like the later ones by Abraham Briloff in the 1970s, the application of cognitive psychology to the field of creative accounting, etc., though each with its own variation, contribution, and disagreement, in *Sec. 2.2.1*; different variants of the protoptye, as shown in many studies since the original U.S. Generally Accepted Accounting Principles [FAS 141] on goodwill—like the later ones such as "U.S. GAAP [FAS 142]," "International Financial Reporting Standards," etc., though each with its own variation, contribution, and disagreement, in *Sec. 2.2.2*; different variants of the protoptye, as shown in many studies since the original EMAS Regulation [EMAS I] in 1993—like the later ones such as EMAS II in 2001, EMAS III in 2010, etc., though each with its own variation, contribution, and disagreement, in *Sec. 2.3.1*; different variants of the protoptye, as shown in many studies since the original works by D. R. Scott and Thorstein Veblen on institutional economics—like the later ones by Takatera and Sawabe, though with their own variation, contribution, and disagreement, in *Sec. 2.3.2*; different variants of the protoptye, as shown in many studies since the original theory of internal auditing by

Lawrence Sawyer —like the later ones such as "the Sarbanes-Oxley Act of 2002," the revision by the IIA in 2010, etc., though each with its own variation, contribution, and disagreement, in *Sec. 2.4.1*; different variants of the protoptye, as shown in many works since the original concept of VAT by Wilhelm von Siemens in 1918—like the later ones by Maurice Lauré in 1954, the EU in 2006, etc., though each with its own variation, contribution, and disagreement, in *Sec. 2.4.2*; different variants of the protoptye, as shown in many studies since the original "accounting structure" built over the last few hundred years—like the later ones such as the accounting reform suggested by Jeremy Nicholls of the SROI Network, the "Constant Purchasing Power model" proposed by the "International Accounting Standards Board," etc., though each with its own variation, contribution, and disagreement, in *Sec. 2.5.1*; different variants of the protoptye, as shown in many works since the original Committee on Accounting Procedure set by the AICPA in 1939—like the later ones such as "the Financial Accounting Standards Board" in 1973, "the Governmental Accounting Standards Board" in 1984, etc., though each with its own variation, contribution, and disagreement, in *Sec. 2.5.2*; different variants of the protoptye, as shown in many studies since the original "control self-assessment…developed by Gulf Canada in 1987"—like the later ones such as the "control self-assessment practices" by "the Federal Deposit Insurance Corporation," "the Canadian Deposit Insurance Corporation," etc., though each with its own variation, contribution, and disagreement, in *Sec. 3.2.1*; different variants of the protoptye, as shown in many studies since the original work on mental accounting by Richard Thaler —like the later ones by J. Cross, H. Shefrin, etc., though each with its own variation, contribution, and disagreement, in *Sec. 3.2.2*; different variants of the protoptye, as shown in many studies since the original use of "Hubbert peak theory" for sustainabilty measurement "in relation to peak oil production"—like the later uses of "Hubbert peak theory" for sustainabilty measurement in relation to "other natural resources" like "natural gas," "coal," etc., though each with its own variation, contribution, and disagreement, in *Sec. 3.3.1*; different variants of the protoptye, as shown in many studies since the original idea by Marsh in 1864, "when he challenged the idea that Earth's natural resources are unbounded by pointing out changes in soil fertility in the Mediterranean"—like the later ones by authors like Osborn [1948], Vogt [1948], Leopold [1949], etc., though each with its own variation, contribution, and disagreement, in *Sec. 3.3.2*; different variants of the protoptye, as shown in many studies since the original works on OOP "invoking 'objects' and 'oriented' in the modern sense of object-oriented programming…at MIT in the late 1950s and early 1960s"—like the later ones which "have emerged that are primarily object-oriented, but that are also compatible with procedural methodology," like "Python and Ruby", etc., though each with its own variation, contribution, and disagreement, in *Sec. 3.4.1*; different variants of the protoptye, as shown in many works since the original "Foreign Corrupt Practices Act [FCPA] of 1977" for internal control in the U.S.—like the later ones such as "the Sarbanes–Oxley Act of 2002," the guideliens by "the AICPA, IIA, and ACFE" in 2008, etc., though each with its own variation, contribution, and disagreement, in *Sec. 3.4.2*; different variants of the protoptye, as shown in many works since the original work on accounting ethics by Luca Pacioli in 1494—like the later ones "expanded by government groups, professional organizations, and independent companies" like "the International Financial Reporting Standards [IFRS]," "the United States Generally Accepted Accounting Principles [GAAP]," etc., though each with its own variation, contribution, and disagreement, in *Sec.3.5.1*; and different variants of the protoptye, as shown in many studies since the original work on "convergence" which "took the form of harmonization, the reduction of differences between the various accounting standards used internationally," before the 1990s —like the later ones in the 2000s by the FASB, the IASB, etc., though each with its own variation, contribution, and disagreement, in *Sec. 3.5.2*).

And the reverse direction also holds true.

Again, it should be stressed, however, that there are different shades of gray (or different degrees of truth) in the two opposites, and the classification is not necessarily mutually exclusive either. In fact, even when some combinations of the two occur, they only end up sharing the same dialectic relationship, but in a different degree.

Also, there are some other relationships (as a kind of family resemblance) which have something in common with the principle, but they are not exactly the same but only more or less comparable, which varies from case to case.

Good examples of family resemblance in relation to the principle are archetypicality-descendedness, origin-descendedness, prototypicality-duplication, archetypicality-mutation, prototypicality-copyness, ancestry-descendedness, origin-reenactment, prototypicality-recreation, and the like.

15th Thesis: The Change-Constancy Principle

The fifteenth thesis concerns the change-constancy principle (on the alteration of things) in the category of "process" in existential dialectics, which is first named in *BCPC* (and were worked out in my books like *FHC*, *FCD*, and *FPHC)*, in that change occurs over time, although constancy is also allowed.

Old players fade away, and new ones emerges, with ever new causes and ever new forms—and asymmetry undergoes changes over time too, so does symmetry. There is no change without constancy—and vice versa.

For instance, on the one hand, there is change (e.g., the ever new ways to study fraudulent activities, as shown in the application of cognitive psychology to the field of creative accounting, in *Sec. 2.2.1*; the ever new ways to study accounting, as shown in the work on goodwill, in *Sec. 2.2.2*; the ever new ways to study accounting, as shown in the research on environmental auditing like EMAS, in *Sec. 2.3.1*; the ever new ways to study accounting, as shown in the work on space-time in accounting, in *Sec. 2.3.2*; the ever new ways to study accounting, as shown in the work on "internal auditing," in *Sec. 2.4.1*; the ever new ways to study accounting, as shown in the implementation of VAT aforediscussed, in *Sec. 2.4.2*; the ever new ways to study accounting, as shown in the work on accounting principles, in *Sec. 2.5.1*; the ever new ways to study accounting, as shown in the work on US GAAP, in *Sec. 2.5.2*; the ever new ways to study accounting, as shown in the work on "control self-assessment," in *Sec. 3.2.1*; the ever new ways to study accounting, as shown in the research on mental accounting, in *Sec. 3.2.2*; the ever new ways to study accounting, as shown in the work on carbon accounting, in *Sec. 3.3.1*; the ever new ways to study accounting, as shown in the work on the E P & L accounting, in *Sec. 3.3.2*; the ever new ways to study bookkeeping, as shown in the work on boilerplate codes like OOP, in *Sec. 3.4.1*; the ever new ways to study accounting, as shown in the work on internal control, in *Sec. 3.4.2*; the ever new ways to study accounting, as shown in the work on accounting ethics, in *Sec. 3.5.1*; and the ever new ways to study accounting, as shown in the work on the convergence of accounting standards, in *Sec. 3.5.2*).

On the other hand, there is constancy (e.g., the ever constant existence of different problems with the ever new ways to study fraudulent activities, as shown in the problems with the application of cognitive psychology to the field of creative accounting pointed out by the critics, in *Sec. 2.2.1*; the ever constant existence of different problems with the ever new ways to study accounting, as shown in the problems with the work on goodwill pointed out by the critics, in *Sec. 2.2.2*; the ever constant existence of different problems with the ever new ways to study accounting, as shown in the problems with the research on environmental auditing like EMAS pointed out by the critics, in *Sec. 2.3.1*; the ever constant existence of different problems with the ever new ways to study accounting, as shown in the problems with te work on space-time in accounting pointed out by the critics, in *Sec. 2.3.2*; the ever constant existence of different problems with the ever new ways to study accounting, as shown in the problems with the work on "internal auditing" pointed out by the critics, in *Sec. 2.4.1*; the ever constant existence of different problems with the ever new ways to study accounting, as shown in the problems with the implementation of VAT pointed out by the critics, in *Sec. 2.4.2*; the ever constant existence of different problems with the ever new ways to study accounting, as shown in the problems with the work on accounting principles pointed out by the critics, in *Sec. 2.5.1*; the ever constant existence of different problems with the ever new ways to study accounting, as shown in the problems with the work on US GAAP pointed out by the critics, in *Sec. 2.5.2*; the ever constant existence of different problems with the ever new ways to study accounting, as shown in the problems with the work on "control self-assessment" pointed out by the critics, in *Sec. 3.2.1*; the ever constant existence of different problems with the ever new ways to study accounting, as shown in the problems with the research on mental accounting pointed out by the critics, in *Sec. 3.2.2*; the ever constant existence of different problems with the ever new ways to study accounting, as shown in the problems with the work on carbon accounting pointed out by the critics, in *Sec. 3.3.1*; the ever constant existence of different problems with the ever new ways to study accounting, as shown in the problems with the work on the E P & L accounting pointed out by the critics, in *Sec. 3.3.2*; the ever constant existence of different problems with the ever new ways to study bookkeeping, as shown in the problems with the work on boilerplate codes like OOP pointed out by the critics, in *Sec. 3.4.1*; the ever constant existence of different problems with the ever new ways to study accounting, as shown in the problems with the work on internal control pointed out by the critics, in *Sec. 3.4.2*; the ever constant existence of different problems with the ever new ways to study accounting, as shown in the problems with the work on accounting ethics pointed out by the critics, in *Sec. 3.5.1*; and the ever constant existence of different problems with the ever new ways to study accounting, as shown in the problems with the work on the convergence of accounting standards pointed out by the critics, in *Sec. 3.5.2*).

And the reverse direction also holds true.

Again, it should be stressed, however, that there are different shades of gray (or different degrees of truth) in the two opposites, and the classification is not necessarily mutually exclusive either. In fact, even when some combinations of the two occur, they only end up sharing the same dialectic relationship, but in a different degree.

Also, there are some other relationships (as a kind of family resemblance) which have something in common with the principle, but they are not exactly the same but only more or less comparable, which varies from case to case.

Good examples of family resemblance in relation to the principle are dynamicness-staticness, instability-stability, and the like.

16th Thesis: The Order-Chaos Principle

The sixteenth thesis is called the the order-chaos principle (on the pattern of things) in the category of "process" in existential dialectics (as first worked out in *FC*), in that both order and chaos are vital in the process of change in the world.

The preference for order is biased, since it does not give sufficient attention to the vital role of chaos in the transformation of the world (without somehow reducing it for the understanding of order). There is no order without chaos—and vice versa.

The scientific search for order in the world is often a hidden bias in its ontological obsession with order, since chaos is often treated as the "bad" guy, with order as the "good" guy (for the end goal of science).

Neither order nor chaos is the final end of the world, and one is not to be treated as the means for the other in the transformation of things. Both are fundamental in their recurrent dialectical interactions with each other over time, without reducing one for the other.

Order has its different forms (e.g., ascending vs. descending order in mathematics, and feeding vs. bleeding order in morphology), just as chaos has its own versions (e.g., as understood differently in chaos theory vs. complexity theory).

For instance, on the one hand, there is order (e.g., the more or less orderly view on acounting, if looked from the sole vantage point of "acceptable accounting practices" as "a matter of personal integrity," in *Sec. 2.2.1*; the more or less orderly view on "reputation," if looked from the sole vantage point of a specific "model" of reputation by its advocates, in *Sec. 2.2.2*; the more or less orderly view on EMAS, if looked from the sole vantage point of of its official regulations, in *Sec. 2.3.1*; the more or less orderly view on accounting, if looked from the sole vantage point of "accounting order out of chaos" by Takatera and Sawabe, in *Sec. 2.3.2*; the more or less orderly view on the history of internal auditing, if looked from the sole vantage point of the implementation of "the Sarbanes-Oxley Act of 2002" at the time, in *Sec. 2.4.1*; the more or less orderly view on VAT, if looked from the sole vantage point of the advocates, in *Sec. 2.4.2*; the more or less orderly view on accounting princples, if looked from the sole vantage point of the advocates of the 3 main ones aforediscussed, in *Sec. 2.5.1*; the more or less orderly view on accounting standards, if looked from the sole vantage point of the well-established US GAAP, in *Sec. 2.5.2*; the more or less orderly view on "risk," if looked from the sole vantage point of by the advocates for "control self-assessment," who may "understate risk by not identifying extreme downside risk," in *Sec. 3.2.1*; the more or less orderly view on decision making about money, if looked from the sole vantage point of the principle of "fungibility" by the advocates, in *Sec. 3.2.2*; the more or less orderly view on the study of carbon accounting, if looked from the sole vantage point of the approach which is only "growth-based," in *Sec. 3.3.1*; the more or less orderly view on the E P & L accounting, if looked from the sole vantage point of the advocates, in *Sec. 3.3.2*; the more or less orderly view on OOP, if looked from the sole vantage point of "objects" and "nouns" by the advocates, in *Sec. 3.4.1*; the more or less orderly view on internal control, if looked from the sole vantage point by the advocates that "internal control can provide reasonable...assurance that the objectives of an organization will be met," in *Sec. 3.4.2*; the more or less orderly view on acounting standards, if looked from the sole vantage point of "the International Financial Reporting Standards [IFRS]" which "are used by over 115 countries" around the world and are "principle-based," in *Sec. 3.5.1*; and the more or less orderly view on the convergence of accounting standards, if looked from the sole vantage point of "comparability" by the advocates, in *Sec. 3.5.2*).

On the other hand, there is chaos (e.g., the more or less chaotic view on accounting, if looked from the multiple, conflicting vantage points of both the advocates of creative accounting and the critics, especially in regard to the problems aforecited, such that they do not add up to much of anything coherent, in *Sec. 2.2.1*; the more or less chaotic view on "reputation," if looked from the multiple "often conflicting models of reputation," according to Kevin Money and Carola Hillenbrand, such that they do not add up to much of anything coherent, especially when compounded with different problems aforediscussed, in *Sec. 2.2.2*; the more or less chaotic view on EMAS, if looked from the multiple, conflicting vantage points of both the advocates and the critics, especially in regard to the problems aforecited, such that they do not add up to much of anything coherent, in *Sec. 2.3.1*; the more or less chaotic view on accounting, if looked from the vantage point of "noise-induced" chaos, when "the micro level noise is sometimes amplified in the system to the extent that it qualitatively alters the behavior of the whole system when chaos apparently dominates at elementary level interactions," according to Takatera and Sawabe, in *Sec. 2.3.2*; the more or less chaotic view on the history of internal auditing, if looked from the multiple, conflicting vantage points as revealed in the inconsistent history of the implementation of internal auditing aforedescribed, in *Sec. 2.4.1*; the more or less chaotic view on VAT, if looked from the multiple, conflicting vantage points of both the advocates and the critics, especially in regard to the problems aforecited, such that they do not add up to much of anything coherent, in *Sec. 2.4.2*; the more or less chaotic view on accounting princples, if looked from the multiple, conflicting vantage points of both the advocates and the critics, especially in regard to the different problems aforecited, such that they do not add up to much of anything coherent, in *Sec. 2.5.1*; the more or less chaotic view on accounting standards, if looked from the disorderly vantage point of "the Asian financial meltdown in the late 1990s" which "has been partially attributed due to the lack of detailed accounting standards," as "giant firms…were able to take advantage of their poorly devised accounting standards to cover up immense debts and losses, which yielded a collective effect that eventually led the whole region into financial crisis," in *Sec. 2.5.2*; the more or less chaotic view on "risk," if looked from the disorderly vantage point of "extreme downside risk," which is "a highly improbable event that would have catastrophic consequences if it occurred," in *Sec. 3.2.1*; the more or less chaotic view on decision making about money, if looked from the multiple, conflicting vantage points of those who mess up their purchases because of different "mental accounts" based on illusions aforecited, in *Sec. 3.2.2*; the more or less chaotic view on the study of carbon accounting, if looked from the multiple, conflicting vantage points of both the advocates and the critics, especially in regard to the problems aforecited, such that they do not add up to much of anything coherent, in *Sec. 3.3.1*; the more or less chaotic view on the E P & L accounting, if looked from the multiple, conflicting vantage points of both the advocates and the critics, especially in regard to the problems aforecited, such that they do not add up to much of anything coherent, and "further complicating the process, these decisions are often spatial, always multi-objective, and based on uncertain data, models, and estimates," in *Sec.*

3.3.2; the more or less chaotic view on OOP, if looked from the multiple, conflicting vantage points of both the advocates and the critics, especially in regard to the problems aforecited, such that they do not add up to much of anything coherent, in *Sec. 3.4.1*; the more or less chaotic view on internal control, if looked from the multiple, conflicting vantage points of both the advocates and the critics, especially in regard to the problems aforecited, such that they do not add up to much of anything coherent, especially since internal control does not provide "absolute" assurance that "the objectives of an organization will be met" and also "has no control over external factors which can affect the performance of an organization," in *Sec. 3.4.2*; the more or less chaotic view on acounting standards, if looked from the multiple, conflicting vantage points of both the advocates and the critics, especially in regard to different problems aforecited, such that they do not add up to much of anything coherent, in *Sec. 3.5.1*; and the more or less chaotic view on the convergence of accounting standards, if looked from the multiple, conflicting vantage points of both the advocates and the critics, especially since the critics like Shyam Sunder dismissed "the link between convergence and comparability" as "overblown," together with other problems, such that all of these views do not add up to much of anything coherent, in *Sec. 3.5.2*).

And the reverse direction also holds true.

Once more, it should be stressed, however, that there are different shades of gray (or different degrees of truth) in the two opposites, and the classification is not necessarily mutually exclusive either. In fact, even when some combinations of the two occur, they only end up sharing the same dialectic relationship, but in a different degree.

Also, there are some other relationships (as a kind of family resemblance) which have something in common with the principle, but they are not exactly the same but only more or less comparable, which varies from case to case.

Good examples of family resemblance in relation to the principle are lawfulness-disorder, order-disorder, pattern-chaos, linearity-nonlinearity, unity-disunity, linearity-irreversalness, sequence-chaos, etc.

17th Thesis: The Slowness-Quickness Principle

The seventeenth thesis concerns the slowness-quickness principle (on the speed of change) in the category of "process" in existential dialectics (which was first worked out in *FIA*), in that both slowness and quickness co-exist, with their own internal tension, to the extent that each fights for its own relevance with the other, in accordance to the perspectives of nature, the mind, culture, and society, without one being the victor and the other being the vanquished in the long haul. There is no slowness without quickness—and vice versa.

For instance, on the one hand, there is slowness (e.g., the relatively slower speed for "an auditor to detect…sophisticated fraud that is covert and much harder to detect," in *Sec. 2.2.1*; the relatively slower speed of "reputation recovery," which is "long and arduous" in our time, in *Sec. 2.2.2*; the relatively slower speed of auditing with the use of "more flexible protocols that rely on the expertise and knowledge of experienced auditors and source documents [regulations, permits, etc.] directly" for more technical reviews, in *Sec. 2.3.1*; the relatively slower speed of "the value of money" to change throughout accounting periods if "the purchasing power of money" is assumed to be "constant" or more or less so, in *Sec. 2.3.2*; the relatively slower speed of internal auditing within an organization to complete, if it is very broad in scope, not only covering "an organization's governance, risk management and management controls" but also "conducting proactive fraud audits to identify potentially fraudulent acts; participating in fraud investigations under the direction of fraud investigation professionals, and conducting post investigation fraud audits to identify control breakdowns and establish financial loss," in *Sec. 2.4.1*; the relatively slower speed for "carousel fraud" to occur, if VAT is not implemented at all or only trivially implemented, in *Sec. 2.4.2*; the relatively slower speed for businesses to conduct transactions nowadays, in the absence of any accounting conventions like the 3 accounting principles, in *Sec. 2.5.1*; the relatively slower speed for the U.S. to move "from US GAAP to the International Financial Reporting Standards," because "the provisions of which differ considerably from GAAP" so "progress has been slow and uncertain," in *Sec. 2.5.2*; the relatively slower speed for those in "control self-assessment" to ever detect "extreme downside risk," in *Sec. 3.2.1*; the relatively slower speed of those "young geniuses with the foresight to save some of it [birthday money]," to ever buy "the most outlandish and impractical things we could," in *Sec. 3.2.2*; the relatively slower speed of carbon dioxide emissions after the implementation of different projects to curb them under the Kyoto Protocol, in *Sec. 3.3.1*; the relatively slower speed of "wild bees" to "provide partial or complete pollination services" in the presence of "intensified agricultural practices" which "can quickly erode pollination services through the loss of species," in California's agricultural region as studied by Kremen [2005], in *Sec. 3.3.2*; the relatively slower speed of a programmer to finish writing codes when he has to rely on the need for boilerplate manual codes without any "metaprogramming," "convention over configuration," and "model-driven engineering," in *Sec. 3.4.1*; the relatively slower speed of accounting scandals to occur, in the presence of "internal control" which "plays an important role in the prevention and detection of fraud," in *Sec. 3.4.2*; the relatively slower speed of corruption and fraud to occur, in the presence of accounting reforms like "education" and "regulations" as described earlier, in *Sec. 3.5.1*; and the relatively slower speed of communication among "investors, companies, auditors, and other participants" if handicapped by "incomparability" among different accounting standards, in *Sec. 3.5.2*).

On the other hand, there is quickness (e.g., the relatively quicker speed for "an auditor to detect error," in *Sec. 2.2.1*; the relatively quicker speed of "reputation loss" in our time, since "the 'stumble rate' among companies has risen exponentially over the past five years" and "79% of the world's most admired companies have lost their number one positions in industries in that time period," in *Sec. 2.2.2*; the relatively quicker speed of auditing with the use of "large, highly detailed and prescriptive protocols" for "easier reviews," in *Sec. 2.3.1*; the relatively quicker speed of "the value of money" to change throughout accounting periods if "the purchasing power of money" is not assumed to be constant and can change rapidly, in response to the "volatile market prices" in "exernal transactions," in *Sec. 2.3.2*; the relatively quicker speed of internal auditing within an organization to complete, if it is narrow in scope and covers only "an organization's governance, risk management and management controls" but nothing else, in *Sec. 2.4.1*; the relatively quicker speed for "carousel fraud" to occur, if it is to be implemented, "because of its particular mechanism of collection, VAT becomes quite easily the target of specific frauds like carousel fraud," in *Sec. 2.4.2*; the relatively quicker speed for businesses to conduct transactions nowadays, with the help of accounting conventions like the 3 accounting principles, in *Sec. 2.5.1*; the relatively quicker speed for "the Securities and Exchange Commission" to simply "issue...a preliminary 'roadmap'" in 2008 to move "from US GAAP to the International Financial Reporting Standards," in *Sec. 2.5.2*; the relatively quicker speed for those in "control self-assessment" to detect "normal" risk, in *Sec. 3.2.1*; the relatively quicker speed of most "kids" to buy "the most outlandish and impractical things" if given "birthday money," in *Sec. 3.2.2*; the relatively quicker speed of carbon dioxide emissions before the implementation of different projects to curb them under the Kyoto Protocol, in *Sec. 3.3.1*; the relatively quicker speed of "wild bees" to "provide partial or complete pollination services" in the absence of any "intensified agricultural practices," in California's agricultural region as studied by Kremen [2005], in *Sec. 3.3.2*; the relatively quicker speed of a programmer to finish writing codes when the need for boilerplate manual code can be reduced by way of "metaprogramming," "convention over configuration," and "model-driven engineering," in *Sec. 3.4.1*; the relatively quicker speed of accounting scandals to occur, in the absence of any "internal control" or any government oversight agencies like the Securities and Exchange Commission in the United States, in *Sec. 3.4.2*; the relatively quicker speed of corruption and fraud to occur, in the absence of any accounting reforms like "education" and "regulations" as described earlier, together with the institutional indulgence of "largest CPA firms" to spend "on advertisements" which "rose from US$4 million in the 1980s to more than $100 million in the 2000s," in *Sec. 3.5.1*; and the relatively quicker speed of communication among "investors, companies, auditors, and other participants" if helped by "comparability" in the convergence of accounting standards, in *Sec. 3.5.2*).

And the reverse direction also holds true.

Again, it should be stressed, however, that there are different shades of gray (or different degrees of truth) in the two opposites, and the classification is not necessarily mutually exclusive either. In fact, even when some combinations of the two occur, they only end up sharing the same dialectic relationship, but in a different degree.

Also, there are some other relationships (as a kind of family resemblance) which have something in common with the principle, but they are not exactly the same but only more or less comparable, which varies from case to case.

Good examples of family resemblance in relation to the principle are inconvenience-convenience, passiveness-activenessness, gradualness-abruptness, deceleration-acceleration, difficulty-ease, nonresilience-resilience, patience-impatience, deliberativeness-impulsiveness, and the like.

18th Thesis: The Expansion-Contraction Principle

The eighteenth thesis is called the expansion-contraction principle in the category of "process" in existential dialectics, in that entities in the world can both expand in some ways and contract in other ones, as part of their nature. There is no expansion without contraction—and vice versa.

This principle, although not so explicitly called, was already used in my previous works on different topics (e.g., the theory of floating consciousness in *FCD* and *FPHC*, the union of the unions in *BWT*, and the cyclical progression of system fragmentation and integration in *FCD*).

For instance, on the one hand, there is expansion (e.g., the relatively more developed ability to analysis "external influences" among "behavioral psychologists," in *Sec. 2.2.1*; the relative spread of "purchase accounting" nowadays, in *Sec. 2.2.2*; the relative spread of the influence of EMAS III nowadays, in *Sec. 2.3.1*; the relative spread of the "going-concern" in the accrual accounting system, in *Sec. 2.3.2*; the relative increase of independence if done by an "external auditor," in *Sec. 2.4.1*; the relative increase of "national government" revenue in India due to VAT, in *Sec. 2.4.2*; the relatively more developed ability to deal with "those transactions that can be expressed using monetary units" by those favoring "the principle of the unit of measure," in *Sec. 2.5.1*; the relative spread of "the International Financial Reporting Standards" nowadays, in *Sec. 2.5.2*; the relative spread of "control self-assessment" since 1987, in *Sec. 3.2.1*; the relative increased perception of money as "play money" when it is "money you can afford to lose," in *Sec. 3.2.2*; the relative spread of "sustainable" economic models nowadays, in *Sec. 3.3.1*; the relative spread of "environmental awareness" nowadays, in *Sec. 3.3.2*; the relative spread, nowadays, of high-level mechanisms like "metaprogramming," "convention over configuration," and "model-driven engineering" to reduce the need for boilerplate manual codes nowadays, in *Sec. 3.4.1*; "inflated" revenues in the accounting scandal involving Bristol-Myers Squibb in 2002, in *Sec. 3.4.2*; the relative expansion of "education" and "regulations" against corruption and fraud nowadays, in *Sec. 3.5.1*; and the relative spread of the convergence of accounting standards nowadays, in *Sec. 3.5.2*).

On the other hand, there is contraction (e.g., the relatively less developed ability to analyze "internal processes" among "behavioral psychologists," in *Sec. 2.2.1*; the relative decline of "pooling-of-interests accounting" nowadays, in *Sec. 2.2.2*; the relative decline of the influence of EMAS I nowadays, in *Sec. 2.3.1*; the relative decline of the "pre-going concern" in the accrual accounting system, in *Sec. 2.3.2*; the relative decrease of independence if done by "internal auditor," in *Sec. 2.4.1*; the relative decrease of "many sub-national government" revenues in India due to VAT, in *Sec. 2.4.2*; the relatively less developed ability to deal with "other transactions which cannot be so easily expressed using monetary units, and good examples include "carbon trading," the costs and benefits of "religious salvation," etc. among those favoring "the principle of the unit of measure," in *Sec. 2.5.1*; the relative decline of the US GAAP nowadays, as it is in the process of moving to IFRS, in *Sec. 2.5.2*; the relative decline of those "external auditors" nowadays who still want to "ignore...the benefits of control self-assessment," in *Sec. 3.2.1*; the relative decreased perception of money as "play money" when it is "safety capital," in *Sec. 3.2.2*; the relative decline of "growth-based" economic models nowadays, in *Sec. 3.3.1*; the relative decline of the practice of the "tragedy of the commons" nowadays, in *Sec. 3.3.2*; the relative decline, nowadays, of those types of computer proamming where "the programmer must write a lot of code to do minimal jobs," like the dependence on boilerplate manual codes, in *Sec. 3.4.1*; "understated" earnings in the accounting scandal involving Freddie Mac in 2002, in *Sec. 3.4.2*; the relative decline, nowadays, of the institutional indulgence of "largest CPA firms" to spend "on advertisements" which "rose from US$4 million in the 1980s to more than $100 million in the 2000s," in *Sec. 3.5.1*; and the relative decline, nowadays, of individual country accounting standards with no interaction with each other, in *Sec. 3.5.2*).

And the reverse direction also holds true.

Again, it should be stressed, however, that there are different shades of gray (or different degrees of truth) in the two opposites, and the classification is not necessarily mutually exclusive either. In fact, even when some combinations of the two occur, they only end up sharing the same dialectic relationship, but in a different degree.

Also, there are some other relationships (as a kind of family resemblance) which have something in common with the principle, but they are not exactly the same but only more or less comparable, which varies from case to case.

Good examples of family resemblance in relation to the principle are conquest-autarky, rise-fall, spread-shrink, extendingness-shorteningness, widening-narrowing, construction-deconstruction, expansion-destruction, construction-demolition, creation-destruction, expansion-contraction, expansion-constriction, expansion-decline, growth-breakdown, formation-decay, composition-decomposition, insertion-deletion, addition-division, multiplication-division, overstatedness-understatedness, multiplication-subtraction, and so forth.

19th Thesis: The Optimality-Nonoptimality Principle

The nineteenth thesis concerns the optimality-nonoptimality principle (on the nature of highly favorable options) in the category of "process" in existential dialectics, which was first proposed in *FPHMORPH*, in that both optimality and non-optimality are important, in relation to the mind, nature, culture, and society, albeit in different ways. There is no optimality without nonoptimality—and vice versa.

There are five clarifications to remember, as shown in the fields of morphology (e.g., "perfection" in *FPHMORP*), data analysis (e.g., "bounded rationality" in *FPHO*, *FFHDA* and *FPHPROP*), probability (e.g., "decision making" in *FPHO* and *FPHPROP*), economics (e.g., "negative externalities" in *FCD* and *BCPC*), and technology (e.g., "post-capitalism" in *FCD*), just to cite five well-documented sources:

- Optimality (in terms of highly favorable options) is not perfect and can be realized in different degrees.
- "Maximization" does not necessarily entail "optimization," as shown in the problem of "negative externalities" in economics.
- Non-optimality can come in all shapes and sizes, in different degrees, like "maximizing" in economics vs. "satisficing" in sociology.
- Optimality can be more feasible in the longer term of technological improvement.
- Optimality should not be confused with the beauty of simplicity, since the more optimal a process is, the more complicated (not necessarily the simpler) it can be.

In any event, on the one hand, there is optimality (e.g., the search for highly favorable options, like the quest for systematic "accounting reform" by the International Accounting Standards Board, in *Sec. 2.2.1*; the search for highly favorable options, like the quest for a comprehensive account for "reputation building," like "the stakeholder theory," which says "corporations should be run for the benefit of all 'stakeholders,' not just the shareholders," in *Sec. 2.2.2*; the search for highly favorable options, like the quest for the global eco-management with the introduction of EMAS III in 2010, in *Sec. 2.3.1*; the search for highly favorable options, like the quest for "certainty" in the accrual accounting system, when it "systematically exclude[s] uncertainties from the accrual process" and "the accrual accounting system deals with uncertain phenomena in a deterministic manner," in light of its assumptions on space and time aforediscussed, in *Sec. 2.3.2*; the search for highly favorable options, like the quest for the comprehensive "integrated audit," where "auditors, in addition to an opinion on the financial statements, must also express an opinion on the effectiveness of a company's internal control over financial reporting, in accordance with PCAOB Auditing Standard No. 5," in *Sec. 2.4.1*; the search for highly favorable options, like the quest for the more efficient way of implementing VAT, such that "the tax revenue is used for productive spending or has positive externalities'—in other words, governments may do more than simply consume the tax income," in *Sec. 2.4.2*; the search for highly favorable options, like the quest for the comprehensive reform of accounting principles, such as the proposal by Jeremy Nicholls of the SROI Network, in *Sec. 2.5.1*; the search for highly favorable options, like the quest for the specification of "constraints" in US GAAP to minimize costs and achieve better efficiency and effectiveness, in *Sec. 2.5.2*; the search for highly favorable options, like the quest for the comprehensive way "to reduce or eliminate the risk of error," as in "control self-assessment," in *Sec. 3.2.1*; the search for highly favorable options, like the quest for "the comprehensive account frame" as a way to deal with mental accounting problems, in *Sec. 3.2.2*; the search for highly favorable options, like the quest for sustainability measurement like the use of "Hubbert peak theory…in relation to peak oil production," in *Sec. 3.3.1*; the search for highly favorable options, like the quest for a comprehensive "environmental profit and loss" [E P&L] accounting which can provide "a company's monetary valuation and analysis of its environmental impacts including its business operations and its supply chain from cradle-to-gate," so as to give "a more holistic view of a company's performance," in *Sec. 3.3.2*; the search for highly favorable options, like the quest for the best "design patterns" which "can be used" as "any general, repeatable solution to a commonly occurring problem in software design," in *Sec. 3.4.1*; the search for highly favorable options, like the quest for the comprehensive three internal controls to "play..an important role in the prevention and detection of fraud," in *Sec. 3.4.2*; the search for highly favorable options, like the quest for the most comprehensive accounting reform against corruption and fraud "in July 2010 when President Obama signed 'The Dodd-Frank Wall Street Reform and Consumer Protection Act'" which "covers a broad range of changes," in *Sec. 3.5.1*; and the search for highly favorable options, like the quest for the most extensive "comparability" in the convergence of accounting standards by the FASB and IASB, in *Sec. 3.5.2*).

On the other hand, there is non-optimality (e.g., the existence of non-optimal alternatives to optimality, like the argument for more realistic studies by the critics, who pointed out different problems with "accounting reform" aforediscussed, in *Sec. 2.2.1*; the existence of non-optimal alternatives to optimality, like the argument for more realistic studies by the critics, who pointed out different problems with reputation capital and management aforediscussed, in *Sec. 2.2.2*; the existence of non-optimal alternatives to optimality, like the argument for more realistic studies by the critics, who pointed out different problems with EMAS aforecited, in *Sec. 2.3.1*; the existence of non-optimal alternatives to optimality, like the argument for more realistic studies by the critics like Takatera and Sawabe, who pointed out the problems with the "fictional" assumptions" on space and time in the accrual accounting system, in *Sec. 2.3.2*; the existence of non-optimal alternatives to optimality, like the argument for more realistic studies by the critics, who pointed out different problems with auditing aforecited, in *Sec. 2.4.1*; the existence of non-optimal alternatives to optimality, like the argument for more realistic studies by the critics, who pointed out different problems with VAT as discussed earlier, in *Sec. 2.4.2*; the existence of non-optimal alternatives to optimality, like the argument for more realistic studies by the critics, who pointed out different problems with accounting reform as viewed by different sides in the debate, in *Sec. 2.5.1*; the existence of non-optimal alternatives to optimality, like the argument for more realistic studies by the critics, who pointed out different problems with US GAAP aforediscussed, in *Sec. 2.5.2*; the existence of non-optimal alternatives to optimality, like the argument for more realistic studies by the critics, who pointed out different problems with "control self-assessment" aforecited, in *Sec. 3.2.1*; the existence of non-optimal alternatives to optimality, like the argument for more realistic studies by the critics, who pointed out different problems with any solutions to mental accounting, since "optimization is not feasible [too costly]," because "repairing one problem may create another," in *Sec. 3.2.2*; the existence of non-optimal alternatives to optimality, like the argument for more realistic studies by the critics, who pointed out different problems with sustainability measurement aforecited, in *Sec. 3.3.1*; the existence of non-optimal alternatives to optimality, like the argument for more realistic studies by the critics, who pointed out different problems with the E P & L accounting aforecited, in *Sec. 3.3.2*; the existence of non-optimal alternatives to optimality, like the argument for more realistic studies by the critics, who pointed out different problems with OOP and its "design patterns" aforedisucussed, in *Sec. 3.4.1*; the existence of non-optimal alternatives to optimality, like the argument for more realistic proposals by the critics, who pointed out different problems with internal control aforedescribed, in *Sec. 3.4.2*; the existence of non-optimal alternatives to optimality, like the argument for more realistic works by the critics, who pointed out different problems with accounting reforms aforecited, in *Sec. 3.5.1*; and the existence of non-optimal alternatives to optimality, like the argument for more realistic studies by the critics, who pointed out different problems with the convergence of accounting standards, in *Sec. 3.5.2*).

And the reverse direction also holds true.

Again, it should be stressed, however, that there are different shades of gray (or different degrees of truth) in the two opposites, and the classification is not necessarily mutually exclusive either. In fact, even when some combinations of the two occur, they only end up sharing the same dialectic relationship, but in a different degree.

Also, there are some other relationships (as a kind of family resemblance) which have something in common with the principle, but they are not exactly the same but only more or less comparable, which varies from case to case.

Good examples of family resemblance in relation to the principle are optimality-satisfaction, optimality-maximization, optimality-heuristicness, idealness-realisticness, and the like.

20th Thesis: The Simultaneity-Nonsimultaneity Principle

The twentieth thesis concerns the simultnaneity-nonsimultaneity principle (on the sequence of time) in the category of "process" in existential dialectics, which is first proposed in *FPHPHON*, in that both simultaneity and nonsimultaneity exist—in relation to the mind, nature, culture, and society, albeit in different ways. There is no simultaneity without nonsimultaneity—and vice versa.

There are the following important aspects of the relationship between simultaneity and nonsimultaneity to be clarified:

- An event B can occur after the occurrence of another event A in a causal situation. But A and B can also occur simultaneously in a different situation, and this is important, because, in quantum mechanics (as addressed in *FPHPT*), for instance, particles move so fast that, for all practical purposes, A and B can be treated as occurring simultaneously.
- In conventional wisdom, A and its negation ~A are different and do not coexist simultaneously. But in certain circumstances, it is also important to consider A and ~A simultaneously, as shown in the interactions between two opposites in the nature vs. nature debate (as addressed in *BNN*) and in the ritual/speech coevolution debate as well as the coarticulation debate (as addressed in *FPHPHON*).

- Simultaneity does not have to be absolute, as it can be more or less so for two entities to occur simultaneously.
- On one end of the line is exteme simultaneity and on the other end is extreme nonsimultaneously, and between the two are different degrees or gradations of simultaneity or nonsimultaneity.

In any event, on the one hand, there is simultaneity (e.g., the occurrence of different things more or less at the same time, such as the case about governments which "have the authority to determine what are accepted accounting principles, while using questionable accounting practices themselves," in *Sec. 2.2.1*; the occurrence of different things more or less at the same time, such as the simultaneous co-existence of both "cognitive" and "affective" elements of "reputation," in *Sec. 2.2.2*; the occurrence of different things more or less at the same time, such as the simultaneous co-existence of organizaitons which implement EMAS and those which do not, in *Sec. 2.3.1*; the occurrence of different things more or less at the same time, such as the simultaneous co-existence of two sub-systems of the accrual accounting system, that is, "external transaction" and "internal transaction," in *Sec. 2.3.2*; the occurrence of different things more or less at the same time, such as the simultaneous co-existence of different audits nowadays, like "financial audits," "information systems audits," etc., in *Sec. 2.4.1*; the occurrence of different things more or less at the same time, such as the simultaneous co-existence of VAT-free countries like the U.S. and countries with VAT like those in the EU, in *Sec. 2.4.2*; the occurrence of different things more or less at the same time, such as the simultaneous co-existence of the "Historical Cost model" and the "Constant Purchasing Power model" in our time, in *Sec. 2.5.1*; the occurrence of different things more or less at the same time, such as the simultaneous co-existence of US GAAP and IFRS nowadays, in *Sec. 2.5.2*; the occurrence of different things more or less at the same time, such as the simultaneous co-existence of different approaches nowadays, like "control self-assessment" and "auditing," in *Sec. 3.2.1*; the occurrence of different things more or less at the same time, such as the simultaneous co-existence of different cognitive biases in people with "mental accounting," in *Sec. 3.2.2*; the occurrence of different things more or less at the same time, such as the simultaneous co-existence of three different "dimensions" of sustainability aforecited, in *Sec. 3.3.1*; the occurrence of different things more or less at the same time, such as the more or less simultaneous announcements in November 2011, by PUMA in regard to its "overall E P&L" on "air pollution, land conversion and waste" and by the PPR Group in regard to "a Group E P&L" which "would be implemented across its Luxury and Sport & Lifestyle brands by 2015," in *Sec. 3.3.2*; the occurrence of different things more or less at the same time, such as the simultaneous co-existence of "state [data] and behavior [code]" when they are integrated by "an object-oriented system" into an "object," in *Sec. 3.4.1*; the occurrence of different things more or less at the same time, such as the simultaneous co-existence of different statutes nowadays, like "the Foreign Corrupt Practices Act [FCPA] of 1977" and "the Sarbanes–Oxley Act of 2002," in *Sec. 3.4.2*; the occurrence of different things more or less at the same time, such as the simultaneous co-existence of different accounting standards, like the U.S. GAAP and the London-based IFRS, in *Sec. 3.5.1*; and the occurrence of different things more or less at the same time, such as the simultaneous co-existence of different accounting standards nowadays, like the US GAAP and the IFRS, in *Sec. 3.5.2*).

On the other hand, there is nonsimultaneity (e.g., the occurrence of something after something else, such as the term "Creative Accounting" which was first used in 1968, before Abraham Briloff wrote against it in the 1970s, in *Sec. 2.2.1*; the occurrence of something after something else, such as the "causes" of reputation and its subsequent "consequences," in *Sec. 2.2.2*; the occurrence of something after something else, such as the development of EMAS III in 2010 after that of EMAS II in 2001, in *Sec. 2.3.1*; the occurrence of something after something else, such as the "pre-going concern" period before the "going concern" period, in *Sec. 2.3.2*; the occurrence of something after something else, such as the historical development of "financial audits" before that of "information systems audits" in our time, in *Sec. 2.4.1*; the occurrence of something after something else, such as the use of VAT in Australia in 2000, after Canada introduced its own version of VAT in 1991, in *Sec. 2.4.2*; the occurrence of something after something else, such as the creation of the "Constant Purchasing Power model" by the "International Accounting Standards Board" in 1989 after the establishment of the "Historical Cost model" several hundred years ago, in *Sec. 2.5.1*; the occurrence of something after something else, such as the creation of US GAAP prior to the creation of IFRS, in *Sec. 2.5.2*; the occurrence of something after something else, such as the creation of "control self-assessment" in 1987, long after the existence of "auditing," in *Sec. 3.2.1*; the occurrence of something after something else, such as the diagnosis of those with cognitive biases in "mental accounting" before it is recommended that they can receive treatment afterwards from a CBM [Cognitive Bias Modification] program, in *Sec. 3.2.2*; the occurrence of something after something else, such as the existence of the approach which is more "growth-based," before the existence of the new approach by "measuring environmental-economic-social interrelationships," in *Sec. 3.3.1*; the occurrence of something after something else, such as "the first-ever E P&L" which "was conducted on 2010 data and released in two phases," well before the decision by the PPR Group in regard to "a Group E P&L" which "would be implemented across its Luxury and Sport & Lifestyle brands by 2015," in *Sec. 3.3.2*; the occurrence of something after something else, such as the development of "object-oriented programming…at MIT in the late 1950s and early 1960s," before it was recently "refined," like "the use of design patterns, design by contract, and modeling languages," in *Sec. 3.4.1*; the occurrence of something after something else, such as the passing of the "Foreign Corrupt Practices Act [FCPA] of 1977" before the passing of "the Sarbanes–Oxley Act of 2002," in *Sec. 3.4.2*; the occurrence of something after something else, such as the decision in 2008 by the Securities and Exchange Commission in the U.S. to "issue…a preliminary 'roadmap' that may lead the U.S. to abandon GAAP in the future…and to join…the London-based IFRS," after "more than 100 countries around the world" had already joined, in *Sec. 3.5.1*; and the occurrence of something after something else, such as the effort for the convergence of accounting standards by the FASB and IASB, after the existence of the US GAAP and the IFRS, in *Sec. 3.5.2*).

And the reverse direction also holds true.

Again, it should be stressed, however, that there are different shades of gray (or different degrees of truth) in the two opposites, and the classification is not necessarily mutually exclusive either. In fact, even when some combinations of the two occur, they only end up sharing the same dialectic relationship, but in a different degree.

Also, there are some other relationships (as a kind of family resemblance) which have something in common with the principle, but they are not exactly the same but only more or less comparable, which varies from case to case.

Good examples of family resemblance in relation to the principle are contemporaneity-successiveness, simultaneity-sequence, simultaneity-consecutiveness, and the like.

21st Thesis: The Isolation-Interaction Principle

The twenty-first thesis concerns the isolation-interaction principle (on the interaction among entities) in the category of "process" in existential dialectics, which was first proposed in *FPHACCO*, in that both isolation and interaction are important, in relation to the mind, nature, culture, and society, albeit in different ways. There is no isolation without interaction—and vice versa.

There are some clarifications to remember, as shown in different case studies already addressed in my previous works (e.g., "interlanguage" in *FPHPHON,* "system integration and fragmentation" in *FCD*, "sound change" in *FPHPHON,* and "mini-systems" in *FHC*), just to cite a few well-documented sources:

- Isolation and interactions are not perfect and can be realized in different degrees.
- Isolation and interaction take time to develop.
- Interaction can have different outcomes over time (e.g., "merging," "split," "loss," etc. in sound change).
- The isolation-interaction pair should not be confused with the absoluteness-relativeness pair, although both concern the issue of autonomy, but there are different opposites in relation to autonomy (in the contexts of method, process, etc.).

In any event, on the one hand, there is isolation (e.g., the state of a situation that is separate from others, like a creative accountant before being audited by an auditor, in *Sec. 2.2.1*; the state of a situation that is separate from others, such as "the good reputation of a company and its existings products" before any "reputation transfer" to "new markets and new products," in *Sec. 2.2.2*; the state of a situation that is separate from others, such as the separation of organizations in the non-industrial sector from the jurisdiction of EMAS in 1995, in *Sec. 2.3.1*; the state of a situation that is separate from others, such as the isolation or independence of the "demarcated space" in the "going concern" period, away from "external transactions," in *Sec. 2.3.2*; the state of a situation that is separate from others, such as the formal separation of the role of "internal auditing" from "corporate governance" in the older days, when "internal auditing activity as it relates to corporate governance has in the past been generally informal, accomplished," in *Sec. 2.4.1*; the state of a situation that is separate from others, such as the immunity or separation of goods and services from any VAT in Australia prior to 2000, in *Sec. 2.4.2*; the state of a situation that is separate from others, such as the isolation of the "Historical Cost model" from any accounting reform, prior to the creation of the "Constant Purchasing Power model" by the "International Accounting Standards Board," in *Sec. 2.5.1*; the state of a situation that is separate from others, such as the separation of US GAAP from IFRS prior to 2008, in *Sec. 2.5.2*; the state of a situation that is separate from others, such as the isolation of "auditing" from "control self-assessment" in the older days, when "external auditors ignored the benefits of control self-assessment," in *Sec. 3.2.1*; the state of a situation that is separate from others, such as the isolated mindset of someone with mental accounting, when he treats tax refund as "a sort of windfall whose eventual disposal means nothing to the broader financial picture" of his life, in *Sec. 3.2.2*; the state of a situation that is separate from others, such as the separation of "environmental" and "social" factors from the study of "economic factors" in the older days, before the construction of carbon accounting and sustainability accounting, in *Sec. 3.3.1*; the state of a situation that is separate from others, such as the "tragedy of the commons" in the older days, without "environmental awareness," in *Sec. 3.3.2*; the state of a situation that is separate from others, such as the separate construction of computer programs as "code" and "data" in the older days, prior to the emergence of OOP, in *Sec. 3.4.1*; the state of a situation that is separate from others, such as the separation of internal control in companies from any legal requirement in 18th century America, in *Sec. 3.4.2*; the state of a situation that is separate from others, such as the isolation of US GAAP from the London-based International Financial Reporting Standards in the older days, in *Sec. 3.5.1*; and the state of a situation that is separate from others, such as the isolated existence of the US GAAP in the older days without any interaction with the IFRS, in *Sec. 3.5.2*).

On the other hand, there is interaction (e.g., the influence of entities on one another, like the interaction between a creative accountant and an auditor in the process of an audit, in *Sec. 2.2.1*; the influence of entities on one another, such as the "reputation transfer" of "the good reputation of a company and its existings products" to "new markets and new products" elsewhere, in *Sec. 2.2.2*; the influence of entities on one another, such as the diverse relationships between different organizations in the non-industrial sector and EMAS in 2010, in *Sec. 2.3.1*; the influence of entities on one another, such as the close connection between the function and scope of accounts and external transactions in the "pre-going concern" period, in *Sec. 2.3.2*; the influence of entities on one another, such as the more "formal" interaction between "internal auditing" and "corporate governance" nowadays, since "the IIA has advocated more formal evaluation of Corporate governance, particularly in the areas of board oversight of enterprise risk, corporate ethics, and fraud," in *Sec. 2.4.1*; the influence of entities on one another, such as the influence of VAT on goods and services in Australia after 2000, in *Sec. 2.4.2*; the influence of entities on one another, such as the influence of the "Constant Purchasing

Power model" on accounting, after it was proposed by the "International Accounting Standards Board," in *Sec. 2.5.1*; the influence of entities on one another, such as the impact of IFRS on US GAAP after 2008, which aims to move to IFRS by 2014 or later in *Sec. 2.5.2*; the influence of entities on one another, such as the integration of "control self-assessment" into "auditing" nowadays, in *Sec. 3.2.1*; the influence of entities on one another, such as the situation of someone with mental accounting under "a CBM [Cognitive Bias Modification] program," when he learns how to integrate tax refund with "the broader financial picture" of his life, in *Sec. 3.2.2*; the influence of entities on one another, such as the inclusion of "environmental" and "social" factors into the study of "economic factors" in the new approach by "measuring environmental-economic-social interrelationships," in *Sec. 3.3.1*; the influence of entities on one another, such as the "environmental awareness" nowadays, which includes or considers "the ecosystem services a business depends on to operate in addition to the cost of direct and indirect negative impacts on the environment," in *Sec. 3.3.2*; the influence of entities on one another, such as the integration in "an object-oriented system" of "the two ['code' and 'data'] using the concept of an 'object,'" so that "an object has state [data] and behavior [code]," in *Sec. 3.4.1*; the influence of entities on one another, such as the legal requirement, under the Sarbanes-Oxley Act of 2002, of companies "to perform a fraud risk assessment and assess related controls" in the U.S., in *Sec. 3.4.2*; the influence of entities on one another, such as the steady integration of US GAAP into the London-based International Financial Reporting Standards nowadays, after the decision in 2008 by the Securities and Exchange Commission in the U.S. to "issue…a preliminary 'roadmap' that may lead the U.S." to eventually join the international standards, in *Sec. 3.5.1*; and the influence of entities on one another, such as the effort to integrate the US GAAP into the IFRS, as part of the project for the convergence of accounting standards by the FASB and IASB, in *Sec. 3.5.2*).

And the reverse direction also holds true.

Again, it should be stressed, however, that there are different shades of gray (or different degrees of truth) in the two opposites, and the classification is not necessarily mutually exclusive either. In fact, even when some combinations of the two occur, they only end up sharing the same dialectic relationship, but in a different degree.

Also, there are some other relationships (as a kind of family resemblance) which have something in common with the principle, but they are not exactly the same but only more or less comparable, which varies from case to case.

Good examples of family resemblance in relation to the principle are isolation-exchange, isolation-intercourse, autonomy-trade, aloneness-interrelationship, autarky-commerce, insulation-interaction, secludedness-interrelationship, segregation-interaction, and the like.

In Relation to Agency

The next eight theses are in relation to agency in existential dialectics, as shown below.

22nd Thesis: The Theory-Praxis Principle

The twenty-second thesis is the theory-praxis principle (on the duality of knowledge) in the category of "agency" in existential dialectics (which was first worked out in *FPHE*), in that, if there is theoretical construction, there is likewise its practical application, both technical and normative. There is no theory without praxis—and vice versa.

For instance, on the one hand, there is theory (e.g., the theoretical construction of professional ethics against creative accounting by Abraham Briloff, in *Sec. 2.2.1*; the theoretical construction of the 3 levels of "reputation-based decisions" in the field of psychology, in *Sec. 2.2.2*; the theoretical construction of "voluntary environmental management instrument" in the interdisciplinary field of environmental studies and accounting, in *Sec. 2.3.1*; the theoretical construction of "chaos theory" in mathematics, in *Sec. 2.3.2*; the theoretical construction of "internal auditing" by Lawrence Sawyer, in *Sec. 2.4.1*; the theoretical construction of "VAT" by the German industrialist Dr. Wilhelm von Siemens in 1918, in *Sec. 2.4.2*; the theoretical construction of "purchasing power" in the field of economics, in *Sec. 2.5.1*; the theoretical construction of "a common global language for business affairs" like "IFRS" in the field of finance and accounting, in *Sec. 2.5.2*; the theoretical construction of "a framework to support the analysis and evaluation of control processes by operational staff" by Gulf Canada in the 1980s, in *Sec. 3.2.1*; the theoretical construction of "cognitive bias" in the field of psychology, in *Sec. 3.2.2*; the theoretical construction of the "pluralised concept of knowledge" in the field of philosophy, in *Sec. 3.3.1*; the theoretical construction of "ecosystem" in the field of ecology, in *Sec. 3.3.2*; the theoretical construction of "inheritance" in social sciences, in *Sec. 3.4.1*; the theoretical construction of "internal control procedures" in the field of accounting, in *Sec. 3.4.2*; the theoretical construction of "ethical principles" in the field of ethics, in *Sec. 3.5.1*; and the theoretical construction of "fair value" in the field of economics, in *Sec. 3.5.2*).

On the other hand, there is praxis (e.g., what Abraham Briloff "advocated has been forced on the industry in the wake of the Enron scandal," in *Sec. 2.2.1*; the practical application of the 3 levels of "reputation-based decisions" in the field of psychology to the field of accounting, like "reputation management," in *Sec. 2.2.2*; the practical application of "voluntary environmental management instrument" in the interdisciplinary field of environmental studies and accounting to the political realm, like the use of EMAS by the European Commission, in *Sec. 2.3.1*; the practical application of chaos theory" in mathematics to the field of accounting, as shown in the work by Takatera and Sawabe, in *Sec. 2.3.2*; the practical application of "internal auditing" by "the IIA," which, in 2010, "began advocating for the broader role internal auditing should play in the corporate arena, in keeping with the IPPF's philosophy" envisioned by Sawyer, in *Sec. 2.4.1*; the practical application of "VAT" to the EU countries, which implemented it in the 2000s, in *Sec. 2.4.2*; the practical application of "purchasing power" in the field of economics to the field of accounting, like the "Constant Purchasing Power model" by "International Accounting Standards Board," in *Sec. 2.5.1*; the practical application of "a common global language for business affairs" like "IFRS" in the field of finance and accounting to countries around the world, since, "as of August 2008, more than 113 countries around the world" use it, in *Sec. 2.5.2*; the practical application of "a framework to support the analysis and evaluation of control processes by operational staff" to the public sphere like its application by "the Federal Deposit Insurance Corporation and the Canadian Deposit Insurance Corporation," in *Sec. 3.2.1*; the practical application of the idea of "cognitive bias" in the field of psychology to the field of accouunting, like the work on "mental accounts," in *Sec. 3.2.2*; the practical application of the "pluralised concept of knowledge" in the field of philosophy to the field of accounting, like the view that "knowledges…can well be used to understand better the status of knowledge produced by carbon accounting: carbon accounting produced a version of understanding of carbon emissions. Other carbon accountants would produce other results," in *Sec. 3.3.1*; the practical application of "ecosystem" in the field of ecology to the field of accounting, like the E P & L accounting, in *Sec. 3.3.2*; the practical application of "inheritance" in social sciences to the field of computing, like "object inheritance" in OOP, in *Sec. 3.4.1*; the practical application of "internal control procedures" in the field of accounting to the public sphere, like the adoption of "internal control procedures" under "the

Foreign Corrupt Practices Act (FCPA) of 1977 and the Sarbanes–Oxley Act of 2002" in the U.S., in *Sec. 3.4.2*; the practical application of "ethical principles" in the field of ethics to the field of accounting, like "accounting ethics," in *Sec. 3.5.1*; and the practical application of "fair value" in the field of economics to the realm of the convergence of accounting standards by the FASB and IASB, as shown in their "fair value option" project in 2007, in *Sec. 3.5.2*).

And the reverse direction also holds true.

Again, it should be stressed, however, that there are different shades of gray (or different degrees of truth) in the two opposites, and the classification is not necessarily mutually exclusive either. In fact, even when some combinations of the two occur, they only end up sharing the same dialectic relationship, but in a different degree.

Also, there are some other relationships (as a kind of family resemblance) which have something in common with the principle, but they are not exactly the same but only more or less comparable, which varies from case to case.

Good examples of family resemblance in relation to the principle are discovery-application, knowledge-action, invention-innovation, theory-intentionality, discourse-action, knowledge-interest, knowledge-motivation, universality-particularity, potentiality-actuality, theory-practice, and so forth.

23rd Thesis: The Convention-Novelty Principle

The twenty-third thesis is the convention-novelty principle (on the nature of creative thinking) in the category of "agency" in existential dialectics (which was first worked out in *FPHCT*), in that, if there is conventional wisdom, there is likewise novel challenge, to the extent that both convergent and divergent thinking are part of life. There is no convention without novelty—and vice versa.

As srummarized from *FPHCT* (almost verbatim here), there are (a) "creative techniques" and (b) "creative traits," which, when satisfied—in relation to the larger context of the mind, nature, society, and culture—can be used to enhance creative works.

In addition, creative thinking has its own possibilities and limits (in relation to invention), just as it has its own promises and pitfalls (in relation to innovation)— as already analyzed in *FPHCT*).

In the end, creative thinking has its own desirability and dark sides (as also already analyzed in *FPHCT*).

With these clarifications in mind—there are good empirical examples for the convention-novelty princple.

For instance, on the one hand, there is convention (e.g., the conventional wisdom about the role of governments "to determine what are accepted accounting principles," in *Sec. 2.2.1*; the conventional wisdom about "reputation capital and management," in *Sec. 2.2.2*; the conventional wisdom about "the value of large, highly detailed and prescriptive protocols" as an audit tool for "easier review," in *Sec. 2.3.1*; the conventional wisdom about "the status of an organization as an independent economic entity" in the "accrual income accounting," in *Sec. 2.3.2*; the conventional wisdom that, "due to constraints, an audit seeks to provide only reasonable assurance that the statements are free from material error," in *Sec. 2.4.1*; the conventional wisdom about using "VAT" to "avoid…the cascade effect of sales tax," in *Sec. 2.4.2*; the conventional wisdom about the 3 accounting principles, in *Sec. 2.5.1*; the conventional wisdom about the need for US GAAP since 1939, in *Sec. 2.5.2*; the conventional wisdom about "external auditing" for risk management prior to the 1980s, in *Sec. 3.2.1*; the conventional wisdom about "mental accounting" as undesirable, as it often leads to cognitve biases, in *Sec. 3.2.2*; the conventional wisdom about the approach to accounting which is more "growth-based," in *Sec. 3.3.1*; the conventional wisdom about the "tragedy of the commons" in the older days, in *Sec. 3.3.2*; the conventional wisdom about the separate construction of computer programs as "code" and "data" in the older days, in *Sec. 3.4.1*; the conventional wisdom about internal control to play "an important role in the prevention and detection of fraud," in *Sec. 3.4.2*; the conventional wisdom about US GAAP as the accounting standards to be used in the U.S., which are "ruled-based," in *Sec. 3.5.1*; and the conventional wisdom about the need for the convergence of accounting standards by the FASB and IASB, in *Sec. 3.5.2*).

On the other hand, there is novelty (e.g., the alternative novel challenge to the conventional wisdom about the role of governments "to determine what are accepted accounting principles"—by the new idea that "governments "us[e] questionable accounting practices themselves," in *Sec. 2.2.1*; the alternative novel challenge to the conventional wisdom about "reputation capital and management"—by the new idea that the obsession with reputation can be a psychological problem instead, "exaggerated in importance due to the fragile nature of the human ego," in *Sec. 2.2.2*; the alternative novel challenge to the conventional wisdom about "the value of large, highly detailed and prescriptive protocols" as an audit tool" for "easier review"—by the alternative idea of a "more flexible protocols that rely on the expertise and knowledge of experienced auditors and source documents [regulations, permits, etc.] directly" for technical review, in *Sec. 2.3.1*; the alternative novel challenge to the conventional wisdom about "the status of an organization as an independent economic entity" in the "accrual income accounting"—by the new idea that the independent status is fictional on the basis of its assumptions on space and time, as pointed out by Takatera and Sawabe, in *Sec. 2.3.2*; the alternative novel challenge to the conventional wisdom that, "due to constraints, an audit seeks to provide only reasonable assurance that the statements are free from material error"—by the new idea that the standard is no longer deemed sufficient for some people, since "recently, the argument that auditing should go beyond just true and fair is gaining momentum," in *Sec. 2.4.1*; the alternative novel challenge to the conventional wisdom about using "VAT" to "avoid...the cascade effect of sales tax"—by the new idea that the implementation of VAT can lead to "carousel fraud," in *Sec. 2.4.2*; the alternative novel challenge to the conventional wisdom about the 3 accounting principles—by the new idea that they are out-dated and need to be reformed, as discussed earlier, in *Sec. 2.5.1*; the alternative novel challenge to the conventional wisdom about the need for US GAAP since 1939 —by the new idea in the late 2000s about the U.S. to abandon US GAAP and move to IFRS by 2014 or later, in *Sec. 2.5.2*; the alternative novel challenge to the conventional wisdom about "external auditing" for risk management prior to the 1980s—by the new idea of using "control self-assessment" in 1987, in *Sec. 3.2.1*; the alternative novel challenge to the conventional wisdom about "mental accounting" as undesirable—by the new idea of the critics that it does not give sufficient attention to the "good" side of mental accounting [but its "bad" side], etc., in *Sec. 3.2.2*; the alternative novel challenge to the conventional wisdom about the approach to accounting which is more "growth-based"—by the new approach on the basis of "measuring environmental-economic-social interrelationships," in *Sec. 3.3.1*; the alternative novel challenge to the conventional wisdom about the "tragedy of the commons" in the older days—by the new idea of the "environmental awareness" nowadays, which includes "the ecosystem services a business depends on to operate in addition to the cost of direct and indirect negative impacts on the environment," in *Sec. 3.3.2*; the alternative novel challenge to the conventional wisdom about the separate construction of computer programs as "code" and "data" in the older days—by the new idea of integrating in "an object-oriented system" "the two ['code' and 'data'] using the concept of an 'object,'" so that "an object has state [data] and behavior [code]," in *Sec. 3.4.1*; the alternative novel challenge to the conventional wisdom about internal control to play "an important role in the prevention and detection of fraud"—by the new idea that internal control does not provide "absolute" assurance that "the objectives of an organization will be met" and also "has no control over external factors which can affect the performance of an organization," in *Sec. 3.4.2*; the alternative novel challenge to the conventional wisdom about US GAAP as the accounting standards to be used in the U.S., which are "ruled-based"—by the new idea of the Securities and Exchange Commission in the U.S. to "issue...a preliminary 'roadmap'" in 2008 "that may lead the U.S." to eventually join the international standards IFRS which are "principle-based," in *Sec. 3.5.1*; and the alternative novel challenge to the conventional wisdom about the need for the convergence of accounting standards by the FASB and IASB—by the new idea of Shyam Sunder, who "has called the link between convergence and comparability 'overblown'" in his questioning of the project for the convergence of accounting standards, in *Sec. 3.5.2*).

And the reverse direction also holds true.

Again, it should be stressed, however, that there are different shades of gray (or different degrees of truth) in the two opposites, and the classification is not necessarily mutually exclusive either. In fact, even when some combinations of the two occur, they only end up sharing the same dialectic relationship, but in a different degree.

Also, there are some other relationships (as a kind of family resemblance) which have something in common with the principle, but they are not exactly the same but only more or less comparable, which varies from case to case.

Good examples of family resemblance in relation to the principle are convergence-divergence (or convergent thinking vs. divergent thinking), normalness-nonnormalness, conformity-nonconformity, familiarity-freshness, inertia-newness, equilibrium-disturbance, and so on.

24th Thesis: The Evolution-Transformation Principle

The twenty-fourth thesis refers to the evolution-transformation principle (on the multiple kinds of agency). in the category of "agency" in existential dialectics (which was first worked out in *FAE* and then in other books of mine).

This principle—and the symmetry-asymmetry principle, for instance—are both about agency. More precisely, the word "agency," in a formal definition, refers to "a person or thing through which power is exerted or an end is achieved." (MWD 2007) It therefore does not have to necessarily involve an intelligent lifeform.

Because of this dual meaning in agency, the evolution-transformation principle is more concerned with *the kind of agency*, that is, both about the *evolution* in the state of nature (e.g., an object of natural beauty) and the *transformation* in the world of intelligent lifeforms (e.g., a work of art). There is no evolution without transformation—and vice versa.

In classical Darwinian evolutionary theory (as more detailedly analyzed in *BNN*), evolution is "blind." But in the human world, change often takes place because of the conscious intervention of humans in transforming society and culture, just to cite two instances.

And the transformative part of the principle precisely refers to the other dimension in the dual meaning of agency, in giving technology (as invented by intelligent lifeforms like humans and, soon, post-humans) a major role to play in the change of the world, which is something that I extensively analyzed in *FHC* in the context of the technophilic lifeworld, espeically though not exclusively since modern times. The same theme is also elaborated in many other books of mine.

For instance, on the one hand, there is evolution (e.g., the natural evolution of humans to count things in the state of nature in a primitive way, in Sec. *2.2.1*; the natural evolution of humans to hold certain opinions about others in the state of nature, in Sec. *2.2.2*; the natural evolution of humans to deal with the environment in the state of nature in a primitive way, in Sec. *2.3.1*; the natural evolution of humans to experience space and time in the state of nature, in Sec. *2.3.2*; the natural evolution of humans to "see" or "hear" things in the state of nature in a primitive way, in Sec. *2.4.1*; the natural evolution of humans to consume things in the state of nature in a primitive way, in Sec. *2.4.2*; the natural evolution of humans to measure things in the state of nature in a primitive way, in Sec. *2.5.1*; the natural evolution of humans to exchange things with each other in the state of nature in a primitive way, in Sec. *2.5.2*; the natural evolution of humans to commit errors in the state of nature, in Sec. *3.2.1*; the natural evolution of humans to experience illusions in the state of nature, in Sec. *3.2.2*; the natural evolution of humans to encounter the "natural occurrences" of "carbon dioxide," in Sec. *3.3.1*; the natural evolution of humans to encounter "ecosystem services" in the state of nature in a primitive way, in Sec. *3.3.2*; the natural evolution of humans to encounter "objects" in the state of nature, in Sec. *3.4.1*; the natural evolution of humans to make mistakes, either intentionally or unintentionally, in the state of nature, in Sec. *3.4.2*; the natural evolution of humans to behave badly sometimes in the state of nature, in Sec. *3.5.1*; and the natural evolution of humans to count and report things in the state of nature in a primitive way in Sec. *3.5.2*).

On the other hand, there is transformation (e.g., the technical transformation of human ability to count by the invention of "creative accounting," in Sec. *2.2.1*; the technical transformation of human ability to to hold certain opinions about others by the invention of "reputation management," in Sec. *2.2.2*; the technical transformation of human ability to deal with the environment by the invention of EMAS, in Sec. *2.3.1*; the technical transformation of human ability to experience space and time by the invention of different assumptions on space and time in the modern accrual accounting system, in Sec. *2.3.2*; the technical transformation of human ability to "see" or "hear" things by the invention of "internal auditing," with the word "audit" from "Latin word 'audire' which means 'to hear,'" in Sec. *2.4.1*; the technical transformation of human ability to consume by the invention of a consumption tax known as "VAT," in Sec. *2.4.2*; the technical transformation of human ability to measure things by the invention of accounting principles like "the unit of measure," in Sec. *2.5.1*; the technical transformation of human ability to exchange things with each other by the invention of US GAAP for a "common…language for business affairs" in the U.S., in Sec. *2.5.2*; the technical transformation of human ability to cope with errors by the invention of "control self-assessment." in Sec. *3.2.1*; the technical transformation of human ability to cope with illusions by the invention of "CBM [Cognitive Bias Modification] program," in Sec. *3.2.2*; the technical transformation of human ability to deal with "carbon dioxide" by the invention of "carbon accounting," in Sec. *3.3.1*; the technical transformation of human ability to deal with "ecosystem services" by the invention of the E P & L accounting, in Sec. *3.3.2*; the technical transformation of human ability to deal with "objects" by the invention of OOP, in Sec. *3.4.1*; the technical transformation of human ability to deal with making mistakes, either intentionally or unintentionally, by the invention of internal control, in Sec. *3.4.2*; the technical transformation of human ability to deal with bad behaviors by the invention of "accounting ethics," in Sec. *3.5.1*; and the technical transformation of human ability to deal with counting and reporting by the invention of "the convergence of accounting standards" for "comparability" nowadays, in Sec. *3.5.2*).

And the reverse direction also holds true.

Again, it should be stressed, however, that there are different shades of gray (or different degrees of truth) in the two opposites, and the classification is not necessarily mutually exclusive either. In fact, even when some combinations of the two occur, they only end up sharing the same dialectic relationship, but in a different degree.

Also, there are some other relationships (as a kind of family resemblance) which have something in common with the principle, but they are not exactly the same but only more or less comparable, which varies from case to case.

Good examples of family resemblance in relation to the principle are inorganicness-volition, naturalness-technologicalness, naturalness-nonnaturalness, nonwillingness-willingness, inorganicness-motivation, evolution-creation, naturalness-artificialness, and so on.

25th Thesis: The Symmetry-Asymmetry Principle

The twenty-fifth thesis is labeled as the symmetry-asymmetry principle (on the relationships among existents) in the category of "agency" in existential dialectics (which was also already worked out in my previous books), in that there is no asymmetry without symmetry—and vice versa.

For instance, the Same can be symmetric and asymmetric towards the Others. But in case of asymmetry (as analyzed in *BDPD*), oppression and self-oppression can occur. So, when the Same is asymmetric towards the Others, the Same can also be relatively asymmetric towards itself in self-oppression, just as the Others can be likewise towards themselves. The subsequent oppressiveness is dualistic, as much by the Same against the Others and itself, as by the Others against the Same and themselves. Both oppression and self-oppression can be achieved by way of downgrading differences between the Same and the Others and of accentuating them.

This is true, even though not all forms of asymmetry have to be about oppression and self-oppression.

In addition, from Chapter Three of *FPHG*, symmetry is not perfect, to be understood in an approximate sense under many life circumstances. With this clarification in mind, hereafter are some empirical examples.

For instance, on the one hand, there is symmetry (e.g., the co-existence of governments and ordinary citizens, in *Sec. 2.2.1*; the co-existence of different methods in the field of accounting, like "purchase accounting" and "pooling-of-interests accounting," in *Sec. 2.2.2*; the co-existence of different entities, like the European Commission, companies, etc., in *Sec. 2.3.1*; the co-existence of "certain" and "uncertain" phenomena in the world, in *Sec. 2.3.2*; the co-existence of different auditors, like "internal auditors," "external auditors," etc., in *Sec. 2.4.1*; the co-existence of different entites like the government and the consumers, in *Sec. 2.4.2*; the co-existence of different accounting principles, like the "Constant Purchasing Power model" and the

"Historical Cost model," in *Sec. 2.5.1*; the co-existence of different organizations like "the American Institute of Certified Public Accountants [AICPA]" and its members like individual business enterprises, in *Sec. 2.5.2*; the co-existence of different entities like governments and individual businesses, in *Sec. 3.2.1*; the co-existence of both "good" and "bad" sides of mental accounting, in *Sec. 3.2.2*; the co-existence of different entities, like companies and governments, in *Sec. 3.3.1*; the co-existence of different entities like animals and humans, in *Sec. 3.3.2*; the co-existence of different features of OOP, like "Inheritance, Object, Class, Encapsulation, Method, Message Passing, Polymorphism, Abstraction," etc., in *Sec. 3.4.1*; the co-existence of different entities like government oversight agencies and individual companies, in *Sec. 3.4.2*; the co-existence of different entities like the Italian court system and individual companies like Parmalat, in *Sec. 3.5.1*; and the co-existence of different entities like the standard-setters FASB and IASB, and individual companies, in *Sec. 3.5.2*).

On the other hand, there is asymmetry (e.g., governments "have the authority to determine what are accepted accounting principles"—but ordinary citizens do not have this authority, in *Sec. 2.2.1*; in practice, "since 2001, U.S. Generally Accepted Accounting Principles [FAS 141] no longer allows the pooling-of-interests method," but the "purchase accounting" is allowed, in *Sec. 2.2.2*; the European Commission has the authority to set up EMAS for companies to comply—but companies do not have the authority to set up EMAS for the European Commission to comply, in *Sec. 2.3.1*; the accrual accounting system favors "certainty" and "systematically exclude[s] uncertainties from the accrual process," so "the accrual accounting system deals with uncertain phenomena in a deterministic manner," in a misleadingly "fictional" way, according to Takatera and Sawabe, in *Sec. 2.3.2*; "external auditors" have more "independence" or the power of distancing than "internal auditors," in *Sec. 2.4.1*; the government can tax the consumers with VAT—but the consumers cannot tax the government, in *Sec. 2.4.2*; the "Historical Cost model" is "still used in most accounting systems"—but the "Constant Purchasing Power model" is not widely used yet, in *Sec. 2.5.1*; AICPA issues codes of professional ethics for individual business enterprises to follow—but individual business enterprises do not have the power to issue codes of professional ethics for AICPA to follow, in *Sec. 2.5.2*; "the United Kingdom government" in 1992 "required that the directors of a company should report on the effectiveness of the company's system of internal control in each annual report" —but individual companies do not have the authority to require the UK government how or what to do about its annual reports, in *Sec. 3.2.1*; in mainstream literature, "a question that has not received much attention is whether mental accounting is good for us"—but the literature has received much attention in regard to the "bad" side of mental accounting like cognitive biases which led to irresponsible purchases, in *Sec. 3.2.2*; "in June 2012, the UK coalition government announced the introduction of mandatory carbon reporting, requiring around 1,100 of the UK's largest listed companies to report their greenhouse gas emissions every year"—but these companies do not have the authority to require the British government to report its greenhouse gas emissions to them, in *Sec. 3.3.1*; humans have damaged the environment in a much more severe way than animals, since "it has been reported that approximately 40-50% of Earth's ice-free land surface has been heavily transformed or degraded by anthropogenic activities, 66% of marine fisheries are either overexploited or at their limit, atmospheric CO_2 has increased more than 30% since the advent of industrialization, and nearly 25% of Earth's bird species have gone extinct in the last two thousand years," in *Sec. 3.3.2*; according to Deborah Armstrong, in descending order of popularity, the features of OOP are: "Inheritance, Object, Class, Encapsulation, Method, Message Passing, Polymorphism, Abstraction" in *Sec. 3.4.1*; government oversight agencies like the Securities and Exchange Commission in the United States have the authority to prosecute individual companies for corporate misdeeds—but individual companies do not have the authority to prosecute government oversight agencies for their own misdeeds, in *Sec. 3.4.2*; the Italian court system sentenced the CEO of Parmalat, Calisto Tanzi, to "10 years in prison for fraud relating to the collapse of the dairy group"—but Tanzi cannot sentence the Italian court system to "10 years in prison," in *Sec. 3.5.1*; and the standard-setters like the FASB and the IASB can set up standards for the convergence of accounting standards for individual companies—but individual companies cannot set up standards for the FASB and the IASB to follow, in *Sec. 3.5.2*).

And the reverse direction also holds true.

Again, for the last time, it should be stressed, however, that there are different shades of gray (or different degrees of truth) in the two opposites, and the classification is not necessarily mutually exclusive either. In fact, even when some combinations of the two occur, they only end up sharing the same dialectic relationship, but in a different degree.

Also, there are some other relationships (as a kind of family resemblance) which have something in common with the principle, but they are not exactly the same but only more or less comparable, which varies from case to case.

Good examples of family resemblance in relation to the principle are balance-extremity, reflexiveness-unreflexiveness, equality-inequality, harmoniousness-unharrmoniousness, center-periphery, compatibility-incompatibility, proportion-disproportion, balance-unbalance, and something like that.

26th Thesis: The Softness-Hardness Principle

The twenty-sixth thesis refers to the softness-hardness principle (on the force of change) in the category of "agency" in existential dialectics (which was first worked out in *ALD*), in that any change by an agent, be it organic (like humans) or non-organic (like natural objects), can occur in a forceful (aggressive) or gentle (pacific) way, which can come in all shapes and sizes, of course. There is no softness without hardness—and vice versa.

For instance, on the one hand, there is softness (e.g., the defense of the application of "cognitive psychology" to accounting by the advocates aforecited, in *Sec. 2.2.1*; the defense of "reputation management" by the advocates aforecited, in *Sec. 2.2.2*; the defense of EMAS by the advocates aforecited, in *Sec. 2.3.1*; the defense of "the status of an organization as an independent economic entity" by those for the "accrual accounting system," in *Sec. 2.3.2*; the defense of "internal auditing" by the advocates like Sawyer aforecited, in *Sec. 2.4.1*; defense of VAT by the advocates like Wilhelm von Siemens, in *Sec. 2.4.2*; the defense of the 3 accounting principles by the advocates aforecited, in *Sec. 2.5.1*; the defense of US GAAP by the advocates like "the Governmental Accounting Standards Board" and "the Public Company Accounting Oversight Board," in *Sec. 2.5.2*; the defense of "control self-assessment" by Gulf Canada, in *Sec. 3.2.1*; the defense of using CBM programs to treat those with cognitive biases in mental accounting by the advocates aforecited, in *Sec. 3.2.2*; the defense of carbon accounting by the advocates aforecited, in *Sec. 3.3.1*; the defense of the E P & L accounting by the advocates aforecited, in *Sec. 3.3.2*; the defense of OOP by the advocates aforecited, in *Sec. 3.4.1*; the voluntary compliance of individual companies with the Securities and Exchange Commission in the United States in regard to internal control, in *Sec. 3.4.2*; the defense of Parmalat by the advocates like Tanzi, in *Sec. 3.5.1*; and the defense of the convergence of accounting standards by the FASB and the IASB, in *Sec. 3.5.2*).

On the other hand, there is hardness (e.g., the critique of the application of "cognitive psychology" to accounting, as shown in the problems pointed out by the critics aforecited, in *Sec. 2.2.1*; the critique of "reputation management," as shown in the problems pointed out by the critics aforecited, in *Sec. 2.2.2*; the critique of EMAS, as shown in the problems pointed out by the critics aforecited, in *Sec. 2.3.1*; the critique of "the status of an organization as an independent economic entity" by those in "transaction cost economics," in *Sec. 2.3.2*; the critique of "internal auditing," as shown in the problems pointed out by the critics aforecited, in *Sec. 2.4.1*; the critique of VAT, as shown in the problems pointed out by the critics aforecited, in *Sec. 2.4.2*; the critique of the 3 accounting principles, as shown in the problems pointed out by the critics aforecited, in *Sec. 2.5.1*; the critique of US GAAP, as shown in the problems pointed out by the critics aforecited, as there is a call for it to move to IFRS by 2014 or later, in *Sec. 2.5.2*; the critique of control self-assessment," as shown in the problems pointed out by the critics aforecited, in *Sec. 3.2.1*; the critique of using CBM programs to treat those with cognitive biases in mental accounting, as shown in the problems pointed out by the critics aforecited, in *Sec. 3.2.2*; the critique of carbon accounting, as shown in the problems pointed out by the critics aforecited, in *Sec. 3.3.1*; the critique of the E P & L accounting, as shown in the problems pointed out by the critics aforecited, in *Sec. 3.3.2*; the critique of OOP, as shown in the problems pointed out by the critics aforecited, in *Sec. 3.4.1*; the prosecution of individual companies by the Securities and Exchange Commission in the United States for corporate misdeeds, in *Sec. 3.4.2*; the critique of Parmalat, as shown in the conviction of Tanzi by the Italian court system aforecited, in *Sec. 3.5.1*; and the critique of the convergence of accounting standards, as shown in the problems pointed out by the critics aforecited, in *Sec. 3.5.2*).

And the reverse direction also holds true.

Again, it should be stressed, however, that there are different shades of gray (or different degrees of truth) in the two opposites, and the classification is not necessarily mutually exclusive either. In fact, even when some combinations of the two occur, they only end up sharing the same dialectic relationship, but in a different degree.

Also, there are some other relationships (as a kind of family resemblance) which have something in common with the principle, but they are not exactly the same but only more or less comparable, which varies from case to case.

Good examples of family resemblance in relation to the principle are peacefulness-violence, cooperation-competition, pacificity-aggression, dovishness-hawkishness, reward-punishment, peace-war, friendliness-unfriendliness, kindness-hostility, love-hatred, acceptance-discrimination, admiration-belittlement, inclusion-exclusion, cooperation-predation, like-dislike, courteousness-violence, polishedness-harshness, integration-rivalry, bribery-killing, femininity-masculinity, absorption-resistance, amicality-animosity, and the like.

27th Thesis: The Seriousness-Playfulness Principle

The twenty-seventh thesis refers to the seriousness-playfulness principle (on the extent of seriousness) in the category of "agency" in existential dialectics (which is first worked out in *FPHH*), in that, if there is seriousness, there is also playfulness, especially for any agent with some kind of intelligence life, be it about humans, animals, or, later, post-humans. There is no seriousness without playfulness—and vice versa.

Lest any misunderstanding occurs, the word "playfulness" here should not be confused with other terms like "laughter," "smile," "joke," "tickleness," and "laugh-like vocalization" (in some animals), for instance.

Being playful can come in all shapes and sizes and therefore is not necessarily tied up with any of these terms, alhough it can result from any of them, or a combination of them, or something else altogether.

This principle, together with some other principles (especially though not exclusively, like the formalness-informalness principle), have important implications for what constitutes rationality, be it about "rationality of application" at the practical level or "rationality of knowledge" at the meta-theoretical level (as summarized in *Sec. 4.20* of *FPHH*).

For instance, on the one hand, there is seriousness (e.g., the serious business of those who try to alter financial reports for creative accounting, in *Sec. 2.2.1*; the serious business of those who try to understand the nature of "reputation," in *Sec. 2.2.2*; the serious business of those who try to understand the nature of eco-management and audit scheme, in *Sec. 2.3.1*; the serious business of those who try to analyze the new accounting system known as the "accrual accounting system," in *Sec. 2.3.2*; the serious business of those who try to find out the appropriate "role" of auditing in "internal control," in *Sec. 2.4.1*; the serious business of those who try to implement the VAT, in *Sec. 2.4.2*; the serious business of those who try to evaluate the "historical cost principle," in *Sec. 2.5.1*; the serious business of those who tried to set up the necessary concepts for US GAAP in the older days, in *Sec. 2.5.2*; the serious business of those who try to understand the nature of "control self-assessment," in *Sec. 3.2.1*; the serious business of those who try to understand the nature of mental accounting, in *Sec. 3.2.2*; the serious business of those who research on sustainability measurement, in *Sec. 3.3.1*; the serious business of those who try to understand the nature of ecosystem services, in *Sec. 3.3.2*; the serious business of those who try to understand the features of OOP, in *Sec. 3.4.1*; the serious business of those who try to understand the nature of internal control, in *Sec. 3.4.2*; the serious business of those who try to understand the reasons of ethical failures in accounting, in *Sec. 3.5.1*; and the serious business of those who work on the convergence of accounting standards, in *Sec. 3.5.2*).

On the other hand, there is playfulness (e.g., the playful part of those who try to alter financial reports for creative accounting, when they play around with different ways over the decades, like "unsuitable revenue recognition," "inappropriate accruals and estimates of liabilities," "excessive provisions," "generous reserve accounting," "intentional minor breaches of financial reporting requirements that aggregate to a material breach," etc., in *Sec. 2.2.1*; the playful part of those who try to understand "reputation," when they play around with different possible levels of analysis over the last decades, like the "epistemic," the "pragmatic–strategic," the "memetic," etc., in *Sec. 2.2.2*; the playful part of those who try to understand the nature of eco-management and audit scheme, when they play around over the last few decades with different implementation steps for organizations to follow, as shown in the 7 examples aforecited, in *Sec. 2.3.1*; the playful part of those who who try to analyze the new accounting system known as the "accrual accounting system," when they play around with different assumptions over the last few centuries, like the 2 assumptions on time and the 2 assumptions on space aforediscussed, in *Sec. 2.3.2*; the playful part of those who try to find out the appropriate "role" of auditing in "internal control," when they play around over the decades with possible "core objectives" like the 4 ones aforecited, in *Sec. 2.4.1*; the playful part of those who try to implement the VAT, when they play around with different options and possibilities over the decades, like the 2 types of the method of collection and the 2 types of the timing of collection, in *Sec. 2.4.2*; the playful part of those who try to evaluate the "historical cost principle," when they play around with different possible advantages and disadvantages over the centuries, like the 3 advantages and 3 disadvantages aforecited, in *Sec. 2.5.1*; the playful part of those who tried to set up the necessary concepts for US GAAP in the older days, when they play around with different "assumptions," "principles," and "constraints" over the decades, as discussed earlier, in *Sec. 2.5.2*; the playful part of those who try to understand the nature of "control self-assessment," when they play around over the decades with different basic methodologies, like the 6 ones aforecited, in *Sec. 3.2.1*; the playful part of those who try to understand the nature of mental accounting, when they play around with its different possibilities over the decades, like the 9 types aforedescribed, in *Sec. 3.2.2*; the playful part of those who research on sustainability measurement, when they play around over the decades with different quantitative methods like "benchmarks," "indices," "metrics," "auditing," etc., in *Sec. 3.3.1*; the playful part of those who try to understand the nature of ecosystem services, when they play around with different definitions over the decades, as shown in the comflicting definitions aforedescribed, in *Sec. 3.3.2*; the playful part of those who try to understand the features of OOP, when they play around with different possibilities over the decades, like the 5 features by Benjamin C. Pierce, the 4 features by John C. Mitchell, and the 3 features by Michael Lee Scott, in *Sec. 3.4.1*; the playful part of those who try to understand the nature of internal control, when they play around with different definitions over the decades, like the 3 types aforediscussed, in *Sec. 3.4.2*; the playful part of those who try to understand the reasons of ethical failures in accounting, when they play around with different possibiliites over the decades, like the 9 possible reasons aforecited, in *Sec. 3.5.1*; and the playful part of those who work on the convergence of accounting standards, when they play around with different convergence projects over the decades, like the 8 examples aforecited, in *Sec. 3.5.2*).

And the reverse direction also holds true.

Again, it should be stressed, however, that there are different shades of gray (or different degrees of truth) in the two opposites, and the classification is not necessarily mutually exclusive either. In fact, even when some combinations of the two occur, they only end up sharing the same dialectic relationship, but in a different degree.

Also, there are some other relationships (as a kind of family resemblance) which have something in common with the principle, but they are not exactly the same but only more or less comparable, which varies from case to case.

Good examples of family resemblance in relation to the principle are seriousness-jokingness, seriousness-humorousness, seriousness-wittiness, seriousness-nonseriousness, Apollo-Dionysus, seriousness-restingness, seriousness-relaxedness, criticalness-imaginativeness, seriousness-clownishness, and so forth.

28th Thesis: The Activeness-Inactiveness Principle

The twenty-eighth thesis refers to the activeness-inactiveness principle (on the extent of activeness) in the category of "agency" in existential dialectics (which was first worked out in *FPHTRAN*), in that, if there is activeness, there is also inactiveness, especially (though not exclusively) for any agent with some kind of intelligence life, be it about humans, animals, or, later, post-humans. There is no activeness without inactiveness—and vice versa.

Being active can come in all shapes and sizes, just as a person (or an animal) can be inactive because he (or it) is tired, old, injured, sick, sleepy (or hibernating), uninterested, in a state of passive inertia (or disposition), etc.—and an object can be inactive because it is at rest, out of work (or order), etc.

This principle, together with some other principles (especially though not exclusively, like the formalness-informalness principle and seriousness-playfulness principle), have important implications for what constitutes rationality, be it about "rationality of application" at the practical level or "rationality of knowledge" at the meta-theoretical level (as summarized in *Sec. 4.20* of *FPHH*).

For instance, on the one hand, there is activeness (e.g., the relative activeness of those executives in Enron in the 1990s who engaged in earnings management, in *Sec. 2.2.1*; the relative activeness of "purchase accounting" nowadays, which is allowed to use, under the U.S. Generally Accepted Accounting Principles, in *Sec. 2.2.2*; the relative activeness of EMAS III nowadays, in *Sec. 2.3.1*; the relative activeness of those who use the accrual accounting system nowadays, in *Sec. 2.3.2*; the relative activeness of those auditors when it is "broad" in scope, that is, not only covering "an organization's governance, risk management and management controls" but also "conducting proactive fraud audits to identify potentially fraudulent acts; participating in fraud investigations under the direction of fraud investigation professionals, and conducting post investigation fraud audits to identify control breakdowns and establish financial loss," in *Sec. 2.4.1*; the relative activeness of those governments in countries with VAT to implement it, in *Sec. 2.4.2*; the relative more activeness of the "Historical Cost model," because it is "still used in most accounting systems," in *Sec. 2.5.1*; the relative activeness of those who try to implement IFRS around the world, in *Sec. 2.5.2*; the relative activeness of those auditors who make good use of "control self-assessment" nowadays, in *Sec. 3.2.1*; the relative activeness of those who can make more decisions about money which are smart and responsible, without mental accounting, in *Sec. 3.2.2*; the relative activeness of those countries in the EU which have participated in "Emissions Trading Schemes" under the Kyoto Protocol, so they are active in trading emissions, in *Sec. 3.3.1*; the relative activeness of those who act in accordance to the environmental awareness nowadays, because this is increasingly encouraged, in *Sec. 3.3.2*; the relative activeness of those programmers who have to rely on the need for boilerplate manual codes without any "metaprogramming," "convention over configuration," and "model-driven engineering," in *Sec. 3.4.1*; the relative activeness of internal controllers when they investigate an organization for "misusing or misdirecting funds," "overstating revenues," "understating expenses," "overstating the value of corporate assets," "underreporting the existence of liabilities," etc., in *Sec. 3.4.2*; the relative activeness of those like Tanzi in Parmalat who actively engaged in fraudulent activities when he was the CEO, in *Sec. 3.5.1*; and the relative activeness of those who work on the convergence of accounting standards, since it is demanded around the world, in *Sec. 3.5.2*).

On the other hand, there is inactiveness (e.g., the relative inactiveness of those executives in Enron after the Enron scandal, when they were sentenced to prison, so that they could no longer engage in creative accounting, in *Sec. 2.2.1*; the relative inactiveness of the "pooling-of-interests method," which is no longer allowed nowadays, due to the U.S. Generally Accepted Accounting Principles [FAS 141] since 2001, in *Sec. 2.2.2*; the relative inactiveness of EMAS I nowadays, in *Sec. 2.3.1*; the relative inactiveness of those who still want to do accounting in the old way, when "the scope and function of accounts were limited…to recording decisions rendered by the market" during the "pre-going concern" period, because these accountants are not in much demand anymore, in *Sec. 2.3.2*; the relative inactiveness of those auditors, when they are "narrow" or "minimal" in scope, like covering a small specific task like "security audit" only, in *Sec. 2.4.1*; the relative inactiveness of those "missing traders" in "carousel fraud" to comply with the implementation of VAT, in *Sec. 2.4.2*; the relative inactiveness of the "Constant Purchasing Power model," because it is still not used in most accounting systems, in *Sec. 2.5.1*; the relative inactiveness of those who want to keep US GAAP as it was, since there is a call for it to move to IFRS by 2014 or later, in *Sec. 2.5.2*; the relative inactiveness of those auditors who still "ignored the benefits of control self-assessment" nowadays, in *Sec. 3.2.1*; the relative inactiveness of those who suffer from "cognitive vulnerability" in mental accounting, so that they cannot engage in smart, responsible decision making about money, in *Sec. 3.2.2*; the relative inactiveness of those countries like China, India and the United States, which have not participated in "Emissions Trading Schemes" under the Kyoto Protocol, so they are not active in trading emissions, in *Sec. 3.3.1*; the relative inactiveness of those who act in accordance to the "tragedy of the commons" nowadays, because this is increasingly condemned, in *Sec.*

3.3.2; the relative inactiveness of those programmers who reduce the need for boilerplate manual codes by way of "metaprogramming," "convention over configuration," and "model-driven engineering," in *Sec. 3.4.1*; the relative inactiveness of internal controllers, when they have no control over external factors which can affect the performance of an organization, in *Sec. 3.4.2*; the relative inactiveness of those like Tanzi in Parmalat who later spent "10 years in prison for fraud," so he could not engage in corporate misdeeds anymore, in *Sec. 3.5.1*; and the relative inactiveness of those who still work in the old-fashioned way of individual country accounting standards which have no interaction with the IFRS, as they become more and more out-of-date, in *Sec. 3.5.2*).

And the reverse direction also holds true.

Again, it should be stressed, however, that there are different shades of gray (or different degrees of truth) in the two opposites, and the classification is not necessarily mutually exclusive either. In fact, even when some combinations of the two occur, they only end up sharing the same dialectic relationship, but in a different degree.

Also, there are some other relationships (as a kind of family resemblance) which have something in common with the principle, but they are not exactly the same but only more or less comparable, which varies from case to case.

Good examples of family resemblance in relation to the principle are alertness-unalertness, activeness-decrepidness, alertness-sleepiness, activeness-tiredness, alertness-unwareness, alertness-uninterestedness, soberness-inactiveness, alertness-inactiveness, soberness-unawareness, awakenness-hibernatedness, activeness-passiveness, and so on.

29th Thesis: The Selfness-Otherness Principle

The twenty-ninth thesis refers to the selfness-otherness principle (on the magnitude of self-centeredness) in the category of "agency" in existential dialectics (which is first worked out in *FPHCA*), in that, if there is the tendency to be mostly centered on one's self or group, there is likewise the tendency to be mostly centered on others, especially (though not exclusively) for any agent with some kind of intelligence life, be it about humans, animals, or, later, post-humans. There is no selfness without otherness—and vice versa.

Being self-centered and being other-centered do not have to be mutually exclusive, as the two can come in all shapes and sizes, or in different degrees. And other-centeredness does not entail desirability, any more than self-centeredness implies non-desirability, as both can go either way.

In addition, whether self-centeredness constitutes a moral alternative or, conversely, an alternative to morality is of course debatable (as addressed in my other book titled *Beyond Ethics to Post-Ethics*).

For instance, on the one hand, there is selfness (e.g., the tendency to be mostly centered on one's self or group, as shown in those accountants who engage in earnings management for private gains, in *Sec. 2.2.1*; the tendency to be mostly centered on one's self or group, as shown by those who are obsessed with their "reputation," in *Sec. 2.2.2*; the tendency to be mostly centered on one's self or group, as shown in the implementation of EMAS by some organizations for the "benefits" as cited before, in *Sec. 2.3.1*; the tendency to be mostly centered on one's self or group, as shown by those individuals in the older days who acted solely as the "results" of their "subjective judgments" and interests in the market, prior to the rise of modern accounting conventions, in *Sec. 2.3.2*; the tendency to be mostly centered on one's self or group, as shown by those who commit "fraudulent activities" and are discovered when audited by an auditor, in *Sec. 2.4.1*; the tendency to be mostly centered on one's self or group, as shown by those who commit "carousel fraud" for their personal gains, in *Sec. 2.4.2*; the tendency to be mostly centered on one's self or group, as shown in the defense of the continued use of the "Historical Cost model" in most accounting systems," in spite of the problems aforecited, in *Sec. 2.5.1*; the tendency to be mostly centered on one's self or group, as shown in the bureaucratic inertia of US GAAP, since, "although the SEC has a stated goal of moving from US GAAP to the International Financial Reporting Standards,...progress has been slow and uncertain," in *Sec. 2.5.2*; the tendency to be mostly centered on one's self or group, as shown in the problem of greed in "financial scandals, notably the collapse of Robert Maxwell's publishing empire," which later "led to the increased adoption of control self-assessment in the UK," in *Sec. 3.2.1*; the tendency to be mostly centered on one's self or group, as shown in those who received "birthday money" from relatives and "eagerly rushed out and bought the most outlandish and impractical things" for themselves, in *Sec. 3.2.2*; the tendency to be mostly centered on one's self or group, as shown in the finding that "it is unrealistic to expect business to voluntarily commit the resources required for full sustainable accounting implementation," in *Sec. 3.3.1*; the tendency to be mostly centered on one's self or group, as shown by those who act in accordance to the "tragedy of the commons," in *Sec. 3.3.2*; the tendency to be mostly centered on one's self or group, as shown in the criticism by Paul Graham

[2013], who "suggested that OOP's popularity within large companies is due to 'large [and frequently changing] groups of mediocre programmers," in *Sec. 3.4.1*; the tendency to be mostly centered on one's self or group, as shown in the corporate misdeeds by those in the accounting scandal involving Freddie Mac aforecited, in *Sec. 3.4.2*; the tendency to be mostly centered on one's self or group, as shown in the corruption and fraud by Tanzi in Parmalat, in *Sec. 3.5.1*; and the tendency to be mostly centered on one's self or group, as shown by those in the U.S. who still prefer the US GAAP over the IFRS, in *Sec. 3.5.2*).

On the other hand, there is otherness (e.g., the tendency to be mostly centered on others, as shown in the activism of Abraham Briloff, "who for years wrote a column for Barron's that constantly analyzed breaches of ethics and audit professionalism among CPA firms" and "much of what he advocated has been forced on the industry in the wake of the Enron scandal," as an advancement in the field of accounting reform, in *Sec. 2.2.1*; the tendency to be mostly centered on others, as shown in the academic analysis of both sides of the debate on "reputation capital and management," as an advancement in the field of accounting for humanity, in *Sec. 2.2.2*; the tendency to be mostly centered on others, as shown in the academic analysis of both sides of the debate on EMAS, as an advancement in the knowledge of eco-management and audit scheme for humanity, in *Sec. 2.3.1*; the tendency to be mostly centered on others, as shown in the academic analysis of both sides of the debate on the evolution of accounting conventions in history, as an advancement in the knowledge of space-time in income accounting for humanity). (S. Takatera 2000: 790, in *Sec. 2.3.2*; the tendency to be mostly centered on others, as shown in the academic analysis of both sides of the debate on "internal auditing, " as an advancement in the knowledge of accounting for humanity, in *Sec. 2.4.1*; the tendency to be mostly centered on others, as shown in the academic analysis of both sides of the debate on VAT, as an advancement in the knowledge of taxation and accounting for humanity, in *Sec. 2.4.2*; the tendency to be mostly centered on others, as shown in the academic analysis of both sides of the debate on accounting principles, as an advancement in the knowledge of accounting for humanity, in *Sec. 2.5.1*; the tendency to be mostly centered on others, as shown in the academic analysis of both sides of the debate on accounting standards, as an advancement in the knowledge of accounting standards for humanity, in *Sec. 2.5.2*; the tendency to be mostly centered on others, as shown in the academic analysis of both sides of the debate on "control self-assessment," as an advancement in the knowledge of accounting for humanity, in *Sec. 3.2.1*; the tendency to be mostly centered on others, as shown in the academic analysis of both sides of the debate on mental accounting, as an advancement in the knowledge of accounting and psychology for humanity, in *Sec. 3.2.2*; the tendency to be mostly centered on others, as shown in the academic analysis of both sides of the debate on sustainability accounting, as an advancement in the knowledge of accounting for humanity, in *Sec. 3.3.1*; the tendency to be mostly centered on others, as shown in the preaching of "the environmental awareness" nowadays, which "is rapidly improving in our contemporary world" in relation to nature, in *Sec. 3.3.2*; the tendency to be mostly centered on others, as shown in the academic analysis of both sides of the debate on OOP, as an advancement in the knowledge of computing and bookkeeping for humanity, in *Sec. 3.4.1*; the tendency to be mostly centered on others, as shown in the academic analysis of both sides of the debate on internal control, as an advancement in the knowledge of accounting for humanity, in *Sec. 3.4.2*; the tendency to be mostly centered on others, as shown in the academic analysis of both sides of the debate on accounting ethics, as an advancement in the knowledge of professional ethics for humanity, in *Sec. 3.5.1*; and the tendency to be mostly centered on others, as shown in the academic analysis of both sides of the debate on the convergence of accounting standards, as an advancement in the knowledge of the evolution of accounting standards for humanity, in *Sec. 3.5.2*).

And the reverse direction also holds true.

Again, it should be stressed, however, that there are different shades of gray (or different degrees of truth) in the two opposites, and the classification is not necessarily mutually exclusive either. In fact, even when some combinations of the two occur, they only end up sharing the same dialectic relationship, but in a different degree.

Also, there are some other relationships (as a kind of family resemblance) which have something in common with the principle, but they are not exactly the same but only more or less comparable, which varies from case to case.

Good examples of family resemblance in relation to the principle are egoisticness-altruisticness, selfishness-unselfishness, selfness-selflessness, egocentricness-othercentricness, parasiticalness-helpfulness, take-give, and so on.

In Relation to Outcome

The next six theses are in relation to outcome in existential dialectics, as shown below.

30th Thesis: The Regression-Progression Principle

The thirtieth thesis is called the regression-progression principle (on the direction of history) in the category of "outcome" in existential dialectics (which was also already worked out in my previous books), in that neither the cyclical nor the linear views are adequate for explaining many phenomena at all levels. There is no regression without progression—and vice versa.

History progresses to more advanced forms, but with a regressive touch. Examples include no freedom without unfreedom, no equality without inequality, and no civilization without barbarity. This is not an inevitable law, but merely a highly likely empirical trend.

For instance, on the one hand, there is regression (e.g., the regression made by the fight against creative accounting, as shown in the problems pointed out by the critics, in *Sec. 2.2.1*; the regression made by "reputation management," as shown in the problems pointed out by the critics, in *Sec. 2.2.2*; the regression made by EMAS, as shown in the problems pointed out by the critics, in *Sec. 2.3.1*; the regression made by the work on space-time in income accounting, as shown in the problems pointed out by the critics, in *Sec. 2.3.2*; the regression made by "internal auditing," as shown in the problems pointed out by the critics, in *Sec. 2.4.1*; the regression made by VAT, as shown in the problems pointed out by the critics, in *Sec. 2.4.2*; the regression made by the 3 accounting principles, as shown in the problems pointed out by the critics, in *Sec. 2.5.1*; the regression made by US GAAP, as shown in the problems pointed out by the critics, in *Sec. 2.5.2*; the regression made by "control self-assessment practices," as shown in the problems pointed out by the critics, in *Sec. 3.2.1*; the regression made by the CBM program, as shown in the problems pointed out by the critics, in *Sec. 3.2.2*; the regression made by carbon accounting, as shown in the problems pointed out by the critics, in *Sec. 3.3.1*; the regression made by the E P & L accounting, as shown in the problems pointed out by the critics, in *Sec. 3.3.2*; the regression made by OOP, as shown in the problems pointed out by the critics, in *Sec. 3.4.1*; the regression made by internal control, as shown in the problems pointed out by the critics, in *Sec. 3.4.2*; the regression made by accounting ethics, as shown in the problems pointed out by the critics, in *Sec. 3.5.1*; and the regression made by the convergence of accounting standards, as shown in the problems pointed out by the critics, in *Sec. 3.5.2*).

On the other hand, there is progress (e.g., the progress made by the fight against creative accounting, as shown in the contribution to accounting reform aforecited, in *Sec. 2.2.1*; the progress made by "reputation management," as shown in the contribution to the "transfer" of "the good reputation of a company and its existings products to new markets and new products" aforecited, in *Sec. 2.2.2*; the progress made by EMAS, as shown in the contribution to a global scheme for eco-management and audit scheme aforecited, in *Sec. 2.3.1*; the progress made by the work on space-time in income accounting, as shown in the contribution to the understanding of the evolution of accounting conventions aforecited, in *Sec. 2.3.2*; the progress made by "internal auditing," as shown in the contribution to the improvement of the financial performance of businesses aforecited, in *Sec. 2.4.1*; the progress made by VAT, as shown in the contribution to the increase of government revenue aforecited, in *Sec. 2.4.2*; the progress made by the 3 accounting principles, as shown in the contribution to the facilitation of business transactions aforecited, in *Sec. 2.5.1*; the progress made by US GAAP, as shown in the contribution to a common langauge for business affairs in the U.S. aforecited, in *Sec. 2.5.2*; the progress made by "control self-assessment practices," as shown in the contribution to the reduction of the "risk of error" aforecited, in *Sec. 3.2.1*; the progress made by the CBM program, as shown in the contribution to the treatment of cognitive biases aforecited, in *Sec. 3.2.2*; the progress made by carbon accounting, as shown in the contribution to the reduction of carbon emissions aforecited, in *Sec. 3.3.1*; the progress made by the E P & L accounting, as shown in the contribution to the understanding of "the ecosystem services a business depends on to operate in addition to the cost of direct and indirect negative impacts on the environment" aforecited, in *Sec. 3.3.2*; the progress made by OOP, as shown in the contribution to the understanding of boilerplate codes aforecited, in *Sec. 3.4.1*; the progress made by internal control, as shown in the contribution to "the prevention and detection of fraud," in *Sec. 3.4.2*; the progress made by accounting ethics, as shown in the contribution to the prevention of corruption and fraud aforecited, in *Sec. 3.5.1*; and the progress made by the convergence of accounting standards, as shown in the contribution to "comparability" aforecited, in *Sec. 3.5.2*).

And the reverse direction also holds true.

Again, it should be stressed, however, that there are different shades of gray (or different degrees of truth) in the two opposites, and the classification is not necessarily mutually exclusive either. In fact, even when some combinations of the two occur, they only end up sharing the same dialectic relationship, but in a different degree.

Also, there are some other relationships (as a kind of family resemblance) which have something in common with the principle, but they are not exactly the same but only more or less comparable, which varies from case to case.

Good examples of family resemblance in relation to the principle are undesirability-desirability, risk-opportunity, badness-goodness, cost-benefit, dysfunctionality-functionality, cyclicity-linearity, failure-success, and so on.

31st Thesis: The Same-Difference Principle

The thirty-first thesis refers to the same-difference principle (on the metamorphosis of change) in the category of "outcome" in existential dialectics (which was worked out in *ALD* and other books of mine), in that an entity, as it evolves over time, can be both different from and similar to its opposing alternatives and does not have to be solely more different from them over time. There is no similarity without difference—and vice versa.

Opposites are not absolute in a black-or-white fashion; so, an entity can become relatively more similar to (or more different from) its opposite over time.

This is further constrained by another principle, that is, the symmetry-asymmetry principle about the relationships among existents under the category about agency in existential dialectics, in that if there is symmetry (equality) between two entities, there is likewise asymmetry (inequality) emerging in a different way.

For instance, on the one hand, there is similarity in outcome (e.g., the contribution to the molding and control of beliefs, values, and behaviors, regardless of whether this be done by way of creative accounting, or by way of accounting reform, in *Sec. 2.2.1*; the contribution to the molding and control of beliefs, values, and behaviors, regardless of whether this be done by way of the "stakeholder theory," or by way of the "shareholder theory," on building reputation, in *Sec. 2.2.2*; the contribution to the molding and control of beliefs, values, and behaviors, regardless of whether this be done by way of "highly detailed and prescriptive protocols," or by way of "more flexible protocols," in *Sec. 2.3.1*; the contribution to the molding and control of beliefs, values, and behaviors, regardless of whether this be done by way of using the accrual accounting system, or by way of using chaos theory, in *Sec. 2.3.2*; the contribution to the molding and control of beliefs, values, and behaviors, regardless of whether this be done by way of appealing to "reasonable assurance," or by way of going beyond just "reasonable assurance," in *Sec. 2.4.1*; the contribution to the molding and control of beliefs, values, and behaviors, regardless of whether this be done by way of VAT, or by way of carousel fraud, in *Sec. 2.4.2*; the contribution to the molding and control of beliefs, values, and behaviors, regardless of whether this be done by way of the continuation of the existing 3 accounting principles, or by way of the proposal of new accounting principles as discussed earlier, in *Sec. 2.5.1*; the contribution to the molding and control of beliefs, values, and behaviors, regardless of whether this be done by way of US GAAP, or by way of IFRS, in *Sec. 2.5.2*; the contribution to the molding and control of beliefs, values, and behaviors, regardless of whether this be done by way of person-to-person "interview techniques," or by way of IT-focused "customised questionnaires," in *Sec. 3.2.1*; the contribution to the molding and control of beliefs, values, and behaviors, regardless of whether this be done by way of the observance of the principle of "fungibility," or by way of the appeal to "mental accounts," in *Sec. 3.2.2*; the contribution to the molding and control of beliefs, values, and behaviors, regardless of whether this be done by way of "Emissions Trading," or by way of "the Verified Emission Reductions," in *Sec. 3.3.1*; the contribution to the molding and control of beliefs, values, and behaviors, regardless of whether this be done by way of the "tragedy of the commons," or by way of the "environmental awareness," in *Sec. 3.3.2*; the contribution to the molding and control of beliefs, values, and behaviors, regardless of whether this be done by way of "OOP," or by way of "procedural programming," in *Sec. 3.4.1*; the contribution to the molding and control of beliefs, values, and behaviors, regardless of whether this be done by way of corporate misdeeds committed by creative accountants, or by way of prosecution against individual companies by government oversight agencies, in *Sec. 3.4.2*; the contribution to the molding and control of beliefs, values, and behaviors, regardless of whether this be done by way of corporate misdeeds by Tanzi, or by way of the law enforcement by the Italian court system against Tanzi, in *Sec. 3.5.1*; and the contribution to the molding and control of beliefs, values, and behaviors, regardless of whether this be done by way of individual country accounting standards, or by way of the convergence of accounting standards, in *Sec. 3.5.2*).

On the other hand, there is difference in outcome (e.g., the contribution to the molding and control of beliefs, values, and behaviors by way of creative accounting for a relatively more amoral lifeworld—but the contribution to the molding and control of beliefs, values, and behaviors by way of accounting reform for a relatively more moral lifeworld, in *Sec. 2.2.1*; the contribution to the molding and control of beliefs, values, and behaviors by way of the "stakeholder theory" on building reputation for a relatively broader view of management—but the contribution to the molding and control of beliefs, values, and behaviors by way of the "shareholder theory" on building reputation for a relatively more narrow view of management, in *Sec. 2.2.2*; the contribution to the molding and control of beliefs, values, and behaviors by way of "highly detailed and prescriptive protocols" for a relatively more novice worldview—but the contribution to the molding and control of beliefs, values, and behaviors by way of "more flexible protocols" for a relatively more experienced worldview, in *Sec. 2.3.1*; the contribution to the molding and control of beliefs, values, and behaviors by way of using the accrual accounting system for a relatively more deterministic worldview—but the contribution to the molding and control of beliefs, values, and behaviors by way of using chaos theory for a relatively more uncertain worldview, in *Sec. 2.3.2*; the contribution to the molding and control of beliefs, values, and behaviors by way of appealing to "reasonable assurance" for a relatively looser professional standard in the business lifeworld—but the contribution to the molding and control of beliefs, values, and behaviors by way of going beyond just "reasonable assurance" for a relatively stricter professional standard in the business lifeworld, in *Sec. 2.4.1*; the contribution to the molding and control of beliefs, values, and behaviors by way of VAT for a relatively more legalistic lifeworld—but the contribution to the molding and control of beliefs, values, and behaviors by way of carousel fraud for a relatively more underground lifeworld, in *Sec. 2.4.2*; the contribution to the molding and control of beliefs, values, and behaviors by way of the continuation of the existing 3 accounting principles for a relatively more conservative lifeworld—but the contribution to the molding and control of beliefs, values, and behaviors by way of the proposal of new accounting principles for a relatively more liberal lifeworld, in *Sec. 2.5.1*; the contribution to the molding and control of beliefs, values, and behaviors by way of US GAAP for a relatively more regional lifeworld—but the contribution to the molding and control of beliefs, values, and behaviors by way of IFRS for a relatively more global lifeworld, in *Sec. 2.5.2*; the contribution to the molding and control of beliefs, values, and behaviors by way of person-to-person "interview techniques" for a relatively more people-interactive lifeworld—but the contribution to the molding and control of beliefs, values, and behaviors by way of IT-focused "customised questionnaires" for a relatively more impersonal lifeworld, in *Sec. 3.2.1*; the contribution to the molding and control of beliefs, values, and behaviors by way of the observance of the principle of "fungibility" for a relatively more realistic lifeworld—but the contribution to the molding and control of beliefs, values, and behaviors by way of the appeal to "mental accounts" for a relatively more illusory lifeworld, in *Sec. 3.2.2*; the contribution to the molding and control of beliefs, values, and behaviors by way of "Emissions Trading," for a relatively mandatory lifeworld—but the contribution to the molding and control of beliefs, values, and behaviors by way of "the Verified Emission Reductions" for a relatively more voluntary lifeworld, in *Sec. 3.3.1*; the contribution to the molding and control of beliefs, values, and behaviors by way of the "tragedy of the commons" for a relatively more individualist lifeworld—but the contribution to the molding and control of beliefs, values, and behaviors by way of the "environmental awareness" for a relatively more collectivist lifeworld, in *Sec. 3.3.2*; the contribution to the molding and control of beliefs, values, and behaviors by way of "OOP" for a relatively more object-oriented worldview—but the contribution to the molding and control of beliefs, values, and behaviors by way of "procedural programming" for a relatively more procedural worldview, in *Sec. 3.4.1*; the contribution to the molding and control of beliefs, values, and behaviors by way of corporate misdeeds committed by creative accountants for a relatively more selfish lifeworld—but the contribution to the molding and control of beliefs, values, and behaviors by way of prosecution by government oversight agencies against individual companies for a relatively more legalistic lifeworld, in *Sec. 3.4.2*; the contribution to the molding and control of beliefs, values, and behaviors by way of corporate misdeeds by Tanzi for a relatively more amoral lifeworld—but the contribution to the molding and control of beliefs, values, and behaviors by way of the law enforcemeny by the Italian court system against Tanzi for a relatively more moral lifeworld, in *Sec. 3.5.1*; and the contribution to the molding and control of beliefs, values, and behaviors by way of individual country accounting standards for a relatively more nationalistic lifeworld—but the contribution to the molding and control of beliefs, values, and behaviors by way of the convergence of accounting standards for a relatively more globalized lifeworld, in *Sec. 3.5.2*).

And the reverse direction also holds true.

Again, it should be stressed, however, that there are different shades of gray (or different degrees of truth) in the two opposites, and the classification is not necessarily mutually exclusive either. In fact, even when some combinations of the two occur, they only end up sharing the same dialectic relationship, but in a different degree.

Also, there are some other relationships (as a kind of family resemblance) which have something in common with the principle, but they are not exactly the same but only more or less comparable, which varies from case to case.

Good examples of family resemblance in relation to the principle are homogeneity-heterogeneity, we-they, similarity-contrast, assimilation-dissimilation, and so forth.

32nd Thesis: The Stability-Reaction Principle

The thirty-second thesis is called the stability-reaction principle (on the feedback mechanism for further changes) in the category of "outcome" in existential dialectics (which was first worked out in *FPHCHEM*), in that once an outcome becomes stabilized over time (which varies from case to case), there is resistance to it, such that it undergoes further changes in its multifaceted interactions with the outside world (e.g., with other entities, energy, space, and so on).

There is no stability without reaction (to stability for further changes)—and vice versa.

Works on chemical reactions (as in *FPHCHEM*) and feedback mechanisms for learning in adaptive management (as in *BNR*), just to name a few of them, have contributed to the understanding of stability and reaction.

For instance, on the one hand, there is stability in outcome (e.g., the relatively more stable condition of "breaches of ethics and audit professionalism among CPA firms" in the older days, in *Sec. 2.2.1*; the relatively more stable condition of using "pooling-of-interests accounting" in the older days, in *Sec. 2.2.2*; the relatively more stable condition of EMAS I in the 1990s, in *Sec. 2.3.1*; the relatively more stable condition of the "pre-going concern" in the older days, in *Sec. 2.3.2*; the relatively more stable condition of the Sarbanes-Oxley Act of 2002 at the time, in *Sec. 2.4.1*; the relatively more stable condition of the spread of VAT in modern times, in *Sec. 2.4.2*; the relatively more stable condition of the 3 accounting principles as developed in the past few hundred years, in *Sec. 2.5.1*; the relatively more stable condition of setting up US GAAP in the older days, in *Sec. 2.5.2*; the relatively more stable condition of "external auditing" in the older days pior to the creation of "control self-assessment," in *Sec. 3.2.1*; the relatively more stable condition of the use of mental accounting among some people, in *Sec. 3.2.2*; the relatively more stable condition of "growth-based" economic models in the older days, in *Sec. 3.3.1*; the relatively more stable condition of the practice of the "tragedy of the commons" in the older days, in *Sec. 3.3.2*; the relatively more stable condition of "object-oriented software design methods" in the 1950s and 1960s, in *Sec. 3.4.1*; the relatively more stable condition of "the Foreign Corrupt Practices Act [FCPA] of 1977 in the older days, in *Sec. 3.4.2*; the relatively more stable condition of accounting scandals in the last few decades of the 20th century, in *Sec. 3.5.1*; and the relatively more stable condition of the US GAAP in the older days, in *Sec. 3.5.2*).

On the other hand, there is reaction in outcome (e.g., "breaches of ethics and audit professionalism among CPA firms" in the older days then led to further changes later on, when Abraham Briloff "for years wrote a column for Barron's that constantly analyzed breaches of ethics and audit professionalism among CPA firms" and "much of what he advocated has been forced on the industry in the wake of the Enron scandal," in *Sec. 2.2.1*; the use of "pooling-of-interests accounting" in the older days then led to further changes later on, when, "since 2001, U.S. Generally Accepted Accounting Principles no longer allows the pooling-of-interests method," in *Sec. 2.2.2*; EMAS I in the 1990s then led to further changes later on, when nowadays, there is EMAS III, in *Sec. 2.3.1*; the "pre-going concern" in the older days then led to further changes later on, when it was superseded by the "going concern" in modern times, in *Sec. 2.3.2*; the Sarbanes-Oxley Act of 2002 then led to further changes later on, when in 2010, there was the revised interpretation by the IIA, in *Sec. 2.4.1*; the spread of VAT in modern times then led to further changes later on, when nowadays, there is "carousel fraud," in *Sec. 2.4.2*; the 3 accounting principles as developed in the past few hundred years then led to further changes later on, when nowadays there is the call for accounting reform to create new accounting principles to replace the old ones, in *Sec. 2.5.1*; US GAAP in the older days then led to further changes later on, when nowadays, there is a call for it to move to IFRS by 2014 or later, in *Sec. 2.5.2*; "external auditing" in the older days then led to further changes later on, when "control self-assessment" was created in 1987 as another approach, in *Sec. 3.2.1*; the use of mental accounting among some people try then led to further changes later on, when nowadays, some propose the treatment of cognitive biases in mental accounting with a CBM program, in *Sec. 3.2.2*; "growth-based" economic models in the older days then led to further changes later on, when nowadays, there is the alternative idea of "sustainable" economic models, in *Sec. 3.3.1*; the practice of the "tragedy of the commons" in the older days then led to further changes later on, when nowadays, there is the "environmental awareness," in *Sec. 3.3.2*; "object-oriented software design methods" in the 1950s and 1960s then led to further changes later on, when nowadays, they are "refined," with "the use of design patterns, design by contract, and modeling languages [such as UML]," in *Sec. 3.4.1*; "the Foreign Corrupt Practices Act [FCPA] of 1977 in the older days then led to further changes later on, when it was later refined by "the Sarbanes–Oxley Act of 2002," in *Sec. 3.4.2*; accounting scandals in the last few decades of the 20th century then led to further changes later on, when nowadays, there are different accounting reforms like "education" and "regulations" aforecited, in *Sec. 3.5.1*; the US GAAP in the older days then led to further changes later on, when nowadays, there is a call for the reduction of "the differences between the US GAAP and the IFRS for the ultimate convergence of accounting standards, in *Sec. 3.5.2*).

And the reverse direction also holds true.

Again, it should be stressed, however, that there are different shades of gray (or different degrees of truth) in the two opposites, and the classification is not necessarily mutually exclusive either. In fact, even when some combinations of the two occur, they only end up sharing the same dialectic relationship, but in a different degree.

Also, there are some other relationships (as a kind of family resemblance) which have something in common with the principle, but they are not exactly the same but only more or less comparable, which varies from case to case.

Good examples of family resemblance in relation to the principle are stability-forwardness, normalcy-feedback, balance-forwardness, balance-reaction, consolidation-modification, normalcy-feedback, stability-breakup, rest-motion, and the like.

33rd Thesis: The Functionality-Nonfunctionality Principle

The thirty-third thesis is called the functionality-nonfunctionality principle (on the presence of function) in the category of "outcome" in existential dialectics (which is first worked out in *FPHCA*), in that, if there is the presence of action for which something or someone is fitted, there is likewise the relative lack of action for which something or someone is fitted. There is no functionality without nonfunctionallity—and vice versa.

A *function* should not be confused with an *alternative function,* which is just another kind of function, just as *nonfunction* should not be confused with *dysfunction* (which has more to do with malfunction and is thus related more to the regression-progression principle).

In addition, being functional (or useful) and being nonfunctional (or useless) do not have to be mutually exclusive, as the two can come in all shapes and sizes, or in different degrees. And what is functional at one time may become nonfunctional at another time, and vice-versa.

For instance, on the one hand, there is functionality in outcome (e.g., the presence of action for which something or someone is fitted, as shown in the role of governments "to determine what are accepted accounting principles," which serves the function of maintaining professional ethics in the business world, in *Sec. 2.2.1*; the presence of action for which something or someone is fitted, as shown in "reputation management," which serves the function of increasing the "benefit" of "stakeholders," for instance, in *Sec. 2.2.2*; the presence of action for which something or someone is fitted, as shown in the creation of EMAS, which serves the function of providing a way to do eco-management and audit scheme, in *Sec. 2.3.1*; the presence of action for which something or someone is fitted, as shown in the creation of the "accrual accounting system," which serves the function of providing an "internal calculative logic" of accounting which is "structurally separated from the external logic of the market," in *Sec. 2.3.2*; the presence of action for which something or someone is fitted, as shown in the appeal to "reasonable assurance," which serves the function of providing a realistic standard of professional ethics under the condition of "constraints," in *Sec. 2.4.1*; the presence of action for which something or someone is fitted, as shown in the creation of VAT, which serves the function of increasing government revenue, in *Sec. 2.4.2*; the presence of action for which something or someone is fitted, as shown in the establishment of the 3 accounting principles, which serves the function of providing an accounting groundwork for business transactions in the last few hundred years, in *Sec. 2.5.1*; the presence of action for which something or someone is fitted, as shown in the creation of US GAAP, which serves the function of providing a common language for business affairs in the U.S., in *Sec. 2.5.2*; the presence of action for which something or someone is fitted, as shown in the creation of "control self-assessment," which serves the function to reduce the "risk of error," in *Sec. 3.2.1*; the presence of action for which something or someone is fitted, as shown in the proposal of the mental accounting theory, which serves the function of explainng how "a person subjectively frames a transaction in their mind" and "will determine the utility they receive or expect," in *Sec. 3.2.2*; the presence of action for which something or someone is fitted, as shown in the creation of "carbon accounting," which serves the function of providing a "sustainable" economic model to reduce carbon emissions, in *Sec. 3.3.1*; the presence of action for which something or someone is fitted, as shown in the creation of the E P & L accounting, which serves the function of providing an evaluation of a company's environmental impacts, so as to reduce "the cost of direct and indirect negative impacts on the environment," in *Sec. 3.3.2*; the presence of action for which something or someone is fitted, as shown in the development of OOP, which serves the function of dealing with the problem of bolierplate codes in computing, in *Sec. 3.4.1*; the presence of action for which something or someone is fitted, as shown in the creation of "internal control," which serves the function to "prevent…and detect…fraud," in *Sec. 3.4.2*; the presence of action for which something or someone is fitted, as shown in the construction of accounting ethics, which serves the function of preventing corruption and fraud in the profession of accounting, in *Sec. 3.5.1*; and the presence of action for which something or someone is fitted, as shown in the convergence of accounting standards, which serves the function of providing "comparability" and other benefits aforecited, in *Sec. 3.5.2*).

On the other hand, there is nonfunctionality in outcome (e.g., the relative lack of action for which something or someone is fitted, as shown in the problems with the role of governments "to determine what are accepted accounting principles," since they also use "questionable accounting practices themselves," such that they are not as functional as the advocates would like us to believe, in *Sec. 2.2.1*; the relative lack of action for which something or someone is fitted, as shown in different problems with "reputation management" aforecited, such that it is not as functional as the advocates would like us to believe, in *Sec. 2.2.2*; the relative lack of action for which something or someone is fitted, as shown in the problems with EMAS, such that it is not as functional as the advocates would like us to believe, in *Sec. 2.3.1*; the relative lack of action for which something or someone is fitted, as shown in the problems with the "accrual accounting system" in regard to its fictional assumptions, such that it is not as functional as the advocates would like us to believe, to be further accounted for by an alternative model known as "accounting order out of chaos" by Takatera and Sawabe, in *Sec. 2.3.2*; the relative lack of action for which something or someone is fitted, as shown in the problems with the appeal to "reasonable assurance," since, to the critics, the conventional standard of "reasonable assurance" is no longer sufficient for some people since "recently, the argument that auditing should go beyond just true and fair is gaining momentum," such that it is no longer as functional as the advocates would like us to believe, in *Sec. 2.4.1*; the relative lack of action for which something or someone is fitted, as shown in the different problems with the implementation of VAT, such that it is not as functional as the advocates would like us to believe, in *Sec. 2.4.2*; the relative lack of action for which something or someone is fitted, as shown in the problems with the 3 accounting principles aforediscussed, such that the 3 principles are not as functional as the advocates would like us to believe, in *Sec. 2.5.1*; the relative lack of action for which something or someone is fitted, as shown in the problems with US GAAP aforediscussed, such that it is not as functional as the advocates would like us to believe, in *Sec. 2.5.2*; the relative lack of action for which something or someone is fitted, as shown in the problems with "control self-assessment," such that it is not as functional as the advocates would like us to believe, in *Sec. 3.2.1*; the relative lack of action for which something or someone is fitted, as shown in the problems with the mental accounting theory aforecited, such that it is not as functional as the advocates would like us to believe, in *Sec. 3.2.2*; the relative lack of action for which something or someone is fitted, as shown in the problems with "carbon accounting" aforecited, such that it is not as functional as the advocates would like us to believe, in *Sec. 3.3.1*; the relative lack of action for which something or someone is fitted, as shown in the problems with the E P & L accounting, such that it is not as functional as the advocates would like us to believe, in *Sec. 3.3.2*; the relative lack of action for which something or someone is fitted, as shown in the problems with OOP aforecited, such that it is not as functional as the advocates would like us to believe, in *Sec. 3.4.1*; the relative lack of action for which something or someone is fitted, as shown in the problems with "internal control" aforediscussed, such that it is not as functional as the advocates would like us to believe, in *Sec. 3.4.2*; the relative lack of action for which something or someone is fitted, as shown in the criticism that there have been business scandals in spite of the existence of accounting ethics, such that it is not as functional as the advocates would like us to believe, in *Sec. 3.5.1*; and the relative lack of action for which something or someone is fitted, as shown in the problems with the convergence of accounting standards, such that it is not as functional as the advocates would like us to believe in *Sec. 3.5.2*).

And the reverse direction also holds true.

Again, it should be stressed, however, that there are different shades of gray (or different degrees of truth) in the two opposites, and the classification is not necessarily mutually exclusive either. In fact, even when some combinations of the two occur, they only end up sharing the same dialectic relationship, but in a different degree.

Also, there are some other relationships (as a kind of family resemblance) which have something in common with the principle, but they are not exactly the same but only more or less comparable, which varies from case to case.

Good examples of family resemblance in relation to the principle are usefulness-nonusefulness, valueness-nonvalueness, usefulness-uselessness, and the like.

34th Thesis: The Intentionality-Nonintentionality Principle

The thirty-fourth thesis is called the intentionality-nonintentionality principle (on the planning for outcomes) in the category of "outcome" in existential dialectics (which is first worked out in *FPHW*), in that, if there is the planning or design of something for a certain outcome, there is likewise the relative absence of planning or design of something for a certain outcome. There is no intentionality without nonintentionallity—and vice versa.

An intention for an outcome does not have to be certain, so the outcome can even be counter-intentional in being the opposite of what was originally planned. In this sense, counter-intentionality is a special case of non-intentionality; in other words, there are unintented consequences which can be just the opposite of what was originally planned.

An intention for an outcome is mostly relevant to intelligent life, but there are non-intelligent life and non-living matter in the universe too, so nonintentionality is also important.

In addition, an intention does not have to be obvious or straightforward, as William Shakspeare once famously said: "I must be cruel only to be kind."

Especially, works on unintended consequences (like *FPHAERO*) have contributed to the understanding of nonintentionality.

In any event, on the one hand, there is intentionality in outcome (e.g., the planning or design of something for a certain outcome, as shown in the original intention of Ulric Neisser who, in 1967, "proposed a psychology type focusing on the mind's internal processes as they impact behavior," that is, "cognitive psychology," so as to counter the influence of "behavioral psychology," in *Sec. 2.2.1*; the planning or design of something for a certain outcome, as shown in the original intention of the advocates of the "stakeholder theory" to build reputation for the benefit of all concerned, in *Sec. 2.2.2*; the planning or design of something for a certain outcome, as shown in the original intention of those to design "highly detailed and prescriptive protocols" for "easier review," so as to become institutionalized as the preferred audit tool, in *Sec. 2.3.1*; the planning or design of something for a certain outcome, as shown in the original intention of those in the older days who acted solely as the "results" of their "subjective judgments" and interests, prior to the rise of modern accounting conventions, in *Sec. 2.3.2*; the planning or design of something for a certain outcome, as shown in the original intention of those advocating for "the implementation in the United States of the Sarbanes-Oxley Act [SOX] of 2002" to enhance "the profession's exposure and value," in *Sec. 2.4.1*; the planning or design of something for a certain outcome, as shown in the original intention of the German industrialist Dr. Wilhelm von Siemens to propose VAT as a way to increase government revenue, in *Sec. 2.4.2*; the planning or design of something for a certain outcome, as shown in the original intention of those who proposed the "Historical Cost model" in the older days so as to provide an accounting groundwork for the facilitation of business transactions, in *Sec. 2.5.1*; the planning or design of something for a certain outcome, as shown in the original intention of the "American Institute of Certified Public Accountants" to work on some accounting standards for the U.S., later known as US GAAP, in *Sec. 2.5.2*; the planning or design of something for a certain outcome, as shown in the original intention of Gulf Canada to introduce control self-assessment to reduce "risk," in *Sec. 3.2.1*; the planning or design of something for a certain outcome, as shown in the original intention of those who proposed the CBM program so as to cure those with cognitive biases, in *Sec. 3.2.2*; the planning or design of something for a certain outcome, as shown in the original intention of M. King Hubbert to propose "Hubbert peak theory" to analyze "oil production," in *Sec. 3.3.1*; the planning or design of something for a certain outcome, as shown in the original intention of Marsh in 1864 to challenge "the idea that Earth's natural resources are unbounded" for better environmental awareness, in *Sec. 3.3.2*; the planning or design of something for a certain outcome, as shown in the original intention of designing OOP to offer an alternative approach to procedural programming to deal with boilerplate codes, in *Sec. 3.4.1*; the planning or design of something for a certain outcome, as shown in the original intention of the creation of internal control for "the prevention and detection of fraud," in *Sec. 3.4.2*; the planning or design of something for a certain outcome, as shown in the original intention of those to construct professional ethics to combat corruption and fraud, in *Sec. 3.5.1*; and the planning or design of something for a certain outcome, as shown in the original intention of the FASB and IASB to focus on the convergence of accounting standards, in *Sec. 3.5.2*).

On the other hand, there is nonintentionality in outcome (e.g., the relative absence of planning or design of something for a certain outcome, such that the outcome can even be counter-intentional in being the opposite of what was originally planned, as shown in the rejection of "cognitive psychology" by "some critics" nowadays who "find cognitive psychology lacking in validity," because it is a daunting task to prove "the mind's internal processes," so "behavioral psychologists" for the analysis of "external influences" still remains influential in our time—contrary to the original intention of Neisser, in *Sec. 2.2.1*; the relative absence of planning or design of something for a certain outcome, such that the outcome can even be counter-intentional in being the opposite of what was originally planned, as shown in the critique that the obsession with reputation is a psychological problem instead, "a human fault, exaggerated in importance due to the fragile nature of the human ego"—contrary to the original intention of the advocates of the "stakeholder theory" to build reputation for the benefit of all concerned, in *Sec. 2.2.2*; the relative absence of planning or design of something for a certain outcome, such that the outcome can even be counter-intentional in being the opposite of what was originally planned, as shown in the alternative proposal nowadays to use "more flexible protocols that rely on the expertise and knowledge of experienced auditors and source documents [regulations, permits, etc.] directly"—contrary to the original intention of those who design to design "highly detailed and prescriptive protocols" for "easier review," so as to become institutionalized as the preferred audit tool, in *Sec. 2.3.1*; the relative absence of planning or design of something for a certain outcome, such that the outcome can even be counter-intentional in being the opposite of what was originally planned, as shown in the rise of modern accounnting conventions, which involved "individual actions that are often interpreted as results of the subjective judgments of individual actors" but "these norms are achieved without any explicit human intentions or interventions or design," with the "paradox of individual actions resulting unintentionally in the formation of social institutions," according to Takatera and Sawabe—contrary to the original intention of those in the older days who acted solely as the "results" of their "subjective judgments" and interests, in *Sec. 2.3.2*; the relative absence of planning or design of something for a certain outcome, such that the outcome can even be counter-intentional in being the opposite of what was originally planned, as shown in the setback when "the focus by internal audit departments of publicly traded companies on SOX related financial policy and procedures derailed progress made by the profession in the late 20th century toward Larry Sawyer's vision for internal audit"—contrary to the original intention of those advocating for "the implementation in the United States of the Sarbanes-Oxley Act of 2002" to enhance "the profession's exposure and value," in *Sec. 2.4.1*; the relative absence of planning or design of something for a certain outcome, such that the outcome can even be counter-intentional in being the opposite of what was originally planned, as shown in the different problems with the implementation of VAT, like the decrease of sub-national government revenues, tax evasion via "carousel fraud," etc.—contrary to the original intention of the German industrialist Dr. Wilhelm von Siemens to propose VAT as a way to increase government revenue, in *Sec. 2.4.2*; the relative absence of planning or design of something for a certain outcome, such that the outcome can even be counter-intentional in being the opposite of what was originally planned, as shown in the creation of "the Constant Purchasing Power model" as "an alternative basic accounting model" to the "Historical Cost model" by the "International Accounting Standards Board" in 1989—contrary to the original intention of those who proposed the "Historical Cost model" in the older days so as to provide an accoutning groundwork for the facilitation of business transactions, in *Sec. 2.5.1*; the relative absence of planning or design of something for a certain outcome, such that the outcome can even be counter-intentional in being the opposite of what was originally planned, as shown in the call nowadays to abandon US GAAP by adopting IFRS by 2014 or later—contrary to the original intention of the "American Institute of Certified Public Accountants" to work on US GAAP, in *Sec. 2.5.2*; the relative absence of planning or design of something for a certain outcome, such that the outcome can even be counter-intentional in being the opposite of what was originally planned, as shown in the problem with "control self-assessment," since it does not give sufficient attention to "downside risk," etc.—contrary to the original intention of Gulf Canada to introduce control self-assessment to reduce "risk," in *Sec. 3.2.1*; the relative absence of planning or design of something for a certain outcome, such that the outcome can even be counter-intentional in being the opposite of what was originally planned, as shown in the problem with the CBM program, since it was found that "participants with cognitive vulnerability are not able to maintain their new learning even within a CBM course, with their negative cognitive bias beginning to re-emerge over time"—contrary to the original intention of those who proposed the CBM program so as to cure those with cognitive biases, in *Sec. 3.2.2*; the relative absence of planning or design of something for a certain outcome, such that the outcome can even be counter-intentional in being the opposite of what was originally planned, as shown in the application of Hubbert peak theory to non-oil "natural resources" like "natural gas," "coal," etc.—contrary to the original intention of M. King Hubbert to propose "Hubbert peak theory" to analyze "oil production," in *Sec. 3.3.1*; the relative absence of planning or design of something for a certain outcome, such that the outcome can even be counter-intentional in being the opposite of what was originally planned, as shown in the criticism that the understanding of ecosystem services is culturally relative in a way that "cannot be addressed by methods that use universal scientific parameters to determine ecological structures and functions"—contrary to the original intention of Marsh in 1864 to challenge "the idea that Earth's natural resources are unbounded" for better environmental awareness, in *Sec. 3.3.2*; the relative absence of planning or design of something for a certain outcome, such that the outcome can even be counter-intentional in being the opposite of what was originally planned, as shown in the "study by Potok et al. [1999]" which "has shown no significant difference in productivity between OOP and procedural approaches"—contrary to the original intention of designing OOP to offer an alternative approach to procedural programming to deal with boilerplate codes, in *Sec. 3.4.1*; the relative absence of planning or design of something for a certain outcome, such that the outcome can even be counter-intentional in being the opposite of what was originally planned, as shown in the criticisms nowadays that internal control does not guarantee the realization of objectives, that internal control has no control over external factors which can affect the performance of an organization, that there are competing views of "internal control," that there are limitations in the verifiability of data in internal control, that the problem with accounting scandals is not wholly the fault of corporate criminals, and that the nature of business cycles in the free market makes it easier for accounting scandals to occur—contrary to the original intention of the creation of internal control for "the prevention and detection of fraud," in *Sec. 3.4.2*; the relative absence of planning or design of something for a certain outcome, such that the outcome can even be counter-intentional in being the opposite of what was originally planned, as shown in the different criticisms by the critics that there have been business scandals in spite of the existence of accounting ethics, that there are competing approaches to ethical principles in accounting ethics, that morality can be "relative" [not "universal"], that there are the larger institutional factors which contribute to the occurrence of accounting scandals [not just due to the fault of creative accountants], that different countries have different legal standards and regulations in regard to the enforcement of accounting practices, etc.—contrary to the original intention of those to construct professional ethics to combat corruption and fraud, in *Sec. 3.5.1*; and the relative absence of planning or design of something for a certain outcome, such that the outcome can even be counter-intentional in being the opposite of what was originally planned, as shown in the recent announcement of the IASB chairman Hans Hoogervorst, when he suggested in 2012 that "convergence is no longer a top priority for the IFRS standard-setter"—contrary to the original intention of the FASB and IASB to focus on the convergence of accounting standards, in *Sec. 3.5.2*).

And the reverse direction also holds true.

Again, it should be stressed, however, that there are different shades of gray (or different degrees of truth) in the two opposites, and the classification is not necessarily mutually exclusive either. In fact, even when some combinations of the two occur, they only end up sharing the same dialectic relationship, but in a different degree.

Also, there are some other relationships (as a kind of family resemblance) which have something in common with the principle, but they are not exactly the same but only more or less comparable, which varies from case to case.

Good examples of family resemblance in relation to the principle are purposiveness-nonpurposiveness, intention-accidentalness, awareness-unawareness, design-randomness, etc.

35th Thesis: The Survivability-Nonsurvivability Principle

The thirty-fifth thesis is called the survivability-nonsurvivability principle (on the survival of things) in the category of "outcome" in existential dialectics (which is first worked out in *FPHPROB*), in that, if there are some entities which survive at certain times, there are likewise other entities which do not survive. There is no survivability without nonsurvivability—and vice versa.

In nature, for illustration, there are some animals which survive in hard times, whereas others do not—and this is true, not just among different species but also among different members of the same species (and the phenomenon of "selection" in evolutionary biology has much to say on this, as already addressed in *BNN*).

Of course, the principle is not restricted solely to the animal world, since it can be used to other cases concerning non-animals too (e.g., trees, lakes, etc. in the natural world—or empires, companies, etc. in the human world).

But one should not confuse this principle with the expansion-contraction principle, since those entities which survive, say, can still expand or contract over time (e.g., those animals which survive can be more capable in some activities but not in others, for instance).

Nor should this principle be confused with the stability-reaction principle, in that the condition of survivability or non-survivability may lead to further changes later on (as shown in the emergence of new species after the extinction of some old ones, as already explained in *BNR*).

In any event, on the one hand, there is survivability in outcome (e.g., the relative survival, nowadays, of the accounting reform against creative accounting, in *Sec. 2.2.1*; the relative survival, nowadays, of "purchasing accounting," in *Sec. 2.2.2*; the relative survival, nowadays, of EMAS III, in *Sec. 2.3.1*; the relative survival, nowadays, of the "going concern," in *Sec. 2.3.2*; the relative survival, nowadays, of "the broader role" of "internal auditing" advocated by the IIA since 2010, in *Sec. 2.4.1*; he relative survival, nowadays, of VAT, in *Sec. 2.4.2*; the relative survival, nowadays, of the "Historical Cost model," as it is "still used in most accounting systems," in *Sec. 2.5.1*; the relative survival, nowadays, of IFRS, in *Sec. 2.5.2*; the relative survival, nowadays, of "control self-assessment," in *Sec. 3.2.1*; the relative survival, nowadays, of the mental accounting theory, in *Sec. 3.2.2*; the relative survival, nowadays, of the approach to sustainbility accounting by "measuring environmental-economic-social interrelationships," in *Sec. 3.3.1*; the relative survival, nowadays, of the "environmental awareness," in *Sec. 3.3.2*; the relative survival, nowadays, of the need for high-level mechanism like "metaprogramming," "convention over configuration," and "model-driven engineering," in *Sec. 3.4.1*; the relative survival, nowadays, of "the Sarbanes–Oxley Act of 2002" in the U.S., in *Sec. 3.4.2*; the relative survival, nowadays, of the international standards like IFRS, in *Sec. 3.5.1*; and the relative survival, nowadays, of the efforts for the convergence of accounting standards, in *Sec. 3.5.2*).

On the other hand, there is non-survivability in outcome (e.g., the relative non-survival, nowadays, of those firms like Enron, Worldcom, etc., because of the accounting scandals, in *Sec. 2.2.1*; the relative non-survival, nowadays, of "pooling-of-interests" accounting, in *Sec. 2.2.2*; the relative non-survival, nowadays, of EMAS I, in *Sec. 2.3.1*; the relative non-survival, nowadays, of the "pre-going concern," in *Sec. 2.3.2*; the relative non-survival, nowadays, of the more narrow focus of internal auditing under the Sarbanes-Oxley Act of 2002, before it was reinterpreted by the IIA in 2010, in *Sec. 2.4.1*; the relative non-survival, nowadays, of the older way to implement VAT without addressing the "regressive" impact on the poor, since "to maintain the progressive nature of total taxes on individuals, countries implementing VAT have reduced income tax on lower income-earners as well as instituted direct transfer payments to lower-income groups, resulting in lower tax burdens on the poor," in *Sec. 2.4.2*; the relative non-survival, nowadays, of the "Historical Cost model" among those in the "International Accounting Standards Board" who proposed in 1989 "the Capital Maintenance in Units of Constant Purchasing Power model" as "an alternative basic accounting model" to the "Historical Cost model," in *Sec. 2.5.1*; the relative non-survival, nowadays, of US GAAP, as it is in the process of moving to IFRS by 2014 or later, in *Sec. 2.5.2*; the relative non-survival, nowadays, of the habit of those auditors who "initially ignored the benefits of control self-assessment even though it was effective at providing audit evidence around the 'soft' areas [such as staff morale] that are critical to the effectiveness of internal control systems," in *Sec. 3.2.1*; the relative non-survival, nowadays, of the naïve belief that a CBM program can cure those with cognitive biases, since it is now recognized that "participants with cognitive vulnerability are not able to maintain their new learning even within a CBM course, with their negative cognitive bias beginning to re-emerge over time," in *Sec. 3.2.2*; the relative non-survival, nowadays, of the "growth-based" economic approach, in *Sec. 3.3.1*; the relative non-survival, nowadays, of the practice of the "tragedy of the commons," in *Sec. 3.3.2*; the relative non-survival, nowadays, of the need for boilerplate manual codes, without any use of high-level mechanism like "metaprogramming," "convention over configuration," and "model-driven engineering," such that "the programmer must write a lot of code to do minimal jobs," in *Sec. 3.4.1*; the relative non-survival, nowadays, of the "dual administration" as it existed in "Hellenistic Egypt," "with one set of bureaucrats charged with collecting taxes and another with supervising them," in *Sec. 3.4.2*; the relative non-survival, nowadays, of US GAAP, since the Securities and Exchange Commission in the U.S. "issued a preliminary 'roadmap'" in 2008 "that may lead the U.S." to abandon US GAAP and to join IFRS instead, in *Sec. 3.5.1*; and the relative non-survival, nowadays, of individiual country accounting standards like the US GAAP, since it is to be eventually abandoned by joining the IFRS, under the convergence of accounting standards, in *Sec. 3.5.2*).

And the reverse direction also holds true.

Again, it should be stressed, however, that there are different shades of gray (or different degrees of truth) in the two opposites, and the classification is not necessarily mutually exclusive either. In fact, even when some combinations of the two occur, they only end up sharing the same dialectic relationship, but in a different degree.

Also, there are some other relationships (as a kind of family resemblance) which have something in common with the principle, but they are not exactly the same but only more or less comparable, which varies from case to case.

Good examples of family resemblance in relation to the principle are survival-extinction, existence-nonexistence, life-death, etc.

36th Thesis: The Post-Human Double-Sidedness

And the thirty-sixth thesis is about the role of "post-humans," which I originally proposed in my previous books, starting with the first book titled *The Future of Human Civilization* in 2000 and all others afterwards.

As already pointed out in *Sec. 1.8*, I need to emphasize, as this is something that I used to repeat (*almost verbatim*) from my previous books, two clarifications here about the term "post-human" as a neologism in my works.

Firstly, the word "post-human" here should *not* be confused with another term which looks similar but has a totally different meaning in the literature of postmodernism, namely, "post-humanism"—which constitutes a critique of "humanism" as tradionally understood (especially, though not exclusively, in relation to the idea of progress in science and reason in the Enlightenment project). (WK 2008)

My works reject the project of "postmodernism" and propose the future world of what I orignally called "after-postmodernity" in *FHC* and *FCD*, for instance.

And secondly, the word "post-human" here should also *not* be confused with a similar term which is used to champion the ideology of technology for the future co-existence between humans and cyborgs in "trans-humanism." (WK 2008a)

Instead, my term "post-human" in relation to "posthuman-ism" also rejects "transhumanism" (especially, though not exclusively, in relation to the promises of technology) and refers to something else altogether, that is, the future extinction of humans and its post-human successors in deep space and beyond unto multiverses.

My critique of "transhumanism" was more extensively elabroated in *Sec. 2.4.1* of *BEPE*—just as my critique of different futuristic scenarios of fictional artificial intelligence in science fiction was given in *Sec. 3.5.2* of *FPHCOM*.

With these two clarifications in mind (as summarized in *Table 1.31*)—the post-human double-sidedness in accounting can be addressed in relation to five main directions, based on my *double-sided theory of accounting*, as analysed in this book (together with my visions as already worked out in my previous books).

Firstly, the post-human double-sidedness in accounting will learn from my double-sided theory of accounting, in relation to the dialectic context of no analysis of addition without that of subtraction (and vice versa), to be later transcended, in the context of accounting.

Secondly, the post-human double-sidedness in accounting will learn from my double-sided theory of accounting in relation to the four perspectives of the mind, nature, society, and culture.

Thirdly, the post-human double-sidedness in accounting will learn from my double-sided theory of accounting in relation to the non-privilege of any specific theory over others in the literature and the non-integration of them all (as they are not necessarily compatible with each other).

Fourthly, the post-human double-sidedness in accounting will have to confront the new challenges as posed by the long-term civilizational development of intelligent life in the distant future, both here on this planet Earth and elsewhere in deep space until multiverses.

As this is something that I regularly pointed out in my previous books for background information (and summarize here, *almost verbatim*, with some updates), I already worked out, in my numerous books, what these new challenges will be and provide different original visions to meet them—especially in regard to the following future evolutions:

- of the mind (e.g., "interconnected health care," "supersession computing," "metamorphic humor," "contrarian personality," "hyper-martial body," "hyper-sexual body," "floating consciousness," "hyper-spatial consciousness," "thinking machines," "unfolding unconsciousness," "thinking robots," "genetically altered superior beings," "cyborgs," and others)
- of nature (e.g., "resilient natural resources," "transdisciplinary architecture," "creational chemistry," "multilateral sound," "resettlement geology," "post-cosmology," "the alteration of space-time," "the creation of new matter-energy," "selective geometry," and the like)
- of society (e.g., "post-capitalism," "post-democracy," "authoritarian liberal democracy," "transfigurative waste," "panoramic transportation," "cyclical-progressive migration," "reflective criminology," "interventive-reshaping geography," "detached gambling," "heterodox education," "multifaceted war and peace," "virtual organizations," the movement of "cyclical progression" at both structural and systemic levels, "ambivalent technology," and so forth)
- of culture (e.g., "inclusivist phonology," "fusional morphology," "modificative sports," "inquisitive culinary art," "interactive

semantics," "transdisciplinary performing arts," "comparative-impartial literature," "post-human mind games," "comparative-substitutive religion," "post-ethics," "post-civilization," "transformative aesthetic experience," "contrastive mathematical-logic," and the whatnot)

- in history (e.g., "multifold history," the age of "after-postmodernity," etc.)
- in methodology (e.g., "sophisticated methodologial holism," "critical-dialectic formal science," "complex data analysis," "interpretivist probability," etc.)
- in ontology (e.g., "existential dialectics," "contrastive rationality," etc.)

Of course, the examples (as listed above) are not exhaustive but illustrative, since my numerous books have worked out many other visions (as already roughly summarized in *Sec.1.8* and *Sec. 1.9*). Many, though not all, of my visions on the mind, nature, society, culture, history, methodology, and ontology (as cited above) are summarized in the tables as shown in Chapter One and Chapter Four (especially in *Table 4.43* about my original theories on numerous topics).

And finally, or fifthly, the post-human double-sidedness in accounting will learn from my double-sided theory of accounting in regard to the need to go beyond both addition and subtraction—to be further elaborated in the next section.

Towards the Post-Human Double-Sidedness

This "post-human" double-sidedness to transcend both addition and subtraction (in the context of accounting) can be understood in terms of nine great future transformations of accountiing, in relation to the mind, nature, society, and culture—as explained below (and summarized in *Table 1.2*).

Nine Great Future Transformations of Accounting

The nine great future transformations of accounting (in relation to the mind, nature, society, and culture) can be explained hereafter, with the caveat, however, that the classification in terms of nine is solely illustrative, as there is no objective basis that it must be classified in term of nine (as this is something that I often stressed and repeated in my previous books, albeit in the context of different topics).

The First Great Future Transformation of Accounting

The first great future transformation of accounting in the post-human era concerns what I originally call *new goals,* in relation to the mind.

This search for "new goals" will give future advanced humans and later their successors, namely, the post-humans, the capacity to be motivated to make accounting in a way that current humans cannot.

There are already some indications that the future of accounting is moving toward this direction.

Consider, for illustration, the different "goals" as already seen in the history of accounting, such as the quest for creative deception (*Sec. 2.2.1*), the search for reputation capital (*Sec. 2.2.2*), the search for eco-management (*Sec. 2.3.1*), the search for the understanding of space-time in accounting (*Sec. 2.3.2*), the quest for better financial performance (*Sec. 2.4.1*), the search for government revenue (*Sec. 2.4.2*), the quest for accounting conventions (*Sec. 2.5.1*), the aim for accounting standards (*Sec. 2.5.2*), the quest for control self-assessment (*Sec. 3.2.1*), the prosecution of creative accounting (*Sec. 3.2.2*), the study of carbon accounting (*Sec. 3.3.1*), the quest for the knowledge of "environmental profit and loss" (*Sec. 3.3.2*), the quest for more efficient computing to do bookkeeping (*Sec. 3.4.1*), the quest for better internal control (*Sec. 3.4.2*), the quest for better accounting ethics (*Sec. 3.5.1*), and the quest for the convergence of accounting standards (*Sec. 3.5.2*), for instance.

More importantly, this future quest for new goals should also be understood in relation to my previous works on the future "post-human" supersession of the human mind from the combined perspectives of the mind, nature, society, and culture—with good examples like "supersession computing," "metamorphic humor," "contrarian personality," "hyper-martial body," "hyper-sexual body," "floating consciousness," "hyper-spatial consciousness," "unfolding unconsciousness," "thinking robots," "thinking machines," "genetically altered superior beings," "cyborgs," and others.

For instance, most advanced forms of future intelligent life in deep space until multiverses (e.g., "genetically altered superior beings," "floating consciousness," "hyper-spatial consciousness," "unfolding unconsciousness," etc.) will confront the issues concerning creative accouning (as already discussed in *Sec. 2.2.1*), mental accounting (as already discussed in *Sec. 3.2.2*), artificial intelligence (as already discussed in *Sec. 3.4.1*), and others in the distant future.

In any event, all these are just some illustrative (relatively primitive) examples in our time, of course, but the important point to remember here is that the search for different new goals will further transform accounting at the more advanced level in the post-human future that the human world has never known.

The Second Great Future Transformation of Accounting

The second great future transformation of accounting in the post-human era concerns what I originally call *new means,* in relation to the mind.

This quest for "new means" will give future advanced humans and later their successors, namely, the post-humans, the capacity to find novel options to achieve the goals in regard to accounting in a way that current humans cannot.

There are already some indications that the future of accounting is moving toward this direction.

Consider, for illustration, the different "means" as already seen in the history of accounting, such as the use of "unsuitable revenue recognition" as a means for creative accounting (*Sec. 2.2.1*), the analysis of the psychology of "reputation" as a way to understand "goodwill" (*Sec. 2.2.2*), the use of "environmental audit" as a way for eco-management (*Sec. 2.3.1*), the analysis of "demarcated space" as a way to understand space-time in accounting (*Sec. 2.3.2*), the study of auditing as a way for better financial performance (*Sec. 2.4.1*), VAT as a way to increase government revenue (*Sec. 2.4.2*), the analysis of accounting principles as a way to study accounting conventions (*Sec. 2.5.1*), writing standards into law as a way to create accounting standards (*Sec. 2.5.2*), interview techniques as a way to do control self-assessment (*Sec. 3.2.1*), the study of "birthday money" as a way to understand mental accounting (*Sec. 3.2.2*), sustainability measurement as way to analyze carbon accounting (*Sec. 3.3.1*), the analysis of "ecosystem services" as a way to study E P&L (*Sec. 3.3.2*), the use of computing for bookkeeping (*Sec. 3.4.1*), the use of internal control to minimize corporate misdeeds (*Sec. 3.4.2*), the use of accounting standards as way to enforce ethics (*Sec. 3.5.1*), and the reduction of the differences between the US GAAP and the IFRS as a way to converge accounting standards (*Sec. 3.5.2*), for instance.

All these are just some illustrative (relatively primitive) examples in our time, of course, but the important point to remember here is that the search for different new means will further transform accounting at the more advanced level in the post-human future that the human world has never known.

The Third Great Future Transformation of Accounting

The third great future transformation of accounting in the post-human era concerns what I originally call *new comparisons*, in relation to nature.

This search for "new comparisons" will give future advanced humans and later their successors, namely, the post-humans, the ability to learn from animals, plants, and the universe, through comparison and contrast, about accounting in a way that current humans cannot.

There are already some indications that the future of accounting is moving toward this direction.

Consider, for illustration, the different forms of comparison and contrast as already seen in the history of accounting, such as the comparison and contrast between organizations and nature in regard to environmental performance (*Sec. 2.3.1*), the comparison and contrast between accounting and nature in regard to space-time (*Sec. 2.3.2*), the comparison and contrast between humans and nature in regard to carbon emissions (*Sec. 3.3.1*), and the comparison and contrast between companies and nature in regard to environmental impacts (*Sec. 3.3.2*), for instance.

All these are just some illustrative (relatively primitive) examples in our time, of course, but the important point to remember here is that the search for different new comparisons will further transform accounting at the more advanced level in the post-human future that the human world has never known.

The Fourth Great Future Transformation of Accounting

The fourth great future transformation of accounting in the post-human era concerns what I originally call *new integrations*, in relation to nature.

This search for "new integrations" will give future advanced humans and later their successors, namely, the post-humans, the ability to live and interact with nature for accounting in a way that current humans cannot.

There are already some indications that the future of accounting is moving toward this direction.

Consider, for illustration, the different integrations for accounting as already seen in the history of accounting, such as the integration of environmental performance into the management of private and public organizations (*Sec. 2.3.1*), the integration of "chaos theory" and "complexity theory" in natural science to the study of the evolution of accounting (*Sec. 2.3.2*), the integration of environmental sustainability into the analysis of carbon emissions by nation states, corporations and individuals (*Sec. 3.3.1*), and the integration of "ecosystem services" into E P&L accounting (*Sec. 3.3.2*), for instance.

More importantly, this future quest for new integrations should also be understood in relation to my previous works on the future "post-human" adventures into different dimensions of space-time unto multiverses from the combined perspectives of the mind, nature, society, and culture—with good examples like "the alteration of space-time" (*Table 1.3* and *Table 1.4*), "multiverses" (*Table 1.7* and *Table 1.8*), "hyperspace" (*Table 1.9* and *Table 1.25*), "time travel" (*Table 1.10* and *Table 1.11*), and others.

In addition, in my previous book titled *The Future of Post-Human Geology* (2010), I already envisioned the three great future transformations of geology, namely, "solar geology," "galactic geology," and "cosmic geology," in the context of my "resettlement theory of geology," in that humans and later post-humans will explore and migrate to other planets, solar systems, galaxies, and cosmos in the long haul.

Then, in another book of mine titled *The Future of Post-Human Geography* (2011), I linked the two books together, since geography has much to learn from geology (as in "physical geography," which, as already introduced in *Sec. 1.3* of *FPHGEOG*, relies heavily on the earth sciences) and discussed the search for "new explorations" in the context of the three forms of environments outside the planet Earth, namely, what I originally called "solar geography," "galactic geography," and "cosmic geography"—as a further elaboration of the three great future transformations of geology (in *FPHGEOL*).

Later, in still another book of mine titled *The Future of Post-Human Migration* (2012), I envisioned the future integrations in different forms, namely, what I originally called "solar migration," "galactic migration," "clustery migration," "cosmic migration," and "multiversal migration."

And then, in an additional book of mine titled *The Future of Post-Human Transportation* (2012), I envisioned the future integrations in different forms, namely, what I originally called "solar transportation," "galactic transportation," "clustery transportation," "cosmic transportation," and "multiversal transportation."

This extension is consistent with my other theory, namely, "the theory of the cyclical progression of system integration and fragmentation" in *FCD* (2002)—at the social-systemic level (as already clarified in *Sec. 1.8* and shown below):

Local → Regional → Global →Solar →
Galactic → Clustery…→ Multiversal

All these are just some illustrative (relatively primitive) examples in our time, of course, but the important point to remember here is that the search for different new integrations will further transform accounting at the more advanced level in the post-human future that the human world has never known.

The Fifth Great Future Transformation of Accounting

The fifth great future transformation of accounting in the post-human era concerns what I originally call *new praxes*, in relation to society.

This search for "new praxes" will give future advanced humans and later their successors, namely, the post-humans, the ability to engage in practice (like activism, politics, personal experience, etc.) for the future of accounting in a way that current humans cannot.

There are already some indications that the future of accounting is moving toward this direction.

Consider, for illustration, the different praxes as already seen in history of accounting like the practical application of Abraham Briloff's work to the fight against creative accounting (*Sec. 2.2.1*), the practical application of the 3 levels of "reputation-based decisions" in the field of psychology to the field of "reputation management" (*Sec. 2.2.2*), the practical application of "voluntary environmental management instrument" in the interdisciplinary field of environmental studies and accounting to the political realm like the use of EMAS by the European Commission (*Sec. 2.3.1*), the practical application of chaos theory" in mathematics to the field of accounting (*Sec. 2.3.2*), the practical application of "internal auditing" by "the IIA" in accordance to the IPPF's philosophy" envisioned by Sawyer (*Sec. 2.4.1*), the practical application of "VAT" to the EU countries (*Sec. 2.4.2*), the practical application of "purchasing power" in the field of economics to the field of accounting, like the "Constant Purchasing Power model" (*Sec. 2.5.1*), the practical application of "IFRS" in the field of finance and accounting to countries around the world (*Sec. 2.5.2*), the practical application of "a framework to support the analysis and evaluation of control processes by operational staff" to the public sphere like its application by "the Federal Deposit Insurance Corporation and the Canadian Deposit Insurance Corporation (*Sec. 3.2.1*), the practical application of the idea of "cognitive bias" in the field of psychology to the field of accouunting (*Sec. 3.2.2*), the practical application of the "pluralised concept of knowledge" in the field of philosophy to the field of accounting (*Sec. 3.3.1*), the practical application of "ecosystem" in the field of ecology to the field of E P & L accounting (*Sec. 3.3.2*), the practical application of "inheritance" in social sciences to the field of computing like "object inheritance" in OOP (*Sec. 3.4.1*), the practical application of "internal control procedures" in the field of accounting to the public sphere (*Sec. 3.4.2*), the practical application of "ethical principles" in the field of ethics to the field of accounting, like "accounting ethics" (*Sec. 3.5.1*), and the practical application of "fair value" in the field of economics to the realm of the convergence of accounting standards (*Sec. 3.5.2*), for instance.

All these are just some illustrative (relatively primitive) examples in our time, of course, but the important point to remember here is that the search for different new praxes will further transform accounting at the more advanced level in the post-human future that the human world has never known.

The Sixth Great Future Transformation of Accounting

The sixth great future transformation of accounting in the post-human era concerns what I originally call *new technologies*, in relation to society.

This search for "new technologies" will give future advanced humans and later their successors, namely, the post-humans, the ability to come up with different techniques, materials, and experiments for accounting in a way that current humans cannot.

There are already some indications that the future of accounting is moving toward this direction.

Consider, for illustration, the use of different technologies as already seen in the history of accounting, such as the technique to study "the personalities of folks" in cognitive psychology (*Sec. 2.2.1*), the technique of reputation management (*Sec. 2.2.2*), the technique of audit scheme (*Sec. 2.3.1*), the technique of "artificially partitioning continuous external transactions into accounting periods" (*Sec. 2.3.2*), the technique of "integrated auditing" (*Sec. 2.4.1*), the technique of "invoice-based" VAT collection (*Sec. 2.4.2*), the technique of calculating historical cost (*Sec. 2.5.1*), the technqiue of recording revenue when realized and earned (*Sec. 2.5.2*), interview techniques (*Sec. 3.2.1*), techniques of mental accounting (*Sec. 3.2.2*), the technique to measure carbon footprint (*Sec. 3.3.1*), the technique to measure and monitor the footprint of a company's operations in E P&L accounting (*Sec. 3.3.2*), the technology of OOP (*Sec. 3.4.1*), the technique of "activity categorization" (*Sec. 3.4.2*), the technique of new regulations for accounting reform (*Sec. 3.5.1*), and the technique of doing the Fair Value Option (*Sec. 3.5.2*), for instance.

All these are just some illustrative (relatively primitive) examples in our time, of course, but the important point to remember here is that the search for new technologies will further transform accounting at the more advanced level in the post-human future that the human world has never known.

The Seventh Great Future Transformation of Accounting

The seventh great future transformation of accounting in the post-human era concerns what I originally call *new stratifications*, in relation to society.

This emergence of "new stratifications" will lead future advanced humans and later their successors, namely, the post-humans, to face new inequalities in accounting in a way that current humans cannot.

There are already some indications that the future of accounting is moving toward this direction.

Consider, for illustration, the different inequalities as already seen in the history of accounting, such as governments vs. creative accountants (*Sec. 2.2.1*), the advocates of the pooling-of-interests method" vs. the advocates of the "purchase accounting" (*Sec. 2.2.2*), the European Commission vs. individual companies (*Sec. 2.3.1*), the advocates of "deterministic" approach vs. the advocates of "non-deterministic" approach to the study of accounting (*Sec. 2.3.2*), "external auditors" vs. "internal auditors" in regard to "independence" (*Sec. 2.4.1*), the government vs. the consumers (*Sec. 2.4.2*), the advocates of the "Historical Cost model" vs. the advocates of the "Constant Purchasing Power model" (*Sec. 2.5.1*), AICPA vs. individual business enterprises (*Sec. 2.5.2*), governments vs. individual businesses in regard to internal control (*Sec. 3.2.1*), those for the "good" side vs. those for the "bad" sides of mental accounting (*Sec. 3.2.2*), the UK coalition government vs. the UK's largest listed companies (*Sec. 3.3.1*), animal impact on nature vs. human impact on nature (*Sec. 3.3.2*), "inheritance" vs. "abstraction" in regard to the order of popularity in OOP (*Sec. 3.4.1*), the Securities and Exchange Commission in the United States vs. individual companies (*Sec. 3.4.2*), the Italian court system vs. Parmalat (*Sec. 3.5.1*), and FASB and IASB vs. individual companies (*Sec. 3.5.2*), for instance.

All these are just some illustrative (relatively primitive) examples in our time, of course, but the important point to remember here is that the emergence of new stratifications will further transform accounting at the more advanced level in the post-human future that the human world has never known.

The Eighth Great Future Transformation of Accounting

The eighth great future transformation of accounting in the post-human era concerns what I originally call *new sub-fields*, in relation to culture.

This search for "new sub-fields" will give future advanced humans and later their successors, namely, the post-humans, the ability to explore new specialties (or new games) for accounting in a way that current humans cannot.

There are already some indications that the future of accounting is moving toward this direction.

Consider, for illustration, the different sub-fields as already seen in the history of accounting, such as the sub-field of creative accounting (*Sec. 2.2.1*), the sub-field of reputation management (*Sec. 2.2.2*), the sub-field of eco-management (*Sec. 2.3.1*), the sub-field of accrual accounting (*Sec. 2.3.2*), the sub-field of auditing (*Sec. 2.4.1*), the sub-field of taxation (*Sec. 2.4.2*), the sub-field of accounting principles (*Sec. 2.5.1*), the sub-field of accounting standards (*Sec. 2.5.2*), the sub-field of control self-assessment (*Sec. 3.2.1*), the sub-field of mental accounting (*Sec. 3.2.2*), the sub-field of carbon accounting (*Sec. 3.3.1*), the sub-field of E P&L accounting (*Sec. 3.3.2*), the sub-field of computing accounting (*Sec. 3.4.1*), . the sub-field of internal control (*Sec. 3.4.2*), the sub-field of accounting ethics (*Sec. 3.5.1*), and the sub-field of the history of accounting standards (*Sec. 3.5.2*), for instance.

All these are just some illustrative (relatively primitive) examples in our time, of course, but the important point to remember here is that the search for new sub-fields will further transform accounting at the more advanced level in the post-human future that the human world has never known.

The Ninth Great Future Transformation of Accounting

And the ninth great future transformation of accounting in the post-human era concerns what I originally call *new interpretations*, in relation to culture.

This search for "new interpretations" will give future advanced humans and later their successors, namely, the post-humans, the ability to come up with different evaluations and views about accounting in a way that current humans cannot.

There are already some indications that the future of accounting is moving toward this direction.

Consider, for illustration, the different interpretations as already seen in the history of accounting, such as the argument against creative accounting as "fraud" (*Sec. 2.2.1*), the argument for reputation capital (*Sec. 2.2.2*), the argument for eco-management (*Sec. 2.3.1*), the argument for the use of chaos theory to study accounting (*Sec. 2.3.2*), the argument for internal audit (*Sec. 2.4.1*), the argument for VAT (*Sec. 2.4.2*), the argument for the principle of "unit of measure" (*Sec. 2.5.1*), the argument for Uniform Accounting Standards (*Sec. 2.5.2*), the argument for control the self-assessment (*Sec. 3.2.1*), the argument about cognitive biases in mental accounting (*Sec. 3.2.2*), the argument for sustainability measurement (*Sec. 3.3.1*), the argument for analyzing ecosystem services (*Sec. 3.3.2*), the argument for the use of OOP to do bookkeeping (*Sec. 3.4.1*), the argument for "precision control" (*Sec. 3.4.2*), the argument for new regulations against accounting scandals (*Sec. 3.5.1*), and the argument in favor of the convergence of accounting standards (*Sec. 3.5.2*), for instance.

All these are just some illustrative (relatively primitive) examples in our time, of course, but the important point to remember here is that the search for new interpretations will further transform accounting at the more advanced level in the post-human future that the human world has never known.

Enormous Implications

These nine great future transformations of accounting have enormous implications for some soul-searching questions which have challenged some of the best minds in the history of accounting (in relation to addition and subtractioin).

For illustration, consider the following questions in the history of accounting:

- Ex: to what extent is the profession of accounting the guardians of social interests?
- Ex: in what ways is the historical cost assumption in accounting to be changed for a better understanding of time?
- Ex: to what extent can creative accounting be better used?
- Ex: what exactly are the use and misuse of mental accounting?
- Ex: in what ways can computing be better applied for accounting?
- Ex: is there really a universal accounting ethics?
- Ex: to what extent is objectivity in accounting possible?

- Ex: to what extent is monetization too narrow for measurement in accounting?
- Ex: to what extent can the separate entity assumption in accounting be revised to include broader interests?
- Ex: in what ways are the assumptions on space in accounting be revised for the better?
- Ex: to what extent can accounting standards be raised for better outcomes without too high costs?
- Ex: to what extent is the convergence of accounting standards really desirable?

Questions and disagreements like this will receive new answers in the future history of accounting in the advanced human and later post-human eras, at least in two fundamental ways.

The first fundamental way is to examine the relationships between addition and subtraction—such that their possibility (or impossibility) and desirability (or undesirability) are not to the extent that the respective defenders would like us to believe.

The different views about accounting, as already addressed in *Sec. 1.1*, *Sec. 1.5*, and the rest of the book, are just different ways to understand accounting and its future, with each from its own vantage point.

Each view has its own merits and defects and is preferred over others by different proponents with their own different interests, but none of them is possible (or impossible) and desirable (or undesirable) to the extent that its respective advocates would like us to believe.

And the second fundamental way is to inquire about, and then transcend, in the long haul, the debate about addition and subtraction, such that it will be shifted to different directions, as shown in the nine great future transformations that I have envisioned and outlined above. To transcend them all is to move into these different directions, in conjunction with the various visions of my previous books (as already summarized in *Sec. 1.8* and also earlier in this chapter, with the further help of the tables at this end of Chapter One and this chapter).

Some good examples can be summarized below:

- Ex: to discover the redemptive role of creative accounting
- Ex: to explore the better use of mental accounting
- Ex: to seek a better way to deal with space and time in accounting
- Ex: to create better computing for accounting
- Ex: to better cope with social interests in the profession of accounting

- Ex: to raise accounting standards for better outcomes without too high costs
- Ex: to broaden measurement in accounting beyond monetization
- Ex: to build an accounting ethics beyond universality and relativity
- Ex: to transcend objectivity and subjectivity for a new accounting
- Ex: to balance convergence and divergence in accounting standards
- Ex: to prepare accounting for the future eras of "post-capitalism," "post-democracy," and "post-civilization" towards the post-human direction

Of course, these examples are illustrative, not exhaustive.

Lest any misunderstanding occurs, one must remember that these nine great future transformations of accounting will bring neither utopia nor dystopia, regardless of some optimistic expectations about the use of computing for accounting in our time, just to cite an obvious example—as this is something that I have warned time and again in all of my previous books, *almost verbatim*.

In any event, as I so often stressed in my previous books, it is not up to us current humans to decide whether or not, or to what extent, the post-human future of accounting will be for better and for worse, because it will be judged by the values and beliefs of future post-humans on earth and in deep space unto multiverses—in a way that the entire history of the human world has never known.

Thus is the future of accounting—and of its post-human fate.

Table 4.1. Sophisticated Methodological Holism (Part I)

• "My methodological holism implies the partiality-totality principle in the ontology of existential dialectics (see the table on the partiality-totality principle for summary), which is against the varieties of (a) reductionism and (b) reverse-reductionism, in relation to (i) concept, (ii) theory, (iii) methodology, and (iv) ontology." (*FC*)

• "[M]y methodological holism here is not opposed to methodological individualism but includes it (and, for that matter, other methodologies too)…." (*FPHC*) For this reason (and others too, as summarized hereafter), my version of methodological holism is sophisticated—not vulgar as sometimes used by inapt scholars using the same term. (*FC*)

• "[M]y methodological holism does not democratically presume that all levels are equally valid, as all levels are not created equal. In other words, in relation to issue X, level A may be more relevant than level B, but in relation to Y, level B can be more relevant than level A instead." (*FPHC*) One excellent example of this vulgar democratic presumption is what I called in *BNN* "the compromise fallacy." (*FC*)

• My methodological holism does not presume that a lower level of analysis is more important than a higher level, solely because the former serves as the foundation for the latter—and vice versa, for that matter. One excellent example of this reductionistic presumption is what I called in *FPHST* "the foundation fallacy." (*FPHST*, *FC*)

(continued on next page)

Table 4.1. Sophisticated Methodological Holism (Part II)

• "[M]y methodological holism does not make any a-priori postulation that there must be a definite (and, for that matter, indefinite) number of levels" in any analysis. (*FPHC*) Nor does it dogmatically require that there must be a certain combination of levels of analysis in a given inquiry. (*FC*)

• "[M]ethodological holism, in my usage, does not assume that all levels…can necessarily be integrated, since methodological holism is not aimed to search for the holy grail of 'an integral theory…' (as is the case for Wilber). In other words, it allows that sometimes some levels may experience irreducible gaps between them, to be understood, at best, as empirical correlations, not as causal relations…." (*FPHC*)

• "[D]ifferent levels may overlap and even interact with each other in a given context (but sometimes may not), and the fact that I even proposed different ways of re-classifying the levels (whenever needed) in *FDC* reinforces this point….The dual danger here is either forcefully making different levels interact when they are just different (or, metaphorically speaking, apples and oranges) or inappropriately ignoring their interactions when some situations instead require them." (*FPHC*)

• "[T]o understand different levels from their own (unique) perspectives (as required by my methodological holism) is not the same as trying to reduce them to a preferred level in the process of learning from other levels. This second kind of multidisciplinary work is not genuine and does no justice to the unique complexities and merits inherent at each level." (*FPHC*)

(continued on next page)

Table 4.1. Sophisticated Methodological Holism (Part III)

• "[My] methodological holism walks a fine line between the artificial classification (separation) of levels and the simultaneous incorporation of them, if only for the sake of human scholarly endeavor. It should be reminded that nature does not impose upon itself the academic classification of the levels of analysis as humans have. The enterprise of classification is therefore anthropocentric." (*FPHC*)

• "[M]y methodological holism advocates neither epistemic subjectivism nor epistemic non-subjectivism (e.g., realism, idealism, and historicism), neither epistemic relativism (e.g., subjectivism, historicism) nor epistemic absolutism (e.g., realism, positivism), neither epistemic reductionism nor epistemic emergencism, and neither epistemic objectivism (e.g., realism, idealism) nor epistemic historicism....Neither does methodological holism, in my usage, accept the false meta-conceptual dichotomy between nominalism and realism....These false dichotomies...are to be transcended. In other words, methodological holism does not fully accept epistemic realism, positivism (a form of epistemic idealism), historicism, subjectivism, and reductionism in epistemology and philosophy of science but learns from the strengths and weaknesses of all of the opposing approaches without siding with any of them...." (*FPHC*)

• "Sophisticated methodological holism is subject to the constraints as imposed by the syntax of existential dialectics (e.g., the partiality-totality principle and the predictability-unpredictability principle). Even in predictability, outcomes are subject to uncertainty, the degree of which varies from case to case." (*FC*)

(continued on next page)

Table 4.1. Sophisticated Methodological Holism (Part IV)

- "Sophisticated methodological holism—when applied, especially though not exclusively, as illustrated in my numerous works—can enrich the understanding of reality in some distinctive ways. Here are three examples (as revealed in each of my books)." (*FPHCT*; *FPHL*)
  - —"Firstly, it provides a comprehensive analysis of a subject matter, from which much can be learned about reality," "both in relation to the perspectives of the mind, nature, society, and culture and also in relation to a new classification of the subject matter."
  - —"Secondly, it suggests some visions of the future in relation to the subject matter in question."
  - —"And thirdly, it proposes some insights on meta-theory (e.g., methodology and ontology) in general—with the clear understanding, however, of the dilemma of specific vs. general ontology (as shown in the table on the syntax of existential dialectics in the context of the dilemma of ontology). For this very reason, all of these ways are important, without reducing one into the analysis of another."
- "In the end, my meta-theory (both sophisticated methodological holism and existential dialectics) serves as a foundation to unify all domains of knowledge into *an unified theory of everything* (by way of some ontological principles and the comprehensive perspectives of the mind, nature, society, and culture). This is so, without commiting the sins of reductionism and reverse-reductionism (as often seen in many holistic approaches, with the fad of sysems approach—be it about systems theory, chaos theory, complexity theory, or else—as a most recent notorious example, which I debunked in *The Future of Complexity* and also in *The Future of Post-Human Formal Science*). My distinctive approach makes good use of different schools of thought without favoring any of them nor trying to integrate them (as they are not necessarily compatible with each other), so as to adjust for subjectivity, diversity, conflict, and complexity, for example. In this sense, the word "unified" does not have to mean integrative approach, in a narrow sense, nor systems approach, in a broad one, both of which I reject while learning from them." (*FPHFS*; *FPHS*)

Sources: A summary of *Sec.1.2* in *FPHC*—and also from *BNN*, *FPHST*, *ALD*, *FC*, *FPHCT*, and the rest of all other books of mine. See the books for more detail.

Table 4.2. On Reductionism and Reverse-Reductionism (Part I)

- **The Partiality-Totality Principle**
  —The partiality-totality principle in the ontology of existential dialectics targets against the varieties of reductionism and reverse-reductionism (as already worked out in my previous books).

- **Against the Varieties of Reductionism**
  —*Conceptual Reductionism*
  - Some illustrative instances involve myriad dualities like mind vs. body, self vs. world, democracy vs. non-democracy, and the like (as already addressed in *FHC*, *FPHC*, and *BDPD*, for instance).

  —*Theoretical Reductionism*
  - A fascinating case study concerns what I originally called "the foundation fallacy" in *FPHST*, in any attempt to naively understand space-time from the physical perspective as the foundation and, consequently, to dangerously dismiss other perspectives.
  - In *FAE*, I elaborated further these versions of reductionism in the literature on aesthetics (e.g., form vs. content, representation vs. expression, critics vs. artists, and externalism vs. internalism).
  - In *FIA*, I revealed other forms of reductionism in the literature on information architecture (e.g., the constructivist argument).
  - In *FPHU*, I showed the persistent legacy of reductionism, this time, in the literature on anomalous experience (e.g., the obsession with physics, chemistry, and biology for explaining anomalous experience).
  - In *FPHE*, I examined another case of reductionism in action, in the context of engineering (e.g., technical constraints vs. normative constraints).
  - In *FPHMM*, I elaborated one more version of reductionism, in relation to the three domains of communication (e.g., the competing views on sending, connecting, and receiving).
  - In *FPHCT*, I explored another version of reductionism, in relation to invention and innovation (e.g., the bio-psychological argument vs. the socio-cultural arguments).

(continued on next page)

Table 4.2. On Reductionism and Reverse-Reductionism (Part II)

- **Against the Varieties of Reductionism** ***(cont'd)***
  - —*Theoretical Reductionism (cont'd)*
    - In *FPHG*, I identify another version of reductionism, in relation to infinity, symmetry, and dimensionality (e.g., the Euclidean argument vs. the non-Euclidean arguments).
    - In *FPHUP*, I analyzed another version of reductionism, in relation to density and void (e.g., the engineering argument and the ecology argument).
    - In *FPHL*, I examined another version of reductionism, in relation to structure and context (e.g., the structuralist argument and the contextualist argument).
    - In *PFHO*, I explored another version of reductionism, in relation to communcation, decision-making, and leadership (e.g., the rational-system argument vs. the natural-system argument).
    - In *PFHMA*, I revealed another version of reductionism, in relation to the martial body and spirit (e.g., the spiritual argument vs. the materialist argument vs. the defensive argument).
    - In *PFHS*, I analyzed another version of reductionism, in relation to the sexual body and spirit (e.g., the naturalist argument vs. the constructivist argument).
    - In *PFHLAW*, I examined another version of reductionism, in relation to law (e.g., the necessity argument vs. the contengency argument).
    - In *FPHWP*, I showed another version of reductionism, in relation to war and peace (e.g., the aggressivist argument vs. the pacifist argument).
    - In *BEPE*, I analyzed another version of reductionism, in relation to morality and immorality (e.g., the objectivist argument vs. the non-objectivist argument vs. the skeptical argument).
    - In *BCOS*, I examined another version of reductionism, in relation to the contested beginnings and speculative ends of the universe (e.g., the scientific argument vs. the religious argument vs. the esoteric argument vs. the metaphysical argument).
    - In *FPHP*, I show another version of reductionism, in relation to normality and abnormality (e.g., the natural argument vs. the social argument vs. the cultural argument vs. the mental argument).

(continued on next page)

Table 4.2. On Reductionism and Reverse-Reductionism (Part III)

- **Against the Varieties of Reductionism *(cont'd)***
  —*Theoretical Reductionism (cont'd)*
  - In *FPHGEOL*, I scrutinized another version of reductionism, in relation to statics and dynamics (e.g., the catastrophe argument vs. the uniformity argument vs. the revision argument).
  - In *FPHCHESS*, I showed another version of reductionism, in relation to tactics and strategy (e.g., the natural argument vs. the social argument vs. the cultural argument vs. the mental argument).
  - In *FPHR*, I revealed another version of reductionism, in relation to secularness and sacredness (e.g., the critical argument vs. the skeptical argument vs. the theist argument).
  - In *FPHEDU*, I examined another version of reductionism, in relation to teaching and learning (e.g., the teacher-centered argument vs. the student-centered argument vs. the balanced argument).
  - In *FPHH*, I explored another version of reductionism, in relation to joking and laughing (e.g., the natural argument vs. the social argument vs. the cultural argument vs. the mental argument).
  - In *FPHA*, I proposed another version of reductionism, in relation to sound and silence (e.g., the natural argument vs. the social argument vs. the cultural argument vs. the mental argument).
  - In *FPHGAM*, I proposed another version of reductionism, in relation to risk and caution (e.g., the natural argument vs. the social argument vs. the cultural argument vs. the mental argument).
  - In *FPHCOM*, I worked out another version of reductionism, in relation to hardware and software (e.g., the natural argument vs. the social argument vs. the cultural argument vs. the mental argument).
  - In *FPHLIT*, I proposed another version of reductionism, in relation to fiction and non-fiction (e.g., the natural argument vs. the social argument vs. the cultural argument vs. the mental argument).
  - In *FPHCHEM*, I proposed another version of reductionism, in relation to substances and their changes (e.g., the natural argument vs. the social argument vs. the cultural argument vs. the mental argument).

(continued on next page)

Table 4.2. On Reductionism and Reverse-Reductionism (Part IV)

- **Against the Varieties of Reductionism *(cont'd)***
  —*Theoretical Reductionism (cont'd)*
  - In *FPHDA*, I proposed another version of reductionism, in relation to quantitative and qualitative research (e.g., the natural argument vs. the social argument vs. the cultural argument vs. the mental argument).
  - In *FPHGEOG*, I proposed another version of reductionism, in relation to environments and their interactions (e.g., the natural argument vs. the social argument vs. the cultural argument vs. the mental argument).
  - In *FPHPA*, I proposed another version of reductionism, in relation to the body and its presence (e.g., the natural argument vs. the social argument vs. the cultural argument vs. the mental argument).
  - In *FPHSEM*, I proposed another version of reductionism, in relation to internality and externality (e.g., the natural argument vs. the social argument vs. the cultural argument vs. the mental argument).
  - In *FPHARCH*, I proposed another version of reductionism, in relation to form and function (e.g., the natural argument vs. the social argument vs. the cultural argument vs. the mental argument).
  - In *FPHCRIM*, I proposed another version of reductionism, in relation to heroes and villains (e.g., the natural argument vs. the social argument vs. the cultural argument vs. the mental argument).
  - In *FPHHIST*, I proposed another version of reductionism, in relation to universality and relativity (e.g., the natural argument vs. the social argument vs. the cultural argument vs. the mental argument).
  - In *FPHMIG*, I proposed another version of reductionism, in relation to Sameness, Otherness, and identity (e.g., the natural argument vs. the social argument vs. the cultural argument vs. the mental argument).

(continued on next page)

Table 4.2. On Reductionism and Reverse-Reductionism (Part V)

- **Against the Varieties of Reductionism** ***(cont'd)***
  - *—Theoretical Reductionism (cont'd)*
    - In *FPHTRAN*, I proposed another version of reductionism, in relation to networks and operations (e.g., the natural argument vs. the social argument vs. the cultural argument vs. the mental argument).
    - In *FPHCA*, I proposed another version of reductionism, in relation to ingredients and techniques (e.g., the natural argument vs. the social argument vs. the cultural argument vs. the mental argument).
    - In *FPHW*, I proposed another version of reductionism, in relation to uselessness and usefulness (e.g., the natural argument vs. the social argument vs. the cultural argument vs. the mental argument).
    - In *BNR*, I proposed another version of reductionism, in relation to diversity and discontinuity (e.g., the natural argument vs. the social argument vs. the cultural argument vs. the mental argument).
    - In *FPHSPORT*, I proposed another version of reductionism, in relation to training and winning (e.g., the natural argument vs. the social argument vs. the cultural argument vs. the mental argument).
    - In *FPHPROB*, I proposed another version of reductionism, in relation to objectivity and subjectivity (e.g., the natural argument vs. the social argument vs. the cultural argument vs. the mental argument).
    - In *FPHMORP*, I proposed another version of reductionism, in relation to typologies and rules (e.g., the natural argument vs. the social argument vs. the cultural argument vs. the mental argument).
    - In *FPHHC*, I proposed another version of reductionism, in relation to mind and body (e.g., the natural argument vs. the social argument vs. the cultural argument vs. the mental argument).

(continued on next page)

Table 4.2. On Reductionism and Reverse-Reductionism (Part VI)

- **Against the Varieties of Reductionism *(cont'd)***
  - —*Theoretical Reductionism (cont'd)*
    - In *FPHTHAN*, I proposed another version of reductionism, in relation to death and after-death (e.g., the natural argument vs. the social argument vs. the cultural argument vs. the mental argument).
    - In *FPHVA*, I proposed another version of reductionism, in relation to techniques and spirits (e.g., the natural argument vs. the social argument vs. the cultural argument vs. the mental argument).
    - In *FPHAERO*, I proposed another version of reductionism, in relation to predictability and nonpredictability (e.g., the natural argument vs. the social argument vs. the cultural argument vs. the mental argument).
    - In *FPHPHON*, I proposed another version of reductionism, in relation to generality and specificity (e.g., the natural argument vs. the social argument vs. the cultural argument vs. the mental argument).
    - And in *FPHACCO*, I propose another version of reductionism, in relation to addiction and subtraction (e.g., the natural argument vs. the social argument vs. the cultural argument vs. the mental argument).
  - —*Methodological Reductionism*
    - A good instance concerns the debate between different versions of qualitative and quantitative methods (as already analyzed in *FC* and also *FHC*). In *FPHML*, I examined similar reductionism, this time, in the literature on mathematical logic (e.g., the obsession with consistency, soundness, and completeness). And in *FPHFS*, I also explored the problems of reductionism in the context of formal science (e.g., the analytical argument).
  - —*Ontological Reductionism*
    - An excellent example is the debate between emergentism and reductionism in complexity theory and also in psychology (as elaborated in *FPHC*, in the context of Being and Becoming).

(continued on next page)

Table 4.2. On Reductionism and Reverse-Reductionism (Part VII)

- **Against the Varieties of Reverse-Reductionism**
  —*Conceptual Reverse-Reductionism*
  - Any concept of "art" (e.g., fine art, cave art, outsider art, junk art) is deemed acceptable in postmodernism (as already addressed in Ch.4 of *FHC*).

  —*Theoretical Reverse-Reductionism*
  - There are numerous art and literary theories co-exist. Take the case of literary studies, as there are now Literary Structuralism, Marxist Literary Criticism, New Criticism, Phenomenology, Hermeneutics, Language-Game Literary Criticism, Feminist Literary Criticism, Reception Theory, Reader Response Criticism, Poststructuralism, Semiotics, Pyschoanalytic Literary Criticism, just to cite some well-known ones, with no one being said to be better than any others (as detailedly analyzed in Ch.4 of *FHC*). (S. Raman 1997) In *BNN*, I even introduced "the compromise fallacy" as another good illustration of theoretical reverse-reductionism, in misleadingly treating both genetic and environmental approaches as equally valid.

  —*Methodological Reverse-Reductionism*
  - There is the "anything-goes" mentality in postmodernism (e.g., doing art without praxis, doing art with praxis, and doing art by sublation), as analyzed in Ch.4 of *FHC*. And in *FPHFS*, I also exposed the problems of reverse-reductionism in the context of formal science in relation to systems theory.

  —*Ontological Reverse-Reductionism*
  - There are likewise no privileged ontology, and the door is open for anything in postmodernism (e.g., the equal status of the ontology of Being vs. that of Becoming, as already addressed in Ch.4 of *FHC*—and also in *FPHC)*. In *FAE*, I also introduced another version of reverse-reductionism, that is, "the pluralist fallacy," in the context of understanding aesthetic experience, for instance—although this fallacy has been committed not exclusively in relation to the ontological level (but also at the conceptual, theoretical, and methodological ones).

Sources: From my previous books.

Table 4.3. The Concepton of Existential Dialectics (Part I)

- **Sets and Elements**
  - —Sets
    - Ex: the Same
    - Ex: the Others
  - —Elements
    - Ex: whites in 20th century America (in the set of "the Same")
    - Ex: Iraq during the U.S. invasion in 2003 (in the set of "the Others")

- **Relations, Operations, Functions**
  - —Relations (e.g., "belongs," "equals to," "is greater than")
    - Ex: symmetric interactions within the Same (or the Others)
    - Ex: asymmetric interactions between the Same and the Others
  - —Operations (e.g., "and," "or," "not," "if…then")
    - Ex: if the Same oppresses the Others, it will also oppress itself.
    - Ex: the Same is not the Others.
  - —Functions (e.g., goals)
    - Ex: the Same is hegemonic in relation to the Others.

- **Truth Values**
  - —"1" if True (in Symbolic Logic)
    - Ex: the proposition that imperial Japan was hegemonic to China during WWII
  - —"0" if False (in Symbolic Logic)
    - Ex: the proposition that Grenada invaded France in 2003
  - —"1" & "0" if Both True and False (in Dialectic Logic)
    - Ex: the proposition that the rabbit-duck picture refers to a duck
  - —"~1" & "~0" if Neither True Nor False (or N/A, in Dialectic Logic)
    - Ex: the proposition that God really exists

(continued on next page)

Table 4.3. The Conception of Existential Dialectics (Part II)

- **Axioms and Postulates**
  —Axioms
    - Ex: the reflexive axiom—"any quantity is equal to itself"

  —Postulates
    - Ex: the SSS postulate—"if the three sides of a triangle are congruent to their corresponding parts, then the triangles are congruent"

- **Theorems and Principles**
  —Theorems (and Principles) in Existential Dialectics
    - *In Relation to Agency*
      –#35: The formalness-informalness principle
      –#34: The absoluteness-relativeness principle
      –#33: The partiality-totality principle
      –#32: The predictability-unpredictability principle
      –#31: The explicability-inexplicability principle
      –#30: The fiction-reality principle
      –#29: The cognitiveness-noncognitiveness principle
    - *In Relation to Structure*
      –#28: The finiteness-transfiniteness principle
      –#27: The preciseness-vagueness principle
      –#26: The simpleness-complicatedness principle
      –#25: The openness-hiddenness principle
      –#24: The denseness-emptiness principle
      –#23: The rule-exception principle
    - *In Relation to Process*
      –#22: The prototypicality-variation principle
      –#21: The change-constancy principle
      –#20: The order-chaos principle
      –#19: The slowness-quickness principle
      –#18: The expansion-contraction principle
      –#17: The optimality-nonoptimality principle
      –#16: The simultaneity-nonsimultaneity principle
      –#15: The isolation-interaction principle

(continued on next page)

Table 4.3. The Conception of Existential Dialectics (Part III)

- **Theorems and Principles (*cont'd*)**
  —Theorems (and Principles) in Existential Dialectics
  - *In Relation to Agency*
    –#14: The theory-praxis principle
    –#13: The convention-novelty principle
    –#12: The evolution-transformation principle
    –#11: The symmetry-asymmetry principle
    –#10: The softness-hardness principle
    –#9: The seriousness-playfulness principle
    –#8: The activeness-inactiveness principle
    –#7: The selfness-otherness principle
  - *In Relation to Outcome*
    –#6: The regression-progression principle
    –#5: The same-difference principle
    –#4: The stability-reaction principle
    –#3: The functionality-nonfunctionality principle
    –#2: The intentionality-nonintentionality principle
    –#1: The survivability-nonsurvivability principle

Notes: The categories and examples in each are solely illustrative (not exhaustive). The comparison is also relative (not absolute), nor are they mutually exclusive. As generalities, they allow exceptions.
Sources: Starting from Ch.6 of *BCPC* and also from other books of mine

Table 4.4. The Syntax of Existential Dialectics I: The Principles (Part I)

- **In Relation to Method**
  —*The Formalness-Informalness Principle*
  (On the Formal Requirements of Logical Systems)
  - The formal requirements of a logical system (e.g., consistency, soundness, and completeness) have both usefulness and non-usefulness, to the extent that, if there are formal systems requiring them, there are alternative ones which do not. Thus, it does *not* exclude classical logics but simply goes beyond both classical and non-classical logics, while learning something from each. There is no formalness without informalness—and vice versa.
  - For instance, existential dialectics can make use of both classical logics under certain conditions (e.g., especially, though not exclusively, when they are clear-cut, etc.) and non-classical logics under alternative conditions (especially, though not exclusively, when they are "unknown," "irrelevant," "ambiguous," "possible," with "different degrees of truth," empirically inconsistent in a desirable way, etc.). (WK 2008u)
  - Family resemblance: e.g., logicalness-nonlogicalness, rationality-nonrationality, etc.
  - Sources: From *FPHML*, *FPHH*, *FPHGAM*, *FPHSEM*, and my later books.

(continued on next page)

Table 4.4. The Syntax of Existential Dialectics I: The Principles (Part II)

- **In Relation to Method** ***(cont'd)***

  —*The Absoluteness-Relativeness Principle*

  (On the Multiplicity of Things)

  - There is the multiplicity of things in reality, be they about entities, qualities (or properties), and relationships, such that what is acceptable from one standpoint may not be so from another. For instance, if there is something absolute, there is likewise something relative. There is no absoluteness without relativeness—and vice versa.
  - Both absoluteness and relativeness here are also relevant to different modalities often cited in the literature on ontology, such as possibility (e.g., something "can" happen) and its opposite (e.g., impossibility) in alethic modalities, probability (e.g., something "will" happen) and its opposite (e.g., improbability) in temporal modalities, and necessity (e.g., something "should" happen) and its opposite (e.g., contingency) in alethic modalities.
  - Family resemblance: e.g., independence-dependence, universalness-relativeness, immanence-transcendence, autonomy-dependency, internalness-externalness, etc.
  - Sources: From *FPHK*. See also my later books.

(continued on next page)

Table 4.4. The Syntax of Existential Dialectics I: The Principles (Part III)

- **In Relation to Method** ***(cont'd)***
  —*The Partiality-Totality Principle*
  (On the Relationships between Whole and Parts)
  - The whole is not the sum of the parts. There is no partiality without totality—and vice versa.
  - Any inquiry about a phenomenon in the work is to guard against the varieties of (a) reductionism and (b) reverse-reductionism.
  - Reductionism and reverse-reductionism can be (i) conceptual, (ii) theoretical, (iii) methodological, and (iv) ontological.
  - Works on emergence (as in *FC*), gestalt psychology (as in *FPHGEOL*), the hermeneutic circle (as in *FPHDA*), and mental language (as in *FPHSEM*), to name just a few of them, have contributed to the understanding of whole and parts.
  - Family resemblance: e.g., individualisticness-holisticness, partness-wholeness, analysis-synthesis, etc.
  - Sources: Especially from *FC* & *FPHGEOL*. See also my later books.

(continued on next page)

Table 4.4. The Syntax of Existential Dialectics I: The Principles (Part IV)

- **In Relation to Method *(cont'd)***
  - *—The Predictability-Unpredictability Principle (On the Occurrence of Events)*
    - Both predictability and unpredictability have a major role to play in the occurrence of things, so that neither determinism nor indeterminism wins the centuries-old fight. There is no predictability without unpredictability—and vice versa.
    - There are events which are predictable, just as there are those which are not. Or what is regarded as unpredictable at one point in time may turn out to be predictable later, and, conversely, what is deemed as predictable may turn out to not be so predictable. Even in predictability, outcomes are subject to uncertainty, the degree of which varies from case to case.
    - Works on complexity theory (as in *FC*), quantum mechanics (as in *FPHST* and *FPHG*), subjective (or Bayesian) probability (as in *FPHPROB*), "special causes" (as in *FPHGAM*), measurement uncertainty (as in *FPHPROB*), the butterfly effect (as in *FPHAERO*), and unintended consequences (as in *FPHAERO*), just to name a few of them, have contributed to the understanding of uncertainity.
    - The predictability-unpredictability principle is not to be confused with the absoluteness-relativeness principle; although both can address the issue of probability, the former is concerned with the extent of certainty in prediction, whereas the latter focuses on the relativity of viewpoints.
    - Family resemblance: e.g., sureness-arbitrariness, certainty-uncertainty, deterministicness-indeterministicness, sureness-doubtfulness, likeliness-riskiness, deterministicness-randomness, reliability-unreliability, etc.
    - Sources: Especially from *FC* & *FPHGAM*. See also my later books.

(continued on next page)

Table 4.4. The Syntax of Existential Dialectics I: The Principles (Part V)

- **In Relation to Method** ***(cont'd)***

—*The Explicability-Inexplicability Principle*
(On the Underlying Mechanisms of Things)

- Both explicability and inexplicability are part of the understanding of things. There is no explicability without inexplicability—and vice versa.
- This principle tells us the duality of the research dilemma, in that, if reality can be explained in some ways, it also has its other ways which are not quite explainable, at a given point in time.
- Family resemblance: e.g., underlyingness-regularness, causation-regularness, causation-correlation, relation-nonrelation, explanation-description, explanation-narration, explanation-interpretation, etc.
- Sources: Especially from *FPHU*—and also *FPHC*. See also my later books.

—*The Fiction-Reality Principle*
(On the different faces of Reality)

- If there is reality, there is also fiction, especially for any agent with some kind of intelligence life, be it about humans, animals, or, later, post-humans. There is no reality without fiction—and vice versa.
- The distinction between reality and fiction is not absolute, since there can be a mixture of both. Fiction can have non-fictional elements, just as non-fiction can have fictional ones. So, the distinction is in degree, not in kind.
- Family resemblance: e.g., fiction-nonfiction, fiction-fact, imagination-fact, subjectivity-objectivity, imagination-nonfiction, inaccuracy-accuracy, error-reality, misunderstanding-correctness, arbitrariness-reality, etc.
- Sources: First from *FPHLIT*.

(continued on next page)

Table 4.4. The Syntax of Existential Dialectics I: The Principles (Part VI)

- **In Relation to Method *(cont'd)***

  —*The Cognitiveness-Noncognitiveness Principle*

  (On the Understanding of Meaning)

  - If a view of things is based on reason and evidences, it is also affected by non-cognitive factors (such that it is neither true nor false in the latter case)—and vice versa. In other words, there is no cognitiveness without noncognitiveness—and vice versa.
  - Noncognitive factors can come in all shapes and sizes, and good examples include envy, jealousy, power, nationality, race, gender, age, class, greed, lust, status, faith, anger, sadness, joy, fear, wish, etc.
  - Communication, even in its ideal type, is seldom solely based reason and evidences, as falsely assumed during the Age of the Enlightenment and its continued supporters nowadays. What is understood about something is often twisted on the basis of something else too. In these cases, what is at stake is often neither true nor false (which is something that classical logic has sidelined and ignored).
  - Family resemblance: e.g., proposition-attitude, cognition-emotion, cognition-prescription, justification-nonjustification, recognition-nonrecogntion, reason-perception, theory-extratheory, etc.
  - Sources: First from *BNR*.

(continued on next page)

Table 4.4. The Syntax of Existential Dialectics I: The Principles (Part VII)

- **In Relation to Structure**

—*The Finiteness-Transfiniteness Principle*

(On the Nature of Numbers)

- If there are finite things, there are likewise transfinite ones. There is no finiteness without transfiniteness—and vice versa.
- To avoid confusion, my usage of the word "transfinite" here differs radically from the one used by Cantor (and other mathematicians) for "relative" infinity—and is more limited, in light of the problems confronting any attempt to understand the idea of infinity, be it by intuition, imagination, and conception (as detailedly analyzed in *Sec. 2.2.3* of *FPHG*).
- Instead, by "transfinity," I allow numbers which can be many times larger—or smaller, for that matter—than the finite things that we encounter in daily life, but they do not have to be related to the idea of infinity at all (which may not exist).
- Of course, there may be some *borderline* cases, in which it is not clear whether the number in question is transfinite (in my usage) or simply a mathematical convenience. A good example of a borderline case is the Planck unit of length for "the smallest space possibly measured in nature," which is "less than billionths of trillionths of trillionths of an inch" (or something like 1.6×10^{-35} meters).
- That said—my usage of "transfinity" can also resolve (or better, dissolve) an age-old problem in philosophy known as "Zeno's paradoxes" (as already explained in *Sec. 2.2.3* of *FPHG*).
- Family resemblance: e.g., boundedness-quasiunboundedness, smallness-largeness, boundedness-unboundedness, etc.
- Source: From *FPHG*. See also my later books.

(continued on next page)

Table 4.4. The Syntax of Existential Dialectics I: The Principles (Part VIII)

- **In Relation to Structure** ***(cont'd)***

  —*The Preciseness-Vagueness Principle*

  (On the Refinement of Things)

  - Both preciseness and vagueness are important, not that one is better than the other, but that both are used, in different degrees of preference, in accordance to the contextual application from the perspectives of nature, the mind, culture, and society. Even when both are used in a combination, the dilemma is shifted instead to one of combinational concern. There is no preciseness without vagueness—and vice versa.
  - In relation to taxonomy, preciseness has its taxonomic clarity, just as vagueness has its classificatory flexibility, for instance. And in relation to network, vagueness has its explorative liberty, just as preciseness has its conceptual definitiveness, for instance.
  - Family resemblance: e.g., clarity-ambiguity, directness-indirectness, quantitativeness-qualitativeness, describability-nondescribability, sharpness-roughness, specificity-generality, concreteness-abstractness, specificity-obscurity, translatability-untranslatability, clearcutness-transitionalness, discreteness-mixedness, discreteness-vagueness, detailedness-broadness, etc.
  - Sources: Especially from *FIA* and *FPHMORP*. See also my later books.

(continued on next page)

Table 4.4. The Syntax of Existential Dialectics I: The Principles (Part IX)

- **In Relation to Structure *(cont'd)***
  —*The Simpleness-Complicatedness Principle*
  (On the Interconnection among Things)
  - Both simpleness and complicatedness are vital, without favoring one over the other, in that each is utilized, depending on the basis of the perspectives of nature, the mind, culture, and society. And even when a combination of them is preferred, the dilemma is only shifted to a combinational degree of concern. There is no simpleness without complicatedness—and vice versa.
  - For illustration, in relation to taxonomy, simpleness has its heuristic usefulness, just as complicatedness has its realistic representation, for instance. In relation to network, simpleness has its economical attractiveness, just as complicatedness has its practical reliability, for instance. And in relation to logic, there is the simple zeroth-order logic, just as there is the complicated higher-order logic.
  - Family resemblance: e.g., inflexibility-flexibility, standardization-specialization, imperfectness-perfectness, superficiality-depth, shallowness-deepness, economicalness-elaboratedness, plainness-circumspection, onesidedness-multisidedness, succinctness-duplicatedness, simpleness-sophisticatedness, crudeness-subtleness, straightness-bendedness, etc.
  - Sources: Especially from *FIA* and *FPHPROB*. See also my later books.

(continued on next page)

**Table 4.4. The Syntax of Existential Dialectics I:
The Principles
(Part X)**

- **In Relation to Structure** ***(cont'd)***
  —*The Openness-Hiddenness Principle*
  (On the Detection of Things)
  - Reality has its hidden face, just as it is open to outside view in some other ways. There is no openness without hiddenness—and vice versa.
  - For instance, in the context of anomalous experience, certain aspects of reality can be open for examination (e.g., the experiment with SPECT images, the transmission of telepathemic bit, and the use of metonymies and metaphors). Other aspects, however, remain hidden, and examples include the elusive deeper nature of spiritual reality in the intangible realm, the undetectability of different branched universes, the underlying mechanisms of signals traveling between different folds of physical space-time, and the dependence on language and the bias for science and logic).
  - Family resemblance: e.g., overtness-covertness, publicness-privateness, openness-closedness, transparency-secrecy, openness-biasedness, participation-obstruction, exposition-illusion, surface-depth, etc.
  - Sources: Especially from *FPHU*. See also my later books.

(continued on next page)

**Table 4.4. The Syntax of Existential Dialectics I:
The Principles
(Part XI)**

- **In Relation to Structure *(cont'd)***
  —*The Denseness-Emptiness Principle*
  (On the Distribution of Entities in Space)
  - Both density and void are needed, in relation to the mind, nature, culture, and society, albeit in different ways. There is no denseness without emptiness—and vice versa.
  - For clarity, the term "void" is used here only as an approximation of emptiness (depending on the degree of the lack of density), since, in physics, it is well known that "empty" space is not really empty all the way, because it can be full of energy (e.g., random quantum fluctuations at the sub-atomic level, and, for that matter, dark energy in the universe) and matter (e.g., different versions of sub-atomic particles, and, for that matter, dark matter in the universe).
  - Family resemblance: e.g., fullness-voidness, nearness-farness, concentration-dispersion, denseness-nothingness, presence-absence, focus-unfocus, solidness-void, something-nothing, heaviness-lightness, composition-emptiness, emphasis-disemphasis, etc.
  - Sources: First worked out in *FPHUP*. See also my later books.

(continued on next page)

Table 4.4. The Syntax of Existential Dialectics I: The Principles (Part XII)

- **In Relation to Structure** ***(cont'd)***
  —*The Rule-Exception Principle*
  (On the Extent of Typicality)
  - Both rules and their exceptions exist, in relation to the mind, nature, culture, and society, albeit in different ways. There is no rule without exception—and vice versa.
  - As the old saying goes, "There is an exception to the rule"; however, to avoid misuse, there are the following four important aspects of the relationship between rules and exceptions to be clarified.
  - If there is an exception to the rule, there can be an exception to the.rule itself that there is an exception to the rule, as shown in some cases where no exception is allowed—but this regression can continue further (e.g., there is an exception to the rule that there can be an exception to the rule itself that there is an exception to the rule).
  - To be an exception to the rule does not mean that one can say something nonsensual, either jokingly or seriously, and then expects others to allow an exception.
  - To be an exception to the rule does not mean that one can appeal to an invalid generalization, either literally or figuratively, and then expects others to allow an exception.
  - An exception to a rule means that the exception implies the existence of a rule, which does not apply in the special case under consideration—but the exception does not automatically justify that the rule is desirable, good, right, etc.
  - Family resemblance: e.g., generalization-exception, custom-deviance, procedure-exception, pattern-deviance, typicality-exception, and the like.
  - Sources: First worked out in *FPHTHAN*. See also my later books.

(continued on next page)

Table 4.4. The Syntax of Existential Dialectics I: The Principles (Part XIII)

- **In Relation to Process**

—*The Prototypicality-Variation Principle*

(On the Nature of Duplication)

- If there are prototypes, there are also variants of the copies. There is no prototypicality without variation—and vice versa.
- Prototypes and their copies are not static and may change or die out over time, but new ones may emerge.
- Some copies may have multiple prototypes, not just one.
- It can be very challenging to trace some copies back to their prototypes, under certain difficult historical conditions.
- Copies are not solely identical duplications of prototypes but have their own variations (mutations).
- Family resemblance: e.g., archetypicality-descendedness, origin-descendedness, ancestry-descendedness, prototypicality-duplication, archetypicality-mutation, origin-reenactment, prototypicality-recreation, prototypicality-copyness, and the like.
- Sources: First worked out in *FPHMORP*. See also my later books.

—*The Change-Constancy Principle*

(On the Alteration of Things)

- Change occurs over time, although constancy is also allowed. There is no change without constancy—and vice versa.
- Asymmetry undergoes changes over time, so does symmetry.
- Old players fade away, and new ones emerges, with ever new causes and ever new forms.
- Family resemblance: e.g., dynamicness-staticness, instability-stability, etc.
- Sources: First named in *BCPC*. Especially from *FHC*, *FCD*, and *FPHC*. See also other books of mine (from that point on).

(continued on next page)

Table 4.4. The Syntax of Existential Dialectics I: The Principles (Part XIV)

- **In Relation to Process *(cont'd)***
  —*The Order-Chaos Principle*
  (On the Pattern of Things)
  - Both order and chaos are vital in the process of change in the world. The preference for order is biased, since it does not give sufficient attention to the vital role of chaos in the transformation of the world (without somehow reducing it for the understanding of order). There is no order without chaos—and vice versa.
  - The scientific search for order in the world is often a hidden bias in its ontological obsession with order, since chaos is often treated as the "bad" guy, with order as the "good" guy (for the end goal of science).
  - Neither order nor chaos is the final end of the world, and one is not to be treated as the means for the other in the transformation of things. Both are fundamental in their recurrent dialectical interactions with each other over time, without reducing one for the other.
  - Order has its different forms (e.g., ascending vs. descending order in mathematics, and feeding vs. bleeding order in morphology), just as chaos has its own versions (e.g., as understood differently in chaos theory vs. complexity theory).
  - Family resemblance: e.g., lawfulness-disorder, order-disorder, linearity-nonlinearity, linearity-irreversalness, pattern-chaos, unity-disunity, sequence-chaos, etc.
  - Sources: Especially from *FC*. See also my later books.

(continued on next page)

Table 4.4. The Syntax of Existential Dialectics I: The Principles (Part XV)

- **In Relation to Process *(cont'd)***

  —*The Slowness-Quickness Principle*
  (On the Speed of Change)

  - Both slowness and quickness co-exist, with their own internal tension, to the extent that each fights for its own relevance with the other, in accordance to the perspectives of nature, the mind, culture, and society, without one being the victor and the other being the vanquished in the long haul. Even when both are chosen in other cases, this dilemma is only transferred into something else with a combinational character. There is no slowness without quickness—and vice versa.
  - In relation to taxonomy, quickness has its efficient usability, just as slowness has its aesthetic appeal, for instance. And in relation to network, quickness has its adventurous readiness, just as slowness has its risk-adverse convenience, for instance.
  - Family resemblance: e.g., inconvenience-convenience, passiveness-activenessness, gradualness-abruptness, deceleration-acceleration, difficulty-ease, nonresilience-resilience, patience-impatience, deliberativeness-impulsiveness, etc.
  - Sources: Especially from *FIA*. See also my later books.

(continued on next page)

Table 4.4. The Syntax of Existential Dialectics I:
The Principles
(Part XVI)

- **In Relation to Process *(cont'd)***
  —*The Expansion-Contraction Principle*
  (On the Growth of Things)
  - Entities in the world can both expand in some ways and contract in other ones, as part of their nature. There is no expansion without contraction—and vice versa.
  - For instance, in the context of anomalous experience, one of its most fascinating expressions is none other than the analysis of different causes of the slow but steady expansion (and, furthermore, if put in a different way, unfolding) of unconsciousness.
  - Family resemblance: e.g., conquest-autarky, rise-fall, spread-shrink, extendingness-shorteningness, widening-narrowing, expansion-destruction, expansion-decline, construction-demolition, creation-destruction, construction-deconstruction, expansion-contraction, composition-decomposition, growth-breakdown, formation-decay, expansion-constriction, insertion-deletion, addition-division, multiplication-division, multiplication-subtraction, overstatedness-understatedness, etc.
  - Sources: Especially from *FPHU*. See also my later books.

(continued on next page)

Table 4.4. The Syntax of Existential Dialectics I: The Principles (Part XVII)

- **In Relation to Process *(cont'd)***

  —*The Optimality-Nonoptimality Principle*

  (On the Nature of Highly Favorable Options)

  - Both optimality and non-optimality are important, in relation to the mind, nature, culture, and society, albeit in different ways. There is no optimality without nonoptimality—and vice versa.
  - Optimality (in terms of highly favorable options) is not perfect and can be realized in different degrees.
  - "Maximization" does not necessarily entail "optimization," as shown in the problem of "negative externalities" in economics.
  - Non-optimality can come in all shapes and sizes, in different degrees, like "maximizing" in economics vs. "satisficing" in sociology.
  - Optimality can be more feasible in the longer term of technological improvement.
  - Optimality should not be confused with the beauty of simplicity, since the more optimal a process is, the more complicated (not necessarily the simpler) it can be.
  - Family resemblance: e.g., optimality-satisfaction, optimality-maximization, optimality-heuristicness, idealness-realisticness, and the like.
  - Sources: First worked out in *FPHMORP*. See also my later books.

(continued on next page)

Table 4.4. The Syntax of Existential Dialectics I: The Principles (Part XVIII)

- **In Relation to Process *(cont'd)***
  —*The Simultaneity-Nonsimultaneity Principle*
  (On the Sequence of Time)
  - Both simultaneity and nonsimultaneity are important, in relation to the mind, nature, culture, and society, albeit in different ways. There is no simultaneity without nonsimultaneity—and vice versa.
  - An event B can occur after the occurrence of another event A in a causal situation. But A and B can also occur simultaneously in a different situation, and this is important, because, in quantum mechanics (as addressed in *FPHPT*), for instance, particles move so fast that, for all practical purposes, A and B can be treated as occurring simultaneously.
  - In conventional wisdom, A and its negation ~A are different and do not coexist simultaneously. But in certain circumstances, it is also important to consider A and ~A simultaneously, as shown in the interactions between two opposites in the nature vs. nature debate (as addressed in *BNN*) and in the ritual/speech coevolution debate as well as the coarticulation debate (as addressed in *FPHPHON*).
  - Simultaneity does not have to be absolute, as it can be more or less so for two entities to occur simultaneously.
  - On one end of the line is exteme simultaneity and on the other end is extreme nonsimultaneously, and between the two are different degrees or gradations of simultaneity or nonsimultaneity.
  - Family resemblance: e.g., simultaneity-sequence, simultaneity-consecutiveness, contemporaneity-successiveness, and the like.
  - Sources: First worked out in *FPHPHON*. See also my later books.

(continued on next page)

Table 4.4. The Syntax of Existential Dialectics I: The Principles (Part XIX)

- **In Relation to Process *(cont'd)***
  —*The Isolation-Interaction Principle*
  (On the Interaction among Entities)
  - Both isolaiton and interaction are important, in relation to the mind, nature, culture, and society, albeit in different ways. There is no isolation and interaction—and vice versa.
  - Isolation and interactions are not perfect and can be realized in different degrees.
  - Isolation and interaction take time to develop.
  - Interaction can have different outcomes over time (e.g., "merging," "split," "loss," etc. in sound change).
  - The isolation-interaction pair should not be confused with the absoluteness-relativeness pair, although both concern the issue of autonomy, but there are different opposites in relation to autonomy (in the contexts of method, process, etc.).
  - Family resemblance: e.g., isolation-exchange, isolation-intercourse, autonomy-trade, aloneness-interrelationship, autarky-commerce, insulation-interaction, secludedness-interrelationship, segregation-interaction , and the like.
  - Sources: First worked out in *FPHACCO*. See also my later books.

- **In Relation to Agency**
  —*The Theory-Praxis Principle*
  (On the Duality of Knowledge))
  - This has to do with the duality of knowledge in existential dialectics, in that, if there is theoretical construction, there is likewise its practical application, both technical and normative. There is no theory without praxis—and vice versa.
  - Family resemblance: e.g., discovery-application, knowledge-action, invention-innovation, knowledge-motivation, theory-intentionality, universality-particularity, discourse-action, knowledge-interest, potentiality-actuality, theory-practice, etc.
  - Source: Especially from *FPHE*. See also my later books.

(continued on next page)

Table 4.4. The Syntax of Existential Dialectics I: The Principles (Part XX)

- **In Relation to Agency** ***(cont'd)***
  - *—The Convention-Novelty Principle*
    - *(On the Nature of Creative Thinking)*
  - If there is conventional wisdom, there is likewise novel challenge, to the extent that both convergent and divergent thinking are part of life. There is no convention without novelty—and vice versa.
  - In addition, in this context of the convention-novelty principle, there are, in the absence of better words, what I want to call (a) *creative techniques* and (b) *creative traits*, which, when satisfied—in relation to the larger context of the mind, nature, society, and culture—can be used to enhance creative works. The list of creative techniques and traits (as summarized in *Table 4.1* of *FPHCT*) are not exhaustive, of course, but the examples here constitute a great beginning of understanding the structure of creative thinking.
  - Yet, lest the reader gets carried away by the euphoria about creative thinking in our time, it should be reminded that creative thinking has its own possibilities and limits (as shown in *Table 2.1* of *FPHCT* on invention), just as it has its own promises and pitfalls (as shown in *Table 3.1* of *FPHCT* on innovation). In the end, it should also be stressed that creative thinking has its own desirability and dark sides (as shown in *Table 4.2* of *FPHCT*, *BNN*, and *FPHMIG*).
  - This has important daily implications for an average individual, and the word "average" here allows different degrees of variation from one individual to another in a population, at any given point in history.
  - Family resemblance: e.g., normalness-nonnormalness, conformity-nonconformity, convergence-divergence, familiarity-freshness, inertia-newness, equilibrium-disturbance, etc.
  - Sources: First from *FPHCT*. See also my later books.

(*continued on next page*)

Table 4.4. The Syntax of Existential Dialectics I: The Principles (Part XXI)

- **In Relation to Agency *(cont'd)***
  - —*The Evolution-Transformation Principle (On the Multiple Kinds of Agency)*
    - This principle (and the symmetry-asymmetry principle) are both about *the agency of change*. The word "agency," in a formal definition, refers to "a person or thing through which power is exerted or an end is achieved." (MWD 2007b) It therefore does not have to necessarily involve an intelligent lifeform.
    - Because of this dual meaning in agency, the evolution-transformation principle is more concerned with *the multiple kinds of agency*, that is, both about the *evolution* in the state of nature (e.g., an object of natural beauty) and the *transformation* in the world of intelligent lifeforms (e.g., a work of art, an air-conditioner, etc.). There is no evolution without transformation—and vice versa.
    - And the transformative part of the principle precisely refers to the other dimension in the dual meaning of agency, in giving technology a major role to play in the change of the world, which is something that I extensively analyzed in *FHC* in the context of the technophilic lifeworld, espeically though not exclusively since modern times. The same theme is also elaborated in other books of mine.
    - Family resemblance: e.g., nonwillingness-willingness, naturalness-technologicalness, evolution-creation, naturalness-nonnaturalness, inorganicness-volition, naturalness-artificialness, inorganicness-motivation, etc.
    - Sources: Especially from *FAE*. See also other books of mine.

(*continued on next page*)

Table 4.4. The Syntax of Existential Dialectics I: The Principles (Part XXII)

- **In Relation to Agency** ***(cont'd)***
  —*The Symmetry-Asymmetry Principle*
  (On the Relationships among Existents)
  - There is no symmetry without asymmetry—and vice versa.
  - For instance, the Same can be symmetric and asymmetric towards the Others. But in case of asymmetry, oppression and self-oppression can occur. So, when the Same is asymmetric towards the Others, the Same can also be relatively asymmetric towards itself in self-oppression, just as the Others can be likewise towards themselves.
  - The subsequent oppressiveness is dualistic, as much by the Same against the Others and itself, as by the Others against the Same and themselves.
  - Both oppression and self-oppression can be achieved by way of downgrading differences between the Same and the Others and of accentuating them.
  - This is true, even though not all forms of asymmetry have to be about oppression and self-oppression.
  - In addition, from Chapter Three of *FPHG*, symmetry is not perfect, to be understood in an approximate sense under many life circumstances.
  - Family resemblance: e.g., balance-extremity, reflexiveness-unreflexiveness, balance-unbalance, equality-inequality, center-periphery, compatibility-incompatibility, proportion-disproportion, harmoniousness-unharrmoniousness, etc.
  - Sources: From all my books, starting with *FHC*. First named in *BCPC*. See, for instance, *FCD* for more details on the Same and the Others.

(*continued on next page*)

Table 4.4. The Syntax of Existential Dialectics I: The Principles (Part XXIII)

- **In Relation to Agency** ***(cont'd)***

  —*The Softness-Hardness Principle*
  (On the Force of Change)

  - This has to do with *the force of change* under the category about agency in existential dialectics, in that any change by an agent, be it organic (like humans) or non-organic (like natural objects), can occur in a forceful (aggressive) or gentle (pacific) way, which can come in all shapes and sizes, of course. There is no softness without hardness—and vice versa.
  - Family resemblance: e.g., peacefulness-violence, cooperation-competition, acceptance-discrimination, love-hatred, peace-war, pacificity-aggression, dovishness-hawkishness, reward-punishment, kindness-hostility, admiration-belittlement, inclusion-exclusion, friendliness-unfriendliness, like-dislike, cooperation-predation, courteousness-violence, polishedness-harshness, integration-rivalry, bribery-killing, femininity-masculinity, absorption-resistance, amicality-animosity, etc.
  - Sources: Especially from *ALD*. See also my later books.

(continued on next page)

Table 4.4. The Syntax of Existential Dialectics I: The Principles (Part XXIV)

- **In Relation to Agency** ***(cont'd)***

  —The Seriousness-Playfulness Principle

  (On the Extent of Seriousness)

  - If there is seriousness, there is also playfulness, especially for any agent with some kind of intelligence life, be it about humans, animals, or, later, post-humans. There is no seriousness without playfulness—and vice versa.
  - Lest any misunderstanding occurs, the word "playfulness" here should not be confused with other terms like "laughter," "smile," "joke," "tickleness," and "laugh-like vocalization" (in some animals), for instance.
  - Being playful can come in all shapes and sizes and therefore is not necessarily tied up with any of these terms, alhough it can result from any of them, or a combination of them, or something else altogether.
  - This principle, together with some other principles (especially though not exclusively, like the formalness-informalness principle), have important implications for what constitutes rationality, be it about "rationality of application" at the practical level or "rationality of knowledge" at the meta-theoretical level (as summarized in *Sec. 4.20* of *FPHH*).
  - Family resemblance: e.g., seriousness-jokingness, seriousness-humorousness, seriousness-wittiness, seriousness-nonseriousness, seriousness-relaxedness, seriousness-restingness, Apollo-Dionysus, seriousness-clownishness, criticalness-imaginativeness, etc.
  - Sources: First from *FPHH* & *FPHGAM* See also my later books.

(*continued on next page*)

Table 4.4. The Syntax of Existential Dialectics I: The Principles (Part XXV)

- **In Relation to Agency *(cont'd)***
  - —*The Activeness-Inactiveness Principle*
    (On the Extent of Activeness)
    - If there is activeness, there is also inactiveness, especially (though not exclusively) for any agent with some kind of intelligence life, be it about humans, animals, or, later, post-humans. There is no activeness without inactiveness—and vice versa.
    - Being active can come in all shapes and sizes, as a person can be inactive because he is tired, old, sick, sleepy, uninterested, etc., just as an object can be inactive because it is at rest, out of work, etc.
    - This principle, together with some other principles (especially though not exclusively, like the formalness-informalness principle and seriousness-playfulness principle), have important implications for what constitutes rationality, be it about "rationality of application" at the practical level or "rationality of knowledge" at the meta-theoretical level (as summarized in *Sec. 4.20* of *FPHH*).
      - Family resemblance: e.g., alertness-unalertness, activeness-decrepidness, alertness-sleepiness, awakenness-hibernatedness, activeness-tiredness, alertness-uninterestedness, alertness-unwareness, soberness-unawareness, soberness-inactiveness, alertness-inactiveness, activeness-passiveness, etc.
    - Sources: First from *FPHTRAN*.

(*continued on next page*)

Table 4.4. The Syntax of Existential Dialectics I: The Principles (Part XXVI)

- **In Relation to Agency** ***(cont'd)***
  - —*The Selfness-Otherness Principle*
    (On the Magnitude of Self-Centeredness)
    - If there is the tendency to be mostly centered on one's self or group, there is likewise the tendency to be mostly centered on others, especially (though not exclusively) for any agent with some kind of intelligence life, be it about humans, animals, or, later, post-humans. There is no selfness without otherness—and vice versa.
    - Being self-centered and being other-centered do not have to be mutually exclusive, as the two can come in all shapes and sizes, or in different degrees. And other-centeredness does not entail desirability, any more than self-centeredness implies non-desirability, as both can go either way.
    - In addition, whether self-centeredness constitutes a moral alternative or, conversely, an alternative to morality is of course debatable (as addressed in my other book titled *Beyond Ethics to Post-Ethics*).
    - Good examples of family resemblance in relation to the principle are egoisticness-altruisticness, selfishness-unselfishness, selfness-selflessness, egocentricness-othercentricness, parasiticalness-helpfulness, take-give, and so on.
    - Source: First from *FPHCA*

(*continued on next page*)

Table 4.4. The Syntax of Existential Dialectics I: The Principles (Part XXVII)

- **In Relation to Outcome**
  —*The Regression-Progression Principle*
  (On the Direction of History)
  - Neither the cyclical nor the linear views are adequate for explaining many phenomena at all levels. There is regression without progression—and vice versa.
  - History progresses to more advanced forms, but with a regressive touch. Examples include no freedom without unfreedom, no equality without inequality, and no civilization without barbarity.
  - This is not an inevitable law, but merely a highly likely empirical trend.
  - Family resemblance: e.g., cost-benefit, undesirability-desirability, badness-goodness, risk-opportunity, dysfunctionality-functionality, cyclicity-linearity, failure-success, etc.
  - Sources: From all my books, starting with *FHC*. First named in *BCPC*.

  —*The Same-Difference Principle*
  (On the Metamorphosis of Change)
  - An entity, as it evolves over time, can be both different from and similar to its opposing alternatives and does not have to be solely more different from them over time. There is no similarity without difference—and vice versa.
  - Opposites are not absolute in a black-or-white fashion; so, an entity can become relatively more similar to (or more different from) its opposite over time.
  - Family resemblance: e.g., homogeneity-heterogeneity, we-they, similarity-contrast, assimilation-dissimilation, etc.
  - Sources: Especially from ALD. See also my later books

(*continued on next page*)

Table 4.4. The Syntax of Existential Dialectics I: The Principles (Part XXVIII)

- **In Relation to Outcome** ***(cont'd)***
  —*The Stability-Reaction Principle*
  (On the Feedback Mechanism for Further Changes)
  - Once an outcome becomes stabilized over time (which varies from case to case), there is resistance to it, such that it undergoes further changes in its multifaceted interactions with the outside world (e.g., with other entities, energy, space, and so on).
  - There is no stability without reaction (to stability for further changes)—and vice versa.
  - Family resemblance: e.g., stability-forwardness, balance-reaction, balance-forwardness, rest-motion, normalcy-feedback, stability-breakup, consolidation-modification, etc.
  - Sources: First from *FPHCHEM.*

(*continued on next page*)

Table 4.4. The Syntax of Existential Dialectics I: The Principles (Part XIX)

- **In Relation to Outcome** ***(cont'd)***
  - —*The Functionality-Nonfunctionality Principle (On the Presence of Function)*
    - If there is the presence of action for which something or someone is fitted, there is likewise the relative lack of action for which something or someone is fitted. There is no functionality without nonfunctionallity—and vice versa.
    - A *function* should not be confused with an *alternative function*, which is just another kind of function; just as *nonfunction* should not be confused with *dysfunction* (which has more to do with malfunction and is thus related more to the regression-progression principle).
    - In addition, being functional (or useful) and being nonfunctional (or useless) do not have to be mutually exclusive, as the two can come in all shapes and sizes, or in different degrees. And what is functional at one time may become nonfunctional at another time, and vice-versa.
    - Good examples of family resemblance in relation to the principle are usefulness-nonusefulness, usefulness-uselessness, valueness-nonvalueness, and the like.
    - Source: First from *FPHCA*

(*continued on next page*)

Table 4.4. The Syntax of Existential Dialectics I: The Principles (Part XXX)

- **In Relation to Outcome** ***(cont'd)***
  —*The Intentionality-Nonintentionality Principle*
  (On the Planning for Outcomes)
  - If there is the planning or design of something for a certain outcome, there is likewise the relative absence of planning or design of something for a certain outcome. There is no intentionality without nonintentionallity—and vice versa.
  - An intention for an outcome does not have to be certain, so the outcome can even be counter-intentional in being the opposite of what was originally planned. In this sense, counter-intentionality is a special case of non-intentionality; in other words, there are unintented consequences which can be just the opposite of what was originally planned.
  - An intention for an outcome is mostly relevant to intelligent life, but there are non-intelligent life and non-living matter in the universe too, so nonintentionality is also important.
  - In addition, an intention does not have to be obvious or straightforward, as William Shakspeare once famously said: "I must be cruel only to be kind."
  - Especially, works on unintended consequences (like *FPHAERO*) have contributed to the understanding of nonintentionality.
  - Good examples of family resemblance in relation to the principle are purposiveness-nonpurposiveness, intention-accidentalness, design-randomness, awareness-unawareness, etc.
  - Source: First from *FPHW*

(*continued on next page*)

Table 4.4. The Syntax of Existential Dialectics I: The Principles (Part XXXI)

- **In Relation to Outcome *(cont'd)***
  —*The Survivability-Nonsurvivability Principle*
  (On the Survival of things)
  - If there are some entities which survive at certain times, there are likewise other entities which do not survive. There is no survivability without nonsurvivability—and vice versa. In nature, for illustration, there are some animals which survive in hard times, whereas others do not—and this is true, not just among different species but also among different members of the same species (and the phenomenon of "selection" in evolutionary biology has much to say on this, as already addressed in *BNN*).
  - Of course, the principle is not restricted solely to the animal world, since it can be used to other cases concerning non-animals too (e.g., trees, lakes, etc. in the natural world—or empires, companies, etc. in the human world).
  - But one should not confuse this principle with the expansion-contraction principle, since those entities which survive, say, can still expand or contract over time (e.g., those animals which survive can be more capable in some activities but not in others, for instance).
  - Nor should this principle be confused with the stability-reaction principle, in that the condition of survivability or non-survivability may lead to further changes later on (as shown in the emergence of new species after the extinction of some old ones, as already explained in *BNR*).
  - Good examples of family resemblance in relation to the principle are survival-extinction, life-death, existence-nonexistence, and the like.
  - Source: First from *FPHPROB*

Notes: The examples here are solely illustrative (not exhaustive). The comparison is also relative (not absolute), nor are they mutually exclusive. As generalities, they allow exceptions.

Table 4.5. The Syntax of Existential Dialectics II: The Principles as Short Cuts (Part I)

• The principles should be treated with caution, lest misunderstanding occurs, since they do not constitute rigid dualities (or dichotomies).

• The reason is that each pair in an ontological principle consists of two opposites, which are, however, merely short cuts both for multiple variations and degrees, as well as for different interactions with multiple other entities unlike them. In this light, each pair can end up having hundred (or even thousand, if not more) different versions, which interact with hundred (or even thousand, if not more) other entities. There are two clarifications here.

• Firstly, this conception of shortcuts is not mutually exclusive nor absolute, in that the opposites can come in all shapes and sizes, with different degrees. For instance, by analogy, just as there are different degrees of the two colors "white" and "black"—there are likewise different degrees of the opposites in each ontological pair, to the extent that there can be multiple entities (not only two) interacting with each other in each pair.

• And secondly, this conception of shortcuts do not ignore other possible entities in interacting with the two opposites (with their different versions) in each pair. For instance, by analogy, there are not only the two colors "white" and "black" as opposites, since there are other colors too besides them like "yellow," "green," "purple," or else. The same logic can be applied to each ontological pair, in that they also interact with other entities, not with only two of them (with their different versions).

• Therefore, with these two clarifications in mind—each pair in an ontological principle serves only as an abbreviation for something more complicated and, therefore, although it contains two opposites, it should not be confused as a duality (dualism).

(continued on next page)

Table 4.5. The Syntax of Existential Dialectics II: The Principles as Short Cuts (Part II)

• Existential dialectics rejects any dualism (or dichotomy) as too rigid and instead allows the multiplicity of entities, to the extent that between the two opposites in each pair exist many other alternatives to choose from. They are named in that short form for aesthetic elegance, instead of listing all possible entities between the two opposites in the title.

• To be dialectic is to go beyond any rigid dichotomy and transcend into something different altogether in the long haul. One may be tempted to call the dialectic logic here with a different name like existential "multilectics" (instead of "dialectics"), but this naming is incorrect (or even misleading), for the two reasons aforestated.

• After all, each ontological pair come in all shapes and sizes (with different degrees) and do not exist by themselves but also interact with other entities unlike them. The virtue of revealing an ontological pair is to show how they relate within themselves (in multiple versions) and also interact with others unlike them (also in multiple versions).

• This will be clear in the table on "the dialectic constraints imposed by the principles."

Source: From Ch.1 of *FPHK*

Table 4.6. The Syntax of Existential Dialectics III: The Principles as Family Resemblances

- Each ontological principle is generic, with some other comparable ontological pairs to be put in the same family (like a *family resemblance*). Ludwig Wittgenstein (1953) once suggested the idea of "family resemblance" in explaining different games classified under the same family called *games*.

- Why should, for instance, playing football and chess as playing "games," when it is well understood that football is not the same as chess? The answer is that, although each game is different and has different rules, many of them (though not all) share, more or less, some commonalites (e.g., scoring as necessary for winning).

- And this is so, even though some games share more than some others in any given selection of critieria, and no two games are exactly identical. So, his point here is that there is no essential core which is common to all games, and the best that one can look for is some characteristics which are common to many (but not all) games. (A. Biletzki 2006)

- By the same logic—in the previous section on selection criteria, the flexibility-inflexibility pair can be put in the family resemblance of the simpleness-complicatedness principle, although the two pairs are not exactly identical. Likewise, the directness-indirectness pair can be put in the family resemblance of the preciseness-vagueness principle, although, again, the two pairs are not exactly identical.

- In this sense, which specific pair in a family should be used to designate the name of the family can be at times a bit arbitrary, but with good reason.

- For illustration, in the context of method, the partiality-totality principle can take the different form like individualisticness-holisticness, just as the explicability-inexplicability principle can take the different form like underlyingness-regularness—although each two pairs are not exactly identical and have slightly different meanings and usages. A more comprehensive listing of this family resemblance for all other principles is shown in the table on the syntax of existential dialectics.

Source: From Ch.1 of *FPHK*

Table 4.7. The Syntax of Existential Dialectics IV: The Dialectic Constraints Imposed by the Principles (Part I)

• **Co-Existent and Asymmetric**

—The principles, as they constitute the syntax of existential dialectics, are dialectic in character, such that, when they are applied, they impose dialectic constraints on how reality is to be understood. Consider, say, the symmetry-asymmetry principle as an illustration here, in order to summarize two main characters of the dialectic constraints in question.

—Firstly, to be dialectic here is to go beyond the narrow dichotomies (and, for that matter, any rigid multi-dimensional classificatory scheme), be they about "self" vs. "world," "freedom" vs. "unfreedom," "barbarity" vs. "civilization," "individuality" vs. "communality," and so on.

—One way to do so (to go beyond) is to consider them all in terms of co-existence (without favoring one over the rest). For instance, my theory of "post-civilization" (to be summarized later in the section on the pragmatics of existential dialectics) is to go beyond barbarity and civilization in terms of understanding barbarity and civilization as being co-existent. And the same logic can be said in relation to my theories of "post-democracy," "post-capitalsm," and others (also to be introduced later in the section on the pragmatics of existential dialectics), in regard to freedom vs. unfreedom, equality vs. inequality, communality vs. individuality, spirituality vs. materiality, and so on.

—But to consider them all (in the dichotomies—and, for that matter, in any rigid multi-dimensional classifactory schme) as co-existent is not the same as to imply that the opposites in any classificatory scheme are all equal, since, in accordance to the symmetry-asymmetry principle (as an illustration here), if they are equal in terms of being considered as co-existent, they are asymmetric in terms of being unequal in dialectic interaction (e.g., X can be more dominant than Y in case A, or Y is more dominant than X in case B).

(continued on next page)

Table 4.7. The Syntax of Existential Dialectics IV: The Dialectic Constraints Imposed by the Principles (Part II)

—For this reason, there are different versions of "post-democracy" and "post-capitalism" in my theories. As an illustration, in version I of the theory of post-democracy, freedom is more dominant than equality, whereas in version II of the theory of post-democracy, equality is more so than freedom.

—But this "X more than Y" has to be understood in the context of dialectic logic (not in conventional logic), in that both "X" and "Y" are important in post-democracy (in the context of dialectic logic), but in an asymmetry way. By contrast, in conventional logic, it often favors one over the other—be it in regard to privileging freedom over equality in Fascism, favoring freedom relatively more than equality in Liberal Democracy, or favoring equality relatively more than freedom in Socialist Democracy. In the latter two cases (about Liberal Democracy and Socialist Democracy), the difference between dialectic logic and conventional logic can be one in degree, not in kind—in this sense, albeit not in other senses.

—The same logic can be said about the relatonships between individuality and communality, between spirituality and materiality, and between formal legalism and informal legalism in the different versions of my theory of post-capitalism.

• **Transcendent**

—Secondly, to be dialectic is to go beyond the narrow dichotomies (and, for that matter, any rigid multi-dimensional classificatory scheme) in another way, this time, in a transcendent way, that is, in exploring other possibilities or even other issues not considered within the narrow confines of narrow dichotomies (and, for that matter, any rigid multi-dimensional classificatory scheme).

(continued on next page)

Table 4.7. The Syntax of Existential Dialectics IV: The Dialectic Constraints Imposed by the Principles (Part III)

—As an analogy, to go beyond the narrow color dichotomy of "black" and "white" is not just to choose both "black" and "white" (as in the first meaning) but also to explore other color options (e.g., "green," "purple," "blue," etc.—and, alternatively, "shade," "line," "curve," etc.). By the same logic, to go beyond "democracy" is to transcend democracy (as in version III of the theory of "post-democracy") and to explore other possibilties of lifeforms (e.g., floating consciousness, hyper-spatial consciousness, etc., to live beyond the narrow obsession with freedom and equality).

—This dialectic character of the principles in existential dialectics has important implications for the pragmatics of existential dialectics (as will be clear shortly, in the section on the pragmatics of existential dialectics).

Notes: The examples here are solely illustrative (not exhaustive). The comparison is also relative (not absolute), nor are they mutually exclusive. As generalities, they allow exceptions.
Sources: First explicitly stated in *ALD*. Also from all other books of mine.

Table 4.8. The Syntax of Existential Dialectics V: Further Clarifications (Part I)

- **Pioneering**
  - —Firstly, the total number of ontological principles is unknown, to be discovered later, as our knowledge of the world becomes more advanced.
  - —Consequently, the principles as introduced in my books are not exhaustive, with new ones being added, whenever more of them are discovered in later research.
  - —At least, future generations can pick up where I leave off and continue the discovery.
  - —In this sense, my work should be treated as a pioneering effort for the development of a systematic, comprehensive analysis of a new general ontology for the future of knowledge.

- **Flexible**
  - —Secondly, the principles are not rigidly classified, as they can be reclassified in a different way.
  - —For instance, the preciseness-vagueness principle is classified under the category of "structure" but can be reclassified under the category of "method," although in so doing, it has a different meaning in the context of method.
  - —The same logic applies to the same-difference principle under the category of "outcome," which can be reclassified under the category of "structure," although in so doing, once more, it has a different meaning in the context of structure.

- **Mutually Constraining**
  - —Thirdly, the principles are to be understood together, not that each prinicple is to be analyzed independently of others. After all, the principles are mutually constraining, in that they work together as a whole.

(continued on next page)

Table 4.8. The Syntax of Existential Dialectics V: Further Clarifications (Part II)

- **Selectively Useful**
  —Fourthly, the principles are relevant to all subject matters, but some principles are more useful to some subject matters than others—as implied in the symmetry-asymmetry principle. This is true, even if different studies of the same kind can yield different views about the degree of relevance for each ontological principle, depending on the specific nature of a research in question, needless to say.

- **Anti-Reductionistic**
  —And fifthly, the principles constitute only two levels of analysis, this time, at the ontological and methodological levels—while other levels of analysis (from the perspectives of the mind, nature, society, and culture) are also needed, in order to understand reality in its totality (as explained in the section on "sophisticated methodological holism," so as to avoid the dual dangers of reductionism and reverse-reductionism).

Notes: The examples here are solely illustrative (not exhaustive). The comparison is also relative (not absolute), nor are they mutually exclusive. As generalities, they allow exceptions.
Sources: From my previous books like *FHC*, *FCD*, *FPHC*, *BDPD*, *BCPC*, *BNN*, *FC*, *FAE*, and so on.

Table 4.9. The Syntax of Existential Dialectics VI: The Dilemma of Specific vs. General Ontology (Part I)

• Ontology is often more complicated and imposes some seemingly insurmountable difficulties concerning what constitutes a kind of ontology which can be valid enough to be accepted by the wider intellectual community in question. An excellent example concerns what I want to call *the dilemma of ontology* in relation to the relationship between "specific ontology" and "general ontology." (M. Bunge 1999; R. Corazzon 2007)

• On the one hand, "general ontology," as Mario Bunge (1999) put it, "studies all existents." But, on the other hand, "special ontology studies one genus of thing or process—physical, chemical, biological, social, etc." (M. Bunge 1999; R. Corazzon 2007)

• Consequently, the relationship between the former and the latter is that, as an illustration, "whereas general ontology studies the concepts of space, time, and event, the ontology of the social investigates such general sociological concepts as those of social system, social structure, and social change." (M. Bunge 1999; R. Corazzon 2007)

• With this formal definition in mind, the relationship between the two forms of ontology only poses *the dilemma of ontology*, in that general ontology is possible to the extent that its general contructs must be broad enough to serve as the "lowest common denominator" for all disciplines, but this character of the lowest common denominator is not only hard to find but also its usefulness is limited. (WK 2008b)

• But specific ontology fares no better either, in that its specific constructs primarily serve a specific discipline in question, although other disciplines can appropriate them for their own purposes, but in a different context (especially, though not exclusively, when used as a metaphor from one field to another). For instance, the Kantian category of understanding under the heading of "quantity" constitutes a type of specific ontology, which is specific to the discipline of mathematics, although other disciplines can make use of the categories in their own context.

(continued on next page)

Table 4.9. The Syntax of Existential Dialectics VI: The Dilemma of Specific vs. General Ontology (Part II)

• A solution lies in my proposal of "sophisticated methodological holism," which requires all levels of analysis in any subject matter. See the tables on sophisticated methodological holism, and on reductionism and reverse-reductionism, for more details.

Source: From *Sec. 1.4* of *FPHK*.

Table 4.10. The Syntax of Existential Dialectics VII: Types of Inappropriate Family Resemblances (Part I)

- **Inadequate Family Resemblances**
  - —Ex: consistency-inconsistency: the formalness-informalenss principle
  - —Ex: cognition-noncognition: the formalness-informalness principle

- **Compound Family Resemblances**
  - —Ex: reality-nonreality: the formalness-informalness principles, the absoluteness-relativeness principle
  - —Ex: normality-abnormality: the symmetry-asymmetry principle, the density-emptiness principle, the convention-novelty principle, the same-difference principle
  - —Ex: structure-context: the simpleness-complicatedness principle, the theory-praxis principle
  - —Ex: freedom-unfreedom: the expansion-contraction principle
  - —Ex: morality-immorality: the denseness-emptiness principle, the symmetry-asymmetry, the same-difference principle
  - —Ex: particularness-universalness: the preciseness-vagueness principle, the simpleness-complicatedness principle
  - —Ex: pureness-mixedness: the absoluteness-relativeness principle, the simpleness-complicatedness principle, the same-difference principle
  - —Ex: shortsightedness-foresightedness: the partiality-totality principle, the simpleness-complicatedness principle
  - —Ex: shorttermness-longtermness: the partiality-totality principle, the simpleness-complicatedness principle, the slowness-quickness principle
  - —Ex: smoothness-roughness: the softness-hardness principle, the symmetry-asymmetry principle
  - —Ex: warmth-coldness: the finiteness-transfiniteness principle, the softness-hardness principle.

- **One-Sided Family Resemblances**
  - —Ex: sound-silence: asymmetry (in the symmetry-asymmetry principle)
  - —Ex: vision-blindness: asymmetry (in the symmetry-asymmetry principle)

(continued on next page)

Table 4.10. The Syntax of Existential Dialectics VII: Types of Inappropriate Family Resemblances (Part II)

- **Vague Family Resemblances**
  —Ex: shortness-longness, narrowness-broadness, smallness-largeness, smallness-greatness, microness-macroness, and littleness-muchness: three different possibilities, namely, (a) finiteness (in the finiteness-transfiniteness principle), if both opposites are defined on a finite scale, (b) transfiniteness (in the finiteness-transfiniteness principle) if both opposites are defined on a transfinite scale, and (c) the finiteness-transfiniteness principle if one opposite is defined on a finite scale and the other opposite is defined on a transfinite scale.

- **One-Sided and Compound Family Resemblances**
  —Ex: extroversion-introversion: one-sided, as part of the side (category) of "normality" in the normality-abnormality pair. And compounded, because the normality-abnormality pair is in turn related to the symmetry-asymmetry principle, the density-emptiness principle, the convention-novelty principle, the same-difference principle
  —Ex: organizedness-spontaneousness: one-sided, as part of the side (category) of "normality" in the normality-abnormality pair. And compounded, because the normality-abnormality pair is in turn related to the symmetry-asymmetry principle, the density-emptiness principle, the convention-novelty principle, the same-difference principle
  —Ex: risk-caution: one-sided, as part of the side (category) of "normality" in the normality-abnormality pair. And compounded, because the normality-abnormality pair is in turn related to the symmetry-asymmetry principle, the density-emptiness principle, the convention-novelty principle, the same-difference principle

Notes: The examples here are solely illustrative (not exhaustive). The comparison is also relative (not absolute), nor are they mutually exclusive. As generalities, they allow exceptions.
Source: First summarized in *BEPE* (and updated in later books)

Table 4.11. The Semantics of Existential Dialectics (Part I)

- **Abstract Structure vs. Specific Meanings**
  —The syntax of existential dialectics so understood in terms of ontological principles only gives us the structure of ontology in the world, in an abstract (general) sense. These principles by themselves do not tell us the specific meanings in a given context.
  —In order to grasp the specific meanings of the principles in a given context, it is necessary to study the semantics of existential dialectics. And the analysis of the ontological principles as family resemblances in the section on syntax is only a starting point (and thus overlaps a bit with the semantics of existential dialectics here).
  —With this caveat in mind, the reason that I have often gone in great lengths in my previous books on different subjects is to explain the specific meanings of the principles when applied in different contexts.

- **Specific Meanings in Specific Fields**
  —For instance, in *FPHST*, I used the first three principles (i.e., the change-constancy principle, the regression-progression principle, and the symmetry-asymmetry principle) to propose "the perspectival theory of space-time," for a better way to understand space and time—especially, though not exclusively, in relation to future post-human history (as summarized in *Table 3.6*, *Table 3.7*, *Table 3.8*, and *Table 3.9*). In so doing, I had to introduce concepts and theories specific to the field of physics and other related fields (e.g., "absolute space" and "absolute time" in "classical mechanics" and "relative space-time" in "the theory of relativity").
  —In *BNN*, I also exploited the three principles to propose the "transcendent" approach to the study of genes and memes as a new way to understand the interaction between nature and nurture. In so doing, I had to explore concepts and theories in the world of evolutionary theory (e.g., "mutation," "variation," "adaptation," "selection," and "inheritance" in Darwinian evolutionary theory) and neural biology (e.g., "chromosome," "gene," "DNA," "RNA," "protein," "neuron," "neural network," and "behavior").

(continued on next page)

Table 4.11. The Semantics of Existential Dialectics (Part II)

- **Two Types of Translation (Interpretation)**
  —In the technical jargons of semantics, these exercises in my books perform two essential functions of translation (or interpretation), namely, (a) "intensive" interpretation (about the properties or relations of the principles in existential dialectics), and (b) "extensive" interpretation (about the classes or cases to which the properties and relations of the principles in existential dialectics are applied). (WK 2011oo)

- **Limits of Translation (Interpretation)**
  —But there are limits of translation (or interpretation), and an excellent example is that "there is a problem with the 'objectivity' of translation, as it is relative to different cultural (or sub-cultural) contexts. For instance, one of the still unresolved problems in linguistics is: 'Is there an objective gauge for the quality of translation?'" (*FPHCOM* & WK 2011pp)
  —This is one serious problem about the logic of translation and interpretation that I already addressed in *Sec. 3.5.1* of *FPHCOM (*and also, for that matter, in *FHC*, in the context of hermeneutics).

Notes: The categories and examples in each are solely illustrative (not exhaustive). The comparison is also relative (not absolute), nor are they mutually exclusive. As generalities, they allow exceptions.
Sources: First explicitly stated in *ALD* (and later revised in *FPHCOM*). Also from all other books of mine.

Table 4.12. The Pragmatics of Existential Dialectics (Part I)

- **The Two-Way Street Connecting Theory and Meta-Theory**
  —The pragmatics of existential dialectics is not a one-way street (that is, using the ontological principles for theoretical insights in praxis) but a two-way one, that is, (a) from meta-theory to theory, and (b) from theory to meta-theory.
  —(a) On one side of the street, the ontological principles can inspire some theoretical insights in praxis, that is, in relation to some specific fields.
  —(b) On the other (opposing) side of the street, however, the study of a subject matter in the specific fields in turn reveals some more hitherto unknown ontological principles to be discovered and identified. For this reason, three new principles were added in *FC*, one in *FAE*, and two in *ALD*, on top of the original three in *BCPC*—after some research on the specific subject matters.

- **Direct and Indirect Applications**
  —*Direct*
  - The logic of existential dialectics can shed some theoretical insights on diverse phenomena in the world, and good instances are the pertinent use of the principles of existential dialectics for the theoretical insights on the freedom/unfreedom dialectics, the equality/inequality dialectics, and the wealth/poverty dialectics in my previous works.
  - My latest books like *FPHST* and *BNN* also use the principles to reveal some theoretical insights on the perspectives of space and time (as in *FPHST*) and of nature and nurture (as in *BNN*).

  —*Indirect*
  - The theoretical insights can further be used to reveal other phenomena directly from them (viz., the theoretical insights) and therefore indirectly from the principles themselves. A good instance is the use of the theoretical insights on the freedom/unfreedom and equality/inequality dialectics for the understanding of the civilization/barbarity dialectics.

(continued on next page)

Table 4.12. The Pragmatics of Existential Dialectics (Part II)

- **Direct and Indirect Applications (*cont'd*)**
    - Even in indirect applications, however, a phenomenon under study can still be directly related back to the principles themselves. In the example as cited above, the civilization/barbarity dialectics can be directly related to the principles of existential dialectics without the intermediate role of the freedom/unfreedom and equality/inequality dialectics.

- **Multiple Levels of Application**

  —The theoretical insights can be applied to different levels of analysis, even though in a given example, it may refer to one level only. For instance, in the example concerning the freedom/unfreedom dialectics, it can be used at the structural level (e.g., in relation to the theory of cyclical progression of hegemony), but it can be exploited as well for other levels (e.g., the theory of post-capitalism at the institutional level).

Notes: The categories and examples in each are solely illustrative (not exhaustive). The comparison is also relative (not absolute), nor are they mutually exclusive. As generalities, they allow exceptions.
Sources: From Ch.6 of *BCPC*. See also other books of mine.

Table 4.13. The Freedom-Unfreedom Dialectics (Part I)

- **On Having**
  - *—In Relation to the Technological*
    - (1) if freer from submission to Nature, then less free from ecological degradation (Deep and Social Ecology), even if in a hi-tech form
    - (2) if freer from technological inconvenience / backwardness, then less free from technological control and the loss of privacy
    - (3) if freer from technological (material) backwardness, then less free from the abusive (barbaric) maltreatment of the primitive Others
  - *—In Relation to the Everyday*
    - (1) if freer from abject poverty, then less free from artificial needs/discontents (Frankfurt School)
    - (2) if freer from sensual suppression, then less free from violent sublimation (Freud)
    - (3) if freer from the snobbishness of high culture, then less free from the shabbiness (leveling-off effect) of mass culture (Tocqueville)
    - (4) if freer from the inefficiency of traditional "compassionate economy," then less free from the bondage of a "ruthless [competitive] economy" (Keynes)
    - (5) if freer from anarchy in the state of nature (system fragmentation), then less free from government regulations and controls in system integration

(continued on next page)

Table 4.13. The Freedom-Unfreedom Dialectics (Part II)

- **On Belonging**

  —*In Relation to the Good and the Just*

  (1) if freer from disciplinary society, then less free from society of control (Foucault)

  (2) if freer from the tyranny of one or a few, then less free from the tyranny of the majority (or sometimes, minority veto)

  (3) if freer from elitist decision making, then less free from political gridlock/cleavage

  (4) if freer from arbitrary (discretionary) administration, then less free from bureaucratic irrationality (Weber) and legal trickery (loopholes)

- **On Being**

  —*In Relation to the True*

  (1) if freer from unscientific dogmas, then less free from instrumental abyss (nihilism). Or conversely, if freer from meaninglessness, then less free from dogmas.

  (2) if freer from the bondage of partiality/partisanship (e.g., prejudice, discrimination), then less free from the danger of impartiality and neutrality (e.g., opportunism, unrealisticness, lack of compassion, inaction)

  (3) if freer from making generalizations, then less free from being unable to understand much of anything

  —*In Relation to the Holy*

  (1) if freer from collective conscience, then less free from social loneliness

  (2) if freer from religious absoluteness, then less free from spiritual emptiness

  —*In Relation to the Beautiful/Sublime*

  (1) if freer from artistic non-autonomy, then less free from aesthetic disillusion (deconstruction)

(continued on next page)

Table 4.13. The Freedom-Unfreedom Dialectics (Part III)

Notes: The examples in each category are solely illustrative (not exhaustive), and the comparison is relative (not absolute), nor are they necessarily mutually exclusive. And some can be easily re-classified elsewhere. As generalities, they allow exceptions.
Sources: A reconstruction from Ch.10 of *FCD*, based on *FHC*

Table 4.14. The Equality-Inequality Dialectics (Part I)

- **On Having**
  - —*In Relation to the Technological*
    - (1) if more equal in treating Nature with spiritual unity, then less equal in suppressing the dominant drive to transcend it altogether
  - —*In Relation to the Everyday*
    - (1) if more equal in building social plurality, then less equal in leveling-off effects (e.g., the subsequent relative intolerance of high/intellectual ethos in mass culture industry)
    - (2) if more equal in socioeconomic distribution beyond a certain point, then less equal in efficiency (e.g. resentment, the erosion of work ethics)
    - (3) if more equal in urging an affirmative action program, then less equal in creating victim mentality (in oneself), stigma (from others), reverse discrimination (against the once privileged), and mediocracy (against the more able)

- **On Belonging**
  - —*In Relation to the Good and the Just*
    - (1) if more equal in banning monarchic/oligarchic exclusion, then less equal in producing "the tyranny of the majority" or of "minority veto"
    - (2) if more equal in encouraging participatory decision making, then less equal in inducing political divisiveness (gridlock/cleavage in power blocs) and organizational oligarchy
    - (3) if more equal in institutionalizing a decentralized bureaucracy, then less equal in falling into more territorial/turf politics (intrigues)

(continued on next page)

Table 4.14. The Equality-Inequality Dialectics (Part II)

- **On Being**
  - —In Relation to the Beautiful / Sublime
    - (1) if more equal in accepting diverse styles ("anything goes" mentality), then less equal in artistic good quality (in leveling-off effects against the best)
  - —In Relation to the True
    - (1) if more equal in tolerating multiple viewpoints (no matter how extreme), then less equal in epistemic standards
  - —In Relation to the Holy
    - (1) if more equal in celebrating any cults and sects (no matter how questionable), then less equal in spiritual depth and authenticity

---

Notes: The examples in each category are solely illustrative (not exhaustive), and the comparison is relative (not absolute), nor are they mutually exclusive. And some can be easily reclassified elsewhere. As generalities, they allow exceptions.
Sources: A reconstruction from Ch.10 of *FCD*, based on *FHC*

Table 4.15. The Duality of Oppression in Existential Dialectics: Oppression and Self-Oppression (Part I)

- **From the Same to the Others and Itself**
  - —The Oppression by the Same against the Others
    - *By way of downgrading differences*
      - –Ex: on judiciary caprice for corporate crimes (*Sec.2.2.1.2.1*)
      - –Ex: on the deceptive politics of liberation (*Sec.3.5*)
      - –Ex: on the humanitarian mystique (*Sec.4.4*)
      - –Ex: on the fad of emotional intelligence (*Sec.5.3*)
    - *By way of accentuating differences*
      - –Ex: on the legal sophistry of self-defense (*Sec.2.3*)
      - –Ex: on the legal semantics of proportionality (*Sec.2.4*)
      - –Ex: on the tricky politics of external threat (*Sec.3.4*)
      - –Ex: on the appeal of the Far Right for democracy (*Sec.5.4*)
      - –Ex: on the democratic axis of evil (*Sec.5.5*)
      - –Ex: on the democratic way of brutality and revenge (*Sec.5.6*)
      - –Ex: on democratic autocracy (*Sec.6.4*)
  - —The Oppression by the Same against Itself
    - *By way of downgrading differences*
      - –Ex: on the politics of fear (*Sec.2.2*)
      - –Ex: on the trickery of compassionate conservatism (*Sec.3.2*)
      - –Ex: on the deceptive politics of patriotism (*Sec.3.3*)
    - *By way of accentuating differences*
      - –Ex: on the caprice of due process on domestic suspects (*Sec.2.2*)
      - –Ex: on the false security/freedom dilemma (*Sec.6.5.2*)

(continued on next page)

Table 4.15. The Duality of Oppression in Existential Dialectics: Oppression and Self-Oppression (Part II)

- **From the Others to the Same and Themselves**
  - —The Oppression by the Others against the Same
    - *By way of downgrading differences*
      - –Ex: on judiciary caprice in the reverse direction (*Sec.2.2.1.2.2*)
      - –Ex: on equal pay (*Sec.6.2.1.1*)
      - –Ex: on equal representation (*Sec.6.2.1.2*)
      - –Ex: on affirmative action program (*Sec.6.3.1.1*)
      - –Ex: on same-sex marriage (*Sec.6.3.1.2*)
    - *By way of accentuating differences*
      - –Ex: on sexual harassment (*Sec.6.2.2.1*)
      - –Ex: on physical violence (*Sec.6.2.2.2*)
      - –Ex: on sexual exploitation (*Sec.6.2.2.3*)
  - —The Oppression by the Others against Themselves
    - *By way of downgrading differences*
      - –Ex: on the reverse-class mystique (*Sec.4.2*)
      - –Ex: on the reverse-black mystique (*Sec.4.3*)
      - –Ex: on self-discrimination by downgrading (*Sec.6.3.2.2*)
    - *By way of accentuating differences*
      - –Ex: on self-discrimination by accentuating (*Sec.6.3.2.1*)

Notes: The examples are solely illustrative (not exhaustive), nor are they mutually exclusive. As generalities, they allow exceptions. Also, both forms of oppression co-exist in all of the examples, so the listing of them are only meant in a relative, not absolute, sense.

Source: A summary of the sections (as cited) in Chs.2-6 of *BDPD*. See text for more info and references.

Table 4.16. The Structure of Existential Dialectics I: The Freedom/Unfreedom and Equality/Inequality Dialectics

• Each freedom and equality produces its own unfreedom and inequality, regardless of whether the pair occurs in political society (with the nation-state), in civil society (with some autonomy from the state), or elsewhere (e.g., in the private sphere of individual homes)—and regardless of whether freedom and equality are understood as "negative" or "positive."

• Oppression is dualistic, as much by the Same against the Others and itself, as by the Others against the Same and themselves.

• Both forms of oppression and self-oppression can be achieved by way of downgrading differences (between the Same and the Others) and of accentuating them.

• The relationships are relatively asymmetric between the Same and the Others and relatively symmetric within them. This is true, even when the Same can be relatively asymmetric towards itself in self-oppression, just as the Others can be likewise towards themselves.

• Symmetry and asymmetry change over time, with ever new players, new causes, and new forms.

Notes: The examples in each category are solely illustrative (not exhaustive) nor necessarily mutually exclusive, and the comparison is relative (not absolute). As generalities, they allow exceptions. "Negative" freedom is freedom "from" (e.g., freedom from poverty), whereas "positive" freedom is freedom "to" (e.g., freedom to the state of enlightenment). "Negative" equality is "procedural" equality (e.g., equality of opportunity), while "positive" equality is "substantive" equality (e.g., equality of outcome). Existential dialectics impose constraints on freedom and equality in democracy, non-democracy, and post-democracy. There is no utopia, in the end; even should there be one, dystopia would exist within it.

Sources: From *Table 1.5* of *BDPD*—and also from *FHC*, *FCD*, and *FPHC*

Table 4.17. The Structure of Existential Dialectics II: The Wealth/Poverty Dialectics

• There is no wealth without poverty, just as there is no poverty without wealth.

• The wealth/poverty dialectics occurs in the realms of having, belonging, and being, in relation to the material, relational, and spiritual.

• The wealth/poverty dialectics also expresses itself at the multiple levels of analysis in accordance to methodological holism, be they about the micro-physical, the chemical, the biological, the psychological, the organizational, the institutional, the structural, the systemic, the cultural, and the cosmological.

• The wealth/poverty dialectics is a different manifestation of existential dialectics in general, subject to the principles in its logic of ontology—just as the freedom/unfreedom and equality/inequality dialectics are likewise.

• There is no economic utopia, in the end; even should there be one, dystopia would exist within it.

Notes: The main points here are solely illustrative (not exhaustive) nor necessarily mutually exclusive, and the comparison is relative (not absolute). As generalities, they allow exceptions.
Sources: From *BCPC*. See also *FCD* and *FHC*.

Table 4.18. The Structure of Existential Dialectics III: The Civilization/Barbarity Dialectics

- There is no civilization without barbarity.

- The civilization/barbarity dialectics applies in the four civilizing processes (e.g., the rationalizing process, the pacifying process, the stewardizing process, and the subliming process).

- The civilization/barbarity dialectics is another (different) manifestation of existential dialectics in general, subject to the principles in its logic of ontology—just as the freedom/unfreedom and equality/inequality dialectics and the wealth/poverty dialectics are likewise.

- There is no utopia, in the end; even should there be one, dystopia would exist within it.

Notes: The main points here are solely illustrative (not exhaustive) nor necessarily mutually exclusive, and the comparison is relative (not absolute). As generalities, they allow exceptions.
Sources: From *BCIV*. See also *FCD*, *FHC*, and *BDPD*.

Table 4.19. The Double Sides of Virtual Organizations (Part I)

- **Psychological**
  - —Ex: virtual psychosis
  - —Ex: impersonality and loneliness in quaternary social relations
- **Organizational**
  - —Ex: the race for power and interests
  - —Ex: the world of unequal successes
  - —Ex: the bureaucratic life of its own
  - —Ex: lesser accountability and transparency
- **Economic**
  - —Ex: the economic divides
  - —Ex: the erosive impact of commercialization
- **Political**
  - —Ex: the anti-authoritarian myth
  - —Ex: different power struggles among groups (e.g., Conservative, Reformist, Radical)
- **Structural**
  - —Ex: different localities (e.g., climates, scenery)
  - —Ex: divided domains (e.g., core, peripheral, sub-peripheral)
  - —Ex: substitution effect of social capital
- **Systemic**
  - —Ex: uneven advances in transportation and communications, and geographical migration
- **Cultural**
  - —Ex: conflicting civilizational fabrics (e.g., Confucian, Islamic, Western)
  - —Ex: the bias of the Liberal Democratic agenda
  - —Ex: the addiction to moral fanaticism

(continued on next page)

Table 4.19. The Double Sides of Virtual Organizations (Part II)

Notes: These categories and examples are solely illustrative (not exhaustive), and some of the items can be reclassified somewhere else. Nor are they mutually exclusive. Since they are generalities, exceptions are expected.
Source: From Ch.7 of *FCD*

Table 4.20. Beyond the World of Titans, and the Remaking of the World Order

- **Hyper-Empires**
  - —Ex: The Chinese Union
  - —Ex: The Indian Union
- **Meso-Empires**
  - —Ex: The European Union
  - —Ex: The North American Union
- **Micro-Empires**
  - —Ex: The Latin American Union
  - —Ex: The Middle Eastern Union
- **The Rest of the World**
  - —<u>Odd Powers</u>
    - Ex: Japan
    - Ex: Russia
  - —<u>The Poor Club</u>
    - Ex: The African Union
  - —<u>Ambivalent Regions</u>
    - Ex: Southeast Asia
    - Ex: Oceania
    - Ex: South Asia
    - Ex: Central Asia
    - Ex: Southern/Eastern Europe
    - Ex: North Africa
    - Ex: Central America
    - Ex: Others (e.g., the Korean peninsula)

Source: A summary of Chs.2-5 (of *BWT*)

Table 4.21. The Origins of Authoritarian Liberal Democracy

- **The Geopower of Nature (Ch.4 of *ALD*)**
  —Ex: Power Character and Geographical Strategy
  —Ex: Living Space and Territorial Expansion
  —Ex: Strategic Heartland and Containment

- **The Biopsychology of the Mind (Ch.5 of *ALD*)**
  —Ex: The Bell Curve and Mass Intelligence
  —Ex: Group Analysis and Mass Knowledge
  —Ex: Groupthink and Elite (Mis)calculation

- **The Disciplinary Control of Society (Ch.6 of *ALD*)**
  —Ex: Social Organizations of Ruthlessness
  —Ex: Social Institutions of Greed
  —Ex: Social Structure of Exclusion
  —Ex: Social Systems of Violence

- **The Molding Force of Culture (Ch.7 of *ALD*)**
  —Ex: The Tradition of Conquest
  —Ex: The Rationalization of Unreason

Notes: The examples in each category are solely illustrative (not exhaustive), and they are also relative (not absolute), nor are they mutually exclusive. As generalities, they allow exceptions.
Source: A summary of Chs.4-7 of *ALD*

Table 4.22. The Theory of Post-Democracy I: The Priority of Freedom over Equality (Part I)

- **Differences**
  - *—For the aggressive Lions (the strong Elitists)*
    - •Setting up rank distinctions among unequals (e.g., between inferior humans and superior post-humans, or later among inferior post-humans and superior ones, relatively speaking)
    - •Yearning for being not only distinguished from unequals, but also the first among equals (the best of the very best)
    - •Soul-searching for a high spiritual culture (not the trashy one for the masses). Mass culture is a dirty joke for them.
  - *—For the manipulative Foxes (the weak Counter-Elitists)*
    - •Seeking a gentle hegemony by way of more communitarian concerns (for inferior humans and, later, inferior post-humans)
    - •Being more sympathetic to less formal-legalistic institutions and values

- **Similarities**
  - *—For both Lions and Foxes*
    - •Exploring different spheres of non-human consciousness in the cosmos (something vastly superior than the human one)
    - •Recognizing the democratic illusions (e.g., no freedom without unfreedom, no equality without inequality, or simply no justice without injustice, and vice versa)

(continued on next page)

Table 4.22. The Theory of Post-Democracy I: The Priority of Freedom over Equality (Part II)

Notes: The two callings and examples in each category are solely illustrative (not exhaustive), since there will be many different post-human value ideals in the distant future of post-human civilization. The comparison is also relative (not absolute) towards post-democracy, so this is not just a version of free-market democracy (nor Fascism/Nazism, as shown in the table later on democracy, non-democracy, and post-democracy). Nor are they mutually exclusive. As generalities, they allow exceptions. And the specific forms of post-human post-democratic ideals need to be further developed in future after-postmodern history, as they will be different from the ones we now know. The point here is to solely give an extremely rough picture of a small part of the world to come that we have never known.
Source: From Ch.10 of *FCD*. Refer to text for more info and references.

Table 4.23. The Theory of Post-Democracy II: The Priority of Equality over Freedom

- **Hybrid Versions of**
  —Ex: the Trans-Feminine Calling
  —Ex: the Trans-Sinitic Calling
  —Ex: the Trans-Islamic Calling
  —Ex: the Trans-Outerspace Calling

- **Qualifications**
  —These four versions of post-capitalist value ideals need not automatically be post-democratic, just as capitalism does not necessarily mean democracy. They are two different entities—though closely related.
  —But up to a certain threshold of elevating equality at the farther expense of freedom, the democratic ideals will be overcome and cease to exist.
  —The overcome will not be socialist or communist, but post-democratic with no freedom without unfreedom and no equality without inequality, subject to the constraints of existential dialectics.

Notes: The callings are solely illustrative (not exhaustive), since there will be many different post-human value ideals in the distant future of post-human lifeforms. The comparison is also relative (not absolute), nor are they mutually exclusive. As generalities, they allow exceptions. And the specific forms of post-human post-democratic ideals need to be further developed in future after-postmodern history, as they will be different from the ones we now know. The point here is to solely give an extremely rough picture of a small part of the world to come that we have never known.
Source: From Ch.10 of *FCD*. Refer to text for more info and references.

Table 4.24. The Theory of Post-Democracy III: The Transcendence of Freedom and Equality (Part I)

- **Transcending Freedom in Floating Existence**
  —*Freedom*: seeking an ultimate elimination of the body. Being without the body. The aim is to transcend freedom in the end into a metaphysical state (i.e., beyond the physique).
  —*Unfreedom*: yet facing difficult trade-offs. The sacrifice of bodily existence and its joyfulness. An eternal boredom in floating existence in dark deep space, though with alternative pleasures. There is no free lunch even in the state of transcending freedom.

- **Transcending Equality in the Rivalry of Cosmic Hegemony**
  —*Inequality*: competing to outlast other lifeforms in floating existence, or just marginalizing them for one's hegemonic expansiveness in the rest of the cosmos (and even beyond). Universalism is only for the mediocre.
  —*Equality*: accepting only those of one's rank as equal partners in the vast spacetime for cosmic supremacy. Even here, the aim is to transcend equality into a metaphysical state.

(continued on next page)

Table 4.24. The Theory of Post-Democracy III: The Transcendence of Freedom and Equality (Part II)

Notes: Do not confuse this transcendence of freedom and equality (as one version of post-democracy) with the naïve temptation to transcend the freedom/unfreedom and equality/inequality dialectics. Existential dialectics hold true for freedom and equality in all cultures and societies—past, present, or future (i.e., democracy, non-democracy, and post-democracy), regardless of whether freedom and equality are conventionally understood as "negative" or "positive."

Also, the two features and examples in each are solely illustrative (not exhaustive), since there will be many different post-human value ideals in the distant future of post-human lifeforms. The comparison is also relative (not absolute), nor are they mutually exclusive. As generalities, they allow exceptions. And the specific forms of post-human ideals even for these radically alien floating lifeforms (and others unknown to us) need to be further developed in future after-postmodern history, as they will likely be different from the ones herein illustrated. The point here is to solely give a very rough picture of a small part of the extremely alien world to come that we have never known.

Source: From Ch.10 of *FCD*. Refer to text for more info and references.

Table 4.25. Democracy, Non-Democracy, and Post-Democracy (Part I)

- **Democracy**
  - —*Theoretical Constructs*
    - •The pursuit of freedom and equality (in various degrees), regardless of whether freedom and equality can be understood as "negative" or "positive"
      - (1) more equality than freedom: The relative priority of the good over the right
      - (2) more freedom than equality: The relative priority of the right over the good
  - —*Types*
    - •Only (1): Different versions of communitarian moral universalism
    - •Only (2): Different versions of liberal moral universalism
    - •(1) or (2): Different versions of anarchic (non-nation-state) moral universalism
    - •(1) or (2): Different versions of postmodern moral localism

- **Non-Democracy**
  - —*Theoretical Constructs*
    - •The focus on (1') equality or (2') freedom, but not both, regardless of whether freedom and equality can be understood as "negative" or "positive"
  - —*Types*
    - •Only (1'): Different versions on the Far Left (e.g., Stalinism, Robespierrianism)
    - •Only (2'): Different versions on the Far Right (e.g., Nazism, absolute monarchism)

(continued on next page)

Table 4.25. Democracy, Non-Democracy, and Post-Democracy (Part II)

- **Post-Democracy**
  —*Theoretical Constructs*
  •The priority of (1'') equality over freedom, or (2'') freedom over equality, or (3'') the transcendence of freedom and equality, regardless of whether freedom and equality are "negative" or "positive." In degree, (1'') or (2'') is less than (1') or (2') but more than (1) or (2)—respectively.
  •Like democracy and non-democracy, post-democracy is also subject to the freedom/unfreedom and equality/inequality dialectics (or existential dialectics in general). Unlike them, post-democracy acknowledges the constraints of existential dialectics and no longer value freedom and equality as sacred virtues. There is no utopia, in the end; even were there one, dystopia would exist within it.
  —*Types*
  •(1''): Different versions of trans-Sinitic value ideals
  •(1''): Different versions of trans-feminine value ideals
  •(1''): Different versions of trans-Islamic value ideals
  •(1''): Different versions of trans-outerspace value ideals
  •(2''): Different versions of post-human elitist value ideals
  •(3''): Different versions of the value ideals of floating consciousness (etc.)

(continued on next page)

Table 4.25. Democracy, Non-Democracy, and Post-Democracy (Part III)

Notes: The examples are solely illustrative (not exhaustive), nor are they mutually exclusive. As generalities, they allow exceptions. "Negative" freedom is freedom "from" (e.g., freedom from poverty), whereas "positive" freedom is freedom "to" (e.g., freedom to the state of enlightenment). "Negative" equality is "procedural" equality (e.g., equality of opportunity), while "positive" equality is "substantive" equality (e.g., equality of outcome). Existential dialectics impose constraints on freedom and equality in democracy, non-democracy, and post-democracy, regardless of whether freedom and equality can be understood as "negative" or "positive" in conventional discourse. Therefore, do not confuse the transcendence of freedom and equality in (3'') with the naïve temptation to transcend existential dialectics. There is no utopia, in the end; even should there be one, it would not exist without dystopia embedded within it.
Sources: A summary, based on my previous works, especially Ch.5 of *FHC*, Chs.5-10 of *FCD*, Chs.2-4 of *FPHC*, and Chs.1 & 7 of *BDPD*. The reader should consult the books for more analysis, as this is only a summary here.

Table 4.26. Multiple Causes of the Emergence of Post-Democracy (Part I)

- **At the Micro-Physical Level**
  —Ex: intelligent life without the human physical-chemical system
  —Sources: Ch.7 of *FHC*; Chs.9-10 of *FCD*; Ch.1 of *FPHC*

- **At the Chemical Level**
  —Ex: space radiation and toxins
  —Sources: Ch.7 of *FHC*; Chs.9-10 of *FCD*

- **At the Bio-Psychological Level**
  —Ex: exo-biological evolution in deep space
  —Ex: genetic engineering of new beings
  —Ex: limits of cognitive partiality
  —Ex: illusions of emotional neutrality
  —Ex: human biological inequality
  —Ex: the rise of unfolding unconsciousness
  —Sources: Ch.2 & Chs.9-10 of *FCD*; Ch.7 of *FHC*; Ch.4 of *BCPC*; *FPHU*

- **At the Institutional Level**
  —Ex: the flawed logic of equality
  —Ex: the conflicting nature of governance
  —Sources: Ch. 5 of *FHC*; Chs. 6 & 10 of *FCD*; Ch. 3 of *FPHC*; Chs.2-5 of *BDPD*

- **At the Organizational Level**
  —Ex: e-civic alienation
  —Ex: the dark sides of formal-legalistic routines
  —Sources: Ch.3 of *FHC*; Ch.7 of *FCD*; Ch.3 of *FPHC*

- **At the Structural Level**
  —Ex: ever new forms of inequities, at home and abroad
  —Ex: the emergence of China, women, and Islam as major actors
  —Sources: Chs.5-6 of *FHC*; Chs.7, 9 & 10 of *FCD*; Chs.4-5 of *BDPD*

(continued on next page)

Table 4.26. Multiple Causes of the Emergence of Post-Democracy (Part II)

- **At the Cultural Level**
  —Ex: freedom/unfreedom dialectics
  —Ex: equality/inequality dialectics
  —Ex: system fragmentation and integration
  —Sources: Ch.5 of *FHC*; Chs. 3, 9 & 10 of *FCD*; Ch.4 of *FPHC*; Ch.1 of *BDPD*; Ch.4 of *BCPC*

- **At the Systemic Level**
  —Ex: space habitats (in zero-gravity) and colonization
  —Ex: ultra advanced future info systems
  —Ex: qualitative demography
  —Sources: Ch.7 of *FHC*; Chs.9 &10 of *FCD*

- **At the Cosmological Level**
  —Ex: the colonization of multiverses
  —Ex: the alteration of space-time and the creation of new matter-energy
  —Ex: the expansion of floating consciousness
  —Ex: the spread of hyper-spatial consciousness
  —Sources: Ch.7 of *FHC*; Chs.9 &10 of *FCD*; Ch.4 of *FPHC*; *FPHST*

Notes: The examples in each category are solely illustrative (not exhaustive), and some of the items can be reclassified somewhere else. Nor are they always mutually exclusive. Since they are generalities, exceptions are expected.
Sources: Especially from *FHC*, *FCD*, *FPHC*, *BCPC*, *BDPD*, *FPHST*, and *FPHU*. See also other books and my perspectives on civilizational holism.

Table 4.27. Some Clarifications about Post-Capitalism and Post-Democracy (Part I)

• The prefix "trans-" in the first category of post-capitalism (with its four versions) refers to something "going beyond" (not "uniting" or "combining"). Ex: *Sec.10.3.3* of *FCD*; *Sec.2.4* & *Sec.4.4* of *FPHC*; *Sec.7.2* of *BCPC*

• Such terms like "post-democracy," "post-capitalism," "post-human elitist," "trans-feminine calling," and the like as used in my works are more for our current intellectual convenience than to the liking of future humans and post-humans, who will surely invent more tasteful neologisms to call their own eras, entities, and everything else, for that matter. But the didactic point here is to use the terms to foretell what the future might be like, not that its eras and entities must be called so exactly and permanently.
Ex: *Sec.11.1* of *FCD*; *Sec.7.2* of *BCPC*

• The four versions in the first category of post-capitalist value ideals need not automatically be post-democratic, just as capitalism does not necessarily mean democracy. They are two different entities—though closely related. But up to a certain threshold of elevating equality at the farther expense of freedom, the democratic ideals will be overcome and cease to exist. The same is true for the post-human elitist calling in the second category of post-capitalism in relation to post-democracy, depending on the extent to which freedom is elevated at the expense of equality.
Ex: *Sec.10.4.3.3* of *FCD*; *Table 3.9* of *FPHC*; *Table 7.6* of *BDPD*

• The comparison in each of the three realms of existence in all forms of post-capitalism is not absolute, but relative. Examples include "communal" vs. "individualistic," and the like.
Ex: Notes in *Table 10.8*, *Table 10.9*, *Table 10.10*, & *Table 10.11* of *FCD*; Chs.2-4 of *FPHC*; *Sec.7.2* of *BCPC*

(continued on next page)

Table 4.27. Some Clarifications about Post-Capitalism and Post-Democracy (Part II)

• The emergence of post-capitalism (and post-democracy, for that matter) has multiple causes (to not be reduced to one or only a few).
Ex: Ch.10 of *FCD*, Chs.2-4 of *FPHC*; *Sec.1.3* & *Sec.7.2* of *BCPC* (or *Table 1.8* & *Table 7.11*)

• The specific forms of post-capitalism (and post-democracy, for that matter) need to be further developed in future after-postmodern history, as they will be different from the ones we now know. The point here is to solely give an extremely rough sketch of a world to come that we have never known.
Ex: *Sec.10.3.3* & *Sec.10.4.3.3* of *FCD*; *Table 10.14* & *Table 10.15* of *FCD*; *Sec.7.2* of *BCPC*

• All forms of post-capitalism are not part of a "teleological law," but of "historical trends" only. The same is also true for all forms of post-democracy.
Ex: *Sec.7.1* of *FHC*; *Sec.9.5.3.2* & *Sec.10.3.4.2* of *FCD*; *Sec.7.2* of *BCPC*

• Post-capitalism is not better than capitalism in an "absolute" sense but only fits in better, on the basis of the historical contingency of culture, society, nature, and the mind in some future eras. The same is true for post-democracy in relation to democracy. The term "better" is historically relative.
Ex: *Sec.10.3.3* of *FCD*; *Sec.1.7* of *BDPD*; *Sec.1.5* of *BCPC*

(continued on next page)

Table 4.27. Some Clarifications about Post-Capitalism and Post-Democracy (Part III)

• All forms of post-capitalism and post-democracy are subject to the constraints of existential dialectics. In the process, the dialectic direction is to go beyond the conventional "either-or" dichotomies (e.g., freedom vs. unfreedom, equality vs. inequality, freedom vs. equality, individuality vs. communality, spirituality vs. materiality, formal legalism vs. informal legalism, etc.). As is true in post-civilization, to go beyond the dichotomies is to acknowledge the co-existence of both in each dichotomy, although the degree of scaling one over the over varies from case to case (e.g., the theory of post-capitalism I, the theory of post-capitalism II, the theory of post-democracy I, the theory of post-democracy II, etc.)—but is not to be extreme in largely favoring one over the other, *on average* (all things considered). There is no utopia to be had in the end; even should there be one, dystopia would exist within it.
Ex: Ch.5 of *FHC*; *Sec.10.4.4.2* of *FCD*; *Sec.1.5* of *BDPD*; *Sec.1.3* of *BCPC*; *BCIV*

• All forms of post-capitalism, however different from each other though they are, share one common feature, in that they all inspire for a higher spiritual culture. The same is also true for post-democracy.
Ex: *Sec.10.3*, *Sec.10.4* & *Sec.10.5* of *FCD*; *Chs.2-4* of *FPHC*; *Sec.7.2* of *BCPC*

• All forms of post-capitalism try to avoid the excess in capitalist consumerism by favoring more basic than artificial needs in having, but the quality and quantity of these "basic" needs will be measured by future standards, not by our current ones. Standards are historically relative.
Ex: *Sec.10.3*, *Sec.10.4* & *Sec.10.5* of *FCD*; Ch.2 of *FPHC*; *Sec.7.2* of *BCPC*

(continued on next page)

Table 4.27. Some Clarifications about Post-Capitalism and Post-Democracy (Part IV)

• All forms of post-capitalism make use of a different degree of political authority with advanced info systems in future history and strives for higher spiritual cultures (especially in the post-human age), while acknowledging the constraints of existential dialectics and no longer valuing free market (as in capitalism) and economic control (as in non-capitalism) as sacred virtues.
Ex: *Sec.10.3.4.2*, *Sec.10.3*, *Sec.10.4* & *Sec.10.5* of *FCD*; Chs.2-4 of *FPHC*; *Sec.1.5* of *BDPD*; *Sec.7.2* of *BCPC*

Notes:: The main points here are solely illustrative (not exhaustive) nor necessarily mutually exclusive, and the comparison is relative (not absolute). As generalities, they allow exceptions. The sections as cited are only illustrative (not exhaustive).
Sources: From *FHC*, *FCD*, *FPHC*, and *BDPD*

Table 4.28. The Theory of Post-Capitalism I.1: By Group— Ex: Spiritual/Communal in the Trans-Feminine Calling

- **More Communal Than Individual**
  —*Sharing*: learning from others, as different ideas mutually enrich
  —*Cooperative*: encouraging a sense of shared leadership and teamwork

- **More Informal-Legalistic Than Formal-Legalistic**
  —*Specific*: listening more from the heart than from the head, to know a person as a concrete, not as an abstract, unit
  —*Affective*: thinking and acting with others on a more affective tone. Business can mix with an emotional touch.
  —*Ascriptive*: hiring (or firing) can be done on the basis of merit (or lack of it), but deep solidarity (sisterhood) is important too.
  —*Particularistic*: making decisions on the basis of cost-benefit analysis, but a given group relationship is vital

- **More Spiritual Than Secular**
  —*Long-Term Looking*: sharing for a long-term relationship (e.g., love, friendship), not just for a short-term gain
  —*Loving/Caring*: showing compassion for the sufferings of others, without quickly blaming and pre-judging
  —*Respectful*: showing acceptance about others' feelings (and thoughts)

Notes: The categories and examples are solely illustrative, since there can be different versions, and the comparison is relative (not absolute), nor are they mutually exclusive. As generalities, they allow exceptions. The specific forms of the trans-feminine version need to be further developed in future after-postmodern history, as they will be different from the ones we now know, since the prefix "trans-" here means going beyond or deconstructing the feminine values, while using them as the inspirational point at the beginning.
Source: From Ch.10 of *FCD*. Refer to text for more info and references.

Table 4.29. The Theory of Post-Capitalism I.2: By Nation-State— Ex: Spiritual/Communal in the Trans-Sinitic Calling

- **More Communal Than Individualistic**
  —*Centralized*: being more top-down in management
  —*Collective*: encouraging more group cooperation
  —*Social*: investing in trust and connection

- **More Informal-Legalistic Than Formal-Legalistic**
  —*Specific*: knowing more of those related or connected
  —*Affective*: behaving in a paternalistic, hierarchical way
  —*Ascriptive*: favoring family members and those related
  —*Particularistic*: building connection (guanxi) as imperative

- **More Spiritual Than Secular**
  —*Expansionist*: diffusing civilizational values (e.g., the superiority complex of civilizationalism)
  —*Holistic*: synthesizing things into a panoramic horizon
  —*Historical*: learning from the lessons of the ancient past
  —*Respectful*: deferential to elders and superiors

Notes: The categories and examples are solely illustrative, since there can be different versions, and the comparison is relative (not absolute), nor are they mutually exclusive. As generalities, they allow exceptions. The specific forms of the trans-Sinitic version need to be further developed in future after-postmodern history, as they will be different from the ones we now know, since the prefix "trans-" here means going beyond or deconstructing the Sinitic values, while using them as the inspirational point at the beginning.
Source: From Ch.10 of *FCD*. Refer to text for more info and references.

Table 4.30. The Theory of Post-Capitalism I.3: By Region— Ex: Spiritual/Communal in the Trans-Islamic Calling

- **More Communal Than Individualistic**
  —*Collective*: building the webs of relationships to bind individuals
  —*Sharing*: cultivating the established "wisdom" through common experience
  —*Cooperative*: stressing harmony, solidarity, and commonality

- **More Informal-Legalistic Than Formal-Legalistic**
  —*Specific*: making efforts to know well the participants (family and larger community) in matters of common concern
  —*Affective*: mixing work with language and ritual on explicit religious (Islamic) ideals, texts, stories, and examples
  —*Ascriptive*: privileging local history and custom on relationships among kinship groups
  —*Particularistic*: preferring an unbiased insider with ongoing connections to all parties

- **More Spiritual Than Secular**
  —*Historical*: learning from the lessons of the past as a source of stability and guidance
  —*Deferential*: showing respect for age, experience, status, and leadership in communal affairs
  —*Honorable*: emphasizing face, dignity, prestige, and fairness
  —*Compassionate*: giving mercy and charity ("Zahah") to others

Notes: The categories and examples are solely illustrative (not exhaustive), and the comparison is relative (not absolute), nor are they mutually exclusive. As generalities, they allow exceptions. The specific forms of the trans-Islamic version need to be further developed in future after-postmodern history, as they will be different from the ones we now know, since the prefix "trans-" here means going beyond or deconstructing the Islamic values, while using them as the inspirational point at the beginning.
Sources: From Ch.10 of *FCD*. Refer to text for more info and references, especially from the works by George Irani (2000) and C. Murphy (September 19, 2001).

Table 4.31. The Theory of Post-Capitalism I.4: By Universe— Ex: Spiritual/Communal in the Trans-Outerspace Calling

- **More Communal Than Individual**
  —*Cooperative*: requiring teamwork in small space habitats
  —*Sharing*: learning from, and enjoying being with, each other in a small group in outer space

- **More Informal-Legalistic Than Formal-Legalistic**
  —*Specific*: knowing more about each other to facilitate living and working together in space, both as fellow astronauts and space-mates
  —*Affective*: being friendly and social to each other as vital to working and living in small space quarters
  —*Ascriptive*: nurturing comaraderie among fellow astronauts as if they are family members over time
  —*Particularistic*: building work relationship with enduring memory in a space mission

- **More Spiritual Than Secular**
  —*Long-Term*: looking beyond selfish materialistic concerns in a precarious space environment with potential life or death
  —*Loving/Caring*: cultivating deep bondage for the success of a long term space mission
  —*Transcendent*: searching for life meaning in outer space

Notes: The calling and examples in each category are solely illustrative (not exhaustive), since there will be many different outer-space value ideals in the distant future of space colonization. The comparison is also relative (not absolute), nor are they mutually exclusive. As generalities, they allow exceptions. And the specific forms of trans-outer-space calling need to be further developed in future after-postmodern history, as they will be different from the ones we now know, since the prefix "trans-" here means going beyond or deconstructing the current outer-space values, while using them as the inspirational point at the beginning. The point here is to solely give an extremely rough picture of a small part of the world to come that we still do not know much about.
Source: From Ch.10 of *FCD*. Refer to text for more info and references.

Table 4.32. The Theory of Post-Capitalism II: Spiritual/Individualistic in the Post-Human Elitist Calling (Part I)

- **More Individualistic Than Communal**
  —Setting up rank distinctions among unequals (e.g., between inferior humans and superior post-humans, or later among inferior post-humans and superior ones, relatively speaking)
  —Yearning for being not only distinguished from unequals, but also the first among equals (the best of the very best)
  —Recognizing the constraints of equality/inequality dialectics (or existential dialectics in general)

- **More Spiritual Than Secular**
  —Soul-searching for a high spiritual culture (not the trashy one for the masses). Mass culture is a dirty joke for them.
  —Exploring different spheres of non-human consciousness in the cosmos (something vastly superior than the human one)
  —Recognizing the constraints of freedom/unfreedom dialectics (or existential dialectics in general)

- **Qualifications**
  —Although post-human elitist post-democracy is comparable to post-human elitist post-capitalism in some respects, the former does not necessarily imply the latter (post-human elitist post-capitalism), just as democracy does not have to entail capitalism. They are two different (though related) entities.
  —But up to a certain threshold of incorporating government intervention with advanced info systems in future civilizations for higher spiritual concerns at the expense of the free market and materialist pursuit, the capitalist ideal will be overcome.
  —The overcome will not be Fascist or feudalistic, but post-capitalist, subject to the constraints of existential dialectics.

(continued on next page)

Table 4.32. The Theory of Post-Capitalism II: Spiritual/Individualistic in the Post-Human Elitist Calling (Part II)

Notes: The calling and examples in each category are solely illustrative (not exhaustive). The comparison is also relative (not absolute), nor are they mutually exclusive. As generalities, they allow exceptions. And the specific forms of post-human elitist post-capitalism need to be further developed in future after-postmodern history, as they will be different from the ones we now know, while using them as the inspirational point at the beginning. The point here is to solely give an extremely rough picture of a small part of the world to come that we still do not know much about.
Sources: From Ch.10 of *FCD* (and also *FPHC*, *BDPD*, and *BCPC*). Refer to the text for more info and references.

Table 4.33. Capitalism, Non-Capitalism, and Post-Capitalism (Part I)

- **Capitalism**
  - *Theoretical Constructs*
    - Allocation of scarce resources among alternative wants largely by free market for competition (whose characteristics in its ideal form include, for instance, no barrier to entry or exit, homogeneity, perfect information, a large number of buyers/sellers, and perfect factor mobility)
    - More formal-legalistic than informal-legalistic, more individualistic than communal, and more material (secular) than spiritual
    - Either (1) minimal government or (2) relatively active government
  - *Types*
    - Only (1): Different versions of market capitalism (e.g., the U.S.)
    - Only (2): Different versions of welfare capitalism (e.g., Sweden)

- **Non-Capitalism**
  - *Theoretical Constructs*
    - Allocation of scarce resources among alternative wants mainly by political authority for policies (which can be regulative, redistributive, symbolic, and participatory)
    - More informal-legalistic than formal-legalistic
    - Either (1') more individualistic (for the elites), often (though not always) for material (secular) concerns, or (2') more communal (for the masses), often (though not always) for spiritual concerns
  - *Types*
    - Only (1'): Different versions on the Right (e.g., Fascist corporate-state economy for the glory of the new Rome, medieval lord-vassal-serf economy for the power of the feudalistic order)
    - Only (2'): Different versions on the Left (e.g., Soviet command economy for the creation of the New Socialist Man)

(continued on next page)

Table 4.33. Capitalism, Non-Capitalism, and Post-Capitalism (Part II)

- **Post-Capitalism**
  - —*Theoretical Constructs*
    - •Allocation of scarce resources among alternative wants largely by political authority with advanced info systems in future civilizations, subject to existential dialectics. In degree of allocating by authority, post-capitalism is more than capitalism but less than non-capitalism.
    - •More spiritual than secular (material)
    - •Either (1'') more individualistic or (2'') more communal
    - •Like capitalism and non-capitalism, post-capitalism is also subject to the freedom/unfreedom and equality/inequality dialectics (or existential dialectics in general). There is no utopia, in the end; even were there one, dystopia would exist within it.
    - •Unlike capitalism and non-capitalism, post-capitalism makes use of a different degree of political authority with advanced info systems in future civilizations and strives for higher-spiritual cultures (especially in the post-human age), while acknowledging the constraints of existential dialectics and no longer valuing free market (as in capitalism) and economic control (as in non-capitalism) as sacred virtues.
  - —*Types*
    - •Only (1''): Different versions of post-human elitist value ideals
    - •Only (2''): Different versions of trans-Sinitic value ideals
    - •Only (2''): Different versions of trans-feminine value ideals
    - •Only (2''): Different versions of trans-Islamic value ideals
    - •Only (2''): Different versions of trans-outerspace value ideals

(continued on next page)

Table 4.33. Capitalism, Non-Capitalism, and Post-Capitalism (Part III)

Notes: The calling and examples in each category are solely illustrative (not exhaustive). The comparison is also relative (not absolute), nor are they mutually exclusive. As generalities, they allow exceptions. And the specific forms of each calling need to be further developed in future after-postmodern history, as they will be different from the ones we now know, while using them as the inspirational point at the beginning. The point here is to solely give an extremely rough picture of a small part of the world to come that we still do not know much about.
Source: From Ch.10 of *FCD*. Refer to the text for more info and references.

Table 4.34. Multiple Causes of the Emergence of Post-Capitalism (Part I)

- **At the Micro-Physical Level**
  —Ex: intelligent life without the human physical-chemical system
  —Ex: mastering of quantum mechanics, electromagnetism, and other fields for the understanding of a broad range of anomalous experiences and the application for artificial intelligence for spiritual quest
  —Sources: Ch.7 of *FHC*; Chs.9-10 of *FCD*; Ch.1 of *FPHC*

- **At the Chemical Level**
  —Ex: space radiation and toxins
  —Sources: Ch.7 of *FHC*; Chs.9-10 of *FCD*

- **At the Bio-Psychological Level**
  —Ex: exo-biological evolution in deep space
  —Ex: genetic engineering of new beings
  —Ex: limits of human cognition
  —Ex: the rise of unfolding unconsciousness
  —Sources: Ch.2 & Chs.9-10 of *FCD*; Ch.7 of *FHC*; *FPHU*

- **At the Institutional Level**
  —Ex: the flawed logic of the free market
  —Ex: the need of a post-autistic economics
  —Sources: Ch.10 of *FCD*

- **At the Organizational Level**
  —Ex: the dark sides of formal-legalistic routines
  —Sources: Ch.3 of *FHC*; Ch.7 of *FCD*; Ch.3 of *FPHC*

- **At the Structural Level**
  —Ex: ever new forms of inequities, at home and abroad
  —Ex: the emergence of China, women, and Islam as major actors
  —Sources: Chs.5-6 of *FHC*; Chs.7, 9 & 10 of *FCD*; Chs.4-5 of *BDPD*

(continued on next page)

Table 4.34. Multiple Causes of the Emergence of Post-Capitalism (Part II)

- **At the Cultural Level**
  —Ex: freedom/unfreedom dialectics
  —Ex: equality/inequality dialectics
  —Sources: Ch.5 of *FHC*; Chs.3 & 10 of *FCD*; Ch.4 of *FPHC*; Ch.1 of *BDPD*

- **At the Systemic Level**
  —Ex: space habitats (in zero-gravity) and colonization
  —Ex: ultra advanced future info systems
  —Ex: qualitative demography
  —Sources: Ch.7 of *FHC*; Chs. 9 & 10 of *FCD*

- **At the Cosmological Level**
  —Ex: the colonization of multiverses
  —Ex: the alteration of space-time and the creation of new matter-energy
  —Ex: the expansion of floating consciousness
  —Ex: the spread of hyper-spatial consciousness
  —Sources: Ch.7 of *FHC*; Chs. 9 & 10 of *FCD*; Ch.4 of *FPHC*; *FPHST*

Notes: The examples in each category are solely illustrative (not exhaustive), and some of the items can be reclassified somewhere else. Nor are they always mutually exclusive. Since they are generalities, exceptions are expected.
Sources: Especially from *FHC*, *FCD*, *FPHC*, *BCPC*, *BDPD*, *FPHST*, and *FPHU*. See also other books and my perspectives on civilizational holism.

Table 4.35. The Theoretical Debate on Civilization

- **The Progressive Theory of Civilization**
  —*Thesis*: The "civilizing" process is "good," as opposed to the "barbarizing" process as something "bad," relatively speaking.
  —*Discourse*: Especially, though not exclusively, in the Enlightenment era and a bit before. Example: Thomas Hobbes—in that the tribes in primitive societies were "savages."

- **The Romantic Theory of Barbarity**
  —*Thesis*: The "civilizing" process is "bad," as opposed to the "barbarizing" process as something "good," relatively speaking.
  —*Discourse*: Especially, though not exclusively, in the Counter-Enlightenment circle. Example: Jean-Jacques Rousseau—in that civilization "corrupts" men, and the "savages" are in fact "noble."

- **The Moderate Theory of Civilization**
  —*Thesis*: The "civilizing" process is "good," but there is a price to pay, especially in systematic (compulsive) self-control.
  —*Discourse*: Especially, though not exclusively, in some late modern and postmodern circles. Example: Norbert Elias—in that social manners become more refined in the civilizing process, but self-control also becomes more systematic.

- **The Theory of Post-Civilization**
  —*Thesis*: The civilizing process is as evil and good as barbarity, and each cannot exist without the other, to be eventually superseded by post-civilization unto the post-human age.
  —*Discourse*: Proposed by Peter Baofu. See the rest of *BCIV* for more analysis.

Source: From *BCIV* on the theoretical debate

Table 4.36. No Freedom Without Unfreedom in the Civilizing Processes (Part I)

- **The Rationalizing Process (at the Level of Culture)**
  - —if freer from the dominance of unreason (as in barbarism) in the civilizing process, then less free from the rationalizing process (be it in the form of the principle of either transcendence or immanence)
  - —if freer from the principle of immanence in the rationalizing process, then less free from the inclination to commit terror in the name of reason and the relative underdevelopment of non-reason (e.g., in relation to yoga and meditation)
  - —if freer from the principle of transcendence in the rationalizing process, then less free from the relative underdevelopment of reason (e.g., in relation to systematic methodology) and the occurrence of oppression in the name of non-reason

- **The Pacifying Process (at the Level of Society)**
  - —if freer from the dominance of pillage (as in savagery) in the civilizing process, then less free from the pacifying process (be it in the form of external control or self-control)
  - —if freer from self-control in the pacifying process, then less free from the temptation of expansionist oppression and rebellious mindset in external control
  - —if freer from external control in the pacifying process, then less free from the gruesome psychological self-torture and conformism in self-control

- **The Stewardizing Process (at the Level of Nature)**
  - —if freer from the dominance of nature (as in the state of nature) in the civilizing process, then less free from the stewardizing process (be it in the form of the stewardship of creation or the covenant with nature)
  - —if freer from the stewardship of creation in the stewardizing process, then less free from material underdevelopment, relatively speaking, and spiritual exclusion in the covenant with nature
  - —if freer from the covenant with nature in the stewardizing process, then less free from ecological degradation and spiritual disconnection from nature in the stewardship of creation

(continued on next page)

Table 4.36. No Freedom Without Unfreedom in the Civilizing Processes (Part II)

- **The Subliming Process (at the Level of the Mind)**
  —if freer from the dominance of spontaneity (as in the wild state of the mind) in the civilizing process, then less free from the subliming process, be it in the form of (cyclical-centric) self-refinement or (linear-centric) self-discipline
  —if freer from (cyclical-centric) self-refinement in the subliming process, then less free from the (linear-centric) self-regimen (as a form of neurosis)
  —if freer from (linear-centric) self-discipline in the subliming process, then less free from the (cyclical-centric) self-torture (equally as a form of neurosis)

Notes: The examples in each category are solely illustrative (not exhaustive), and the comparison is relative (not absolute), nor are they necessarily mutually exclusive. And some can be easily re-classified elsewhere. As generalities, they allow exceptions.
Sources: From *BCIV*. See also *FHC*, *FCD*, *FPHC*, *BDPD*, and *BCPC*.

Table 4.37. No Equality Without Inequality in the Civilizing Processes (Part I)

- **The Rationalizing Process (at the Level of Culture)**
  —if more equal for the role of rationalization in the rationalizing process (of civilizational making), then less equal for that of mythicization (as in barbarism)
  —if more equal for the principle of transcendence in (linear-centric) rationalizing process, then less equal for the principle of immanence
  —if more equal for the principle of immanence in (cyclical-centric) rationalizing process, then less equal for the principle of transcendence

- **The Pacifying Process (at the Level of Society)**
  —if more equal for pacification in civilizational making, then less equal for the institution of pillaging and others (as in savagery)
  —if more equal for external control, relatively speaking, in pacifying process, then less equal for self-control
  —if more equal for self-control, relatively speaking, in pacifying process, then less equal for external-control

- **The Stewardizing Process (at the Level of Nature)**
  —if more equal for stewardship in the stewardizing process (of civilizational making), then less equal for reverent (submissive) existence (as in barbarism)
  —if more equal for the stewardship of creation in (linear- centric) stewardizing process, then less equal for the (cyclical-centric) covenant with nature for harmonious co-existence
  —if more equal for the (cyclical-centric) covenant with nature in the stewardizing process, then less equal for the (linear-centric) stewardship of nature for domination

(continued on next page)

Table 4.37. No Equality Without Inequality in the Civilizing Processes (Part II)

- **The Subliming Process (at the Level of the Mind)**
  —if more equal for the role of reason in the subliming process, then less equal for that of unreason (as in the natural state of wildness)
  —if more equal for the primacy of reason in (linear-centric) subliming process, then less equal for other faculties (e.g., intuition, existential feelings, and analogous thinking) in cyclical-centric one
  —if more equal for the exercise of other faculties (e.g., intuition, existential feelings, and analogous thinking) in cyclical-centric subliming process, then less equal for the role of reason in linear-centric counterpart

Notes: The examples in each category are solely illustrative (not exhaustive), and the comparison is relative (not absolute), nor are they mutually exclusive. And some can be easily reclassified else-where. As generalities, they allow exceptions.
Sources: From *BCIV*. See also *FHC*, *FCD*, *FPHC*, *BDPD*, and *BCPC*.

Table 4.38. Five Theses on Post-Civilization

• Post-civilization no longer treats civilization as good and barbarity as evil (relatively speaking), nor does it nostalgically regard barbarity as good and civilization as evil (relatively speaking again). Civilization is as evil and good as barbarity.

• Post-civilization also no longer accepts the dichotomy between civilization and barbarity. Civilization cannot exist without barbarity. It is no longer necessary to preserve civilization, any more than it is imperative to destroy barbarity. To go beyond civilization and barbarity is to acknowledge the co-existence of both, although the degree of scaling one over the over varies from case to case—but is not to be extreme in largely favoring one over the other, *on average* (subject to the constraints of existential dialectics).

• Post-civilization is thus subject to the constraints of existential dialectics. There is no freedom without unfreedom, and no equality without inequality, for instance. There will be no utopia; even should there be one, there would be dystopia embedded within it.

• Post-civilization will eventually replace civilization (as a form of life settlement), to be dominated by post-capitalist and post-democratic lifeforms here on earth and in deep space (besides other alien lifeforms that we have never known), unto the post-human age in multiverses. Those few post-humans who keep civilization will live in a "post-human civilization," while the rest (the majority), who choose post-civilization, will evolve towards the state of "post-human post-civilization." One therefore should not confuse "post-human civilization" with "post-human post-civilization," as the two are not the same.

• Post-civilization will confront psychosis as a primary problem in the culture of virtuality unto the post-human age, just as civilization has neurosis as a primary one of its own (although both neurosis and psychosis are major problems in both).

Notes: The comparison in each category is relative (not absolute), nor are they necessarily mutually exclusive. And some can be easily re-classified elsewhere. As generalities, they allow exceptions.
Sources: From *BCIV*. See also *FHC*, *FCD*, *FPHC*, *BDPD*, and *BCPC*.

Table 4.39. Barbarity, Civilization, and Post-Civilization

- **The Rationalizing Process (at the Level of Culture)**
  —*Barbarity*
    - More mythicizing than rationalizing, relatively speaking
  —*Civilization*
    - More rationalizing than mythicizing, relatively speaking
  —*Post-Civilization*
    - Beyond the dichotomy, subject to existential dialectics

- **The Pacifying Process (at the Level of Society)**
  —*Barbarity*
    - More pillaging than pacifying, relatively speaking
  —*Civilization*
    - More pacifying than pillaging, relatively speaking
  —*Post-Civilization*
    - Beyond the dichotomy, subject to existential dialectics

- **The Stewardizing Process (at the Level of Nature)**
  —*Barbarity*
    - More revering than stewardizing, relatively speaking
  —*Civilization*
    - More stewardizing than revering, relatively speaking
  —*Post-Civilization*
    - Beyond the dichotomy, subject to existential dialectics

- **The Subliming Process (at the Level of the Mind)**
  —*Barbarity*
    - More impulsing than subliming, relatively speaking
  —*Civilization*
    - More subliming than impulsing, relatively speaking
  —*Post-Civilization*
    - Beyond the dichotomy, subject to existential dialectics

Notes: The comparison in each category is relative (not absolute), nor are they necessarily mutually exclusive. And some can be easily re-classified elsewhere. As generalities, they allow exceptions.
Sources: From *BCIV*. See also *FHC*, *FCD*, *FPHC*, *BDPD*, and *BCPC*.

Table 4.40. Types of Super Civilization in the Cosmos (Part I)

- **Type I**
  —a civilization which gains control of and uses the total energy output "falling on its planet from its sun for interstellar communication" (or, in general, space colonization). For N. Kardashev, who proposed the first three types, human civilization is currently Type Zero (Type O), which is below even Type I, since its present energy consumption for all purposes, let alone for interstellar communication, is still 10,000 times less.

- **Type II**
  —a civilization which gains control of and uses directly the total energy output of its sun for interstellar communication (or, in general, space colonization).

- **Type III**
  —a civilization which gains control of and uses the total energy output of its galaxy for interstellar communication (or, in general, space colonization).

- **Type IV**
  —a civilization which gains control of and uses the total energy output of its cluster of galaxies for interstellar communication (or, in general, space colonization).

- **Type V**
  —a civilization which gains control of and uses the total energy output of its supercluster of galaxies for interstellar communica-tion (or, in general, space colonization).

(continued on next page)

Table 4.40. Types of Super Civilization in the Cosmos (Part II)

- **Type…n**
  —So continues the series in what I call the cyclical progression of hegemony in the cosmos and beyond.

Notes: The Russian astrophysicist Nikolai Kardashev proposed the first three types of super civilization in terms of total energy out-put for interstellar communication. (CSM 1979) I extend his argument further to propose Type IV, Type V, Type VI, and Type…n, in the context of my claim about the cyclical progression of he-gemony in the cosmos and beyond.
Sources: From *Table 9.4* of *FCD*. See *FHC*, *FCD*, and *FPHC* for more info.

Table 4.41. The Civilizational Project from Pre-Modernity to After-Postmodernity (Part I)

	Pre-Modern	*Modern*	*Postmodern*	*After-Postmodern*
Main narratives	•Sacralness •Courtliness •Vitalism •Animism	•Freedom •Equality •Fraternity	•Multiplicity •Hybridization	•Naked contingency •Cyclical progres-sion of hegemony
Main institutions	•Monarchy •Aristocracy •Feudalism •Holy order •Primitivism	•Capitalism •Liberalism •Socialism •Nazism •Fascism	•Capitalism •Liberalism •Postmodern politics of difference	•Post-Capitalism •Post-Democracy •Others
Main techno-logical and economic revolutions	•Agricultural	•Service •Industrial	•Informational	•Biological •Material •Energy •Space •Others

(continued on next page)

Table 4.41. The Civilizational Project from Pre-Modernity to After-Postmodernity (Part II)

	Pre-Modern	*Modern*	*Postmodern*	*After-Postmodern*
Main agents	•Males •Upper strata •Mini-states	•Males •Upper strata •Whites •Empires	•Males •Upper strata •Whites •Others •Supra-states •IO's	•Post-humans •Humans •Others
Main impacts	•Local	•Inter-national	•Global	•Outer-space •Multiverse
Main outcomes	•Towards moderntiy •Rise of linear- & cyclical-centric civiliza-tions	•Towards post-modernity •Dominance of linear-centric civilization	•Towards after-post-moderntiy •Linear-centric civiliza-tion in crisis	•Towards human (& maybe post-human) extinction •Rise of post-civiliza-tion, especially in post-human forms of space-time

Notes: The examples in each category are solely illustrative (not exhaustive) nor necessarily mutually exclusive, and the comparison is relative (not absolute). As generalities, they allow exceptions.

Sources: From *Table 10.16* of *FCD*—and also from *BCIV* on post-civilization (and *FPHST*)

Table 4.42. Civilizational Holism (Part I)

- **At the Micro-Physical Theoretical Level**
  —Ex: Mastering of quantum mechanics, electromagnetism, and other fields for the understanding of a broad range of anomalous experiences and theapplication for artificial intelligence (*Sec.1.4.1* of *FPHC*)

- **At the Chemical Theoretical Level**
  —Ex: Unprecedented expansion of (and violence to) the mind through ever new forms of drugs (and virtual technologies, for that matter) (Ch.9 of *FCD*)

- **At the Biological Theoretical Level**
  —Ex: Humans are not biologically equal, on the basis of race, gender, ethnicity, age, and whatnot. (*Sec.2.6* & Ch.10 of *FCD*; *BNN*) And post-humans will experience the same fate, in an even more amazing way.

- **At the Psychological Theoretical Level**
  —Ex: Human cognitive impartiality and emotional neutrality are quite limited. (*Secs.2.4-2.5* of *FCD*)
  —Ex: Rise of Floating Consciousness (Ch.10 of *FCD*; Chs.1 & 4 of *FPHC*) and Unfolding Unconsciousness (*FPHU*)

- **At the Organizational Theoretical Level**
  —Ex: Administrative colonization of deep space, with less legal-formalism in some corners. (Chs.9-10 of *FCD*)

- **At the Institutional Theoretical Level**
  —Ex: Both capitalism and democracy will not last, to be superseded by different versions of post-capitalism and post-democracy in after-postmodernity. (Ch.10 of *FCD*)

(continued on next page)

Table 4.42. Civilizational Holism
(Part II)

- **At the Structural Theoretical Level**
  —Ex: Social stratification reappears in ever new forms, also with new causes and new players in the cyclical progression of hegemony. (Chs.8-10 of *FCD*)
  —Ex: The world of hyper-empires, and the union of the unions (*BWT*)

- **At the Systemic Theoretical Level**
  —Ex: Outerspace expansion: local → regional → global → solar → galactic → clustery → multiversal (Ch.9 of *FCD*)
  —Ex: Demographic transition: human extinction, and the rise of post-humans (e.g., cyborgs, thinking machines, thinking robots, genetically altered superior beings, floating consciousness, hyper-spatial consciousness) (Ch.4 of *FPHC*; Ch.10 of *FCD*; & Ch.7 of *FHC*)
  —Ex: New technological forces in material sciences, electronic and communication sciences, energy sciences, biosciences, manufacturing and engineering sciences, and space sciences (Ch.10 of *FCD* & Ch.7 of *FHC*)
  —Ex: Systematic dominance towards nature for space colonization (Chs.9-10 of *FCD*; Chs.2 & 7 of *FHC*)

- **At the Cultural Theoretical Level**
  —Ex: The post-human transcendence of freedom and equality (Ch.10 of *FCD*)
  —Ex: Methodological Holism (Ch.1 of *FCD*; Ch.1 of *FPHC*; *Sec.2.1* & *Sec.2.5* of *BCPC*)
  —Ex: The Evolution from Barbarity to Post-Civilization (*BCIV*)

- **At the Cosmological Theoretical Level**
  —Ex: Mastering of dark matter and dark energy, and the exploration of multiverses (Ch.4 of *FPHC*; Ch.10 of *FCD*; & Ch.7 of *FHC*)
  —Ex: Alternation of space-time (*FPHST*)
  —Ex: The emergence of hyper-spatial consciousness (*FPHC*)

(continued on next page)

Table 4.42. Civilizational Holism (Part III)

- **At Other Levels**
  —Ex: Historical: pre-modernity → modernity → postmodernity → after-postmodernity (human distinction, and the rise of post-humans, including floating consciousness, hyper-spatial consciousness, and unfolding unconsciousness) (Ch.7 of *FHC*; Ch.10 of *FCD*; *FPHC*; *FPHU*)

Notes: These examples are solely illustrative (not exhaustive), and some of the items can be reclassified somewhere else. Nor are they always mutually exclusive. Since they are generalities, exceptions are expected. And the comparison is relative, not absolute.
Sources: From *Table 5.1* of *FPHC*—with details from *FHC*, *FCD*, and the rest of my books. Not every aspect in each category as discussed in all my books are presented here, since there are too many issues. For more info, also consult the table on theories on civilizational holism and, of course, the books themselves.

Table 4.43. Theories on Civilizational Holism (Part I)

I. Theories in Relation to Nature

—*At the Chemical Theoretical Level*

• 72. Constructivist Theory of Aerology (Peter Baofu) (*FPHAERO*)

• 71. Creational Theory of Chemistry (Peter Baofu) (*FPHCHEM*)

(• 66). Resilient Theory of Natural Resources (Peter Baofu) (*BNR*)

—*At the Macro-Physical (Cosmological) Theoretical Level*

(• 72.) Constructivist Theory of Aerology (Peter Baofu) (*FPHAERO*)

• 70. Resettlement Theory of Geology (Peter Baofu) (*FPHGEOL*)

• 69. Theory of Post-Cosmology (Peter Baofu) (*BCOS*)

• 68. Theory of Hyper-Spatial Consciousness (Peter Baofu) (Ch.4 of *FPHC*; *FPHG*)

(• 66). Resilient Theory of Natural Resources (Peter Baofu) (*BNR*)

(• 65). Panoramic Theory of Transportation (Peter Baofu) (*FPHTRAN*)

(• 63). Multilateral Theory of Acoustics (Peter Baofu) (*FPHA*)

(• 62). Selective Theory of Geometry (Peter Baofu) (*FPHG*)

(• 61). Perspectival Theory of Space-Time (Peter Baofu) (*FPHST*)

(• 60). Dialectic Theory of Complexity (Peter Baofu) (*FC*)

(• 32). Theory of Floating Consciousness (Peter Baofu) (Ch.10 of *FCD*; Chs.1 & 4 of *FPHC*)

• 67. Theory of the Geopower of Nature (Ch.4 of *ALD*)—refined as the Interventive-Reshaping Theory of Geography (Peter Baofu) (*FPHGEOG*)

(continued on next page)

Table 4.43. Theories on Civilizational Holism (Part II)

I. Theories in Relation to Nature (*cont'd*)

—At the Micro-Physical Theoretical Level

- 66. Resilient Theory of Natural Resources (Peter Baofu) (*BNR*)
- 65. Panoramic Theory of Transportation (Peter Baofu) (*FPHTRAN*)
- 64. Transcendent Theory of Architecture (Peter Baofu) (*FPHARCH*)
- 63. Multilateral Theory of Acoustics (Peter Baofu) (*FPHA*)
- 62. Selective Theory of Geometry (Peter Baofu) (*FPHG*)
- 61. Perspectival Theory of Space-Time (Peter Baofu) (*FPHST*)
- 60. Dialectic Theory of Complexity (Peter Baofu) (*FC*)

II. Theories in Relation to Culture

—At the Cultural Theoretical Level

- 59. Inclusivist Theory of Phonology (Peter Baofu) (*FPHPHON*)
- 58. Ephemeral Theory of the Visual Arts (Peter Baofu) (*FPHVA*)
- 57. Sequential Theory of Thanatology (Peter Baofu) (*FPHTHAN*)
- 56. Fusional Theory of Morphology (Peter Baofu) (*FPHMORP*)
- 55. Modificative Theory of Sports (Peter Baofu) (*FPHSPORT*)
- 54. Inquisitive Theory of Culinary Art (Peter Baofu) (*FPHCA*)
- 53. Interactive Theory of Semantics (Peter Baofu) (*FPHSEM*)
- 52. Transdicsiplinary Theory of Performing Arts (Peter Baofu) (*FPHPA*)

(continued on next page)

Table 4.43. Theories on Civilizational Holism (Part III)

II. Theories in Relation to Culture (*cont'd*)

—*At the Cultural Theoretical Level* (*cont'd*)

- • 51. Comparative-Impartial Theory of Literature (Peter Baofu) (*FPHLIT*)
- • 50. Mediative-Variative Theory of Chess (Peter Baofu) (*FPHCHESS*)
- • 49. Theory of Post-Ethics (Peter Baofu) (*BEPE*)
- • 48. Dualistic Theory of Mass Culture (Peter Baofu) (Ch.2 of *FHC*)
- • 47. Comparative Theory of Religion—also known as the Comparative-Substitutive Theory of Religion (Peter Baofu) (Ch.3 of *FHC*; Ch.9 of *FCD*; Ch.1 of *FPHK*; *FPHR*)
- • 46. Theory of Post-Civilization (Peter Baofu) (*BCIV*)
- • 45. Theory of the Trinity of Modernity to Its After-Postmodern Counterpart (Peter Baofu) (*FHC*; Ch.10 of *FCD*)
- • 44. Transformative Theory of Aesthetic Experience (Peter Baofu) (*FAE*)
- (• 21). Theory of Post-Capitalism (Peter Baofu) (Ch.10 of *FCD*; Chs.2 & 4 of *FPHC*; *BCPC*)
- (• 20). Theory of Post-Democracy (Peter Baofu) (Ch.10 of *FCD*; Chs.3 & 4 of *FPHC*; *BDPD*)
- (• 5). Theory of Methodological Holism (Peter Baofu) (Ch.1 of *FCD*; Ch.1of *FPHC*; *Sec.2.1* & *Sec.2.5* of *BCPC*; *FC*; *FPHK*; etc.)
- (• 4). Theory of Existential Dialectics, or the Holistic Theory of Knowledge (Peter Baofu) (*FHC*; *FCD*; *FPHC*; *BDPD*; *FC*; *FAE*; *ALD; FIA; FPHK*; etc,)
- (• 3). Contrastive Theory of Rationality (Peter Baofu) (*FPHML*)

(continued on next page)

Table 4.43. Theories on Civilizational Holism (Part IV)

III. Theories in Relation to the Mind

—At the Biological Theoretical Level

- • 43. Interconnected Theory of Health Care (Peter Baofu) (*FPHHC*)
- • 42. Theory of Contrastive Advantages (Peter Baofu) (*Sec.2.6* & Ch.10 of *FCD; BNN*)
- (• 32).Theory of Floating Consciousness (Peter Baofu) (Ch.10 of *FCD*; Chs.1 & 4 of *FPHC*)

—At the Psychological Theoretical Level

- • 41. Complex Theory of Data Analysis (Peter Baofu) (*FPHDA*)
- • 40. Supersession Theory of Computing (Peter Baofu) (*FPHCOM*)
- • 39. Metamorphic Theory of Humor (Peter Baofu) (*FPHH*)
- • 38. Contrarian Theory of Personality (Peter Baofu) (*FPHP*)
- • 37. Theory of Virtual Sexuality (Peter Baofu) (*FPHS*)
- • 36. Expansive-Contractive Theory of Martial Arts (Peter Baofu) (*FPHMA*)
- • 35. Multilogical Theory of Learning (Peter Baofu) (*FPHL*)
- • 34. Comprehensive Theory of Creative Thinking (Peter Baofu) (*FPHCT*)
- • 33. Theory of Unfolding Unconsciousness—also known as the Unfolding Theory of Anomalous Experience (Peter Baofu) (*FPHU*)
- • 32. Theory of Floating Consciousness (Peter Baofu) (Ch.10 of *FCD*; Chs.1 & 4 of *FPHC*)
- • 31. Theory of Cognitive Partiality (Peter Baofu) (*Sec.2.4* of *FCD*; *Sec.4.5.1.1* of *BCPC*)

(continued on next page)

Table 4.43. Theories on Civilizational Holism (Part V)

III. Theories in Relation to the Mind (*cont'd*)

—*At the Psychological Theoretical Level* (*cont'd*)

- 30. Theory of Emotional Non-Neutrality (Peter Baofu) (*Sec.2.5* of *FCD*; *Sec.4.5.2* of of *BCPC*)
- 29. Theory of Behavioral Alteration (Peter Baofu) (*Sec.4.5.3* of *BCPC*)

IV. Theories in Relation to Society

—*At the Organizational Theoretical Level*

- 28. Theory of E-Civic Alienation (Peter Baofu) (Ch.7 of *FCD*)
- 27. Combinational Theory of Organization (Peter Baofu) (*FPHO*; Ch.6 of *ALD*)

—*At the Institutional Theoretical Level*

- 26. Double-Sided Theory of Accounting (Peter Baofu) (*FPHACCO*)
- 25. Reflective Theory of Criminality (Peter Baofu) (*FPHCRIM*)
- 24. Detached Theory of Gambling (Peter Baofu) (*FPHGAM*)
- 23. Heterodox Theory of Education (Peter Baofu) (*FPHEDU*)
- 22. Reconstruction Theory of Law (Peter Baofu) (*FPHLAW*)
- 21. Theory of Post-Capitalism (Peter Baofu) (Ch.10 of *FCD*; Chs.2 & 4 of *FPHC*; *BCPC*)
- 20. Theory of Post-Democracy (Peter Baofu) (Ch.10 of *FCD*; Chs.3 & 4 of *FPHC*; *BDPD*)
- 19. Dynamic Theory of Comparative Political Systems (Peter Baofu) (*ALD*)

(• 9.) Theory of the Cyclical Progression of Empire-Building (Peter Baofu) (*BWT*)

(continued on next page)

Table 4.43. Theories on Civilizational Holism (Part VI)

IV. Theories in Relation to Society (*cont'd*)

—*At the Systemic Theoretical Level*

- 18. Transfigurative Theory of Waste (Peter Baofu) (*FPHW*)
- 17. Theory of the Cyclical Progression of Migration (Peter Baofu) (*FPHMIG*)
- (• 70). Resettlement Theory of Geology (Peter Baofu) (*FPHGEOL*)
- 16. Contingent Theory of Urban Planning (Peter Baofu) (*FPHUP*)
- 15. Totalistic Theory of Communication (Peter Baofu) (*FPHMM*; *FCD*; *FHC*)
- 14. Ambivalent Theory of Technology (Peter Baofu) (*FPHE*; *FCD*; *FHC*)
- 13. Multifaceted Theory of War and Peace (Peter Baofu) (Ch.9 of *FCD*; Ch.1 of *FPHK*)
- 12. Theory of Post-Humanity (Peter Baofu) (Ch.7 of *FHC*; Chs.3, & 10 of *FCD*; Chs.1, 3 & 4 of *FPHC*; and other books of mine)
- 11. Theory of the Cyclical Progression of System Integration and Fragmentation (Peter Baofu) (Chs.9-10 of *FCD*)
- 10. Synthetic Theory of Information Architecture (Peter Baofu) (*FIA*)

—*At the Structural Theoretical Level*

- 9. Theory of the Cyclical Progression of Hegemony (Peter Baofu) (Chs.9-10 of *FCD*; Chs.1, 3 & 4 of *FPHC*; *BDPD*)
- 8. Theory of the Cyclical Progression of Empire-Building (Peter Baofu) (*BWT*)

(continued on next page)

Table 4.43. Theories on Civilizational Holism (Part VII)

V. Meta-Theories (in Relation to Theories)

—*At the Methodological Meta-Theoretical Level*

(• 53).Interactive Theory of Semantics (Peter Baofu) (*FPHSEM*)

(• 41). Complex Theory of Data Analysis (Peter Baofu) (*FPHDA*)

• 7. Interpretivist Theory of Probability (Peter Baofu) (*FPHROB*)

• 6. Critical-Dialectic Theory of Formal Science (Peter Baofu) (*FPHFS*)

• 5. Theory of Methodological Holism (Peter Baofu) (Ch.1 of *FCD*; Ch.1of *FPHC*; *Sec.2.1* & *Sec.2.5* of *BCPC*; *FC*; *FPHK*; etc.)

—*At the Ontological Meta-Theoretical Level*

• 4. Theory of Existential Dialectics, or the Holistic Theory of Knowledge (Peter Baofu) (*FHC*; *FCD*; *FPHC*; *BDPD*; *FC*; *FAE*; *ALD; FIA; FPHK*; etc,)

• 3. Contrastive Theory of Rationality (Peter Baofu) (*FPHML*)

VI. Theories in Relation to the Rest

—*At the Meta-Historical Level*

• 2. Multifold Theory of History (Peter Baofu) (*FPHHIST*)

—*At the Empirical-Historical Level*

• 1. Theory of the Evolution from Pre-Modernity to After-Postmodernity (Peter Baofu) (*FHC*; Ch.9-10 of *FCD*; *FPHC*)

Notes: All these theories are my constructions, as some of the main contributions of my grant project on civilization and its future. These examples are solely illustrative (not exhaustive), and some of the items can be reclassified somewhere else. Nor are they always mutually exclusive. Since they are generalities, exceptions are expected.

Sources: From my previous books.

Table 4.44. Three Great Future Transformations of Mind Games

- **Virtual Games**
  - —Ex: online chess
  - —Ex: virtual experience
- **Novel Games**
  - —Ex: new chess variants
  - —Ex: new chess engines
- **Post-Human Mind Games**
  - —Ex: the quest for broader/deeper mental benefits of chess playing
  - —Ex: games designed for the evolution of the mind into different body-less forms (e.g., "floating consciousness," "hyper-spatial consciousness," "unfolding unconsciousness")

Notes: The examples in the categories are solely illustrative (not exhaustive), and the comparison is relative (not absolute), nor are they necessarily mutually exclusive. And some can be easily re-classified elsewhere. As generalities, they allow exceptions.
Sources: From *Sec. 4.16* of *FPHCHESS*. See text for more info. Also, consult *FCD* for strategy and tactics in warfare and *FPHO* for strategy and tactics in organization.

BIBLIOGRAPHY

AccountingForValue.org (ACCO). 2013. "Encouraging a Wider Discussion on the Nature of Accounting Principles" (December 03). <http://www.accounting forvalue.org/>.

AICPA. 2009. "Where Will the SEC Take the IFRS Roadmap? An AICPA Analysis of Comment Letters on the SEC's Proposal" (April 27). <http://www.ifrs.com/updates/aicpa/IFRS_SEC.html>.

_____.2008. "Convergence with International Accounting Standards" (April 29). <http://www.ifrs.com/updates/aicpa/Convergence_Accounting_Std.html>.

Alexander, David; & Anne Britton. 2004. *Financial Reporting*. Stamford, CT: Cengage Learning EMEA.

Alexander, Delroy; Greg Burns, Robert Manor, Flynn McRoberts, & E.A. Torriero. 2002. "The Fall of Andersen" (September 01). <http://www.courant. com/news/nation-world/chi-0209010315sep01,0,6548944.story>.

Álvarez, Gene; & Phil Gledhill. 2010. "A Comprehensive Risk and Control Self-Assessment Methodology." <http://www.risk.net/operational-risk-and-regulation/feature/1900537/comprehensive-risk-control-self-assessment-me thodology>.

_____.2004. *Operational Risk: Practical Approaches to Implementation.*

Ambler, Scott. 1998. "A Realistic Look at Object-Oriented Reuse" (January 01). <http://www.drdobbs.com/a-realistic-look-at-object-oriented-reus/184415 594>.

Anderson, Richard. 2011. "Puma First to Publish Environmental Impact Costs (May 16). *BBC News*. <http://www.bbc.co.uk/news/business-13410397>.

Appelbaum, Steven H.; Kyle J Deguire; & Mathieu Lay. 2005. "The Relationship of Ethical Climate to Deviant Workplace Behaviour." *Corporate Governance*, 5 (4): 43. <http://libproxy.sdsu.edu/login?url=http://proquest.umi.com/pqd web?did=914821791&sid=5&Fmt=4&clientId=17862&RQT=309&VName= PQD>.

Ayling, Joe. 2011. "Supply Chain Weighs on Puma's First Green Accounts" (May 17). <http://www.just-style.com/analysis/supply-chain-weighs-on-pumas-first-green-accounts_id111097.aspx>.

Bakshi, Sunil. 2004. "Control Self-assessment for Information and Related Technology." *Journal of the Information Systems Audit and Control*

Association: 4.

Baofu, Peter. 2014. *The Future of Post-Human Etiology: Towards a New Theory of Cause and Effect*. Volume 2. Hauppauge, NY: Nova Science Publishers, Inc.

_____.2014a. *The Future of Post-Human Etiology: Towards a New Theory of Cause and Effect*. Volume 1. Hauppauge, NY: Nova Science Publishers, Inc.

_____,2014b. *The Future of Post-Human Phonology: Towards a New Theory of Generality and Specificity*. Hauppauge, NY: Nova Science Publishers, Inc.

_____.2014c. *The Future of Post-Human Aerology: Towards a New Theory of Predictaiblity and Nonpredictability*. Volume 2. Hauppauge, NY: Nova Science Publishers, Inc.

_____.2014d. *The Future of Post-Human Aerology: Towards a New Theory of Predictaiblity and Nonpredictability*. Volume 1. Hauppauge, NY: Nova Science Publishers, Inc.

_____.2013. *The Future of Post-Human Thanatology: Towards a New Theory of Death and After-Death*. Berlin, Germany: Logos Verlag Berlin.

_____.2013a. *The Future of Post-Human Health Care: Towards a New Theory of Mind and Body*. Hauppauge, NY: Nova Science Publishers, Inc.

_____.2013b. *The Future of Post-Human Morphology: Towards a New Theory of Typologies and Rules*. Hauppauge, NY: Nova Science Publishers, Inc.

_____.2013c. *The Future of Post-Human Probability: Towards a New Theory of Objectivity and Subjectivity*. Hauppauge, NY: Nova Science Publishers, Inc.

_____.2013d. *The Future of Post-Human Sports: Towards a New Theory of Training and Winning*. Cambridge, England: Cambridge Scholars Publishing.

_____.2013e. *Beyond Natural Resources to Post-Human Resources: Towards a New Theory of Diversity and Discontinuity*. Cambridge, England: Cambridge Scholars Publishing.

_____.2012. *The Future of Post-Human Waste: Towards a New Theory of Uselessness and Usefulness*. Cambridge, England: Cambridge Scholars Publishing.

_____.2012a. *The Future of Post-Human Culinary Art: Towards a New Theory of Ingredients and Techniques*. Cambridge, England: Cambridge Scholars Publishing.

_____.2012b. *The Future of Post-Human Transportation: A Preface to a New Theory of Networks and Operations*. Cambridge, England: Cambridge Scholars Publishing.

_____.2012c. *The Future of Post-Human Migration: A Preface to a New Theory of Sameness, Otherness, and Identity*. Cambridge, England: Cambridge Scholars Publishing.

_____.2012d. *The Future of Post-Human History: A Preface to a New Theory of*

Universality and Relativity. Cambridge, England: Cambridge Scholars Publishing.

_____.2012e. *The Future of Post-Human Criminality: A Preface to a New Theory of Heroes and Villains*. Cambridge, England: Cambridge International Science Publishing, Ltd.

_____.2012f. *The Future of Post-Human Architecture: A Preface to a New Theory of Form and Function*. Cambridge, England: Cambridge International Science Publishing, Ltd.

_____.2012g. *The Future of Post-Human Semantics: A Preface to a New Theory of Internality and Externality*. Cambridge, England: Cambridge Scholars Publishing.

_____.2012h. *The Future of Post-Human Performing Arts: A Preface to a New Theory of the Body and its Presence*. Cambridge, England: Cambridge Scholars Publishing.

_____.2011. *The Future of Post-Human Geography: A Preface to a New Theory of Environments and their Interactions*. Cambridge, England: Cambridge International Science Publishing, Ltd.

_____.2011a. *The Future of Post-Human Data Analysis: A Preface to a New Theory of Quantitative and Qualitative Research*. Cambridge, England: Cambridge International Science Publishing, Ltd.

_____.2011b. *The Future of Post-Human Chemistry: A Preface to a New Theory of Substances and their Changes*. Cambridge, England: Cambridge Scholars Publishing.

_____.2011c. *The Future of Post-Human Literature: A Preface to a New Theory of Fiction and Non-Fiction*. Cambridge, England: Cambridge International Science Publishing, Ltd.

_____.2011d. *The Future of Post-Human Computing: A Preface to a New Theory of Hardware, Software, and the Mind*. Cambridge, England: Cambridge International Science Publishing, Ltd.

_____.2011e. *The Future of Post-Human Gambling: A Preface to a New Theory of Risk and Caution*. Cambridge, England: Cambridge International Science Publishing, Ltd.

_____.2011f. *The Future of Post-Human Acoustics: A Preface to a New Theory of Sound and Silence*. Cambridge, England: Cambridge International Science Publishing, Ltd.

_____.2011g. *The Future of Post-Human Humor: A Preface to a New Theory of Joking and Laughing*. Cambridge, England: Cambridge International Science Publishing, Ltd.

_____.2011h. *Beyond Ethics to Post-Ethics: A Preface to a New Theory of Morality and Immorality*. Charlotte, NC: Information Age Publishing.

______.2011i. *The Future of Post-Human Education: A Preface to a New Theory of Teaching and Learning*. Cambridge, England: Cambridge International Science Publishing, Ltd.

______.2010. *The Future of Post-Human Religion: A Preface to a New Theory of Spirituality*. Cambridge, England: Cambridge International Science Publishing, Ltd.

______.2010a. *The Future of Post-Human Chess: A Preface to a New Theory of Tactics and Strategy*. Cambridge, England: Cambridge International Science Publishing, Ltd.

______.2010b. *The Future of Post-Human Geology: A Preface to a New Theory of Statics and Dynamics*. Cambridge, England: Cambridge International Science Publishing, Ltd.

______.2010c. *Beyond Cosmology to Post-Cosmology: A Preface to a New Theory of Different Worlds*. Cambridge, England: Cambridge International Science Publishing, Ltd.

______.2010d. *The Future of Post-Human Personality: A Preface to a New Theory of Normality and Abnormality*. Cambridge, England: Cambridge International Science Publishing, Ltd.

______.2010e. *The Future of Post-Human War and Peace: A Preface to a New Theory of Aggression and Pacificity*. Cambridge, England: Cambridge Scholars Publishing.

______.2010f. *The Future of Post-Human Formal Science: A Preface to a New Theory of Abstraction and Application*. Cambridge, England: Cambridge Scholars Publishing.

______.2010g. *The Future of Post-Human Law: A Preface to a New Theory of Necessity, Contingency, and Justice*. Cambridge, England: Cambridge Scholars Publishing.

______.2010h. *The Future of Post-Human Sexuality: A Preface to a New Theory of the Body and Spirit of Love Makers*. Cambridge, England: Cambridge Scholars Publishing.

______,2009. *The Future of Post-Human Martial Arts: A Preface to a New Theory of the Body and Spirit of Warriors*. Cambridge, England: Cambridge Scholars Publishing.

______.2009a. *The Future of Post-Human Organization: A Preface to a New Theory of Communication, Decision-Making, and Leadership*. Cambridge, England: Cambridge Scholars Publishing.

______.2009b. *The Future of Post-Human Language: A Preface to a New Theory of Structure, Context, and Learning*. Cambridge, England: Cambridge Scholars Publishing.

______.2009c. *The Future of Post-Human Urban Planning: A Preface to a New Theory of Density, Void, and Sustainability*. Cambridge, England: Cambridge

Scholars Publishing.

_____.2009d. *The Future of Post-Human Geometry: A Preface to a New Theory of Infinity, Symmetry, and Dimensionality*. Cambridge, England: Cambridge Scholars Publishing.

_____.2009e. *The Future of Post-Human Creative Thinking: A Preface to a New Theory of Invention and Innovation*. Cambridge, England: Cambridge Scholars Publishing.

_____.2009f. *The Future of Post-Human Mass Media: A Preface to a New Theory of Communication*. Cambridge, England: Cambridge Scholars Publishing.

_____.2009g. *The Future of Post-Human Engineering: A Preface to a New Theory of Technology*. Cambridge, England: Cambridge Scholars Publishing.

_____.2008. *The Future of Post-Human Mathematical Logic: A Preface to a New Theory of Rationality*. Cambridge, England: Cambridge Scholars Publishing.

_____.2008a. *The Future of Post-Human Knowledge: A Preface to a New Theory of Methodology and Ontology*. Oxford, United Kingdom: Chandos Publishing (Oxford) Ltd.

_____.2008b. *The Future of Post-Human Unconsciousness: A Preface to a New Theory of Anomalous Experience*. Cambridge, England: Cambridge Scholars Publishing.

_____.2008c. *The Future of Information Architecture: Conceiving a New Way to Understand Taxonomy, Network, and Intelligence*. Oxford, United Kingdom: Chandos Publishing (Oxford) Ltd.

_____.2007. *The Rise of Authoritarian Liberal Democracy: A Preface to a New Theory of Comparative Political Systems*. Cambridge, England: Cambridge Scholars Publishing.

_____.2007a. *The Future of Aesthetic Experience: Conceiving a Better Way to Beauty, Ugliness, and the Rest*. Cambridge, England: Cambridge Scholars Publishing.

_____.2007b. *The Future of Complexity: Conceiving a Better Way to Understand Order and Chaos*. London, United Kingdom: World Scientific Publishing Co.

_____.2007c. *Beyond the World of Titans, and the Remaking of World Order: A Preface to a New Logic of Empire-Building*. Cambridge, England: Cambridge Scholars Publishing.

_____.2006. *Beyond Nature and Nurture: Conceiving a Better Way to Understand Genes and Memes*. Cambridge, England: Cambridge Scholars Publishing.

_____.2006a. *The Future of Post-Human Space-Time: Conceiving a Better Way to Understand Space and Time*. New York: Peter Lang Publishing, Inc.

_____.2006b. *Beyond Civilization to Post-Civilization: Conceiving a Better Model of Life Settlement to Supersede Civilization*. New York: Peter Lang

Publishing, Inc.

_____.2005. *Beyond Capitalism to Post-Capitalism: Conceiving a Better Model of Wealth Acquisition to Supersede Capitalism*. New York: The Edwin Mellen Press.

_____.2004. *The Future of Post-Human Consciousness*. New York: The Edwin Mellen Press.

_____.2004a. *Beyond Democracy to Post-Democracy: Conceiving a Better Model of Governance to Supersede Democracy*. Volume 1. New York: The Edwin Mellen Press.

_____.2004b. *Beyond Democracy to Post-Democracy: Conceiving a Better Model of Governance to Supersede Democracy*. Volume 2. New York: The Edwin Mellen Press.

_____.2002. *The Future of Capitalism and Democracy*. Maryland: The University Press of America.

_____.2000. *The Future of Human Civilization*. Volume 1. New York: The Edwin Mellen Press.

_____.2000a. *The Future of Human Civilization*. Volume 2. New York: The Edwin Mellen Press.

Barrow, John D. 1998. *Impossibility: The Limits of Science and the Science of Limits*. Oxford: Oxford University Press.

BBC. 2011. "Puma Completes Environmental Impact Costs" (November 16). <http://www.bbc.co.uk/news/business-15756080>.

_____.2008. "Italian Dairy Boss Gets 10 Years" (December 18). <http://news.bbc.co.uk/2/hi/business/7790803.stm>.

Berton, Lee. 1984. "Advertising Has Hurt Accounting's Ethics: Critics" (May 24). *Chicago Sun-Times*. <http://www.highbeam.com/doc/1P2-1464160.html>.

_____.1984a. "Goal: Ethical Standards for Accounting Practices" (May 24). *Wall Street Journal*. <http://libproxy.sdsu.edu/login?url=http://proquest.umi.com/pqdweb?did=27058245&sid=1&Fmt=3&clientId=17862&RQT=309&VName=PQD>.

Biletzki, Anat; & Anat Matar. 2006. "Ludwig Wittgenstein" (Nov. 17). *Stanford Encyclopedia of Philosophy*. <http://plato.stanford.edu/entries/wittgenstein/>.

BrainyQuote.com (BRAIN). 2014. "Susan George" (January 02). <http://www.brainyquote.com/quotes/keywords/accounting.html>.

Bunge, Mario. 1999. *Dictionary of philosophy*. Amherst, MA: Prometheus Books.

Carruthers, Bruce G., & Wendy Espeland. 1991. "Accounting for Rationality: Double-Entry Bookkeeping and the Rhetoric of Economic Rationality." *American Journal of Sociology*, 97 (1).

Case, K. & R. Fair. 2007. *Principles of Economics*. Upper Saddle River, NJ: Pearson Education.

Central Financial Service (CEN). 2002. "Second Quality Conference for Public Administration in the EU: Note for the File." <http://www.viidakko.fi/innoctus>.

CFO Insight (CFO). 2012. "Convergence of IFRS and US GAAP Is Not Dead, Just Dormant" (December 13). <http://www.cfo-insight.com/reporting-forecasting/accounting/convergence-of-ifrs-and-us-gaap-is-not-dead-just-dormant/>.

Chase, Scott. 1993. "Tachyons" (March). *Physics FAQ*. <http://math.ucr.edu/home/baez/physics/ParticleAndNuclear/tachyons.html>.

Chicago Tribune (CHI). 1907. "Bookkeepers Repel Inference That Their Own Employers Were Aimed At" (February 28). <http://pqasb.pqarchiver.com/chicagotribune/doc/173279169.html?FMT=ABS&FMTS=ABS:AI&type=historic&date=Feb%2028,%201907&author=&pub=Chicago%20Tribune&edition=&startpage=&desc=DISCUSS%20ACCOUNTING%20ETHICS>.

Clikeman, Paul M. 2003. "Educating for the Public Trust." *The CPA Journal*, 73 (8): 80. <http://libproxy.sdsu.edu/login?url=http://proquest.umi.com/pqdweb?di=407305991&sid=3&Fmt=3&clientId=17862&RQT=309&VName=PQD>.

CNN. 2001. "Waste Management Settles" (November 7). <http://money.cnn.com/2001/11/07/news/waste_mgt/index.htm>.

Cohen, Gail. 2013. "How to Apply Cognitive Psychology to Accounting" (December 03). <http://smallbusiness.chron.com/apply-cognitive-psychology-accounting-38759.html>.

Corazzon, Raul. 2007. "Definitions of Ontology: From Nicolai Hartmann to the Present Time." <http://www.formalontology.it/section_4_two.htm>.

Corbett, Dan. 2008. "Introduction to String Theory" (November 23). *ThinkQuest*. <http://library.thinkquest.org/27930/stringtheory1.htm>.

COSO. 2013. "Internal Control—Integrated Framework" (December 31). <http://coso.org/IC-IntegratedFramework-summary.htm>.

Cosgrove, D. E. 1985. *Social Formation and Symbolic Landscape*. Totowa, NJ: Barnes & Noble Books.

Costanza, Robert; Ralph d'Arge; Rudolf de Groot; Stephen Farberk; Monica Grasso; Bruce Hannon; Karin Limburg I ; Shahid Naeem; Robert V. O'Neill; Jose Paruelo; Robert G. Raskin; Paul Suttonkk; & Marjan van den Belt. 1997. "The Value of the World's Ecosystem Services and Natural Capital." *Nature*, 387: 253–260. <http://www.esd.ornl.gov/benefits_conference/nature_paper.pdf>.

Crenshaw, Albert B.; & Brett D. Fromson. 1998. "A Conflict for CPAs?" (March 29). *The Washington Post*. <http://www.washingtonpost.com/wp-srv/business/longterm/ethics/ethics3.htm>.

Cross, Jay. 2013. "8 Examples of 'Mental Accounting' and How to Avoid Them" (August 4). <http://thecollegeinvestor.com/8748/EXAMPLES-OF-MENTAL-ACCOUNTING/>.

Daniel, T. C. et al. 2012. "Contributions of Cultural Services to the Ecosystem

Services Agenda." *Proc. Natl. Acad. Sci.,* 109: 8812–8819. <http://www.pnas.org/content/early/2012/05/18/1114773109.full.pdf+html>.

DeFries, R. S.; J. A. Foley; & G. P. Asner. 2004. "Land-Use Choices: Balancing Human Needs and Ecosystem Function." *Frontiers in Ecology and the Environment*, 2: 249-257.

Dellaportas, Steven. 2006. "Making a Difference with a Discrete Course on Accounting Ethics." *Journal of Business Ethics*, 65 (4): 391–404.

Dey, Iain; & Dominic Rushe. 2009. "Auditors: In the Palm of the Banks?" (January 25). *The Times*. <https://login.thetimes.co.uk/?gotoUrl=http%3A%2F%2Fwww.thetimes.co.uk%2Ftto%2Fbusiness%2Findustries%2Fbanking%2F>.

Dietz, David. 2002. "'Auditors Are Timid.'" *Pittsburgh Post-Gazette* (April 26). <http://news.google.com/newspapers?id=KesNAAAAIBAJ&sjid=QHADAAAAIBAJ&pg=2763,4354640&dq=investors+rely+on+accountants+auditors>.

Dumb.com (DUM). 2013. "Paul Browne" (December 01). <http://www.dumb.com/quotes/accounting-quotes/>.

Duska, Ronald F.; & Brenda Shay Duska. 2003. *Accounting Ethics*. NY: Wiley-Blackwell.

Eccles, Robert; Robert H. Herz; E. Mary Keegan; & David M. H. Phillips. 2001. The *Value Reporting Revolution: Moving Beyond the Earnings Game*. New York: John Wiley.

Elkington, John. 2013. "The Future Quotient" (December 28). <http://www.thecro.com/content/future-quotient>.

Emiliani, Bob. 2007. "Better Thinking Better Results." *CLBM*.

Engel, Terry J.; & Gilbert Joseph. 2007. "Improving Internal and External Audit Coordination 11 (2)." *CSA Sentinel*. <http://www.theiia.org/CSA/index.cfm?iid=537&catid=0&aid=2670>.

Everett, Hugh. 1957. "'Relative State' Formulation of Quantum Mechanics" (July). *Reviews of Modern Physics*, vol. 29, no.3. <http://www.univer.omsk.su/omsk/Sci/Everett/paper1957.html>.

Falkenreck, C.; & R. Wagner. 2011. "The Impact of Perceived Innovativeness on Maintaining a Buyer-Seller Relationship in Health Care Markets: A Cross-Cultural Study." *Journal of Marketing Management*, 3-4 (27): S. 225-242.

FamousQuotesAbout.com (FAM). 2013. "Sue Hales" (December 11). <http://www.famousquotesabout.com/on/Subtraction>.

FASB. 2012. "International Convergence of Accounting Standards—Overview" (April). <http://www.fasb.org/jsp/FASB/Page/SectionPage&cid=1176156245663>.

_____.2012a. "International Convergence of Accounting Standards—A Brief History" (April). <http://www.fasb.org/cs/ContentServer?c=Page&pagename=FASB%2FPage%2FSectionPage&cid=1176156304264>.

_____.2008. "A Roadmap for Convergence between IFRSs and US GAAP—2006-2008 Memorandum of Understanding between the FASB and the IASB." <http://www.fasb.org/cs/ContentServer?c=Document_C&pagename=FASB%2FDocument_C%2FDocumentPage&cid=1176156245558>.

_____.2008a. "Completing the February 2006 Memorandum of Understanding: A Progress Report and Timetable for Completion." <http://www.fasb.org/cs/ContentServer?c=Document_C&pagename=FASB%2FDocument_C%2FDocumentPage&cid=1175801856967>.

Financial Services Authority (FIN). 2011. "Enhancing Frameworks in the Standardised Approach to Operational Risk—Guidance Note." <http://www.fsa.gov.uk/pubs/guidance/guidance11.pdf>.

Firehammer, Reginald. 2005. "Basic Principles of Ontology" (January 05). <http://usabig.com/autonomist/philosophy/ontology2.html>.

Forbes. 2003. "Shaking Steady Freddie" (December 11). <http://www.forbes.com/2003/12/11/cx_aw_1211freddie.html>.

ForConstructionpros.com (FORC). 2013. "Changes to Lease Accounting Standards Could Affect Construction Industry" (June 05). <http://mobile.forconstructionpros.com/news/10956036/changes-to-lease-accounting-standards-could-affect-construction-industry>.

Friedlob, G. Thomas, & Franklin James Plewa. 1996. *Understanding Balance Sheets*. New York: John Wiley & Sons.

Gaines-Ross, Leslie. 2008. *Corporate Reputation: 12 Steps to Safeguarding and Recovering Reputation*. NY: Wiley.

Gamma, Erich; Richard Helm; Ralph Johnson; & John Vlissides. 1995. *Design Patterns: Elements of Reusable Object Oriented Software*. Boston, MA: Addison-Wesley.

German EMAS Advisory Board (GER). 2011."Creating Added Value with EMAS—The Differences between EMAS and ISO 14001." <http://www.emas.de/fileadmin/user_upload/06_service/PDF-Dateien/Creating_Added_Value_with_EMAS.pdf>.

Gewirth, Alan. 1986. "Professional Ethics: The Separatist Thesis." *Ethics*, 96 (2): 282–300. <http://www.jstor.org/discover/10.2307/2381378?uid=3738512&uid=2&uid=4&sid=21103201327417>.

Gott, J. Richard. 2002. *Time Travel in Einstein's Universe*. Boston : Houghton Mifflin.

Gowthorpe, Catherine; & John Blake. 1998. *Ethical Issues in Accounting*. NY: Routledge.

Graham, Paul. 2013. "Why ARC isn't especially Object–Oriented" (December 30). <http://www.paulgraham.com/noop.html>.

Guardian, the (GUA). 2011. "Puma World's First Major Company to Put a Value on its Environmental Impact" (May 16). <http://www.theguardian.com/sustainable-business/puma-value-environmental-impact-biodiversity?intcmp=

239>.

Guest, Peter. 2011. "Puma Reveals Its $124m Ecological Footprint in Groundbreaking New Research" (November 19). <http://www.huffington post.co.uk/2011/11/18/puma-jochen-zeitz-environment-footprint_n_1101126. html>.

Guevarra, Leslie. 2011. "Gucci, YSL to Follow Puma in Revealing Environmental Impact Costs" (November 18). <http://www.greenbiz.com/blog/2011/11/18/ gucci-ysl-follow-puma-revealing-environmental-impact-costs>.

Guth, Alan. 1997. *The Inflationary Universe: The Quest for a New Theory Of Cosmic Origins*. New York: Helix Books.

Hardin, G. 1968. "The Tragedy of the Commons." *Science*, 162: 1243-1248.

Harper, Robert. 2011. "Some Thoughts on Teaching FP." <http://existentialtype.wordpress.com/2011/04/17/some-advice-on-teaching-fp/>.

Healy, P. M.; & J. M. Wahlen. 1999. "A Review of the Earnings Management Literature and its Implications for Standard Setting." *Accounting Horizons* (December): 365–383.

Heeffer, Albrecht. 2009. "On the Curious Historical Coincidence of Algebra and Double-Entry Bookkeeping." <http://logica.ugent.be/albrecht/thesis/FOTFS 2008-Heeffer.pdf>.

Hickey, Rich, 2009. "Are We There Yet?" (November). <http://www.infoq.com/presentations/Are-We-There-Yet-Rich-Hickey>.

IAS. 2009. "Use of IFRSs by Jurisdiction" (March 24). <http://www.iasplus.com/en/resources/ifrs-topics/use-of-ifrs>.

IASB & FASB. 2012. "Update Report to the FSB Plenary on Accounting Convergence" (April). <http://www.financialstabilityboard.org/publications/ r_120420d.pdf>.

IFRS Foundation (IFRS). 2012. "The Move towards Global Standards." <http://www.ifrs.org/Use-around-the-world/Pages/Jurisdiction-profiles.aspx>.

Institute of Internal Auditors (IIA). 1998. I "Professional Practice Pamphlet 98-2 A Perspective on Control Self-Assessment." <http://www.iiajapan.com/pdf/ data/csa/pp98-2.pdf>.

Jackling, Beverly; Barry J. Cooper; Philomena Leung; & Steven Dellaportas. 2007. "Professional Accounting Bodies' Perceptions of Ethical Issues, Causes of Ethical Failure and Ethics Education." *Managerial Auditing Journal*, 22 (9): 928–944.

Jackson, Tim. 2009. *Prosperity without Growth, Economics for a Finite Planet*. London, UK: Routledge.

James, Justin. 2007. "Multithreading is a Verb, not a Noun" (October 01). *Techrepublic.com*. <http://archive.is/HitlN>.

Jeffers, Agatha; Mengyu Wei; & Sidney Askew. 2010. "The Switch from US GAAP to IFRS." *Proceedings of the Northeast Business & Economics Association*: 48–54.

Johnsson, Hans; & Per Erik Kihlstedt. 2005. *Performance-Based Reporting*. NY: John Wiley & Sons.

Joseph, Gilbert W. 2001. "Use of Control Self-Assessment in Audits." *The CPA Journal*. <http://www.allbusiness.com/human-resources/workforce-management/1071523-1.html>.

Journal of Accountacy (JOUR). 2011. "U.K. Government White Paper Calls for Tracking Green Losses, Gains" (June 22). <http://www.journalofaccountancy.com/Web/20114243.htm>.

Kaku, Michio. 1994. *Hyperspace: A Scientific Odyssey Through Parallel Universes, Time Warps, and The Tenth Dimension*. Illustrations by Robert O'Keefe. Oxford: Oxford University Press.

Kindler, E.; & I. Krivy. 2011. "Object-Oriented Simulation of Systems with Sophisticated Control." *International Journal of General Systems*: 313–343.

Kirchhoff, T. 2012. "Pivotal Cultural Values of Nature Cannot be Integrated into the Ecosystem Services Framework." *Proc. Natl. Acad. Sci.*, 109 (46): E3146.

Klewes, Joachim; & Robert Wreschniok. 2010. *Reputation Capital: Building and Maintaining Trust in the 21st Century*.

Kooreman, P. 1997. "'The Labeling Effect of a Child Benefit System." Unpublished working paper, University of Groningen.

Kremen, C. 2005. "Managing Ecosystem Services: What Do We Need to Know about their Ecology?" *Ecology Letters*, 8: 468-479.

Labaton, Stephen. 2002. "Accounting Ethics Panel Members Resign, Rejecting Sec Chief's Plan" (January 23). *Pittsburgh Post-Gazette*. <http://news.google.com/newspapers?id=YuwNAAAAIBAJ&sjid=M3ADAAAAIBAJ&pg=3645,467306&dq=accounting+ethics>.

Lee, Judy: & Lieng-Seng Wee. 2005. "Companies Using Control Self Assessment Don't Really Know their Risk." <http://dragonflyone.com/docs/Dragonfly_Risk_Insights_CSA_A4_s.pdf>.

Leopold, A. 1949. *A Sand County Almanac and Sketches from Here and There*. New York: Oxford University Press.

Lewis, John; & William Loftus. 2008. *Java Software Solutions Foundations of Programming Design*. Pearson Education Inc.

Lippert, I. 2012. "Carbon Classified? Unpacking Heterogeneous Relations Inscribed into Corporate Carbon Emissions." *Ephemera*, 12 (1/2):138–161.

Loeb, Stephen E. 1988. "Teaching Students Accounting Ethics: Some Crucial Issues." *Issues in Accounting Education*, 3: 316–329. <http://libproxy.sdsu.edu/login?url=http://proquest.umi.com/pqdweb?did=756

9002&sid=1&Fmt=7&clientId=17862&RQT=309&VName=PQD>.

Lohmann, L. 2009. "Toward a Different Debate in Environmental Accounting: The Cases of Carbon and Cost–Benefit." *Accounting, Organizations and Society*, 34: 499–534.

MacKenzie, D. 2009. "Making Things the Same: Gases, Emission Rights and the Politics of Carbon Markets." *Accounting, Organizations and Society*, 34 (3-4): 440–455. <http://10.0.3.248/j.aos.2008.02.004>.

Mader, Becca. 2002. "Enron Case Causes Colleges to Champion Accounting Ethics" (May 3). *The Business Journal.* <http://www.bizjournals.com/milwaukee/stories/2002/05/06/focus1.html>.

Marsh, G. P. 1864 (1965). *Man and Nature.* New York: Charles Scribner's Sons.

Mathews, M. R.; & M. Perera. 1991. *Accounting Theory and Development.* London, UK: Chapman and Hall.

McAleer, Neil. 1987. *The Mind-Boggling Universe.* New York: Doubleday & Co.

McCarthy, J.; Paul Abrahams; Daniel Edwards; S. Hart; M. Levin. 1962. *LISP 1.5 Programmer's Manual.* Cambridge, MA: MIT Press.

_____; R. Brayton; D. Edwards; P. Fox; L. Hodes; D. Luckham; K. Maling; D. Park; & S. Russell. 1960. *LISP I Programmers Manual.* Boston, MA: Artificial Intelligence Group, M.I.T. <http://history.siam.org/sup/Fox_1960_LISP.pdf>.

McKenna, Francine. 2011. "Auditors and Audit Reports: Is The Firm's 'John Hancock' Enough?." *Forbes.* <http://www.forbes.com/sites/francinemckenna/2011/07/08/auditors-and-audit-reports-is-the-firms-john-hancock-enough/>.

McNally, J. Stephen. 2007. "Control Self-Assessment: Everybody Pitching in with Internal Controls." <http://www.accountingweb.com/topic/cfo/control-self-assessment-everybody-pitching-internal-controls>.

Merriam-Webster's Collegiate Dictionary Online (MWD). "Addition" (December 01). <http://www.merriam-webster.com/dictionary/addition>.

_____.2013a. "Subtract" (December 01). <http://www.merriam-webster.com/dictionary/subtract>.

_____.2007. "Agency" (March 20). <http://www.m-w.com/dictionary/agency>.

Merritt, Cam. 2013. "What Is Verifiability in Accounting?" (December 31). <http://smallbusiness.chron.com/verifiability-accounting-56715.html>.

Meyer, Bertrand. 1997. *Object-Oriented Software Construction.* NY: Prentice Hall.

Millennium Ecosystem Assessment (MILL). 2005. *Ecosystems and Human Well-Being: Synthesis.* Washington: Island Press. <http://www.unep.org/maweb/documents/document.356.aspx.pdf>.

Mills, Geofrey T. 1993. "Early Accounting in Northern Italy: The Role of Commercial Development and the Printing Press in the Expansion of Double-

Entry from Genoa, Florence and Venice." *Critical Perspectives on Accounting,* 4 (2): 113-140.

Min, Chia-Tern Huey. 2004. "GST in Singapore: Policy Rationale, Implementation Strategy & Technical Design" (October). <http://web.archive.org/web/20071129192831/http://adb.org/Documents/Events/2004/Fourteenth-Tax-Conference/text-chia.pdf>.

Minister of the Economy, Industry and Employment (MIN). 2009. "Les Recettes Fiscales" (October 30). <http://www.performance-publique.budget.gouv.fr/le-budget-et-les-comptes-de-letat/approfondir/les-recettes/les-recettes-fiscales.html>.

Mitchell, John. 2003. *Concepts in Programming Languages*, Cambridge, UK: Cambridge University Press.

Moeller, Robert. 2009. *Brink's Modern Internal Auditing: A Common Body of Knowledge*. Canada: John Wiley & Sons Inc.

New York Times (NEW). 2009. "Enron Shareholders' Move Against Banks Is Rebuffed by Judge" (February 28). *The New York Times*. <http://www.nytimes.com/2003/02/28/business/enron-shareholders-move-against-banks-is-rebuffed-by-judge.html>.

Neward, Ted. 2006. "The Vietnam of Computer Science" (June 26). Interoperability Happens. <http://blogs.tedneward.com/2006/06/26/The+Vietnam+Of+Computer+Science.aspx>.

Nidumolu, Ram; Kevin Kramer, & Jochen Zeitz. 2012. "Connecting Heart to Head." <http://www.ssireview.org/articles/entry/connecting_heart_to_head>.

Nova Online (NO). 2000. "Time Can Vary." <http://www.pbs.org/wgbh/nova/time/thinktime.html>.

O'Curry, S. 1997. "Income Source Effects." Unpublished working paper, DePaul University.

Ojea, E. 2010. "Classifying Ecosystem Services for Economic Valuation: The Case of Forest Water Services." BIOECON Conference, Venice, Italy (September 27–28).

Oldroyd, David. 1995. "The Role of Accounting in Public Expenditure and Monetary Policy in the First Century AD Roman Empire," *Accounting Historians Journal*, 22 (2).

Osborn, F. 1948. *Our Plundered Planet*. Boston, MA: Little, Brown and Company.

Owsen, Dwight M. 1999. "The AICPA's Prosecution of Dr. Abraham Briloff, Some Observations." *In the Public Interest* 27 (2). <http://web.archive.org/web/20041221054447/http://aaahq.org/PublicInterest/newsletr/spring99/item07.htm>.

Patsuris, Penelope. 2002. "The Corporate Scandal Sheet" (August 26). *Forbes*. <http://www.forbes.com/2002/07/25/accountingtracker.html>.

Perks, R. W. 1993. *Accounting and Society*. London, England: Chapman & Hall.

PBS.org (PBS). 1994. "FRONTLINE: How to Steal $500 Million" (June 23). <http://www.pbs.org/wgbh/pages/frontline/programs/transcripts/1304.html>.

Pickett, Spencer K. H.; & Jennifer Pickett. 2010. *The Internal Auditing Handbook*. NY: John Wiley & Sons Ltd.

Pierce, Benjamin. 2002. *Types and Programming Languages*. Cambridge, MA: MIT Press.

Pixley, Francis William. 1990. *Accountancy—Constructive and Recording Accountancy*. London, England: Sir Isaac Pitman & Sons, Ltd.

Potok, Thomas; Mladen Vouk, & Andy Rindos. 1999. "Productivity Analysis of Object-Oriented Software Developed in a Commercial Environment." *Software – Practice and Experience*, 29 (10): 833–847. <http://www.csm.ornl.gov/~v8q/Homepage/Papers%20Old/spetep-%20printable.pdf>.

Porter, M.; & M. Kramer. 2011. "Creating Share Value." *Harvard Business Review*. <http://hbr.org/2011/01/the-big-idea-creating-shared-value>.

Preston, Alistair M.; David J. Cooper; D. Paul Scarbrough; Robert C. Chilton; & J. Edward Ketz, ed. 2006. *Accounting Ethics: Critical Perspectives on Business and Management (Changes in the Code of Ethics of the U.S. Accounting Profession, 1917 and 1988.* London, UK: Routledge.

PricewaterhouseCoopers (PRICE). 2007. "ViewPoint—April 2007" (May 1). <http://www.pwc.ch/user_content/editor/files/publ_ass/pwc_viewpoint_0704_e.pdf>.

QuotationPage.com (QUO). 2013. "Walt Whitman" (December 14). <http://www.quotationspage.com/quote/26914.html>.

Ramamoorti, Sridhar. 2003. "Chapter 1: Internal Auditing: Evolution and Prospects." <https://na.theiia.org/iiarf/Public%20Documents/Chapter%201%20Internal%20Auditing%20History%20Evolution%20and%20Prospects.pdf>.

Raman, Shankar. 1997. Lectures on Literary Theory. Cambridge, MA: MIT (spring).

Reece, Damian. 2004. "Deloitte's John Connolly Faces Call to Resign over Barlow Clowes Link" (January 13). *The Independent* (London).

Rezaee, Zabihollah. 2002. *Financial Statement Fraud: Prevention and Detection*. New York: Wiley.

Rosenwald, Michael S. 2007. "Extreme (Executive) Makeover" (September 10). *The Washington Post*. <http://www.washingtonpost.com/wp-dyn/content/article/2007/09/09/AR2007090901619_pf.html>.

Sangster, Alan. 2007. "The Printing of Pacioli's Summa in 1494: How Many Copies were Printed?" *Accounting Historians Journal*.

Scannell, Kara; & Joanna Slater. 2008. "SEC Moves To Pull Plug On U.S.

Accounting Standards" (August 28). *Wall Street Journal.* <http://online.wsj.com/news/articles/SB121985665095476825>.

Scott, D. 1931. *The Cultural Signicance of Accounts.* New York: Henry Holt and Company.

Scott, Michael Lee. 2006. *Programming Language Pragmatics.* Burlington, MA: Morgan Kaufmann.

Shackle, G. L. S. 1972. *Epistemics and Economics.* London, England: Cambridge University Press

Shefrin, H. H.; & R. Thaler. 1988. "The Behavioral Life-Cycle Hypothesis." *Economic Inquiry*, 26: 609-643.

Shelly, Asaf. 2008. "Flaws of Object Oriented Modeling" (August 22). *Intel Software Network.* <http://software.intel.com/en-us/blogs/2008/08/22/flaws-of-object-oriented-modeling/>.

_____.2008a. "HOW TO: Multicore Programming (Multiprocessing) Visual C++ Class Design Guidelines, Member Functions." <http://support.microsoft.com /?scid=kb%3Ben-us%3B558117>.

Shlain, Leonard. 1991. *Art and Physics: Parallel Visions in Space,Time, and Light.* New York: William Morrow and Co., Inc.

Smith, L. Murphy. 2008. "Luca Pacioli: The Father of Accounting." Texas A&M University. <http://web.acct.tamu.edu/smith/ethics/pacioli.htm>.

Snyder, Stan. 2008. "Understanding Cash and Accrual Basis Accounting." <http://office.microsoft.com/en-us/products/HA101646121033.aspx>.

Somerville, Leigh. 2003. "Accounting Changes Boil Down to Principles vs. Rules" (February 28). *The Business Journal.* <http://www.bizjournals.com/triad/ stories/2003/03/03/focus2.html>.

Stepanov, Alexander. 2010. "STLport: An Interview with A. Stepanov." <http://www.stlport.org/resources/StepanovUSA.html>.

Sunder S, 2011. "IFRS Monopoly: The Pied Piper of Financial Reporting" (February 17). <http://papers.ssrn.com/sol3/papers.cfm?abstract_id=1763 311>.

Sutherland, I. E. 1963. "Sketchpad: A Man-Machine Graphical Communication System." Technical Report No. 296, Lincoln Laboratory, Massachusetts Institute of Technology via Defense Technical Information Center (stinet.dtic.mil).

Swanson, Marianne. 2013. "Security Self-assessment Guide for Information Technology Systems." National Institute of Standards and Technology. <http://www.itl.nist.gov/lab/bulletns/bltnsep01.htm>.

Takatera, Sadao; & Norio Sawabe. 2000. "Time and Space in Income Accounting." *Accounting, Organizations and Society*, 25: 787-798.

TED. 2011. "Pavan Sukhdev: Put a value on nature!" (December).

<http://www.ted.com/talks/pavan_sukhdev_what_s_the_price_of_nature.html>.

TEED. 2010. "The Ecological and Economic Foundation." <http://www.teebtest.org/wp-content/uploads/Study%20and%20Reports/Reports/Ecological%20and%20Economic%20Foundations/TEEB%20Ecological%20and%20Economic%20Foundations%20report/TEEB%20Foundations.pdf>.

Thaler, R. H. 1999. "Mental Accounting Matters." *Journal of Behavioral Decision Making*, 12: 183-206.

_____.1980. "Toward a Positive Theory of Consumer Choice." *Journal of Economic Behavior and Organization*, 1: 39-60.

ThinkExist.com (THI). 2013. "Henri Bergson" (December 03). <http://thinkexist.com/quotes/with/keyword/addition/2.html>.

Tilman, D., C. L. Lehman; & C. E. Bristow. 1998. "Diversity-Stability Relationships: Statistical Inevitability or Ecological Consequence?" *The American Naturalist,* 151: 277-282.

Trofimov, M. 1993. "OOOP—The Third 'O' Solution: Open OOP." *OMG*, 3 (3): 14.

Tysiac, Ken. 2013. "New Mechanisms Eyed by FASB, IASB in Long March toward Global Comparability" (January 10). <http://www.journalofaccountancy.com/News/20137119.htm>.

Vedantam, Shankar. 2007. "Mental Accounting" (May 19). *The Washington Post*. <http://www.washingtonpost.com/wp-dyn/content/article/2007/05/19/AR2007051900316.html>.

Vogt, W. 1948. *Road to Survival*. New York: William Sloan.

Walker, B. H. 1992. "Biodiversity and Ecological Redundancy." *Conservation Biology*, 6: 18-23.

Weber, Max. 1930. *The Protestant Ethic and the Spirit of Capitalism*. Talcott Parsons, tr. New York: Scribner.

Wikipedia (WK). 2013. "Accounting" (December 01). <http://en.wikipedia.org/wiki/Accounting>.

_____.2013a. "Financial Accounting" (December 01). <http://en.wikipedia.org/wiki/Financial_accountancy>.

_____.2013b. "Asset" (December 01). <http://en.wikipedia.org/wiki/Asset>.

_____.2013c. "Liability" (December 01). <http://en.wikipedia.org/wiki/Liability>.

_____.2013d. "Current Liability" (December 01). <http://en.wikipedia.org/wiki/Current_liability>.

_____.2013e. "Quality Audit" (December 01). <http://en.wikipedia.org/wiki/quality_audit>.

_____.2013f. "Debits and Credits" (December 01). <http://en.wikipedia.org/wiki/

Debits_and_credits>.

_____.2013g. “Expense” (December 02). <http://en.wikipedia.org/wiki/Expense>.

_____.2013h. “Income” (December 02). <http://en.wikipedia.org/wiki/Income>.

_____.2013i. “Lean Accounting” (December 02). <http://en.wikipedia.org/wiki/Lean_accounting>.

_____.2013j. “Balance Sheet” (December 02). <http://en.wikipedia.org/wiki/Balance_sheet>.

_____.2013k. “Financial Statement” (December 02). <http://en.wikipedia.org/wiki/Financial_statements>.

_____.2013l. “Bookkeeping” (December 02). <http://en.wikipedia.org/wiki/Bookkeeping>.

_____.2013m. “Constant Purchasing Power Accounting” (December 02). <http://en.wikipedia.org/wiki/Constant_purchasing_power_accounting>.

_____.2013n. “Historical Cost Accounting” (December 02). <http://en.wikipedia.org/wiki/Historical_cost_accounting>.

_____.2013o. “Mental Accounting” (December 02). <http://en.wikipedia.org/wiki/Mental_accounting>.

_____.2013p. “Cognitive Bias Modification” (December 02). <http://en.wikipedia.org/wiki/Cognitive_bias_modification>.

_____.2013q. “Creative Accounting” (December 02). <http://en.wikipedia.org/wiki/Creative_accounting>.

_____.2013r. “Eco-Management and Audit Scheme” (December 02). <http://en.wikipedia.org/wiki/Eco-Management_and_Audit_Scheme>.

_____.2013s. “Carbon Accounting” (December 02). <http://en.wikipedia.org/wiki/Carbon_accounting>.

_____.2013t. “Audit” (December 02). <http://en.wikipedia.org/wiki/Audit>.

_____.2013u. “Boilerplate” (December 02). <http://en.wikipedia.org/wiki/Boilerplate>.

_____.2013v. “Object Oriented Programming” (December 02). <http://en.wikipedia.org/wiki/Object-oriented_programming>.

_____.2013w. “Accounting Ethics” (December 02). <http://en.wikipedia.org/wiki/Accounting_ethics>.

_____.2013x. “Accounting Reform” (December 03). <http://en.wikipedia.org/wiki/Accounting_reform>.

_____.2013y. “Enron Scandal” (December 03). <http://en.wikipedia.org/wiki/Enron_scandal>.

_____.2013z. “Goodwill (Accounting)” (December 03). <http://en.wikipedia.org/wiki/Goodwill_(accounting)>.

_____.2013aa. “Reputation” (December 03). <http://en.wikipedia.org/ wiki/

Reputation>.

_____.2013bb. "Environmental Audit" (December 06). <http://en.wikipedia.org/wiki/Environmental_audit>.

_____.2013cc. "Chaos Theory" (December 06). <http://en.wikipedia.org/wiki/Chaos_theory>.

_____.2013dd. "Complex Systems" (December 06). <http://en.wikipedia.org/wiki/Complex_systems>.

_____.2013ee. "Peregrine Systems" (December 07). <http://en.wikipedia.org/wiki/Peregrine_Systems>.

_____.2013ff. "Internal Audit" (December 08). <http://en.wikipedia.org/wiki/Internal_audit>.

_____.2013gg. "Valued-Added Tax" (December 08). <http://en.wikipedia.org/wiki/ Value-added_tax>.

_____.2013hh. "Missing Trader Fraud" (December 09). <http://en.wikipedia.org/wiki/Missing_trader_fraud>.

_____.2013ii. "Unit of Measure" (December 09). <http://en.wikipedia.org/wiki/Unit_of_measure>.

_____.2013jj. "Standard Accounting Practice" (December 09). <http://en.wikipedia.org/wiki/Standard_accounting_practice>.

_____.2013kk. "International_Financial_Reporting_Standards" (December 10). <http://en.wikipedia.org/wiki/International_Financial_Reporting_Standards>.

_____.2013ll. "US Generally Accepted Accounting Principles" (December 10). <http://en.wikipedia.org/wiki/US_generally_accepted_accounting_princi ples>.

_____.2013mm. "Control Self-Assessment" (December 14). <http://en.wikipedia.org/wiki/Control_self-assessment>.

_____.2013nn. "Carbon Footprint" (December 28). <http://en.wikipedia. org/wiki/Carbon_footprint>.

_____.2013oo. "Sustainability Accounting" (December 28). <http://en.wikipedia.org/wiki/Sustainability_accounting>.

_____.2013pp. "Sustainability Measurement" (December 28). <http://en.wikipedia.org/wiki/Sustainability_measurement>.

_____.2013qq. "Environmental Profit and Loss Account" (December 28). <http://en.wikipedia.org/wiki/Environmental_profit_ and_loss_account>.

_____.2013rr. "Ecosystem Services" (December 29). <http://en.wikipedia.org/wiki/Ecosystem_services>.

_____.2013ss. "Accounting Scandals" (December 31). <http://en.wikipedia.org/wiki/Accounting_scandals>.

_____.2013tt. "Freddie Mac" (December 31). <http://en.wikipedia.org/wiki/

Freddie_Mac>.

_____.2013uu. "Internal Control" (December 31). <http://en.wikipedia.org/wiki/Internal_control>.

_____.2013vv. "Professional Ethics" (December 31). <http://en.wikipedia.org/wiki/Professioal_ethics>.

_____.2013ww. "Parmalat" (December 31). <http://en.wikipedia.org/wiki/Parmalat>.

_____.2013xx. "Convergence of Accounting Standards" (January 01). <http://en.wikipedia.org/wiki/Convergence_of_accounting_standards>.

_____.2011nn. "Design Elements and Principles" (September 30). <http://en.wikipedia.org/wiki/Design_elements_and_principles>.

_____.2011oo. "Interpretation (Logic)" (April 13). <http://en.wikipedia.org/wiki/Interpretation_(Logic)>.

_____.2011pp. "Unsolved Problems in Linguistics" (February 20). <http://en.wikipedia.org/wiki/Unsolved_problems_in_linguistics>.

_____.2008. "Posthumanism" (October 20). <http://en.wikipedia.org/wiki/Posthumanism>.

_____.2008a. "Transhumanism" (October 20). <http://en.wikipedia.org/wiki/Transhumanism>.

_____.2008b. "Lowest Common Denominator" (March 26). <http://en.wikipedia.org/ wiki/Least_common_denominator>.

_____.2008c. "Infinity" (November 11). <http://en.wikipedia.org/wiki/Infinity>.

_____.2008d. "Transfinite Number" (November 10). <http://en.wikipedia.org/wiki/Transfinite_ number>.

_____.2008e. "Quantum Mechanics" (November 26). <http://en.wikipedia.org/wiki/Quantum_mechanics>.

_____.2008f. "Copenhagen Interpretation" (November 26). <http://en.wikipedia.org/wiki/Copenhagen_interpretation >.

_____.2008g. "Many Worlds Interpretation" (November 26). <http://en.wikipedia.org/wiki/Many_Worlds_Interpretation>.

_____.2008h. "Hyperspace (Science Fiction)" (November 23). <http://en.wikipedia.org/wiki/Hyperspace_(science_fiction)>.

_____.2008i. "Time Travel" (November 26). <http://en.wikipedia.org/wiki/Time_travel#Time_dilation>.

_____.2008j. "Wormhole" (November 25). <http://en.wikipedia.org/wiki/Wormhole>.

_____.2007. "Linguistics" (March 04). <http://en.wikipedia.org/wiki/Linguistics>.

Wikiversity (WKV). 2008. "Why 10 Dimensions" (November 23). <ttp://en.wikiversity.org/wiki/Why_10_dimensions>.

Wittgenstein, Ludwig.1953. *Philosophical Investigations*. G. E. M. Anscombe and R. Rhees, eds. G.E.M. Anscombe, tr. Oxford: Blackwell.

Woopidoo.com (WOO). 2013. "Charlie Munger" (December 01). <http://www.woopidoo.com/business_quotes/accounting-quotes.htm>.

Wright, L.; S. Kemp; & I. Williams. 2011. "'Carbon Footprinting': Towards a Universally Accepted Definition." *Carbon Management*, 2 (1): 61-72.

Yegge, Steve. 2006. "Execution in the Kingdom of Nouns" (March 30). <http://steve-yegge.blogspot.com/2006/03/execution-in-kingdom-of-nouns.html>.

INDEX

•A•

•B•

•C•

•D•

•F•

•G•

•H•

•I•

•J•

•K•

•L•

•M•

•P•

•Q•

•R•

•S•

•T•

•U•

•V•

•W•

•X•

•Y•

•Z•

Printed and bound by CPI Group (UK) Ltd, Croydon, CR0 4YY

15/09/2026

14962370-0001